DATE DUE

D0853651

Experiencing Geometry

Bruni

Based on a book by Emma Castelnuovo

Experiencing Geometry

Acknowledgment

The approach to the study of geometry used in this book was inspired by the work of Emma Castelnuovo. Professor Castelnuovo has been teaching at the Scuola Media "Tasso" in Rome, Italy, for many years and has earned international recognition for her innovative work in mathematics education. Her volume, *La Via della Matematica: La Geometria,* published by La Nuova Italia Editrice, formed the basis for this book and was the source of many of the activities and materials adapted and included herein.

Experiencing Geometry

James V. Bruni

Herbert H. Lehman College
City University of New York

Wadsworth Publishing Company, Inc.
Belmont, California

For Emma Castelnuovo, an extraordinary teacher
and a special friend

Per Emma Castelnuovo, un insegnante straodinario
e un'amica speciale

Designer: Wadsworth Design Staff

Mathematics Editor: Don Dellen

Production Editor: Phyllis Niklas

Copy Editor: Don Yoder

Technical Illustrator: John Foster

Photographers: Kenneth R. Levinson (indoors)
Clifton Freedman (outdoors)

Adapted from the original Italian language edition, *La Via della Matematica: La Geometria,* by Emma Castelnuovo, published by La Nuova Italia Editrice, Florence, Italy. Copyright © 1970 by La Nuova Italia Editrice, Florence, Italy.

© 1977 by Wadsworth Publishing Company, Inc., Belmont, California 94002. All rights reserved. No part of this book may be reproduced, stored in a retrieval system or transcribed, in any form or by any means, electronic, mechanical, photocopying, recording or otherwise, without the prior written permission of the publisher.

ISBN-0-534-00422-9
L. C. Cat. Card No. 76-9245
Printed in the United States of America

1 2 3 4 5 6 7 8 9 10—81 80 79 78 77

Contents

4

7

Preface

Dear Reader,

Your study of geometry began long before you opened this book. It began before you entered grade school. In fact, you discovered many geometric ideas even before you began to talk. Your study of geometry began when you were able to recognize "shapes" in your environment. As an infant you developed the ability to recognize many different shapes, like the shadows of persons and things around you. You explored, compared, and transformed shapes with your eyes and with your hands.

Geometry is a special way of thinking about the world. It is a way of organizing ideas about shapes in space. It is a part of mathematics that deals with the study of shapes.

This book is meant to be an informal, intuitive *introduction* to geometry. If you glance through the book, you will find none of the formal theorems or proofs you might expect in a mathematics textbook. You will find numerous concrete models and illustrations intended to be a springboard for the discovery of geometric ideas. Through guided observation, experimentation, and the use of your intuition, you will investigate some fundamental geometric concepts. These experiences can serve as preparation for a more formal, abstract, and logically precise study of different kinds of geometry.

You will be more than a reader. You will be asked to use various materials to make concrete models of the geometric concepts you investigate. This may seem to be unnecessary. Please try it. We all seem to understand mathematical concepts more easily if they are introduced concretely and if we play an *active* role in learning.

I hope you will enjoy the satisfaction of *Experiencing Geometry.*

New York, New York *James V. Bruni*

1 From Points to Polygons

Looking at Shapes

Everywhere we look we see many different kinds of shapes: round shapes, curved shapes, jagged shapes, square shapes—a countless assortment of shapes. Whether directly or indirectly, we are all involved in describing shapes, comparing shapes, or transforming shapes, and making decisions that are influenced by the shape of things. Which key fits this lock? Which pair of shoes do I want to buy? What kind of flower is this? Shape makes a difference.

Imagine you are going to buy a new car. How can considerations about the shape of the car influence your opinion about buying it? You might ask yourself questions like the following: Does the car have an attractive shape? How is its shape the same as or different from other cars I have seen? Is the car too long or too wide for me to manage comfortably? How many passengers will the car hold? What are the shapes of different parts of the car (seats, windows, trunk, doors)? Do their shapes make them more attractive or more functional? Indeed, shape can be very important!

Figure 1-1

Geometry is a branch of mathematics that involves the study of shapes. But how do you study a shape *geometrically?* Look at the cardboard box pictured in Figure 1-1. There are many questions we might ask about the box and many ways to study it. We might be interested in knowing where the box came from, how it can be used, how strong it is, or what material it is made of. Such questions can lead to varied investigations about the box.

What kinds of questions would lead to a geometric investigation of the box? Figure 1-2 offers some possibilities. The questions suggested lead to an exploration of plane and solid geometric

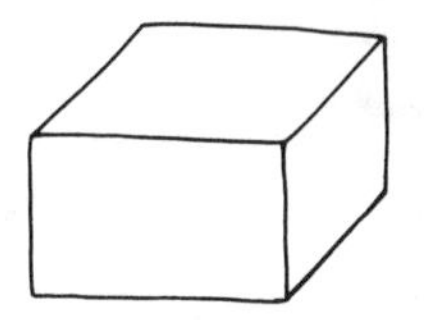

How can you describe the shape of the box?
Which sides are just alike?
How big is the box?
How much does it hold?

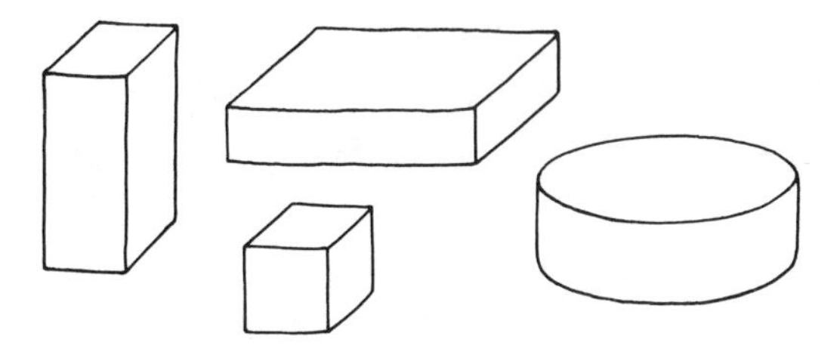

How is the box the same as
or different from other boxes?

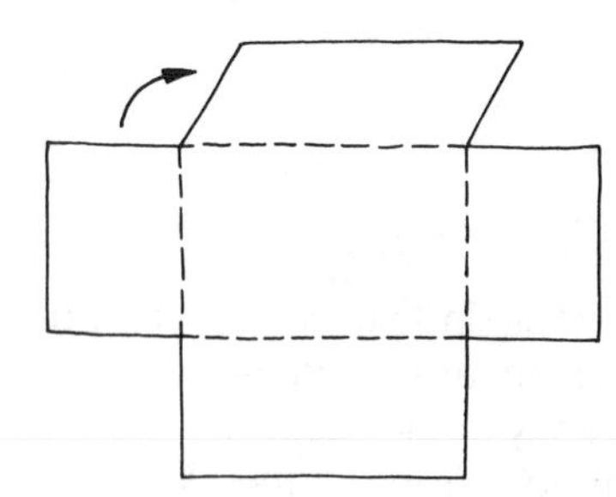

How can you make a box just like this one?
How can you make a box with the same shape,
only larger in size?

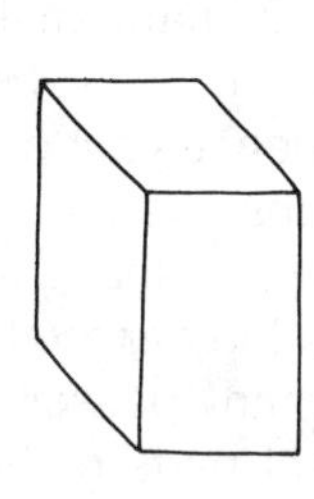

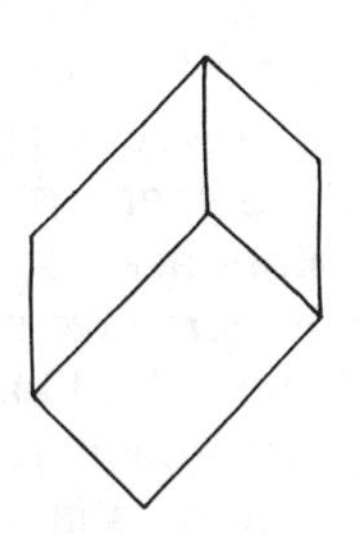

If you turn the box different ways,
does its shape change?
Does its location change?
Does its shadow change?
How?

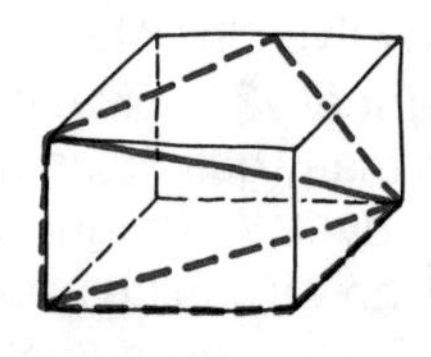

What is the shortest path from one corner
of the top of the box to an opposite corner
on the bottom of the box?
How do you know?

Figure 1-2

Thinking geometrically
about a box

shapes, including the development of such concepts as linear, area, and volume measurement; similarity and congruence; symmetry, and geometric transformations. Throughout this book you will have to respond to questions like these about different kinds of shapes. You will develop a "geometric language" that will be useful in finding out and communicating many things about shapes.

Points in Space

Let us take another look at that cardboard box. How many corners does the box have? Can you find eight? What do you think of as a "corner" of the box?

You can think of a corner of the box as a place where three of its edges meet. The geometric description for that location is a **point.** A point can be thought of as an *exact location or place.* In this case the eight points you are referring to are the places where three edges of the box meet (corners). But imagine other possible points or locations—points *in* the box, points *on* the box, and points *outside* the box. We can think of this **set** or **collection** of all possible points (inside, outside, or on the box) as **space.** Space is the set of *all* possible points—points here, there, and everywhere!

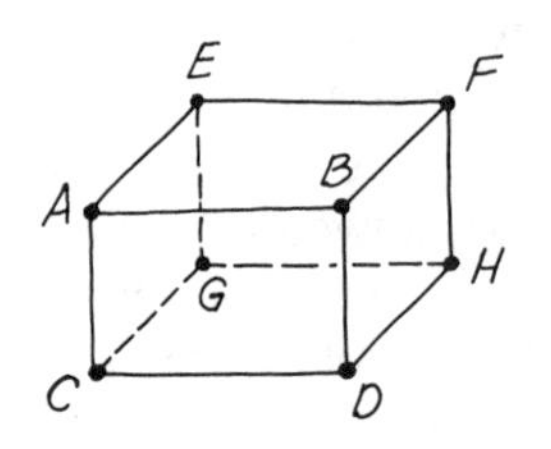

Figure 1-3

We can represent a point as a dot on a piece of paper. In the drawing of the box in Figure 1-3 the corners are represented as dots. The dots are used to picture those eight points in space. Each corner point represented in the diagram of the box is labeled with a letter of the alphabet. This makes it easy to identify a particular point.

Points *A, B, E,* and *F* are on the top of the box. Think of other points on the top of the box, such as all the dots shown in Figure 1-4. These are other places or locations on the top of the box. Imagine any number of other points like these at different places on the top of the box. We can think of the top of the box as a flat surface made up of points.

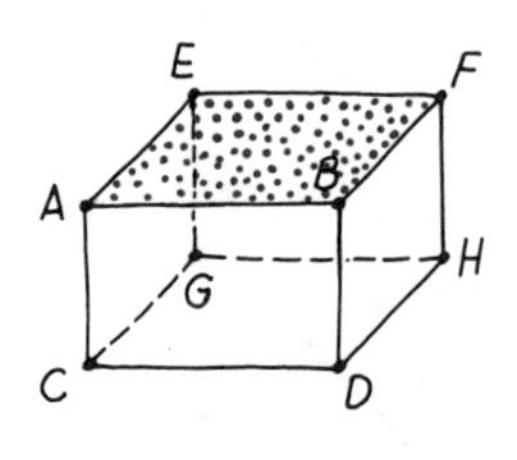

Figure 1-4

Now imagine that the flat surface of the top of the box is extended in all directions indefinitely as suggested by Figure 1-5. We can call that extended flat surface a **plane.** A plane, then, is a special set of points on a flat surface. The top of the box can be thought of as a model of part of a plane. Similarly, each side of the box and the bottom of the box can be considered models of parts of different planes (Figure 1-6).

A plane is a geometric idea. There are no flat surfaces in the real world that extend indefinitely. However, we can think of flat surfaces like the ceiling, floor, or walls of a room as parts of different planes.

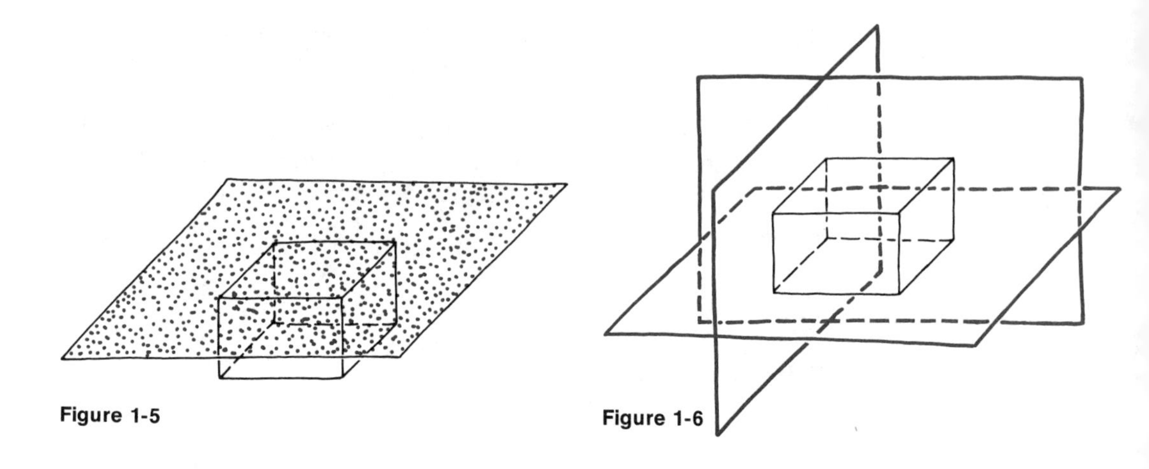

Figure 1-5

Figure 1-6

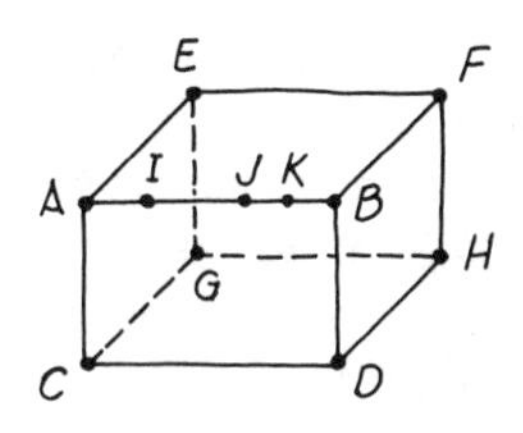

Figure 1-7

Let us consider the two corner points *A* and *B* on the box. Suppose *I*, *J*, and *K* are points along the edge of the box between *A* and *B* (Figure 1-7). Are there other points between *A* and *B* along that edge? There are any number of points between *A* and *B*. Each point represents another place along that edge. We can call the set of all the points along the edge of the box between *A* and *B* (including *A* and *B*) a **line segment.** The symbol used to represent this line segment is $\overline{AB}$ or $\overline{BA}$, where the overbar symbol indicates that we are referring to a line segment and the letters represent the end points of the line segment.

Suppose the line segment from *A* to *B* ($\overline{AB}$) is extended beyond *A* in one direction and beyond *B* in the other direction (Figure 1-8). We now have a model for the geometric idea called a **line.** Unlike a line segment, a line has no end points. Each edge of the box can be thought of as part of a different line (Figure 1-9).

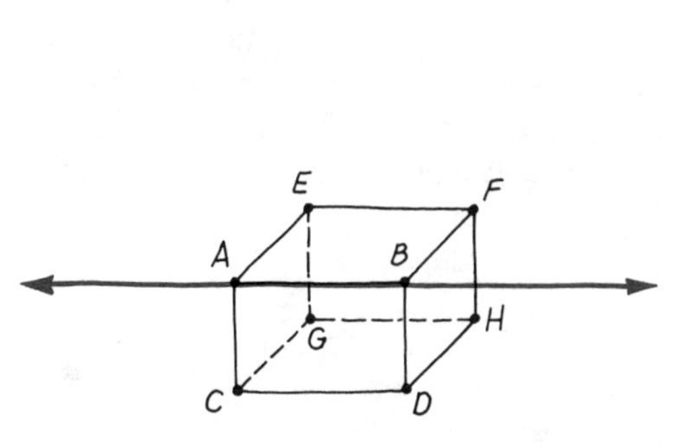

Figure 1-8

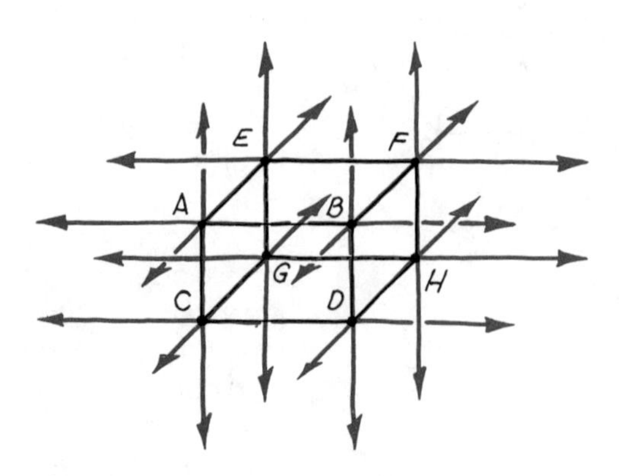

Figure 1-9

The symbol used to represent a line through A and B is $\overleftrightarrow{AB}$ or $\overleftrightarrow{BA}$, where the arrow symbol indicates that we are referring to a line (the arrowheads show that it extends indefinitely in both directions) and the letters indicate that the line includes points A and B. Can you locate the following lines in Figure 1-9: $\overleftrightarrow{AC}$, $\overleftrightarrow{FH}$, $\overleftrightarrow{GH}$, $\overleftrightarrow{AE}$, $\overleftrightarrow{GC}$, $\overleftrightarrow{EF}$, $\overleftrightarrow{DH}$, $\overleftrightarrow{EG}$? Each of these lines represents a special set of points in space.

Figure 1-10

Find a cardboard box like the one used in the previous discussion and make a box model, labeling its corners A, B, C, D, E, F, G, H in the same way the diagram of the box is labeled (Figure 1-10). Which edges of your box correspond to the following line segments: $\overline{AB}$, $\overline{AC}$, $\overline{GH}$, $\overline{EF}$, $\overline{BD}$, $\overline{DH}$, $\overline{CD}$?

Suppose each line segment is extended indefinitely so that you can think of it as a line. Lines $\overleftrightarrow{AB}$ and $\overleftrightarrow{AC}$ can be called **intersecting lines.** They have a point in common. Point A is on $\overleftrightarrow{AB}$ and it is on $\overleftrightarrow{AC}$ at the same time. If you think of $\overleftrightarrow{AB}$ as a set of points and $\overleftrightarrow{AC}$ as another set of points, then point A represents the **intersection** of the two sets (Figure 1-11). That is, point A belongs to both sets of points at the same time. Can you locate other pairs of intersecting lines? What is the point of intersection for each pair of intersecting lines?

Some pairs of lines suggested by the box model have no point of intersection. They do not meet. For example, you can imagine that $\overleftrightarrow{AB}$ and $\overleftrightarrow{CD}$ do not meet, no matter how far they are extended. However, you can think of $\overleftrightarrow{AB}$ and $\overleftrightarrow{CD}$ as being part of the same flat surface or plane. Lines $\overleftrightarrow{AB}$ and $\overleftrightarrow{CD}$ are called **parallel lines.** Similarly, $\overleftrightarrow{BF}$ and $\overleftrightarrow{DH}$ are parallel lines, $\overleftrightarrow{BD}$ and $\overleftrightarrow{FH}$ are parallel, etc. Can you find other pairs of parallel lines?

It is important to note that pairs of lines like $\overleftrightarrow{EG}$ and $\overleftrightarrow{BD}$ are parallel lines, too. They will *never* meet, no matter how far they are extended, and it is possible to imagine a flat surface that includes $\overleftrightarrow{EG}$ and $\overleftrightarrow{BD}$ as indicated in Figure 1-12. Lines $\overleftrightarrow{EG}$ and $\overleftrightarrow{BD}$ are part of that flat surface or plane.

Some pairs of lines suggested by the edges of the box are neither intersecting nor parallel. The pair $\overleftrightarrow{EG}$ and $\overleftrightarrow{CD}$ is an example.

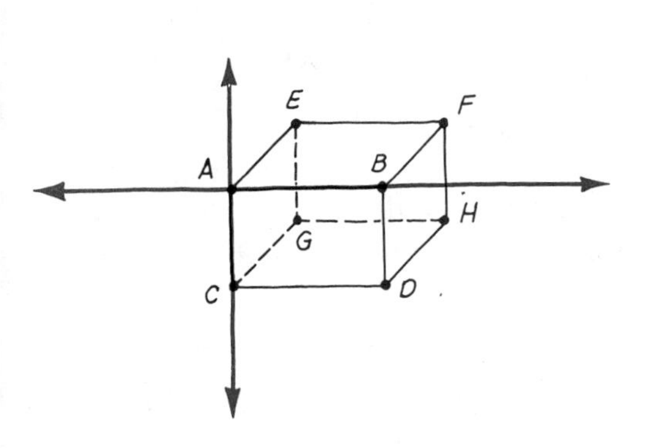

Figure 1-11

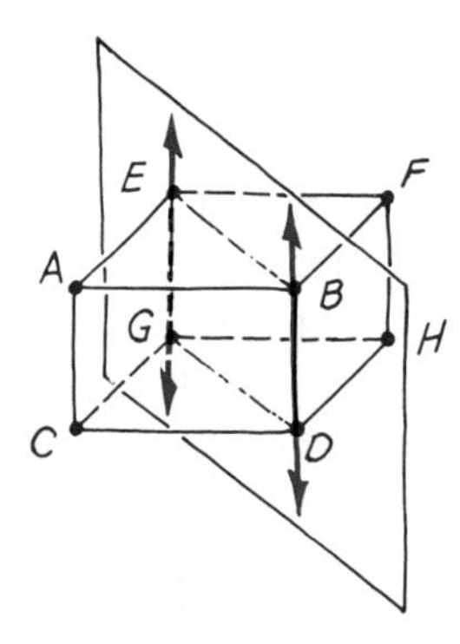

Figure 1-12

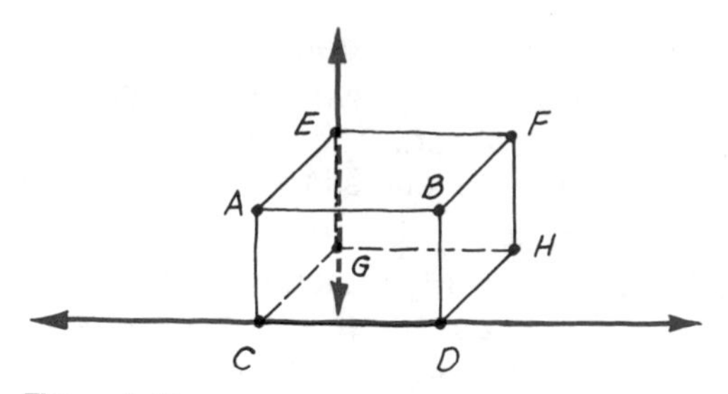

Figure 1-13

Locate these lines on your box model (Figure 1-13). When you extend edges $\overline{EG}$ and $\overline{CD}$ on your model, you will note they do not meet.* Yet it is impossible to find a plane that includes *both* $\overleftrightarrow{EG}$ and $\overleftrightarrow{CD}$. We call $\overleftrightarrow{EG}$ and $\overleftrightarrow{CD}$ **skew lines.** Skew lines *never* meet, and you cannot find a plane that includes both lines. Can you find other pairs of skew lines on your box model (such as $\overleftrightarrow{AC}$ and $\overleftrightarrow{DH}$, $\overleftrightarrow{AE}$ and $\overleftrightarrow{FH}$)?

Measuring Line Segments

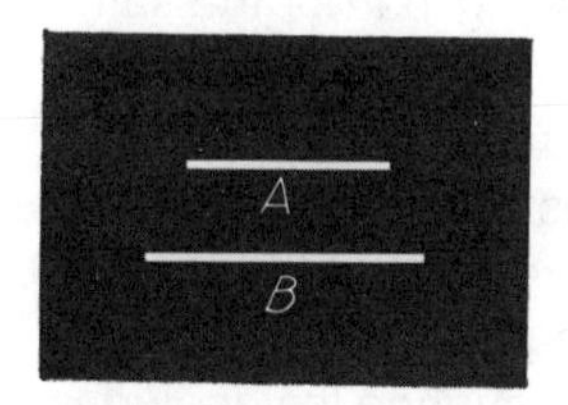

Figure 1-14

In real-life situations it is often important to compare the lengths of different objects. Suppose we want to find out which of the two straws in Figure 1-14 is longer. How would we do it? We could put the straws side by side as in Figure 1-15, and it would be evident that straw *B* is longer than straw *A*. If you imagine that the straws are models of line segments, then the line segment represented by straw *B* is longer than the line segment determined by straw *A*.

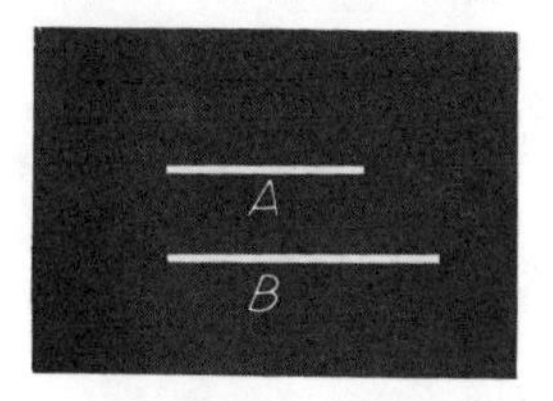

Figure 1-15

But it might be necessary to know how much longer straw *B* is than straw *A*. We could take some other objects, like small paper clips, and find out the length of each straw in terms of the paper clips as in Figure 1-16. We would be using the paper clip as a **unit of measurement.** You can see that it takes about three paper clips to span the length of straw *A* while it takes nearly four to span the length of straw *B*. We might say that straw *B* is "almost one paper clip longer" than straw *A*.

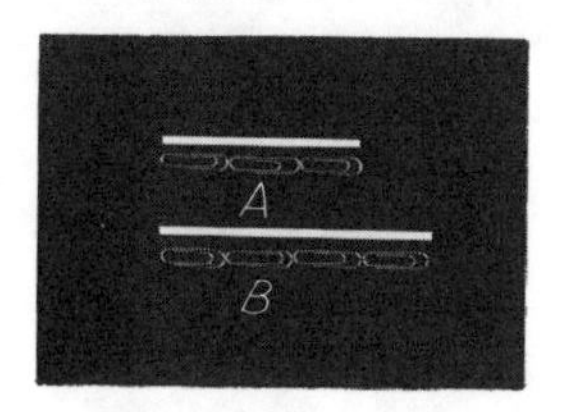

Figure 1-16

It would be better to use some type of unit other than a paper clip to avoid confusion. If you told someone that one straw is about one paper clip longer than another, that information might be very misleading. The person might not know the length of the paper clip you are using as a unit.

**The lines appear to meet on the diagram of the box model since the diagram is an attempt to represent a three-dimensional box on a flat surface. If you locate these lines on an actual box, you will see that they do not meet, no matter how far they are extended.*

Your first impulse might have been to use a ruler. The ruler might be marked off in units called inches or centimeters. Measuring with this type of unit is more useful because others will be familiar with your unit. A unit of measurement that is familiar to most people —like the inch, centimeter, foot, yard, or mile—is called a **standard unit.**

When we measured the lengths of the straws, we were measuring what might be envisioned as the length of the line segments determined by the straws. Similarly, we can imagine that we are measuring line segments when we measure the length or width or height of a room or a building. This type of measurement—where you imagine the object to be measured as a line segment—is called **linear** measurement.

Congruent Line Segments

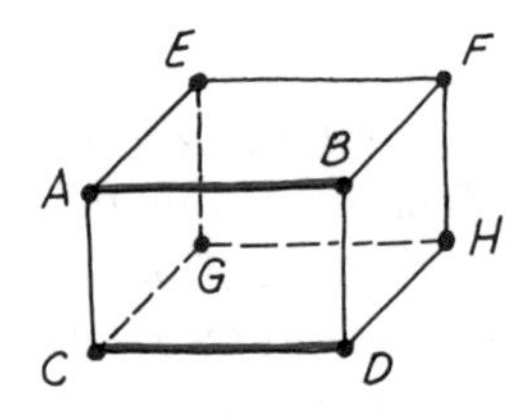

Figure 1-17

Since we can think of the edges of a box as line segments, measuring the length of the edges is another example of linear measurement. If you measure the length of $\overline{AB}$ and then the length of $\overline{CD}$ in Figure 1-17, you will find that they have the same measure.

We can express this information in the following way:

$$m\,\overline{AB} = m\,\overline{CD}$$

This is a mathematical way of saying: "The measure of the line segment with end points A and B is equal to the measure of the line segment with end points C and D."

You might discover the same information by tracing $\overline{CD}$ on a piece of tracing paper and then sliding the tracing paper over the diagram of the box so that $\overline{CD}$ covers $\overline{AB}$ (Figure 1-18). You will see that the two line segments match completely. That is, point C coincides with point A, point D coincides with point B, and all the points between C and D coincide with all the points between A and B. If two geometric figures can be made to coincide or match com-

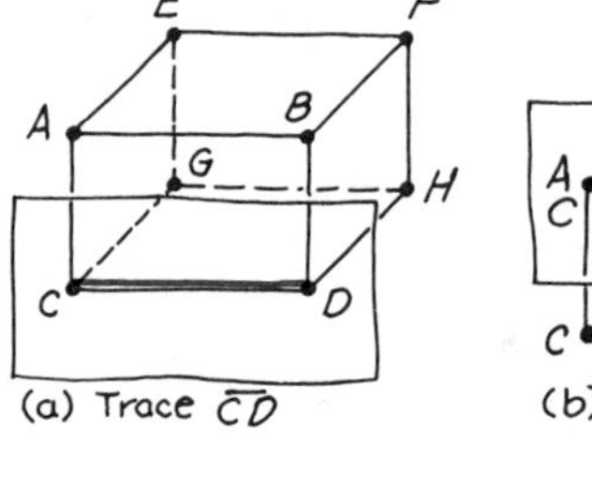

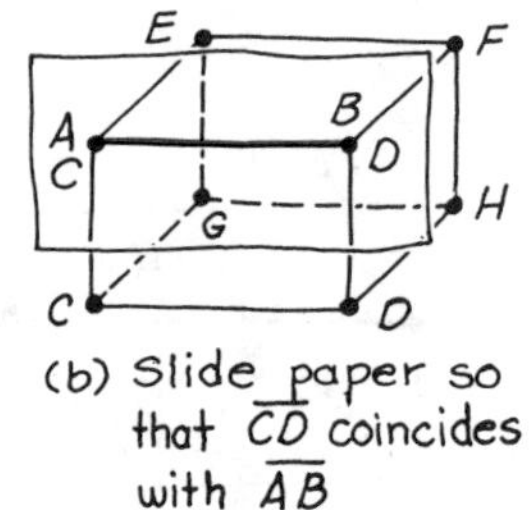

Figure 1-18

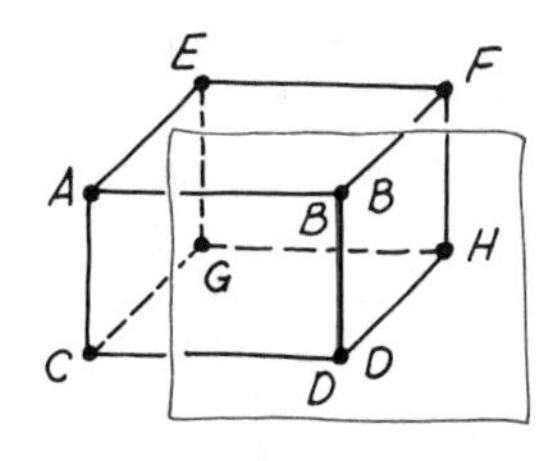

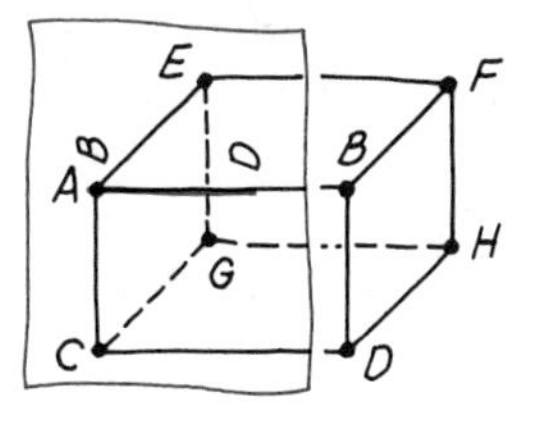

Figure 1-19

pletely, they are said to be **congruent.** Using the symbol $\simeq$ for congruent, we can write

$$\overline{AB} \simeq \overline{CD}$$

Is $\overline{BD}$ congruent to $\overline{AB}$? If you trace line segment $\overline{BD}$ and attempt to match it to $\overline{AB}$, you will find that the two line segments do not coincide completely (Figure 1-19). The two line segments are *not* congruent; they do not have the same length. Can you find other pairs of line segments that are congruent in Figure 1-17? How can you show they are congruent?

Rays

Using the box as a model, we imagined an edge of the box to be a line segment and, if that line segment is extended indefinitely beyond its two end points, we considered that entire set of points as a line (Figure 1-20). If line $\overleftrightarrow{AB}$ is considered as a set of points in space, then $\overline{AB}$ can be thought of as a **subset** of that set. That is, all the points of $\overline{AB}$ are in the set or collection of points that make up $\overleftrightarrow{AB}$. Subset $\overline{AB}$ is a part of the set $\overleftrightarrow{AB}$ (very much like a subcommittee is part of a committee).

We can describe other subsets of the set of points of a line. Consider the following points: all the points of $\overline{AB}$ and all the points that form the extension of $\overline{AB}$ in the B direction (Figure 1-21). This subset of points of $\overleftrightarrow{AB}$ is called a **ray.**

Physically, you might imagine a ray as a beam of light coming from a pinpoint source of illumination. If you paint the glass plate in front of a flashlight bulb black and leave just one small opening for light to come through, the beam of light will look like a ray. Actually, it may seem as if this ray stops at a wall or at the ceiling. A ray, in a geometric sense, is unending; it continues forever in a particular direction.

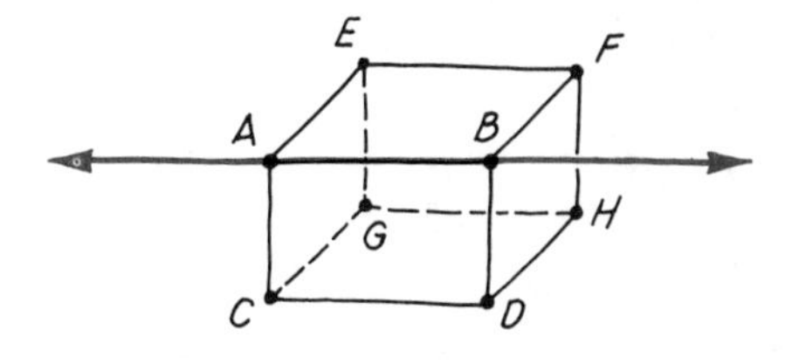

Figure 1-20

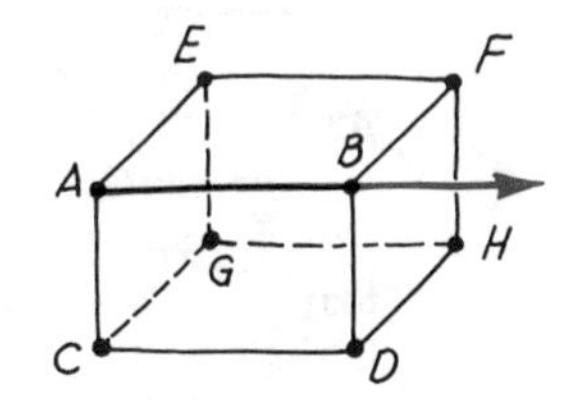

Figure 1-21

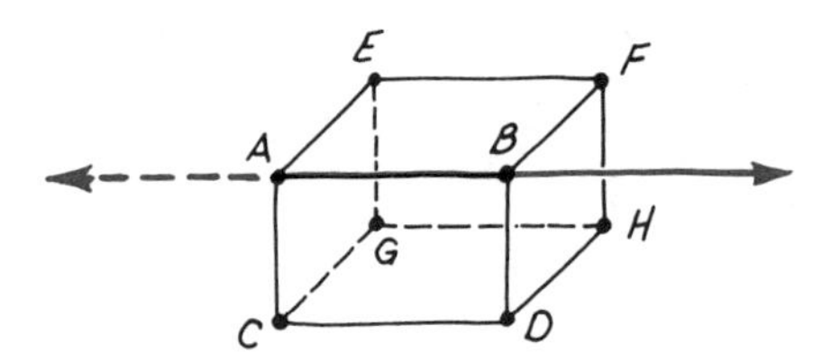

Figure 1-22

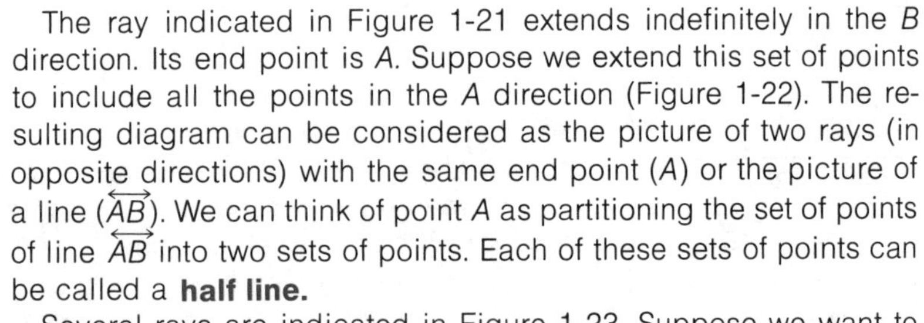

The ray indicated in Figure 1-21 extends indefinitely in the *B* direction. Its end point is *A*. Suppose we extend this set of points to include all the points in the *A* direction (Figure 1-22). The resulting diagram can be considered as the picture of two rays (in opposite directions) with the same end point (*A*) or the picture of a line ($\overleftrightarrow{AB}$). We can think of point *A* as partitioning the set of points of line $\overleftrightarrow{AB}$ into two sets of points. Each of these sets of points can be called a **half line.**

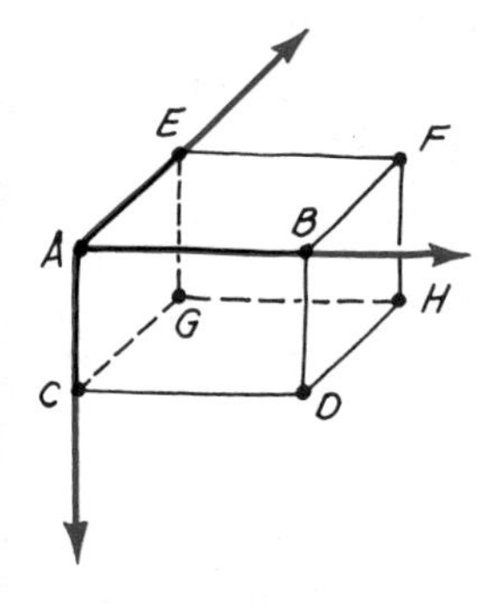

Figure 1-23

Several rays are indicated in Figure 1-23. Suppose we want to refer to one of them. How can we identify it? Calling it "ray *A*" would only lead to confusion, since *A* is the end point for each of these rays. We can avoid confusion by naming the end point of the ray along with some other point on the ray. If we refer to ray *AE*, someone else will know exactly which ray we are talking about.

In mathematical language we could write $\overrightarrow{AE}$. The arrow symbol indicates that we are referring to a ray. It tells us nothing about the direction of the ray. The first letter, *A*, identifies the end point of the ray, and the second letter, *E*, indicates the direction of the ray. How would the other rays in Figure 1-23 be named?

Figure 1-24

Angles

Take the cardboard box and use a knife to cut the top along edge $\overline{AB}$, $\overline{BF}$, and $\overline{FE}$ so that the top of the box is like a hinged door with the hinge along $\overline{AE}$ (Figure 1-24). Imagine that an edge of the lid and the edge of a side are rays as in Figure 1-25. Rays $\overrightarrow{AX}$ and $\overrightarrow{AB}$ are two rays with the common end point *A*. The set of points of $\overrightarrow{AX}$ together with the set of points of $\overrightarrow{AB}$ can be called an **angle.** An angle consists of two rays with a common end point. If each ray is a set of points, then the angle is made up of all the points that either belong to one ray *or* belong to the other ray. This is called the **union** of the two sets of points ($\overrightarrow{AB}$ and $\overrightarrow{AX}$).

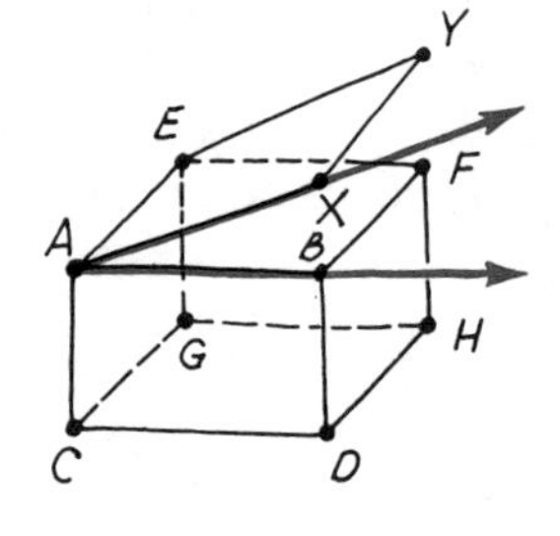

Figure 1-25

Each of the two rays that form an angle is called a **side** of the angle. Rays $\overrightarrow{AB}$ and $\overrightarrow{AX}$ are sides of the angle in Figure 1-25. The

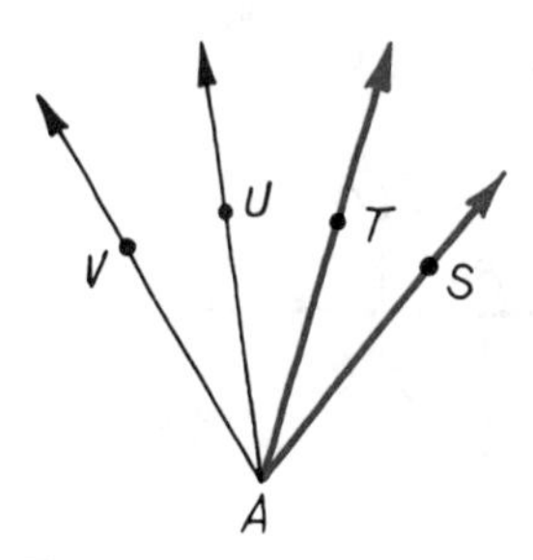

Figure 1-26

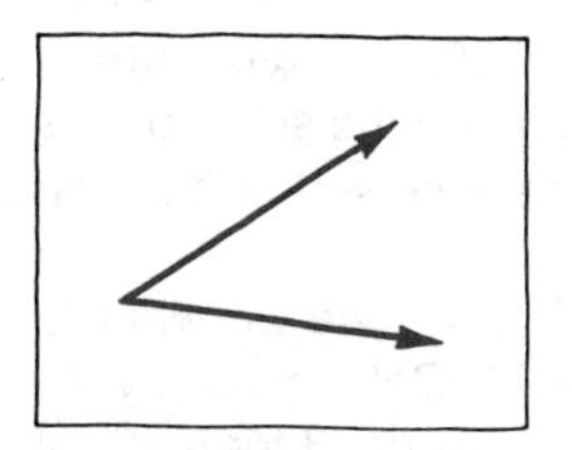

Figure 1-27

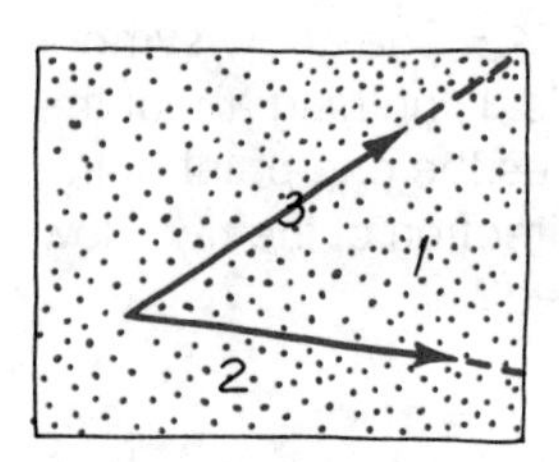

Figure 1-28

common end point, A, is called the **vertex** of the angle. Often, an angle is identified by naming its vertex. This angle might be called angle A. Using the symbol $\angle$ for angle, we can write "$\angle A$."

Sometimes, however, naming an angle by its vertex leads to confusion. Which angle in Figure 1-26 is $\angle A$? We need more information. We can clearly identify the angle we mean by naming three points: a point on one ray, the vertex, and a point on the other ray. The angle outlined in brown in Figure 1-26 might be named $\angle SAT$ or $\angle TAS$. In each case a point on one ray is named, next the vertex, and then a point on the other ray. The vertex is always named by the middle letter. How would you name the other angles in Figure 1-26?

Figure 1-27 is a diagram of an angle on a flat surface. If you visualize that flat surface as a set of points forming a plane, then the plane is partitioned into three subsets of points (Figure 1-28): (1) the subset of points enclosed by the angle (interior points); (2) the subset of points outside the angle (exterior points); and (3) the subset of points that form the angle (all the points on either ray). The points in the plane inside or outside the angle are *not part of the angle*. The angle consists of the two rays only—no interior or exterior points.

Comparing Angles

If we open the lid of the box wider and wider, we can form many different angles as shown in Figure 1-29. How are these angles different? We might respond that some angles form a larger opening than other angles. In Figure 1-30 the lid of the box is opened so far back that its edge and the top edge of the side of the box seem to form a straight line. This type of angle might be called a **straight angle.***

**In many mathematics books two rays with a common end point are not considered an angle if the two rays are part of the same line, as is the case with this straight angle.*

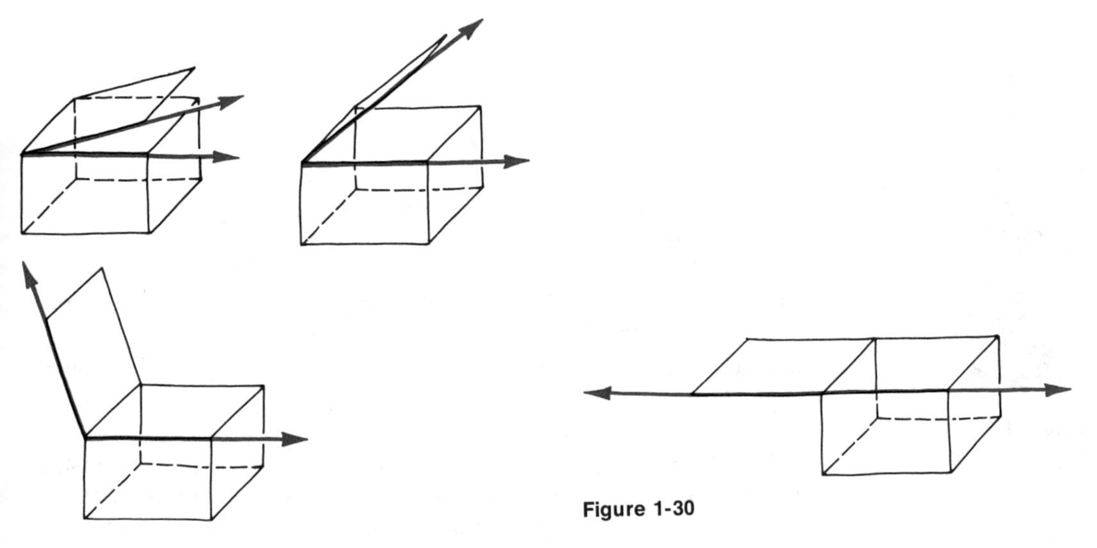

Figure 1-30

Figure 1-29

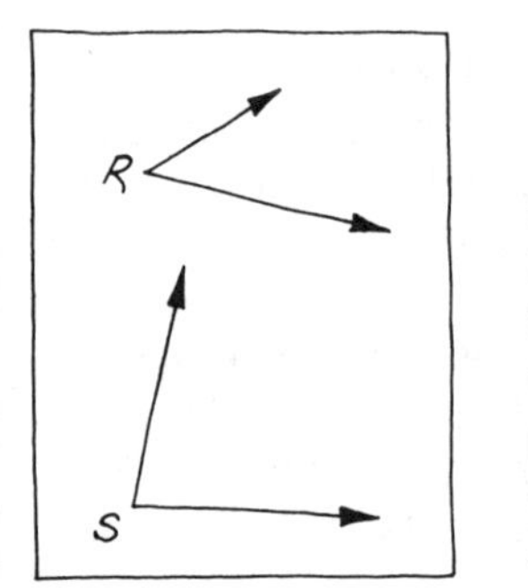

Figure 1-31

How can we compare the amount of opening of one angle with the amount of opening of another angle? Two angles are drawn in Figure 1-31. Which has the greater opening? We might compare the openings of the two angles by following a procedure similar to that used for comparing line segments. We can take some tracing paper and trace $\angle S$ on the paper. Then we slide the paper over the diagram so that the vertex of $\angle S$ coincides with the vertex of $\angle R$ and one side of $\angle S$ coincides with one side of $\angle R$ (Figure 1-32). Note that the other side of $\angle R$ is now located inside $\angle S$. This indicates that $\angle R$ has a smaller opening than $\angle S$. (The lengths of the sides of the angles have nothing to do with the amount of opening. As you know, those sides are rays and rays are unending.)

Two other angles are drawn in Figure 1-33, $\angle T$ and $\angle U$. Which has the bigger opening? If you perform the same test with tracing paper, you will find that both sides of one angle coincide with the corresponding sides of the other (Figure 1-34). This shows that the angles have just as much opening. We might say they are the same size. The two angles, $\angle T$ and $\angle U$, can be called **congruent angles.** Mathematically, we could write $\angle T \simeq \angle U$.

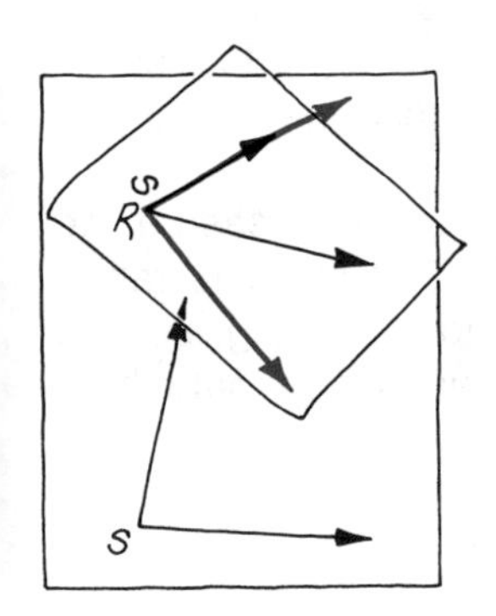

Figure 1-32

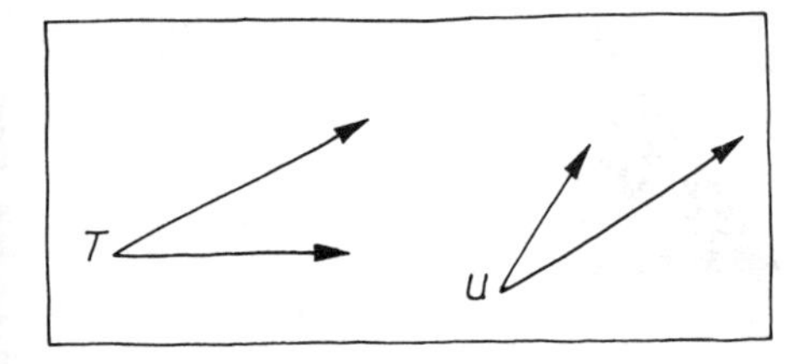

Figure 1-33

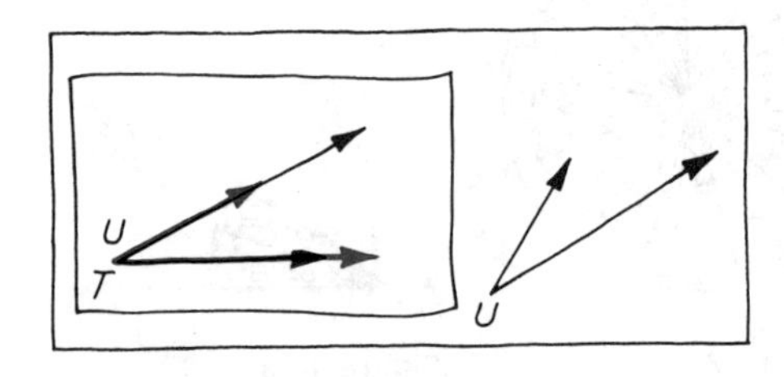

Figure 1-34

Measuring Angles

Figure 1-35

Sometimes it is necessary to know *how much larger* one angle is than another. Then we need some way of measuring the size of each angle. As an example, imagine that slices of pizza represent angles as in Figure 1-35. The two slices of pizza are different sizes. If you consider the angles formed by the two straight sides of each slice of pizza, one angle has more of an opening than the other angle (Figure 1-36). But how much larger is one angle than the other?

A ruler is not going to help us. We are not going to measure along a straight line (linear measurement). Remember that the sides of an angle are rays and cannot be measured since rays are unending. We need some other way of measuring the amount of opening for each angle.

When we attempted to measure the length of straws, we developed a unit of measurement. At first we used a paper clip as our unit. We were able to measure each straw with the paper clip units and to determine how much longer one straw was than the other (in terms of paper clips).

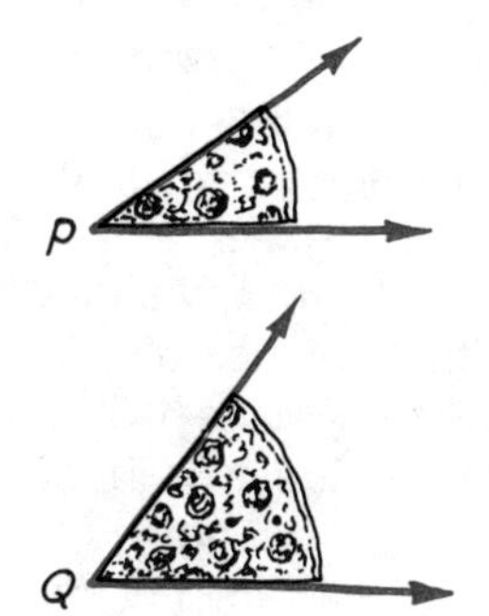

Figure 1-36

Let us develop an appropriate unit for measuring these angles. Suppose we take a whole pizza and cut it into small pieces, all just alike (Figure 1-37). We might use each small piece as a unit for measuring the size of the opening of the angles formed by the larger slices of pizza.

If we cover slice P with small "pizza units," we find that it is completely covered by four pizza units (Figure 1-38). It takes six pizza units to cover slice Q. We can say that the angle formed by slice Q has an opening two pizza units larger than the opening of the angle formed by slice P.

Although we were able to use the pizza unit to measure the angles, this unit is not very convenient. First of all, it is rather messy! Second, someone else might not know how big a pizza unit is. We need a standard unit to measure angles so that others can interpret our measurements.

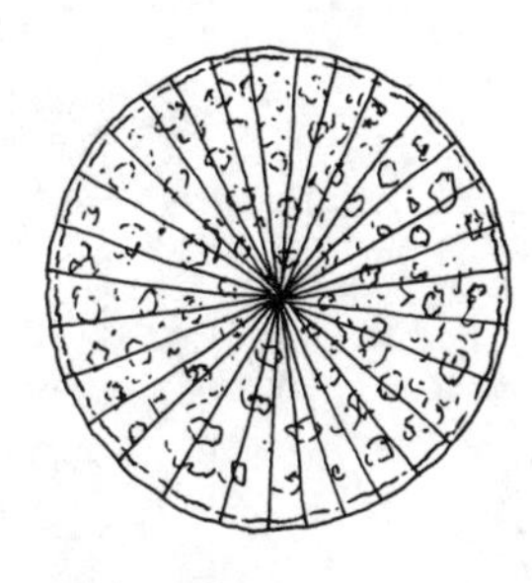

Figure 1-37

Figure 1-38

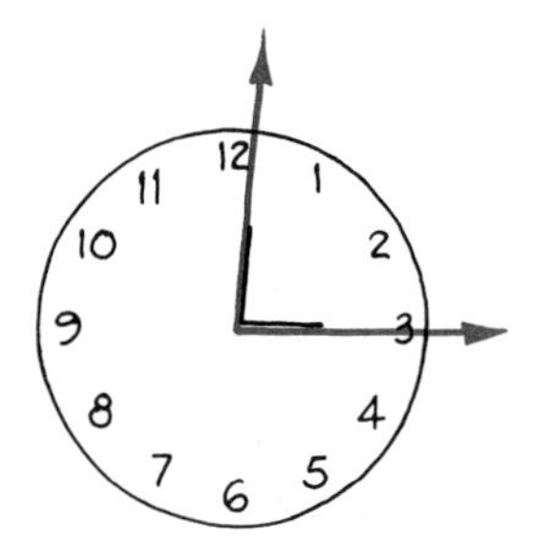

Figure 1-39

The best-known standard unit for measuring angles is the **degree.** Imagine the face of a clock and think of the two hands as rays. Then the hands of the clock form an angle (Figure 1-39). Now imagine a clock with many more hands. Suppose the hands are rays with the center of the clock as their common end point, and position these rays so that they partition the face of the clock by forming 360 congruent angles (angles that are the same size). Then each of these angles is called a degree (Figure 1-40). Like the pizza unit, the degree is used to find the size of angles.

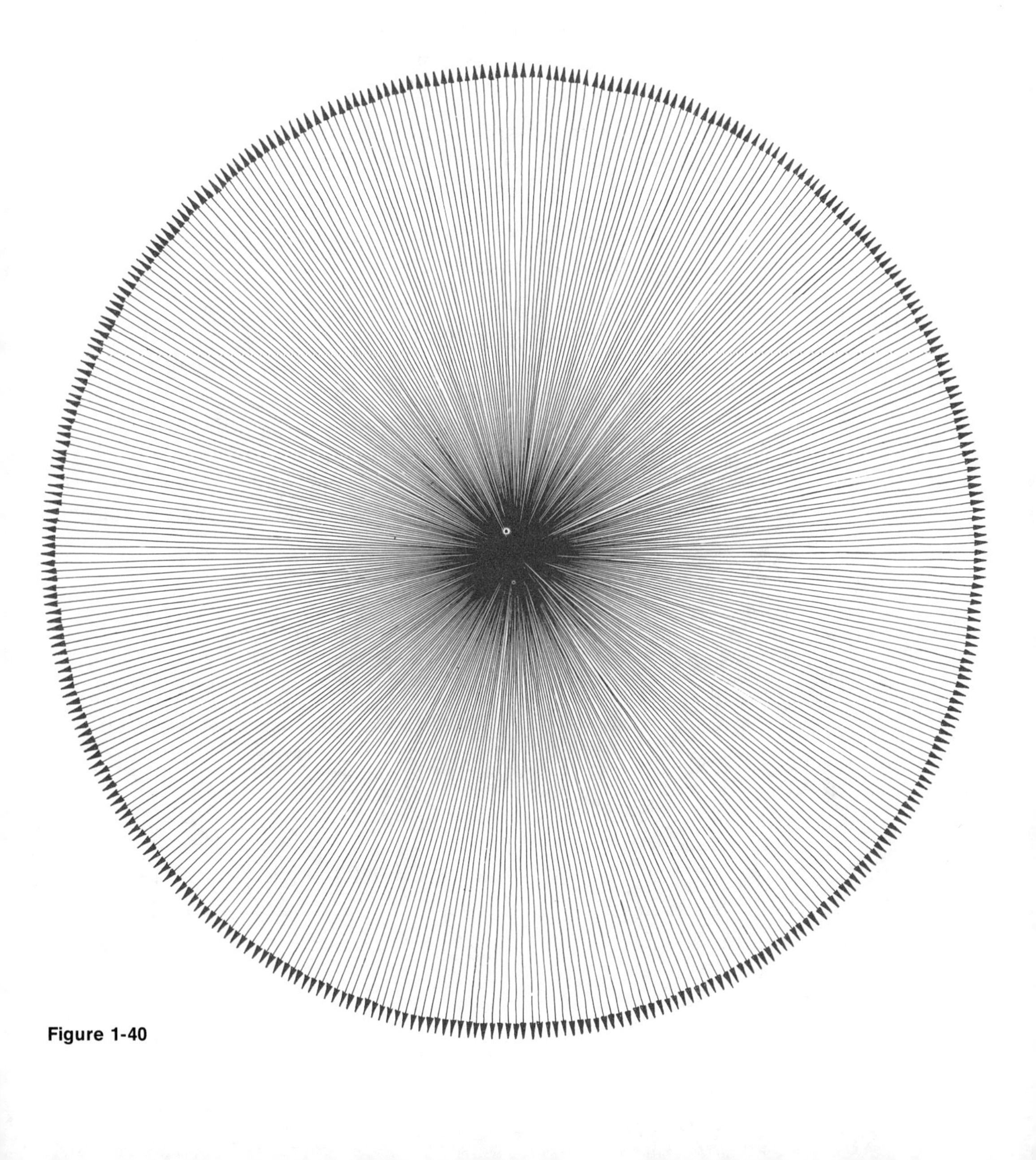

Figure 1-40

Using a Protractor

While the ruler helps us measure line segments, the protractor helps us measure angles (Figure 1-41). Suppose we want to measure $\angle ABC$ (Figure 1-42) in degrees by using a protractor. The following steps would be necessary:

1. Place the protractor over the picture of $\angle ABC$ so that the midpoint M of the bottom of the protractor coincides with the vertex B of the angle.
2. With the midpoint M at vertex B, rotate the protractor so that side $\overrightarrow{BC}$ of the angle coincides with the bottom edge of the protractor.
3. Notice that there are two scales on the protractor: read the one that has a 0° mark coinciding with a side of the angle (in this case, the lower scale).
4. The side $\overrightarrow{BA}$ of the angle intersects the protractor at the 40° mark on the lower scale.

This tells us that it would take 40 of the degree units to cover the interior of $\angle ABC$. In other words, the measure of $\angle ABC$ is 40. We can write this as follows: $m\angle ABC = 40$. The m means "the measure of." We might also say that $\angle ABC$ **measures** 40°.

Figures 1-43 and 1-44 illustrate the use of the protractor to measure other angles. Angle DEF measures 120° while $\angle GHI$ measures 75°. In one case the lower scale is used; in the other case the upper scale is used.

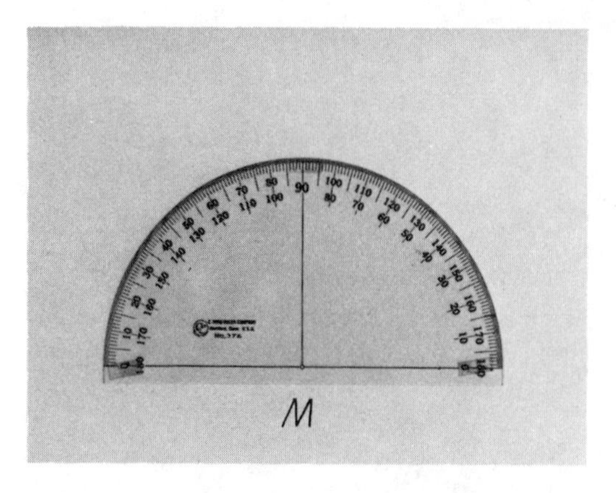

Figure 1-41

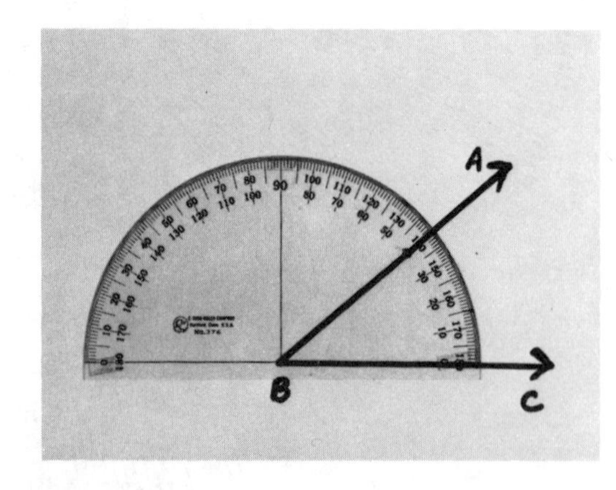

Figure 1-42

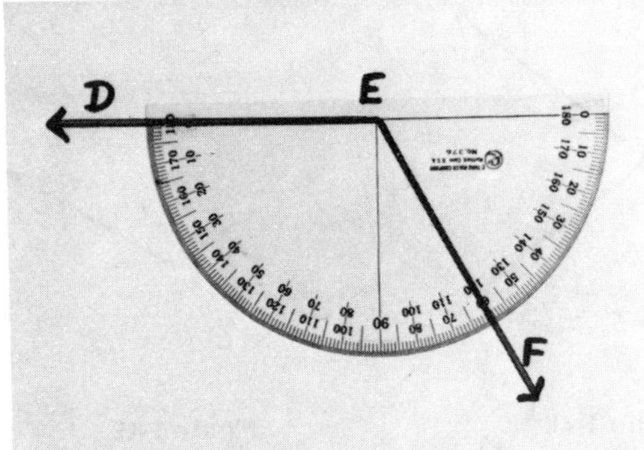

Figure 1-43

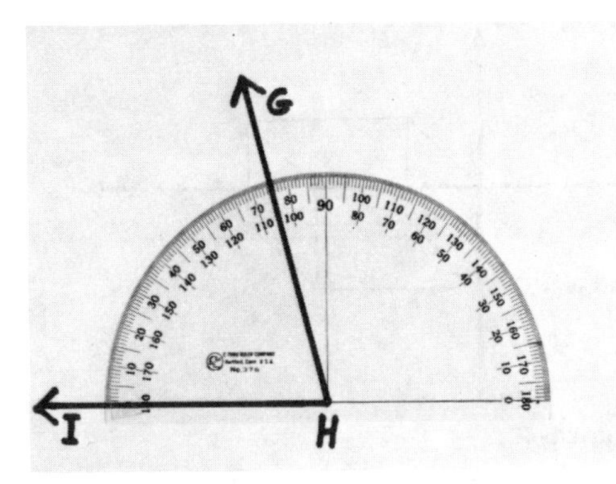

Figure 1-44

Special Angles

You have already been introduced to one special type of angle, the straight angle. The straight angle looks just like a line. Using a protractor, you can verify that a straight angle RST measures 180° (Figure 1-45).

Another special angle is the **right angle.** Suppose you draw another ray in the diagram for $\angle RST$. Draw the ray $\overrightarrow{SV}$ so that the interior of $\angle RST$ is partitioned into two parts that are just alike (Figure 1-46). You have now drawn two angles that are congruent, $\angle RSV$ and $\angle VST$. Because these angles are congruent, they have the same measure. Each angle measures half of 180°, or 90°. Angles RSV and VST are right angles. **All right angles measure 90°.** If you examine the cardboard box we have been using as a model and think of its edges as rays, how many right angles can you find? When two lines intersect and form right angles, they are called **perpendicular** lines. We will also call line segments perpendicular if they can be extended to form perpendicular lines.

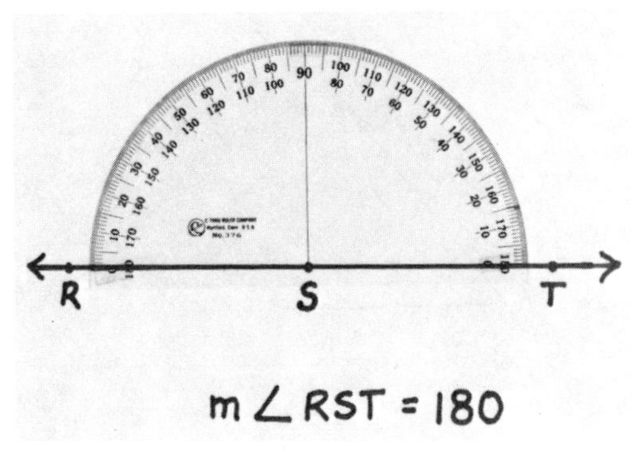

Figure 1-45

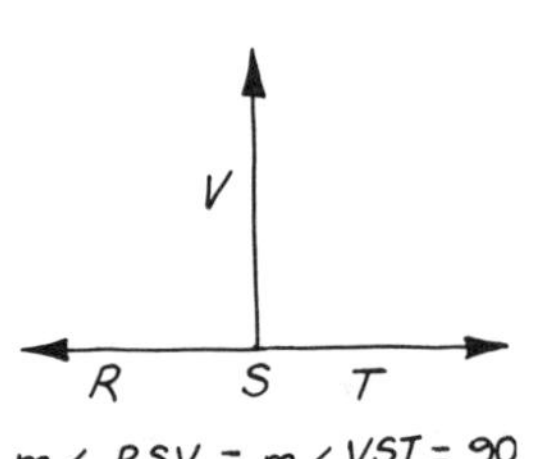

Figure 1-46

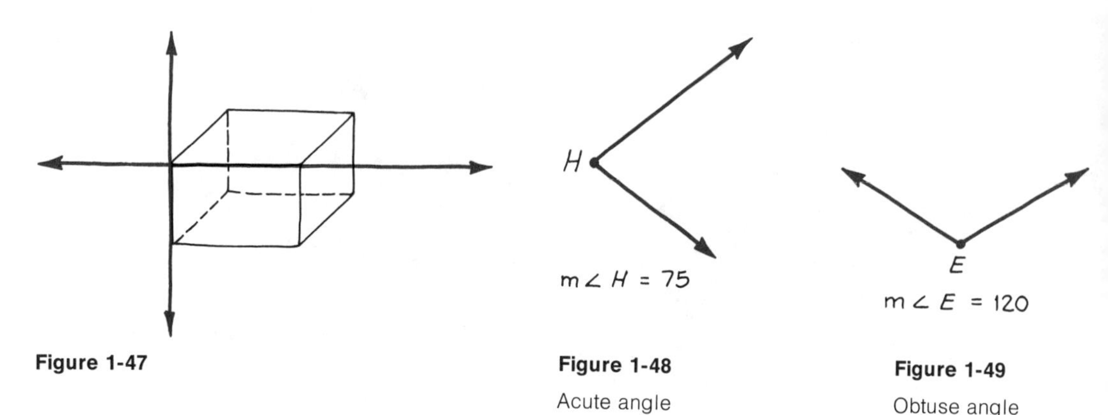

Figure 1-47

Figure 1-48
Acute angle

Figure 1-49
Obtuse angle

L
M N
m∠ M = 40

O
P Q
m∠ P = 50

Figure 1-50

Two edges of the box in Figure 1-47 have been extended to form lines. They form right angles and can be called perpendicular lines. Locate all the pairs of perpendicular lines suggested by the edges of the box.

Any angle whose measure is between 0 and 90° is called an **acute angle.** Angle *H* in Figure 1-48 measures 75°. It can be called an acute angle. Angle *E* in Figure 1-49 measures 120°. It does not qualify as an acute angle, but it has a special name, too—**obtuse angle.** Any angle whose measure is between 90 and 180° is an obtuse angle.

Angles *M* and *P* in Figure 1-50 measure 40 and 50°, respectively. The sum of the number of degrees in both angles is 90. Angles *M* and *P* are called **complementary angles.** Whenever the sum of the number of degrees in two angles is 90, the angles are called complementary angles.

Notice that if we draw $\angle P$ on a piece of tracing paper and slide the drawing over so that vertex *M* coincides with vertex *P* and side $\overrightarrow{PO}$ of $\angle P$ coincides with side $\overrightarrow{MN}$ of $\angle M$ as in Figure 1-51, sides $\overrightarrow{ML}$ and $\overrightarrow{PQ}$ form a right angle. This is because the angles are complementary. A right angle measures 90°, and the sum of the number of degrees in the two angles ($m\angle M + m\angle P$) is 90.

When the sum of the measures in degrees of two angles is 180, the angles are called **supplementary angles.** Angles *T* and *W* in Figure 1-52 are supplementary since the sum of their measures in degrees is 180.

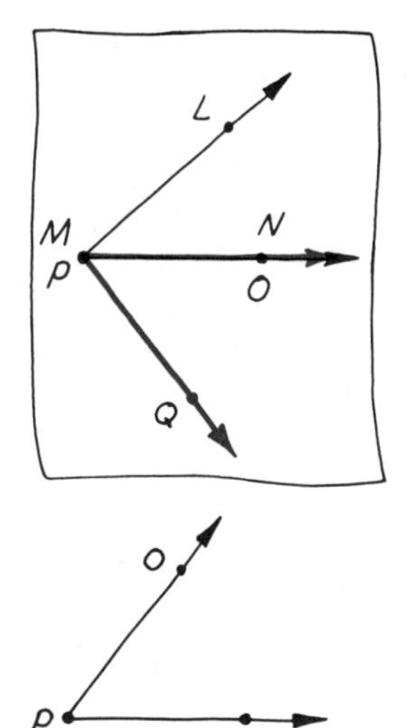

O
P Q

Figure 1-51

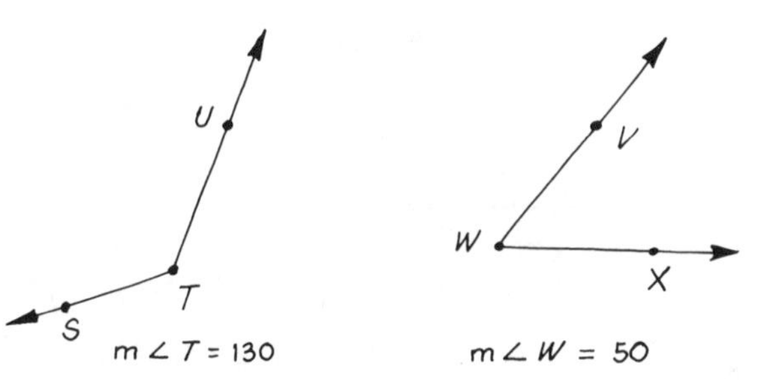

Figure 1-52

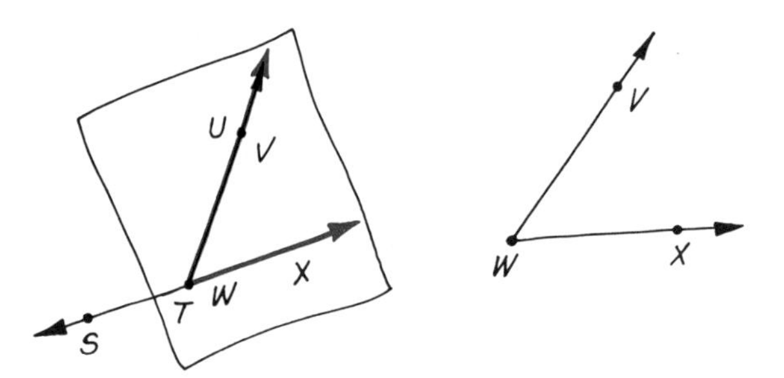

Figure 1-53

If you draw $\angle W$ on tracing paper and move it over so that vertex T coincides with vertex W and side $\overrightarrow{TU}$ of $\angle T$ coincides with side $\overrightarrow{WV}$ of $\angle W$ (Figure 1-53), you can see that $\overrightarrow{TS}$ and $\overrightarrow{WX}$ form a straight angle. A straight angle contains 180°, and the sum of the measures of these two angles ($m\angle T + m\angle W$) in degrees is 180.

Curves

Have you ever taken a piece of paper and pencil and just doodled? If you think of your paper as a plane, then the pictures that resulted from your doodling were drawings of geometric figures on a plane. If the plane is a set of points, then the subset of points covered by your pencil markings forms a **plane figure.**

Take a look at the plane figures illustrated in Figure 1-54. Let us examine the similarities and differences among these figures. By doing this, we will develop a vocabulary of geometric terms useful for describing the doodling.

All the plane figures in Figure 1-54 except (g), (n), and (o) have something in common. What is it? Each can be drawn completely without lifting the pencil from the paper. Any figure that can be drawn without lifting a pencil from a paper can be called a **plane curve.**

You might use string or yarn to make models for curves.* Then, if you can use just *one* piece of string to make a particular figure, the figure is a model of a curve. Notice that the figures in (g), (n), and (o) cannot be formed by using just one piece of string (and in drawing these figures you would have to lift your pencil from the paper a few times). Each of the figures in (g), (n), and (o) is not *a* curve; each can be thought of as being made up of several curves.

*Whenever the word curve is used in this book, we will mean a plane curve.

Figure 1-54

Curves

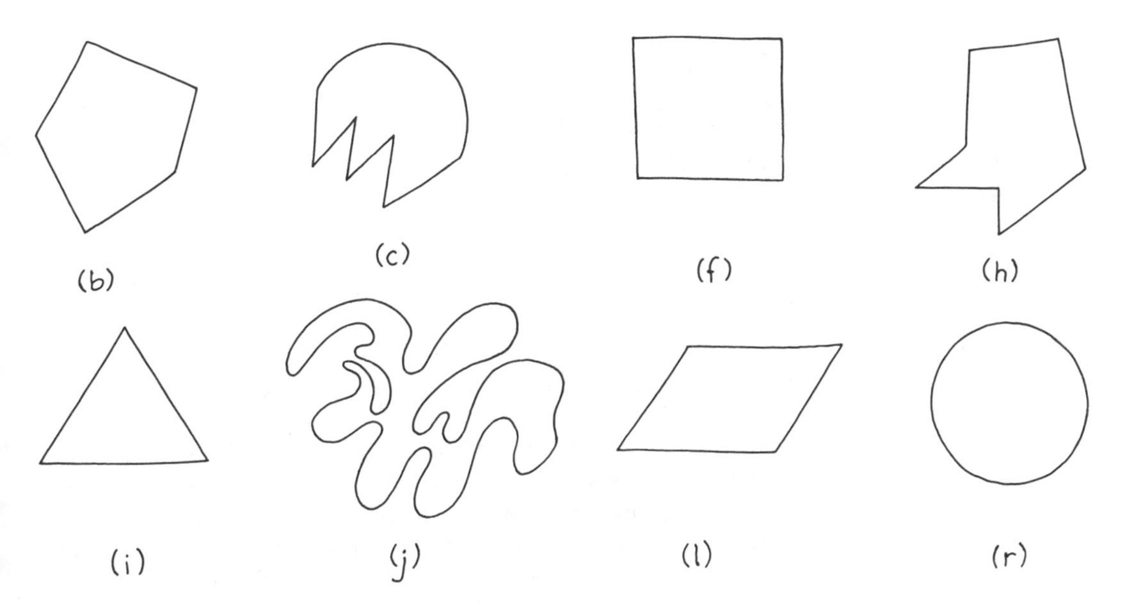

Figure 1-55
Simple closed curves

The mathematical meaning of *curve* differs from its everyday meaning. We usually think of a curve as a figure that is "not straight." But in mathematics a curve *can* be straight, like the line segment in (a). Describing a figure as a curve simply indicates that you can use just *one* piece of string to make a model of the figure.

If the piece of string does not cross itself, then your model is a **simple curve.** In Figure 1-54 (a), (b), (c), (d), (f), (h), (i), (j), (k), (l), (p), and (r) represent simple curves. The curves in (e), (m), and (q) are *not* simple because each curve crosses itself at least once. These nonsimple curves are called **complex curves.**

If you have to join the ends of the piece of string to make a model of a particular curve, then the curve is called a **closed curve.** The curves represented in (b), (c), (e), (f), (h), (i), (j), (l), (m), and (r) are all closed curves. In (a), (d), (k), (p), and (q) you can tell where the ends of the string must be located. These curves are called **open curves.**

Some special curves are both simple and closed at the same time. These curves are called **simple closed curves.** The curves illustrated in Figure 1-55 include all the curves in Figure 1-54 that are simple closed curves.

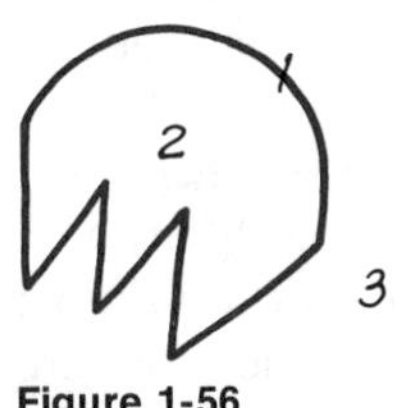

Figure 1-56
Curve (c)

If the simple closed curve (c) shown in Figure 1-55 were a fence and the plane were a field, you would have to cross over the fence to travel from one point inside the fence to another point outside the fence. An important characteristic of a simple closed curve is that it partitions the plane into exactly three subsets of points. If the plane is a set of points, then the simple closed curve partitions the plane into three subsets of points as follows (Figure 1-56): (1) the subset of points forming the curve, (2) the subset

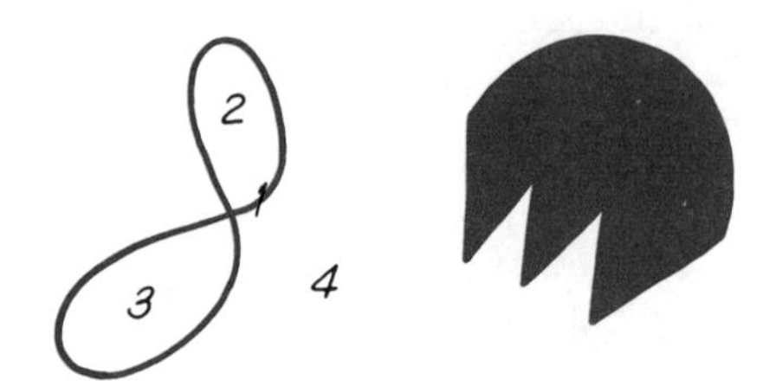

Figure 1-57
Curve (e)

Figure 1-58
Region bounded by curve (c)

of points inside the curve, and (3) the subset of points outside the curve.

Curve (e) is *not* a simple closed curve and does not partition the plane into exactly three sets of points. In this case, the plane is partitioned into four sets of points as indicated in Figure 1-57.

The portion of a plane inside a simple closed curve together with the curve is called a **region** of the plane. The portion of the plane indicated by brown shading in Figure 1-58 is a region. The region contains all the points of the plane located inside curve (c) along with the points of the curve. Curve (c) serves as a **boundary** and is part of this region.

Polygons

Take another look at the simple closed curves illustrated in Figure 1-55. Some of these curves are made up entirely of line segments. They have a special name—**polygons.** A polygon is a simple closed curve formed by the union of line segments. In Figure 1-55 (b), (f), (h), (i), and (l) are polygons. Curve (c) is not a polygon since one part of the curve is not a line segment. Note also that curve (m) in Figure 1-54 does not qualify as a polygon. Although (m) is made up entirely of line segments and is closed, it is *not* a simple curve. To summarize, a figure is a polygon if it is (1) a curve, (2) simple, (3) closed, and (4) made up entirely of line segments.

Let us examine the polygon in Figure 1-59. We can call this figure polygon *ABCDE.* The line segments forming a polygon are called **sides.** Line segments $\overline{AB}$, $\overline{BC}$, $\overline{CD}$, $\overline{DE}$, $\overline{EA}$ are the sides of this polygon. The points where the sides meet are called vertices.* Any two sides that have a common end point are called **adjacent sides.** For example, $\overline{AB}$ and $\overline{BC}$ are adjacent sides and vertex *B* is their common end point. Any two vertices are called **adjacent vertices** if they are the end points of the same side of the polygon.

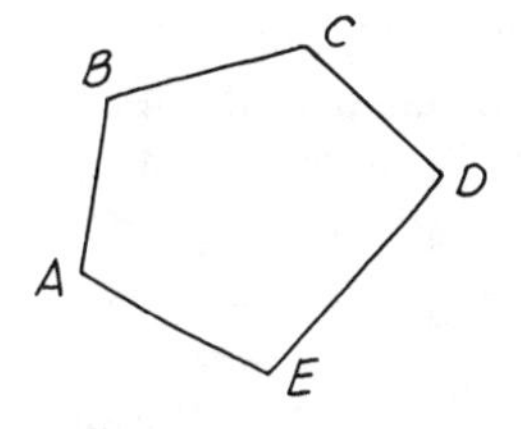

Figure 1-59

**Singular: vertex.*

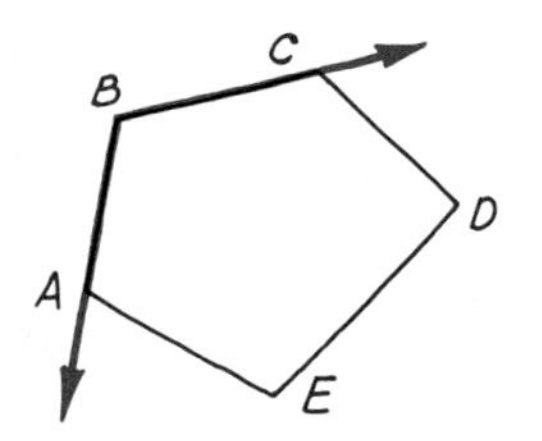

Figure 1-60

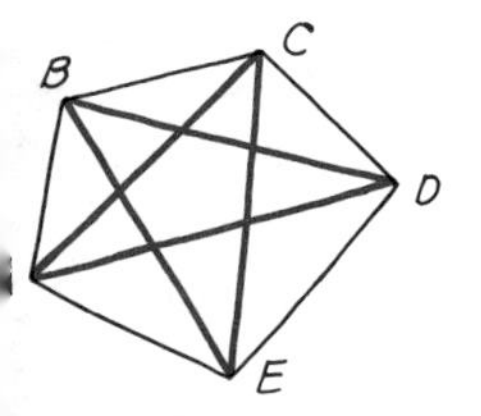

Figure 1-61

Points *A* and *B* are adjacent vertices; *A* and *C* are nonadjacent vertices.

If you imagine that the sides of the polygon are rays, then each pair of adjacent sides would form an angle. In Figure 1-60 the ray along $\overline{BA}$ and the ray along $\overline{BC}$ form an angle (two different rays with a common end point) with vertex at *B*. In this way, each vertex of the polygon might be thought of as a vertex for an angle. These angles are described as the angles formed by the polygon. Therefore, polygon *ABCDE* may be described as having five sides and five angles.

Consider the pairs of nonadjacent vertices in polygon *ABCDE*, such as *A* and *C*, *A* and *D*, and so on. If we form line segments by using these pairs of vertices as end points (Figure 1-61), these line segments are called the **diagonals** of polygon *ABCDE*. A diagonal, then, is a **line segment formed by using two nonadjacent vertices as end points.** Notice that if you used two *adjacent* vertices as end points, the resulting line segment would be a side of the polygon.

All the polygons from the set of simple closed curves (Figure 1-55) are pictured in Figure 1-62. There is something different about the polygon in (h). What is it? You might say that the sides of a few of the angles seem to "cave in." None of the sides of the angles of the other polygons do this. Polygon (h) is called a **nonconvex polygon.** Polygons (b), (f), (i), and (l) are called **convex polygons.**

You can perform a very easy test to determine whether a polygon is convex or nonconvex. Draw all the diagonals of the polygon. If all the diagonals are located inside the polygon (in the region bounded by the polygon), then that polygon is convex. If any diag-

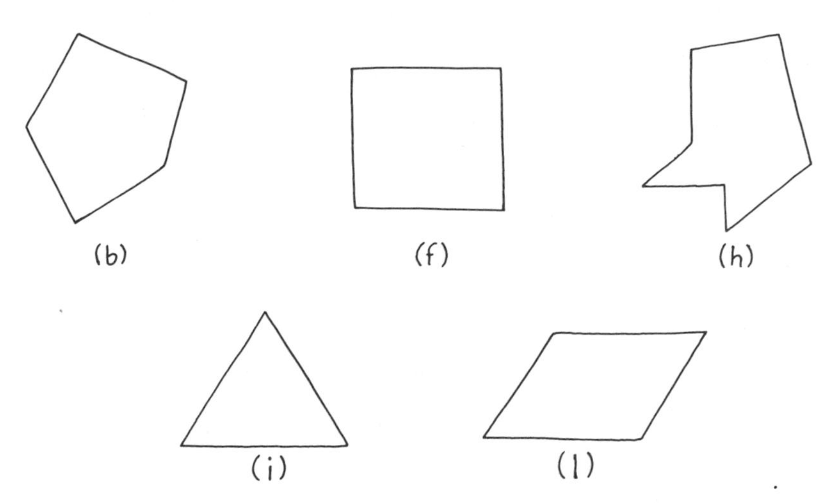

Figure 1-62

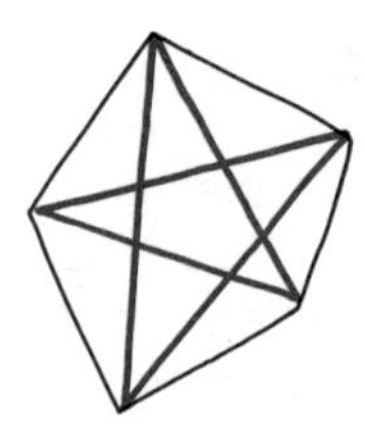

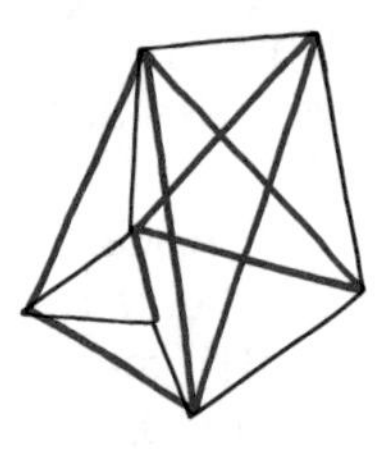

Figure 1-63

onal is located even partly outside the polygon, then the polygon is nonconvex. The test is performed on (b) and (h) in Figure 1-63.

Looking at the polygons in Figure 1-62 once more, note that there is something special about polygons (f) and (i). What is it? In each polygon, all the sides have the same measure. Each polygon has congruent sides. Any polygon with congruent sides is called an **equilateral** (equal-sided) **polygon.**

But polygons (f) and (i) of Figure 1-62 are even more special. They are the most special polygons of all. Not only are all the sides of each of these polygons equal in measure, but all the angles are equal in measure (congruent), too! Polygons (f) and (i) are called **regular polygons.** A polygon can qualify as a regular polygon if it has *both congruent sides and congruent angles.* Is polygon (l) a regular polygon? Why not?

We have discussed ways of describing a curve in the language of geometry and have developed a jargon to help us express geometric ideas. Let us summarize some of these terms by examining a curve and describing it with this new language.

Look at the curve pictured in Figure 1-64. What can we say about that curve? Here are some observations:

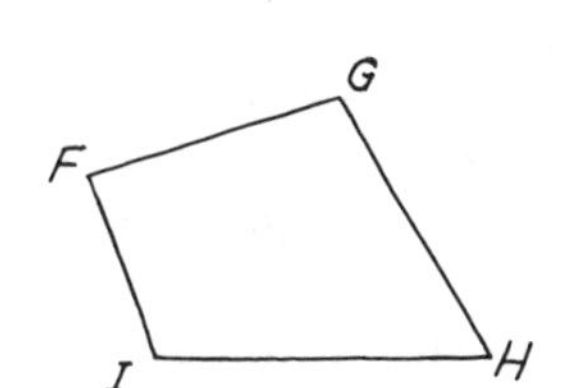

Figure 1-64

1. The figure is a curve (since it can be drawn without lifting a pencil from a paper).
2. It is simple (does not cross itself).
3. It is closed (end points are joined).
4. It is a simple closed curve (since it is simple, closed, and a curve).
5. It is a polygon (since it is a simple closed curve made up entirely of line segments).
6. It is *not* equilateral (since the sides are not congruent).
7. It does *not* have congruent angles.
8. It is *not* a regular polygon (since it does *not* have both congruent sides and congruent angles).

Try to draw some curves of your own. Can you draw the following curves?

1. A simple closed curve that is not a polygon
2. A nonconvex regular polygon
3. An equilateral polygon that is not a regular polygon
4. An open curve that is simple and made up entirely of line segments

Are any of these curves impossible to draw? Why?

More Experiences

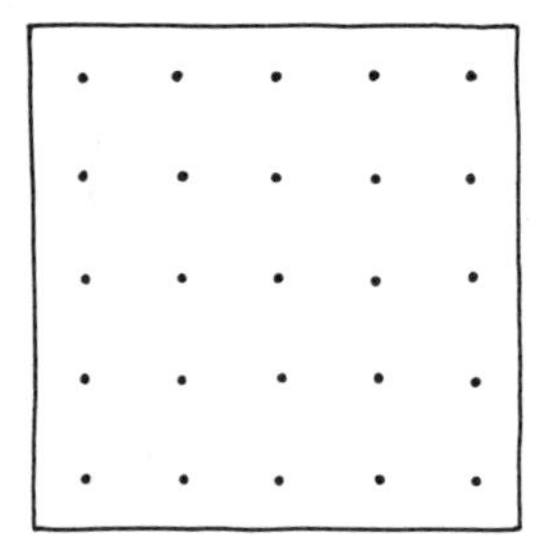

1-1. A **geoboard** is a useful device for exploring geometric relationships. One type of geoboard that is easy to make is a square wooden board with 25 nails or brads arranged in a lattice fashion as shown in the diagram at the left. There are five rows of five nails the same distance apart vertically and horizontally. This type is often called a 5×5 geoboard. Elastic bands of different sizes and colors are used to make shapes on the geoboard. Wooden and plastic geoboards in many sizes are available commercially. Since we will be using geoboards for many activities in this book, you should make or obtain your own.

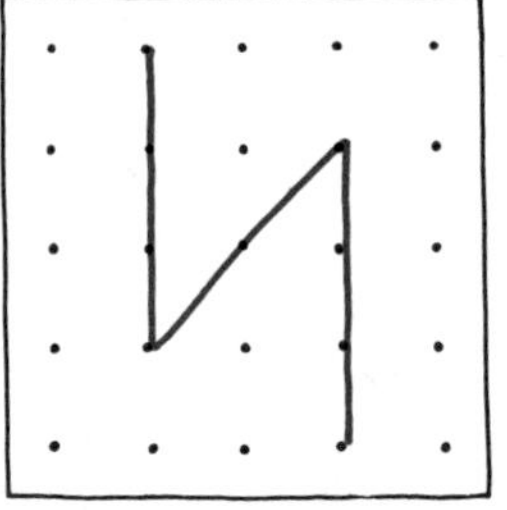

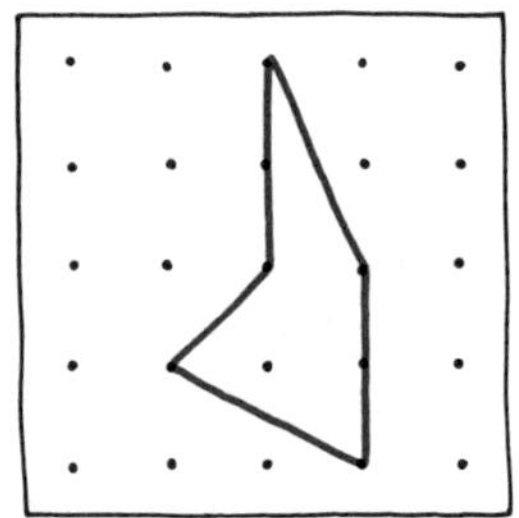

The elastic bands on the geoboards shown at the left form the shapes of a simple open curve and a simple closed curve. Can you make models of the following on a geoboard?

(*a*) Complex closed curve
(*b*) Complex open curve
(*c*) Convex polygon
(*d*) Nonconvex polygon

How many different equilateral polygons can you make on a geoboard? How many different regular polygons can you make? Draw them on paper with 25 dots arranged like the nails on a geoboard.

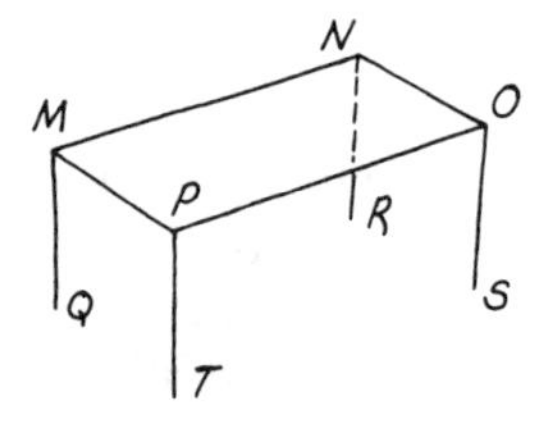

1-2. Suppose the diagram at the left represents the skeletal outline of a table, where *M*, *N*, *O*, and *P* are the corners of the rectangular top and points *Q*, *R*, *S*, and *T* are the end points of the legs. Imagine these same eight points as points in space. How many different lines can you name that include any two of these points? Name them (for example, $\overleftrightarrow{MS}$, $\overleftrightarrow{MT}$, $\overleftrightarrow{MQ}$).

How many pairs of parallel lines can you imagine? Name them. How many pairs of skew lines? Name them. How many pairs of intersecting lines? Name them.

1-3. Estimate the size in degrees of each of the angles shown below. Classify them as acute, obtuse, or right angles. Then measure them with a protractor to find their measures to the nearest degree. Which pairs of angles are complementary? Which pairs are supplementary?

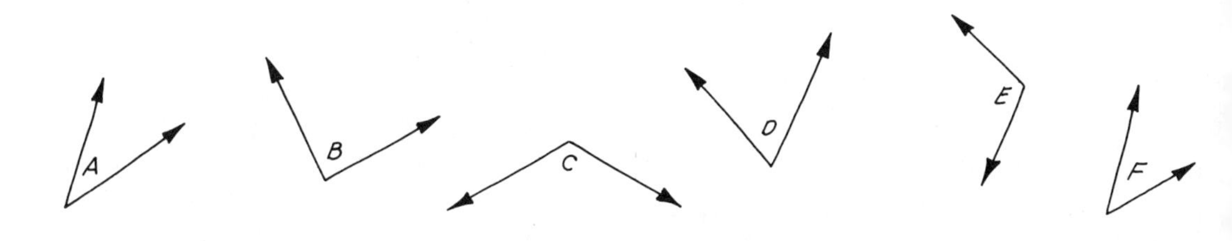

1-4. Which of the following curves can you draw? Draw examples of those that are possible.

(*a*) Complex closed curve
(*b*) Simple open curve
(*c*) Polygon with two sides
(*d*) Nonconvex equilateral polygon
(*e*) Equiangular polygon that is *not* a regular polygon

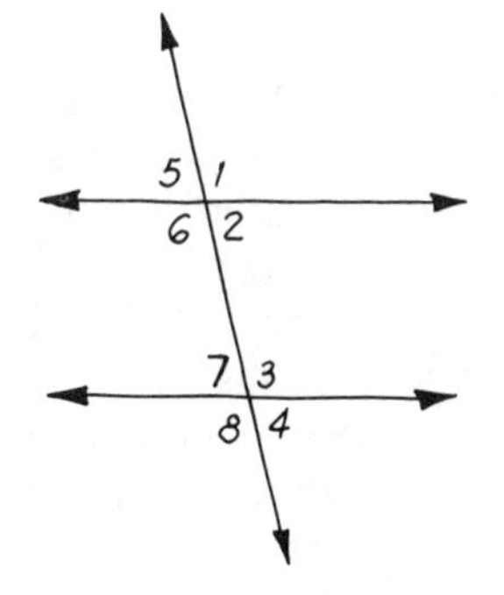

1-5. The diagram at the left represents one line intersecting two parallel lines. Eight angles are formed, labeled 1, 2, 3, . . . , 8. Which angles are congruent? How many pairs of congruent angles can you find? Which angles are supplementary? How many pairs of supplementary angles can you find?

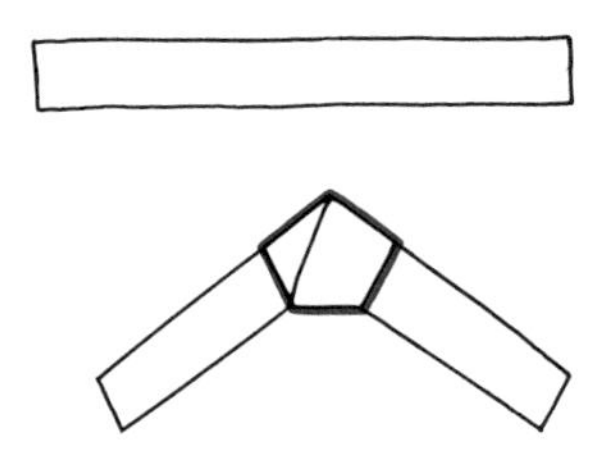

1-6. A very easy way to make a model of a regular pentagon is as follows (see diagram at left). First, cut a long, narrow strip of paper. Then fold the strip carefully into a knot. The knot forms the shape of a regular pentagon. Try to make a regular pentagon in this way.

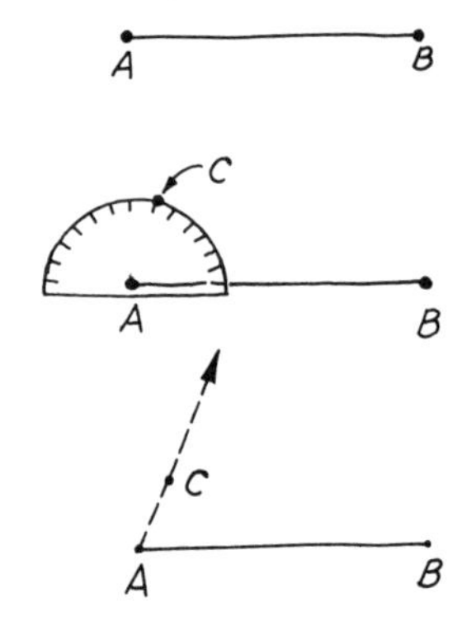

1-7. Besides being a tool for measuring angles, a protractor is useful for constructing angles with a given measure. For example, to construct an angle of 57° you might proceed as follows (see diagram at left):

1. Draw a line segment with end points *A* and *B*.
2. Place the protractor so that its center point coincides with *A* and its baseline coincides with *AB*.
3. Follow the scale that begins with 0 and locate 57°. Make a dot at that point. Call it *C*.
4. Remove the protractor and draw the line segment *AC*. The measure of ∠*BAC* is 57°.

In a similar way, use a protractor to draw models of angles with the following measures: 100°, 12°, 138°. How would you draw an angle of 200°?

2 Triangles

Making Models of Triangles

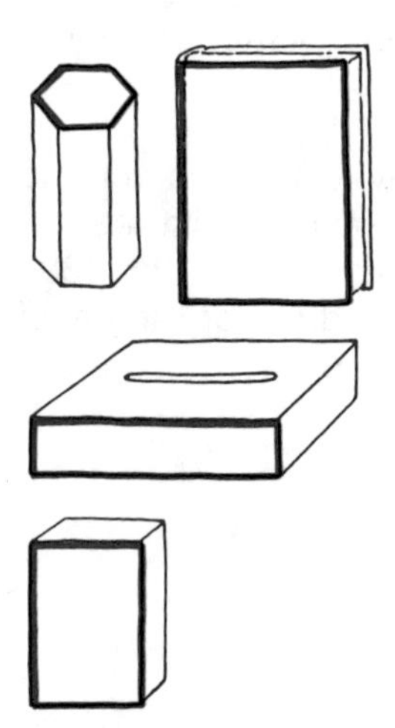

Figure 2-1

Figure 2-2

Many of the shapes we see around us look like polygons. We can imagine polygons to be part of the flat surface of different objects as in Figure 2-1. If a plane is a set of points, then a polygon can be thought of as a subset of the points of a plane.

It is interesting to make models of polygons and then study them. We can create models of polygons with paper, pencil, and straightedge, but we can make more tangible models by using some easily available materials. The following discussion will be much more valuable if you actually make the models suggested and examine them as you read.

The materials used in these activities are homemade cardboard strips and angle-fixers, brass fasteners, and elastic thread (Figure 2-2). You can easily make the strips by mounting the patterns included in Appendix A onto oaktag or lightweight cardboard of some sort and then cutting along the outlines to form strips of 14 different lengths (the number on each strip indicates the distance in centimeters between the end holes), each having a width of 2 cm. The holes are then made with a punch that creates very small holes (preferably $1/8$-in. diameter). The angle-fixers correspond to 30, 45, 60, 75, 90, 105, 120, 135, 150, and 180° angles, as labeled in Appendix B. The brass fasteners have a $1/2$-in. shank. The elastic thread is available in most department stores.

Each strip is a model for a line segment as represented by the line segment drawn on the strip from one end hole to the other. The strips can be joined with brass fasteners as shown in

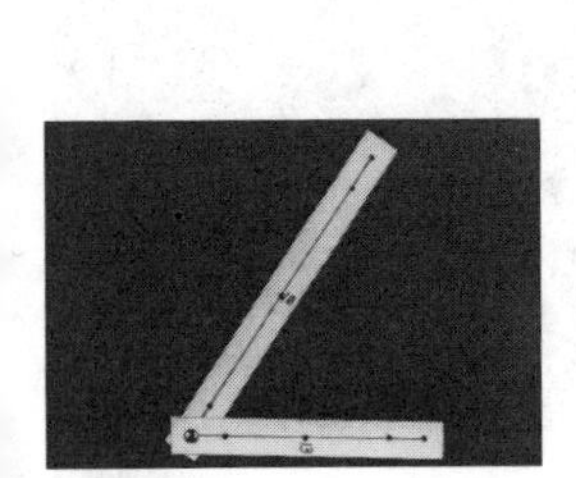

Figure 2-3

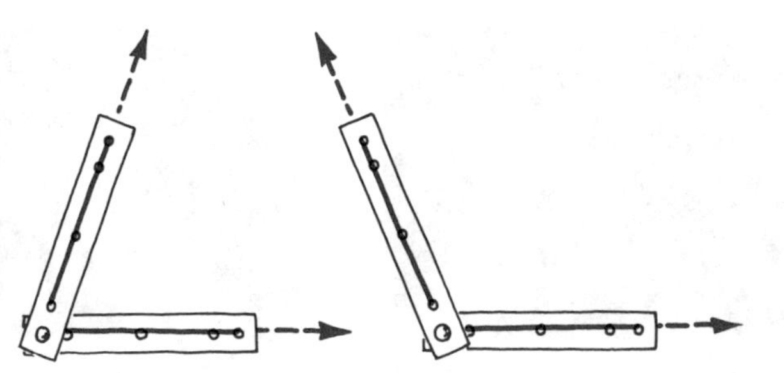

Figure 2-4

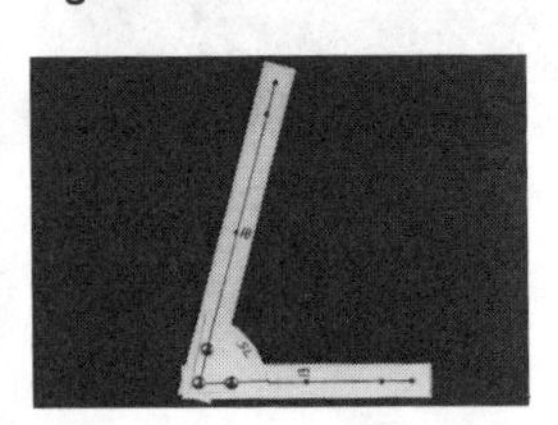

Figure 2-5

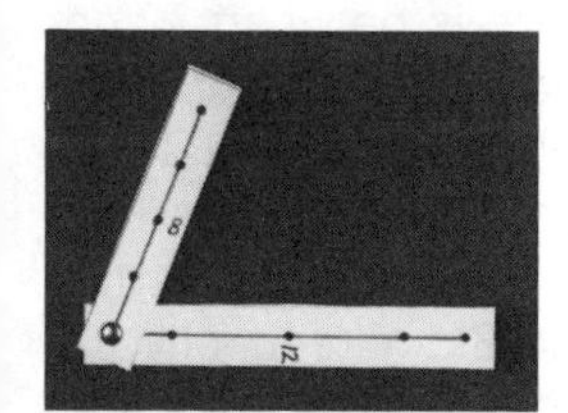

Figure 2-6

Figure 2-3. If you think of the strips as rays, you can imagine two strips joined at an end hole as representing different angles (Figure 2-4). The angle-fixers can be used to hold the strips in place to form a particular angle as shown in Figure 2-5.

Other materials might be used to make similar models of polygons, such as metal strips from erector sets joined with screws and nuts, straws of different lengths joined with pipe cleaners, commercially available plastic geostrips, etc.

Let us make a polygon using the strips. Take any two strips and join them as in Figure 2-6. Notice that this figure does not form a polygon because it is not closed. We will need at least one more side. Attaching a third strip, we can form a three-sided polygon, a **triangle.** A triangle is the polygon with the fewest number of sides.

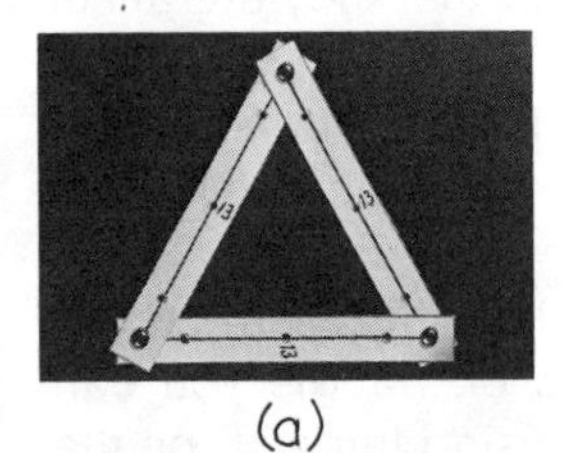

(a)

(b)

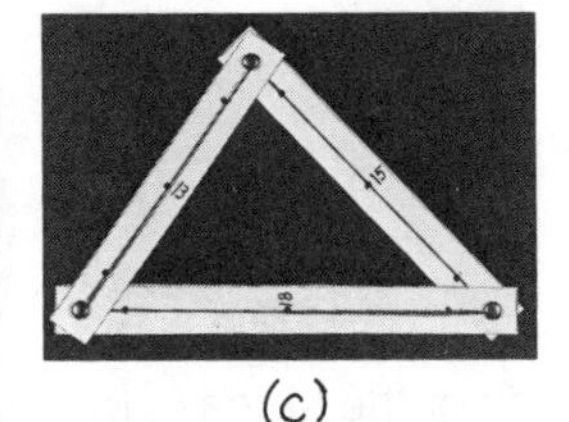

(c)

Figure 2-7

Any three-sided polygon is called a triangle. We can make all sorts of triangles by using the strips. If we make a triangle with three strips of equal length (congruent), then the triangle is an **equilateral** (equal-sided) triangle [Figure 2-7(a)]. If only two strips are congruent and the third is a different size, the resulting triangle is an **isosceles** triangle [Figure 2-7(b)].* With all three strips of different lengths, your triangle would be called a **scalene** triangle [Figure 2-7(c)]. *Any* triangle you make can be described as either equilateral, isosceles, or scalene. If you think of all possible triangles you can make as a set of triangles, then each triangle belongs to either the subset of equilateral triangles, the subset of isosceles triangles, or the subset of scalene triangles.

If you take three strips at random from your collection of strips, will you always be able to make a triangle? Take three strips that are 6, 7, and 15 cm long, respectively, and try to make a triangle. Can you make a triangle? Why not? You will realize that the model does not close (Figure 2-8). Why not?

Suppose you want to make a triangle. You join a 6-cm strip and a 12-cm strip as in Figure 2-9. Try to use strips of the following lengths in centimeters as a third side to form a triangle: 6, 7, 8, 9, 10, 11, 12, 13, 14, 15, 16, 18, 20. When can you make a triangle?

**An isosceles triangle can also be defined as a triangle with at least two congruent sides. Then all equilateral triangles would also qualify as isosceles triangles.*

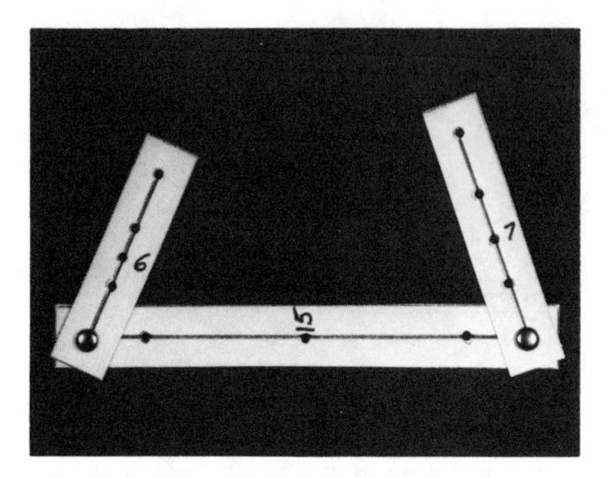

Figure 2-8

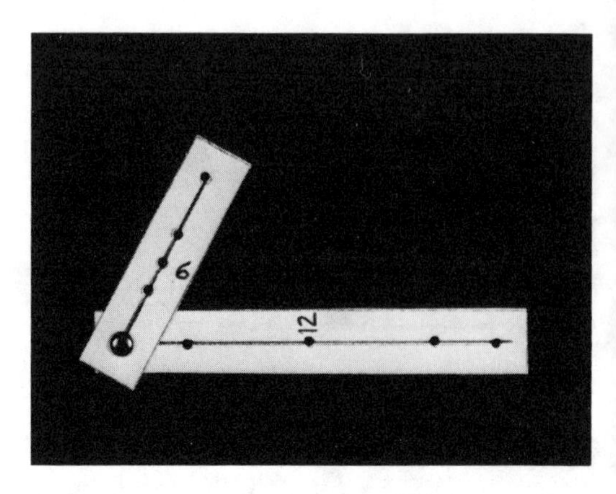

Figure 2-9

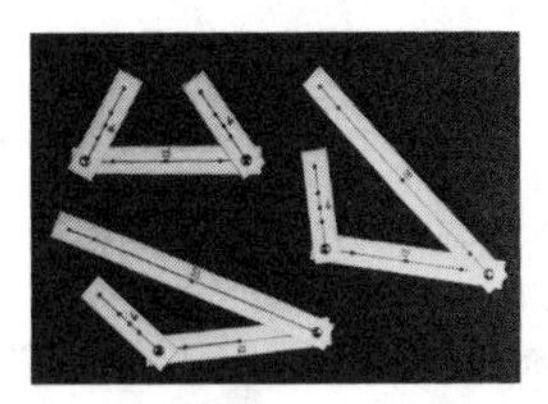

Figure 2-10

In attempting to make these triangles, you will find that it *is* possible to make a triangle when the third side is between 7 and 18 cm long. With two sides of 6 and 12 cm and a third side of 6, 18, or 20 cm, it is *not* possible to form a triangle (Figure 2-10). What can we conclude? Given the lengths of two sides of a triangle, how can we describe the length of a possible third side?

Suppose you take three strips, say 6, 7, and 8 cm, respectively, and make a triangle. Take another set of three strips the same lengths as the first three and make another triangle. Compare the two triangles you have made. You will find that they are alike. You could place one triangle over the other so that each pair of sides the same length is matched (Figure 2-11). The triangles are just the same shape and size; they match or coincide completely (they are congruent).

Figure 2-11

This finding suggests that whenever you take three strips and can make a triangle, there is only one type of triangle you can make with those three strips. Someone else using three strips the same sizes as yours to make a triangle would have to make a triangle just like yours—the same shape and the same size. No other type of triangle is possible with those three strips. We can conclude that all triangles made with a 6-cm, a 7-cm, and an 8-cm strip are congruent.

Angles of a Triangle

Besides having three sides, a triangle also has three angles. The name *triangle* suggests three angles. If you think of the sides of the triangle as rays, then each pair of adjacent sides forms an angle.

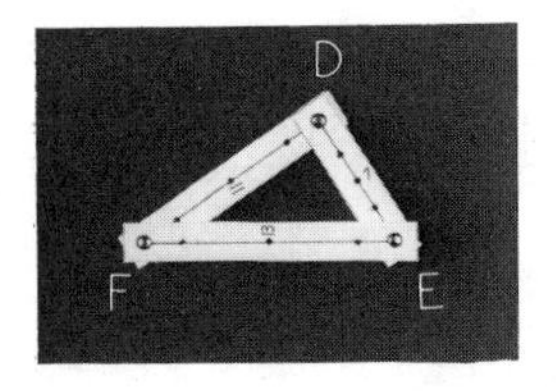

Figure 2-12

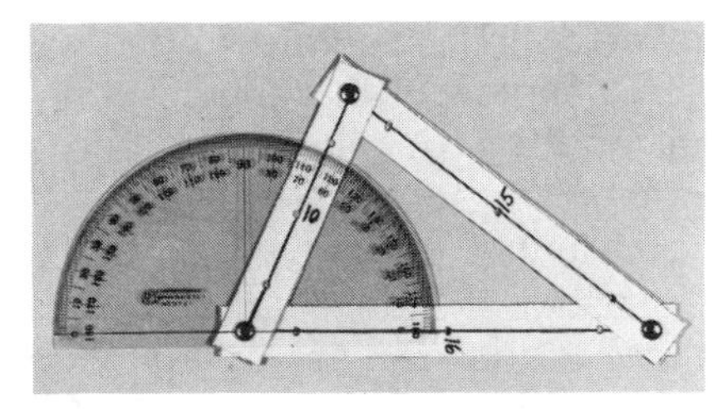

Figure 2-13

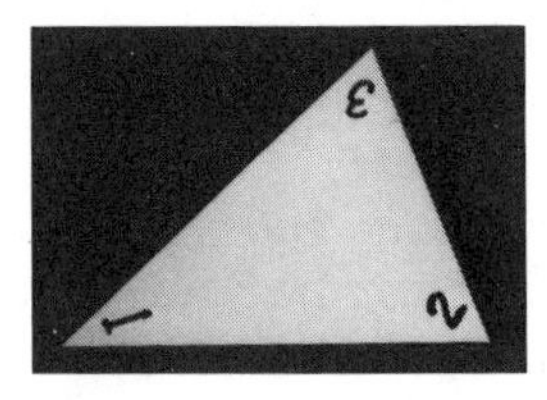

Figure 2-14

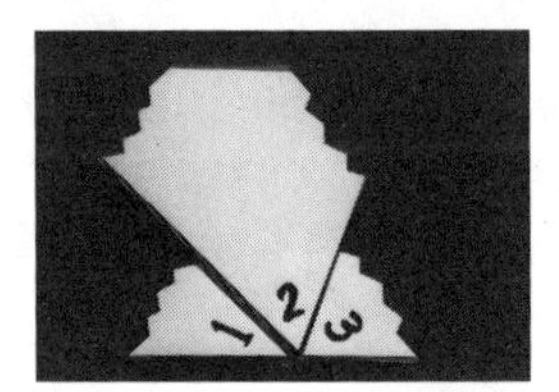

Figure 2-15

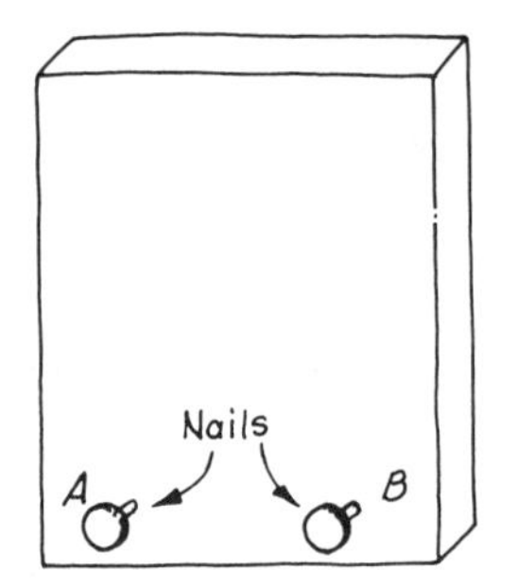

Figure 2-16

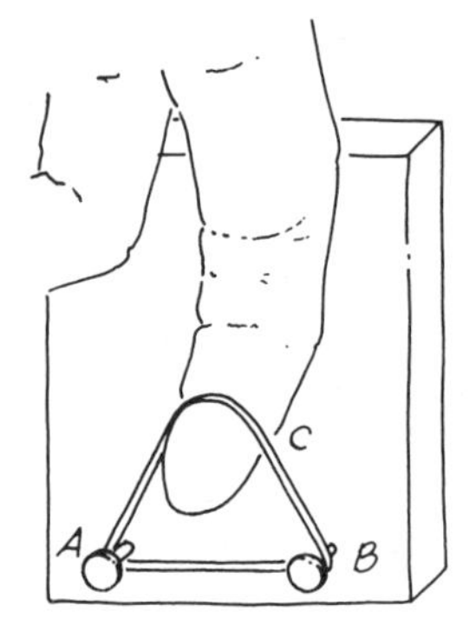

Figure 2-17

Let us take a look at the relationship between the sides and the angles of a triangle. Make a scalene triangle as in Figure 2-12. Which angle in triangle *DEF* is the largest angle? Which side is longest? Which angle is smallest? Which side is shortest? You can verify your estimates by finding the measures of the angles with a protractor and the measures of the sides with a ruler.

Make some other triangles. Which is the largest angle in each triangle? Which is the longest side? Can you find a relationship between the size of the angles and the size of the sides in any triangle?

Make an isosceles triangle. What do you discover about its angles? Make an equilateral triangle. What is true about its angles?

Now let us look at just the angles of a triangle. Make any triangle and carefully measure each angle with a protractor as in Figure 2-13. If you add the number of degrees in all three angles, your sum should be approximately 180°. Make another triangle and find the total number of degrees in its angles. Your total should again be about 180°.

This is a very remarkable property of triangles. **The sum of the number of degrees in each angle of a triangle is always 180.** You can see this better by taking a piece of cardboard cut into a triangular shape and performing the following experiment. Draw a border around the triangular shape. This will serve as a model for a triangle. Label the inside of each angle 1, 2, or 3, respectively, as in Figure 2-14. Now tear or cut the triangle into three parts and rearrange the pieces as in Figure 2-15. We have now moved the angles so that they form one large angle. That large angle is a straight angle and contains 180°. Since combining the angles is like finding the sum of their measures, this indicates that the sum of the number of degrees in the three angles is 180. Try it with other cardboard triangles until you are convinced it always works.

The following experiment is another way of thinking about this phenomenon. Take a piece of wood, and hammer two nails almost entirely into the wood at points *A* and *B* (Figure 2-16). Wrap a large elastic band around the nails so that you form a figure which looks like a line segment. At about the midpoint of that line segment, pull upward on part of the elastic band so that you form another vertex *C* and consequently the model of a triangle, as shown in Figure 2-17. As you pull at point *C*, you make different isosceles triangles. Let us examine the angles of these triangles.

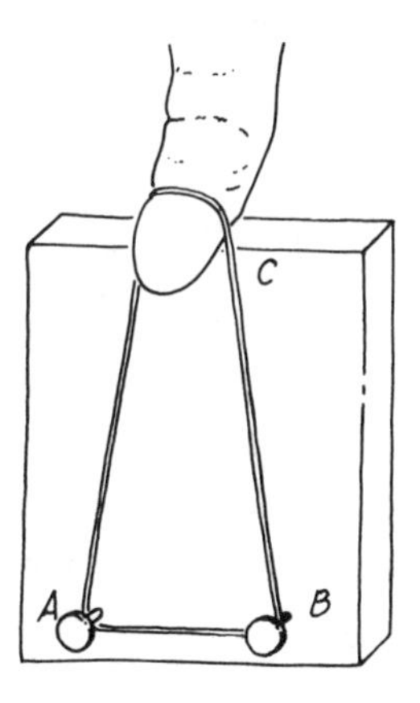

Figure 2-18

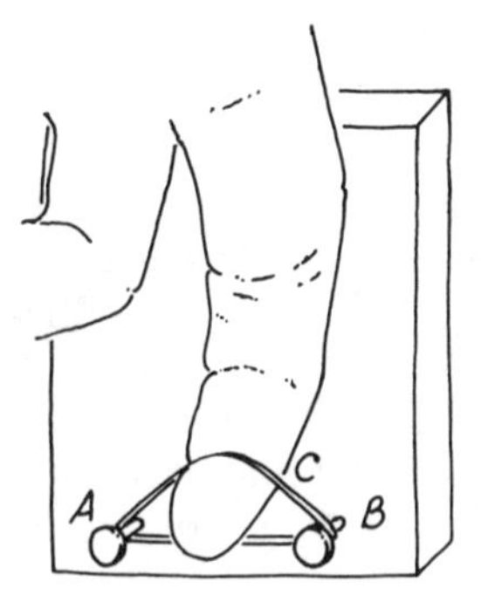

Figure 2-19

At one point the sides are all the same length. We have an equilateral triangle. All the angles of an equilateral triangle are congruent and each contains 60°. The sum of the measures of the three angles is 180°.

Next, pull further upward at C and form other triangles. How does the size of $\angle C$ change as you pull upward? How do the sizes of $\angle A$ and $\angle B$ change? As you pull the elastic upward at C, $\angle C$ gets smaller and smaller while $\angle A$ and $\angle B$ get larger and larger (Figure 2-18). If you can imagine stretching the elastic further and further, in the extreme case $\angle C$ would get closer and closer to 0° in measure and $\angle A$ and $\angle B$ would look more and more like right angles. Again, the sum of the measures of the angles in degrees $(0 + 90 + 90)$ approaches 180.

Now slowly release the elastic band downward and examine the triangles formed. What happens to the angles now? Angle C gets larger and larger while both $\angle A$ and $\angle B$ get smaller. As vertex C gets nearer to side AB, $\angle C$ increases in size until it looks almost like a straight angle of 180° (Figure 2-19). Meanwhile $\angle A$ and $\angle B$ seem to get closer and closer to 0° in measure. Once more the sum of the measures of the angles of the triangles formed $(180 + 0 + 0)$ seems to be fixed at 180. Although this experiment does not *prove* that the sum of the measures of the angles of any triangle is always 180°, it does suggest that this conclusion is reasonable.

If we know that the sum of the measures of the angles of a triangle is always 180°, then we only have to measure two angles of a triangle and we can find the measure of the third angle. For example, if two angles of a triangle measure 30° and 70°, respectively, then the third angle must have a measure of 80°. How do we know each angle of an equilateral triangle must measure 60°?

Classifying Triangles According to Their Angles

Several triangles are shown in Figure 2-20. The triangles in (c) and (d) contain one right angle (an angle that measures exactly 90°). These triangles can be called **right triangles.** The triangles in (b) and (e) each contain one angle that is obtuse (measures between 90 and 180°). These triangles are called **obtuse triangles.** The other triangles in (a), (f), and (g) contain neither an obtuse nor a right angle. All their angles are acute (measure between 0 and 90°). These are called **acute triangles.**

In a right triangle, the sides have special names. The side opposite the right angle is called the **hypotenuse** while the other two perpendicular sides are called **legs** of the right triangle. If

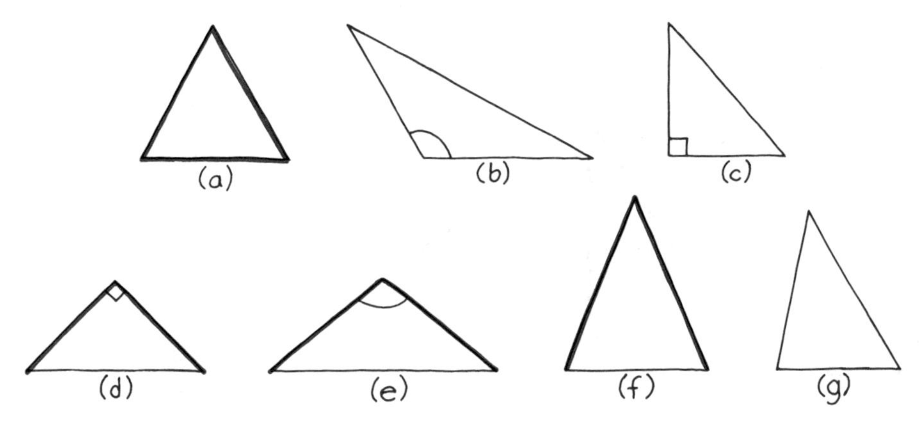

Figure 2-20

you make several right triangles and measure the hypotenuse and legs of each, you will realize that the hypotenuse is always the longest side of any right triangle. This makes sense because the hypotenuse is opposite the largest angle.

Can a triangle have two right angles? Try to make a triangle with two right angles. As you can see in Figure 2-21, you cannot make a three-sided polygon with two right angles. This seems reasonable, since if the sum of the measures of the angles of a triangle is always 180°, then with two 90° angles the other angle would have to be 0° in measure (nonexistent). A triangle can have, at most, one right angle. The other two angles must be acute if the sum of the measures of the three angles is 180°.

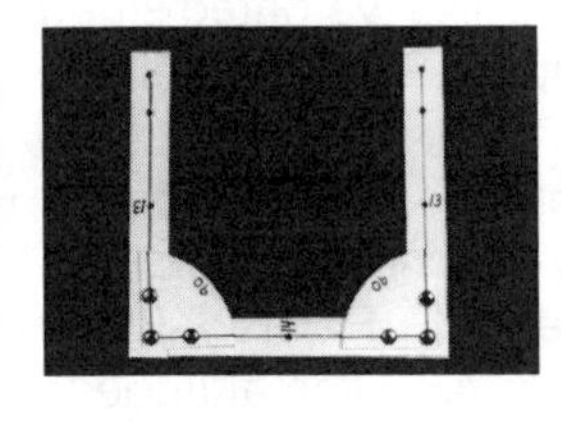

Figure 2-21

Can a triangle have two obtuse angles? Explain your answer.

Earlier in this chapter we considered the set of all possible triangles and we made subsets by classifying the triangles as equilateral, isosceles, or scalene. We can also take the set of triangles and make subsets according to whether a triangle is right, obtuse, or acute. Every triangle will be in either the subset of right triangles, the subset of obtuse triangles, or the subset of acute triangles.

Altitudes of a Triangle

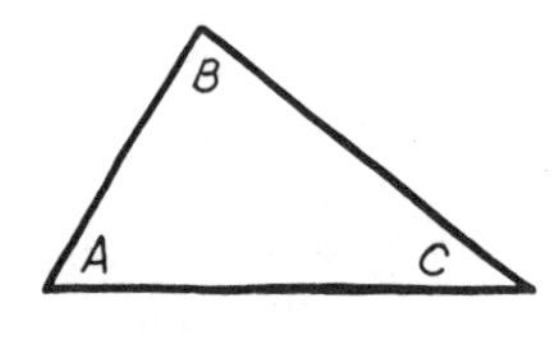

Figure 2-22

Make a triangular shape out of cardboard with a border representing a triangle as shown in Figure 2-22. Call the triangle *ABC* and place it upright. How tall is the triangle? The height of this triangle is the distance from the vertex *B* to the base $\overline{AC}$. But there are any number of line segments that might extend from *B* to base $\overline{AC}$ as in Figure 2-23. Which line segment represents the height?

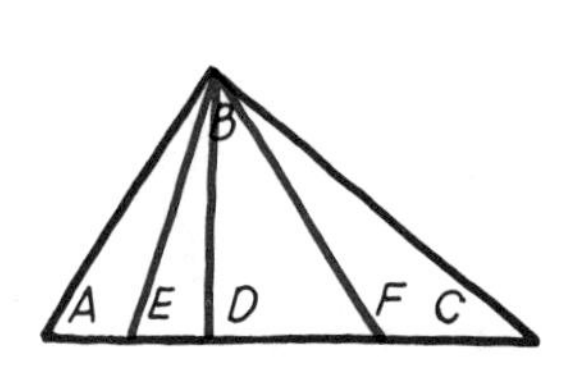

Figure 2-23

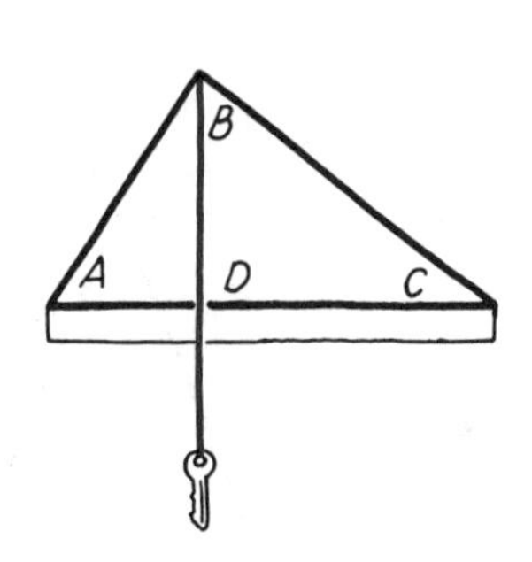

Figure 2-24

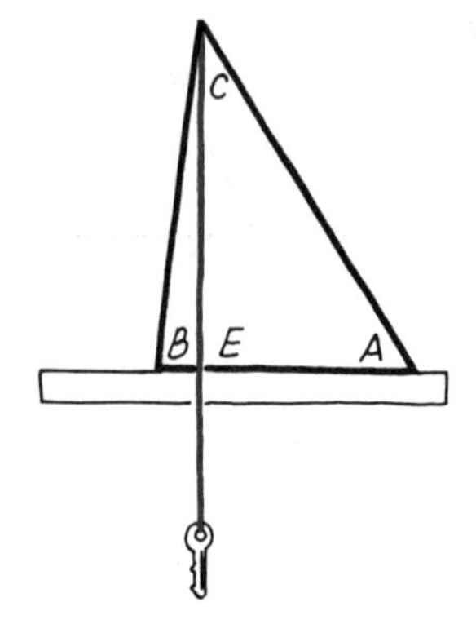

Figure 2-25

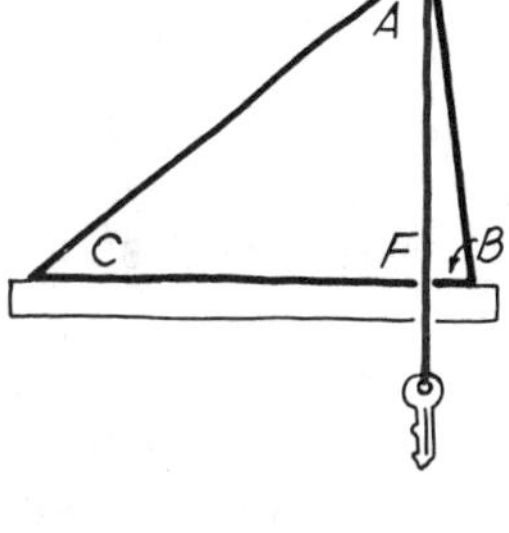

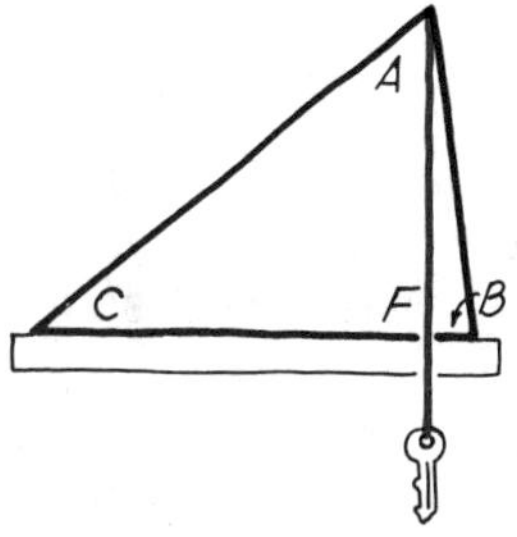

Figure 2-26

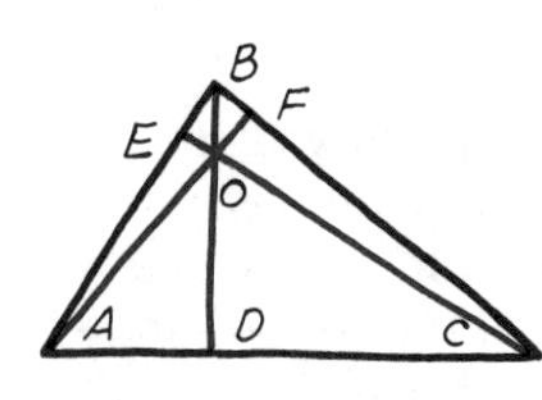

Figure 2-27

Figure 2-28

The height of triangle ABC is determined by the line segment that comes straight down from B to $\overline{AC}$, making right angles with $\overline{AC}$ (perpendicular to AC). This line segment is called an **altitude.** As you can see, $\overline{BD}$ is the altitude for triangle ABC when side $\overline{AC}$ is the base.

A very easy way to locate the altitude of a triangle is to place it upright on a table top and hang a weighted string (a key will do as a weight) from the upper vertex to the base. The weighted string will indicate the altitude. In triangle ABC (Figure 2-24) the altitude to base $\overline{AC}$ is shown to be $\overline{BD}$.

Suppose you turn the triangle counterclockwise so that side $\overline{BA}$ rests on the table. What is the altitude now? Using the weighted string, you can see that the altitude is segment $\overline{CE}$ (Figure 2-25). If you rotate the triangle again, you will find that $\overline{AF}$ is the altitude to base $\overline{CB}$ (Figure 2-26).

If we rotate the triangle once more in the same way as before, we find that the triangle returns to its original position with side $\overline{AC}$ as the base. We have found three altitudes for triangle ABC: $\overline{BD}$ (with base $\overline{AC}$), $\overline{CE}$ (with base $\overline{BA}$), and $\overline{AF}$ (with base $\overline{CB}$). A triangle, therefore, has three altitudes. Each side can be considered the base of the triangle and each base has a corresponding altitude.

What do you notice about the three altitudes? All three altitudes intersect at the same point! (See Figure 2-27.) This point of intersection O of the altitudes is called the **orthocenter** of the triangle.

Sometimes an altitude of a triangle is located outside the triangle. In triangle DEF (Figure 2-28) the weighted string indicates that the altitude (with side $\overline{FE}$ as base) is $\overline{DG}$. If we proceed as before, rotating triangle DEF counterclockwise and finding all three altitudes, will they intersect? If we extend the altitudes, we find that they do intersect: the orthocenter is point O (Figure 2-29). This time the orthocenter is located outside the triangle.

Actually, if you make or draw many different triangles and locate their three altitudes, you will find that the orthocenter can be inside the triangle, outside the triangle, or the same as a vertex of

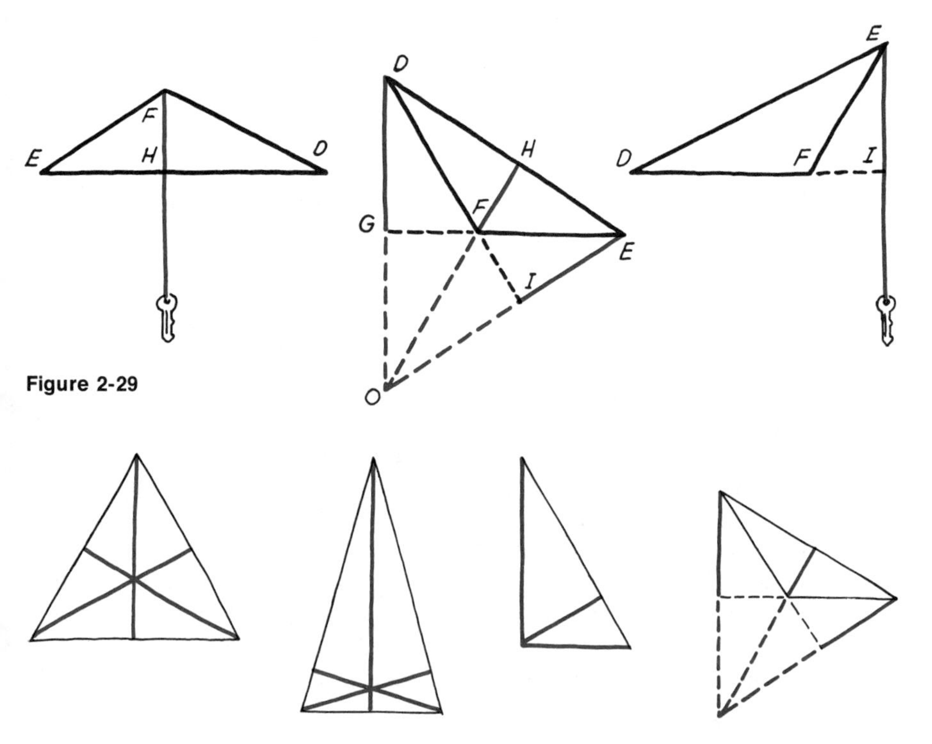

Figure 2-29

Figure 2-30

the triangle (Figure 2-30). However, in *every* triangle the altitudes *always* have a common point of intersection (an orthocenter). Where is the orthocenter for acute triangles? For right triangles? For obtuse triangles? Draw some different triangles and find out.

Medians of a Triangle

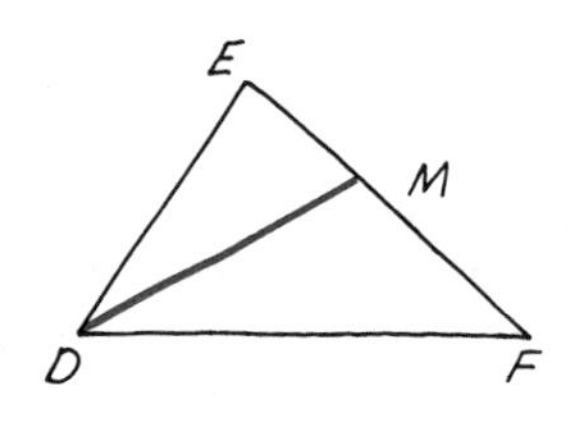

Figure 2-31

A **median** of a triangle is a **line segment that extends from a vertex to the midpoint of the opposite side.** In Figure 2-31, point M is the midpoint of $\overline{EF}$ since $\overline{EM}$ and $\overline{MF}$ are congruent. Segment $\overline{DM}$ is called the median from vertex D to side $\overline{EF}$.

A triangle has three medians. Draw a triangle and locate its three medians as in Figure 2-32. What do you notice about the medians? Will this always be true? Try making other triangles and locating their medians.

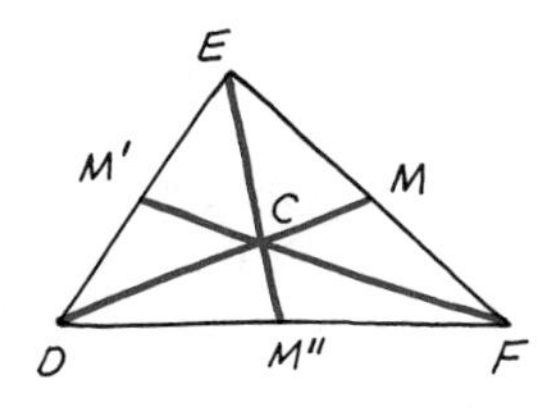

Figure 2-32

Make a cardboard triangle and try to support it with a pencil as in Figure 2-33. If you keep trying, you will find a point on the cardboard inside the triangle where the tip of the pencil can be placed to balance the triangle. That point is called the **centroid** of

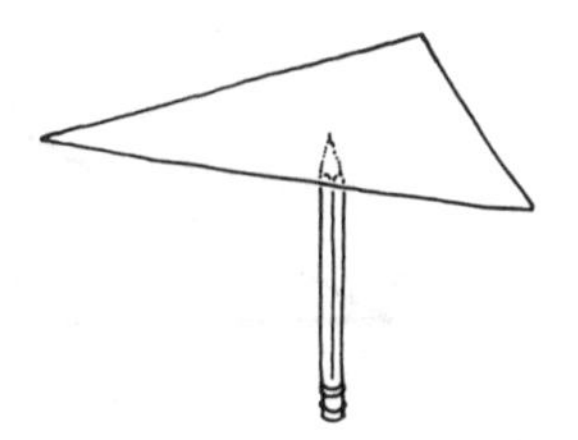

Figure 2-33

Figure 2-34

the triangle. Now draw the three medians of the triangle on your model. Try to balance the triangle by placing the tip of the pencil at the point of intersection of the medians (Figure 2-34). What do you discover? The intersection of the medians is the same point that balanced the triangle before. **The point of intersection of the medians is the centroid of the triangle.**

Make several kinds of triangles out of cardboard and draw the three medians for each triangle. Check to see that the medians meet at one point and that this point is the centroid of each triangle. Is the centroid always inside the triangle?

Bisectors of the Angles

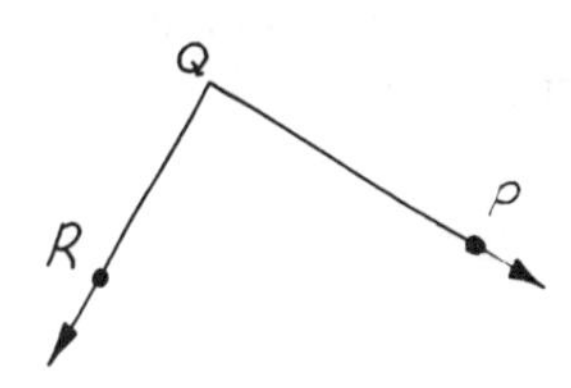

Figure 2-35

Imagine an angle on a plane like $\angle PQR$ in Figure 2-35. Suppose we draw a ray $\overrightarrow{QB}$ in the interior of $\angle PQR$ so that we form two congruent angles (Figure 2-36), $\angle PQB$ and $\angle BQR$. Ray $\overrightarrow{QB}$ is called the **bisector** of $\angle PQR$.

Similarly, in Figure 2-37 imagine that sides $\overline{QP}$ and $\overline{QR}$ of triangle PRQ are rays and that segment $\overline{QB}$ is a ray, too. Also suppose that $\angle PQB$ is congruent to $\angle BQR$. Then we can say that $\overline{QB}$ is on the bisector of $\angle PQR$. In triangle PRQ, we will call $\overline{QB}$ the bisector of $\angle Q$.

Since a triangle has three angles, it will also have three angle bisectors. The three angle bisectors of triangle PRQ are illustrated in Figure 2-38. Notice that they too intersect at one point. This point I is called the **incenter** of the triangle, the point of intersection of the angle bisectors. If you make other triangles and draw the angle

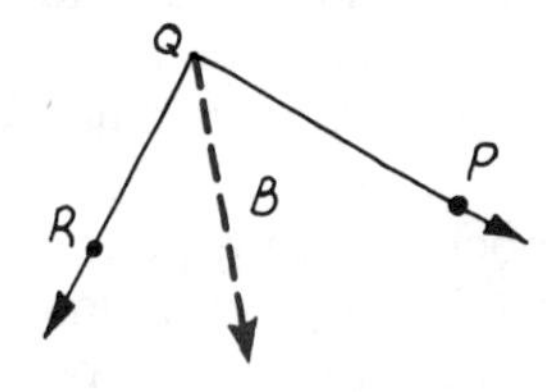

Figure 2-36

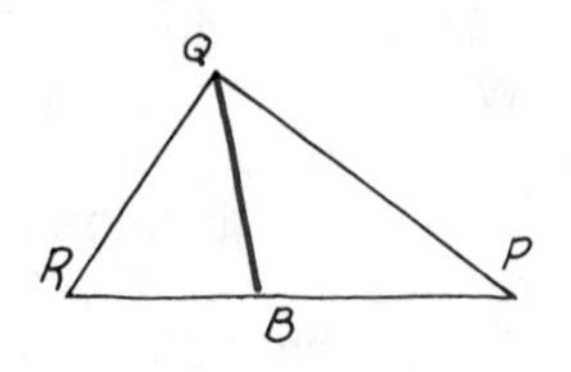

Figure 2-37

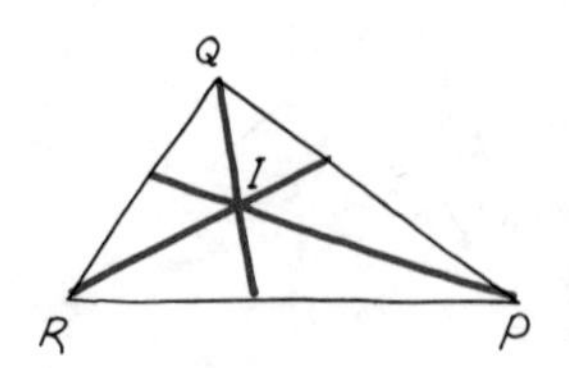

Figure 2-38

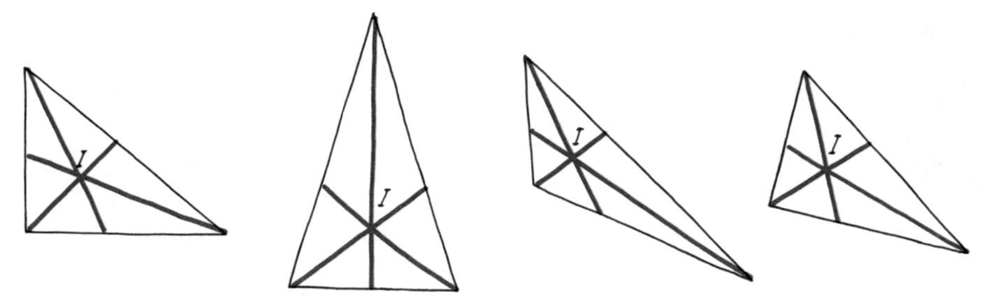

Figure 2-39

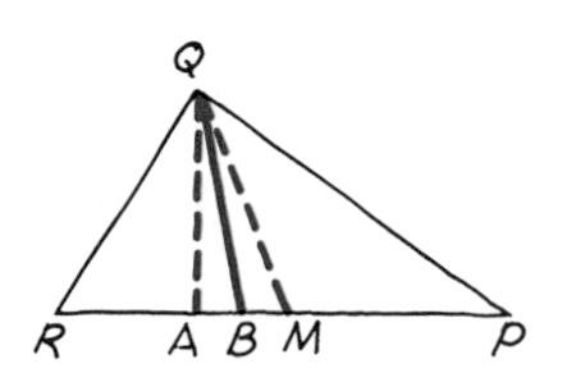

Figure 2-40

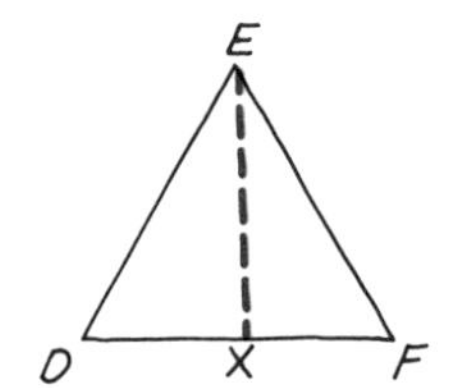

Figure 2-41

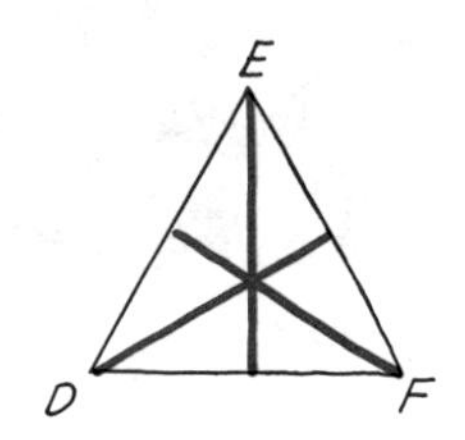

Figure 2-42

bisectors, as in Figure 2-39, you will see that the three bisectors always intersect at one point.*

The altitude, angle bisector, and median from vertex *Q* in triangle *QPR* are shown in Figure 2-40. In this case, they are three different line segments: $\overline{QA}$, $\overline{QB}$, and $\overline{QM}$, respectively. In the special case of an equilateral triangle, as in Figure 2-41, the *same* line segment, $\overline{EX}$, serves as an altitude, angle bisector, and median. Each line segment inside equilateral triangle *DEF* (Figure 2-42) represents an altitude, angle bisector, and median of the triangle *at the same time.* In this special case, the one point of intersection of those line segments is the orthocenter, centroid, and incenter of the triangle.

Rigidity of a Triangle

We have examined many properties of triangles in this chapter. Let us conclude with one more very important property. A triangle is a **rigid** figure. You can appreciate the meaning of this if you do the following experiment.

**We will discuss constructing the bisector of an angle with a compass and straightedge in Chapter 8. For now, you might find it convenient to use a protractor to find out where the bisector should be so that the angles formed are half the size of the angle being bisected.*

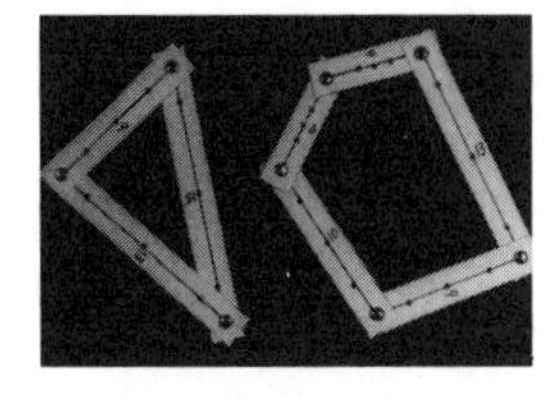

Figure 2-43

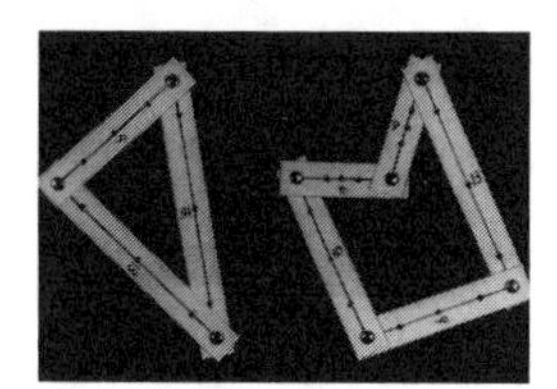

Figure 2-44

Using the strips and brass fasteners, make a triangle and some other polygon as in Figure 2-43. Try to push inward at one or more vertices of the triangle. Do the same thing to the other polygon. You will see that the triangle remains unaltered while the other polygon can be distorted into different shapes. (Figure 2-44). Even though the ends of the strips are not glued together, the triangle keeps its shape; it is rigid. The fact that a triangle is rigid makes it a useful shape in construction. Its rigidity gives strength to structures. How many triangles can you see in Figure 2-45?

Figure 2-45

More Experiences

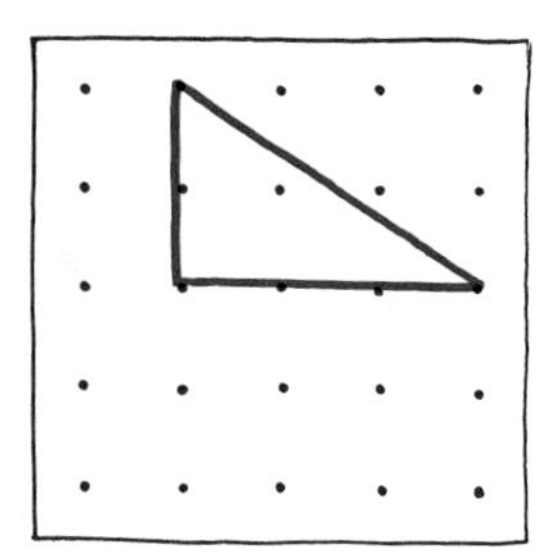

2-1. Which of the following triangular shapes can you make on a geoboard? Draw each of your shapes on 5 × 5 dot paper as in the diagram.

(*a*) Isosceles right triangle
(*b*) Scalene obtuse triangle
(*c*) Scalene acute triangle
(*d*) Equilateral triangle
(*e*) Isosceles acute triangle

Triangle	Length of side 1	Length of side 2	Length of side 3	Possible?
(*a*)	6	7	8	
(*b*)	6	7	15	
(*c*)	8	9	14	
(*d*)	8	9	20	
(*e*)	7	18	6	
(*f*)	8	15	10	
(*g*)	9	18	7	

2-2. Try to draw or make models of triangles with strips that have the measures shown in the table at the left. When is it not possible to make a triangle? Why?

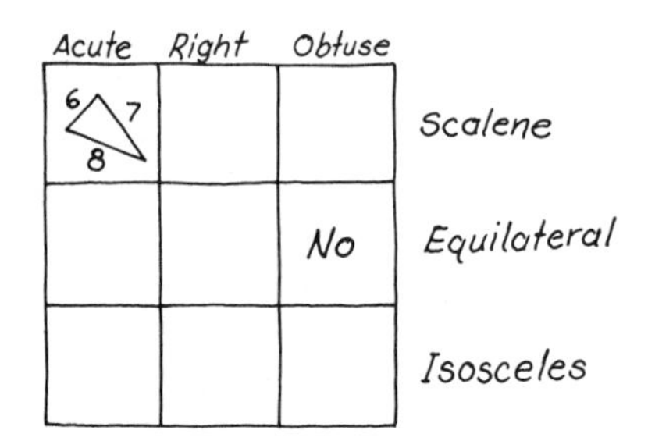

2-3. Each space on the left represents a possible triangle described according to its sides and angles (the triangle indicated is acute and scalene). Which triangles are possible to make? Draw a picture of each one you can make (as shown). Indicate which ones cannot be made (for example, obtuse and equilateral).

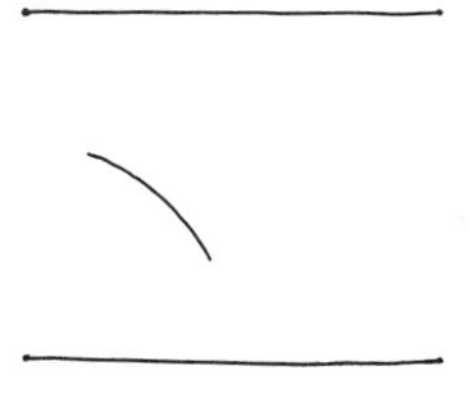

2-4. A compass and ruler are useful for drawing a triangle with specified measurements. If you want to make a triangle with sides 5, 8, and 10 cm, you might proceed as follows:

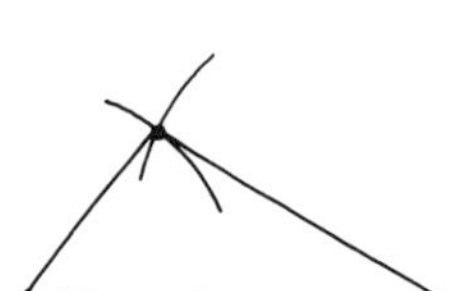

1. Draw a line segment 10 cm long.
2. Open the compass so that the distance between its tips is 5 cm. Then draw an arc with the stationary point of the compass at one end point of the line segment.

3. Similarly, open the compass so that the distance between its tips is 8 cm. Then draw an arc with the stationary point of the compass at the other end point of the line segment.
4. The point of intersection of the arcs and the two end points of the original line segment will be the vertices of the triangle with sides 5, 8, and 10 cm.

How do you know the sides have measures of 5, 8, and 10 cm? What would happen if you continued each arc along the other side of the original line segment?

Use a compass and ruler to draw triangles that have the following measures in inches or centimeters. Which of these triangles is not possible to draw? Why?

(*a*) 6, 8, 9
(*b*) 3, 4, 5
(*c*) 3, 7, 12

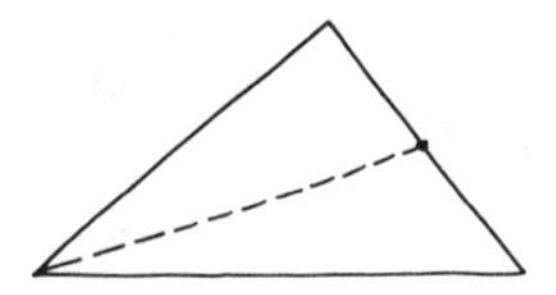

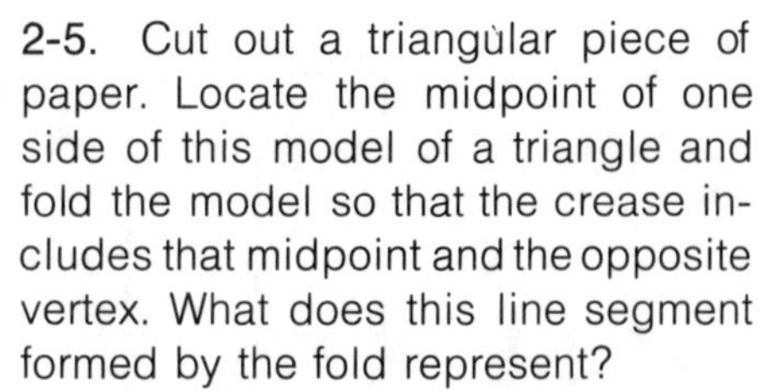

2-5. Cut out a triangular piece of paper. Locate the midpoint of one side of this model of a triangle and fold the model so that the crease includes that midpoint and the opposite vertex. What does this line segment formed by the fold represent?

In the same way, locate the midpoints of the other two sides of this model and fold in the same way. What do you notice about the three folds? Make many different triangular models and repeat the procedure. How will the three folds be related each time?

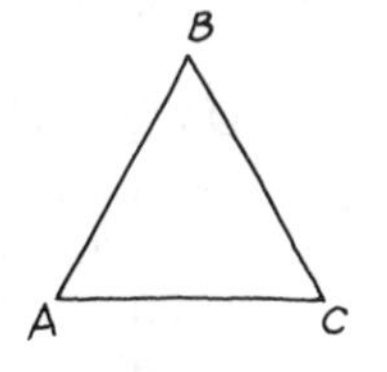

2-6. Draw a triangle like $\triangle ABC$.

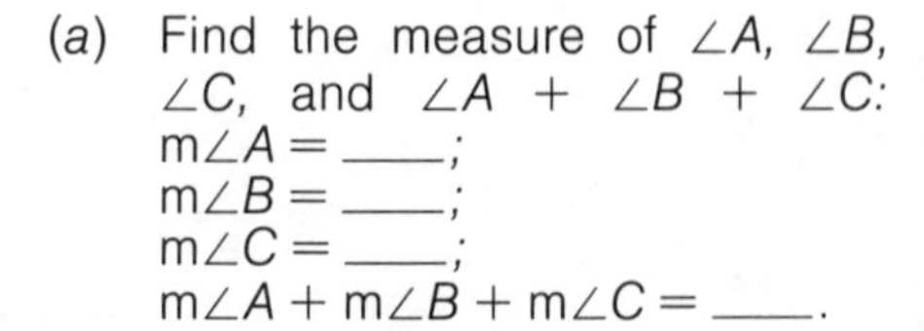

(*a*) Find the measure of $\angle A$, $\angle B$, $\angle C$, and $\angle A + \angle B + \angle C$:
$m\angle A =$ ____;
$m\angle B =$ ____;
$m\angle C =$ ____;
$m\angle A + m\angle B + m\angle C =$ ____.

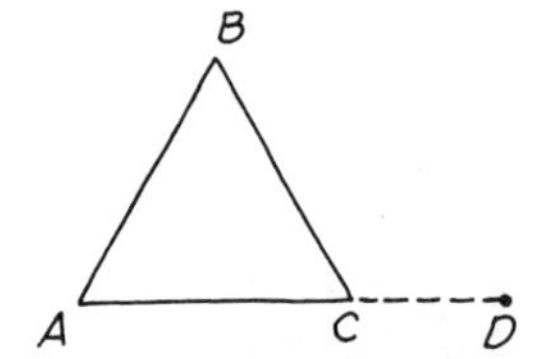

(*b*) Extend $\overline{AC}$ to point *D*. What is the measure of $\angle DCB$? $m\angle DCB =$ ____.

(*c*) Angle *DCB* and $\angle C$ are supplementary angles. How do you know this?

(*d*) What relationship can you find between $m\angle A$, $m\angle B$, and $m\angle DCB$?

(*e*) Suppose $\overline{AB}$ were extended in the *A* direction to *E*. What do you know about $\angle EAC$?

3 Quadrilaterals and Other Polygons

Making Quadrilaterals

Take four strips and make a model of a polygon as in Figure 3-1. This four-sided polygon is called a **quadrilateral.** If you think of the sides of the quadrilateral as rays, how many angles are determined by the sides? Each of the four corners or vertices of the quadrilateral can be considered a vertex of an angle.

Unlike a triangle, a quadrilateral is *not* a rigid figure. If you push inward at one of the vertices, you can transform the quadrilateral into another quadrilateral with a different shape (Figure 3-2).

When we made models of triangles from strips, we found that it was *not* always possible to make a triangle out of any set of three strips. Suppose you take four strips at random. Is it always possible to make a quadrilateral with those four strips? Try it with the following sets of strips with measures in centimeters as indicated: (*a*) 6, 6, 6, 6; (*b*) 6, 7, 9, 14; (*c*) 6, 6, 6, 20. Notice that it is *not* possible to make a quadrilateral for (*c*). Why not? Explain how the lengths of the strips are related to whether a quadrilateral can or cannot be made. Figure 3-3 provides an illustration.

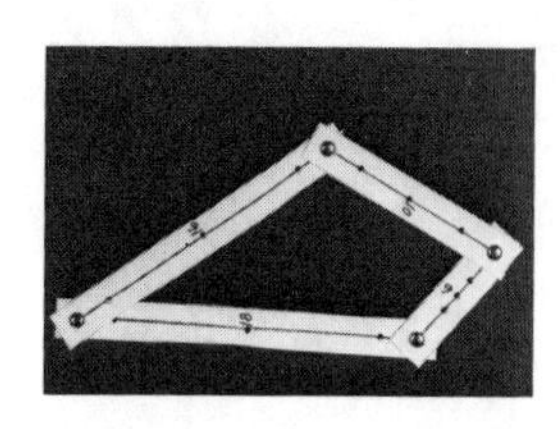

Figure 3-1

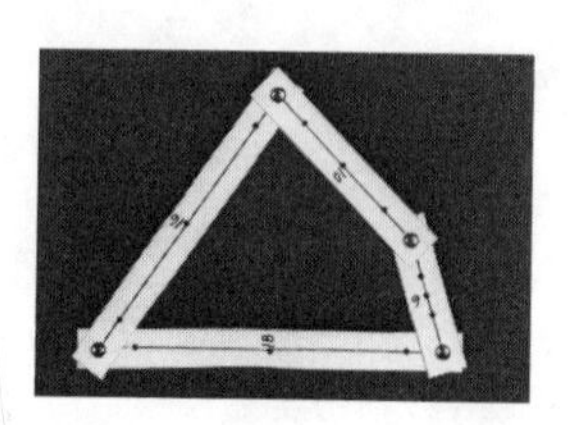

Figure 3-2

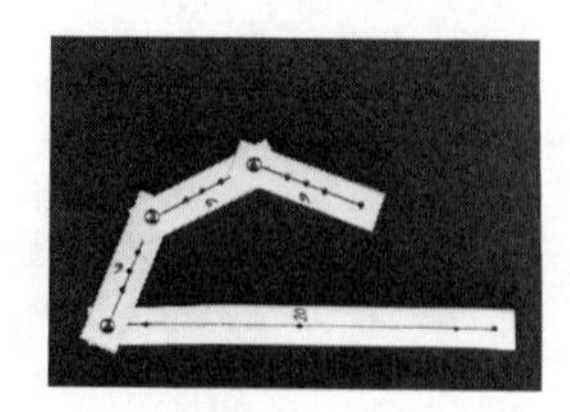

Figure 3-3

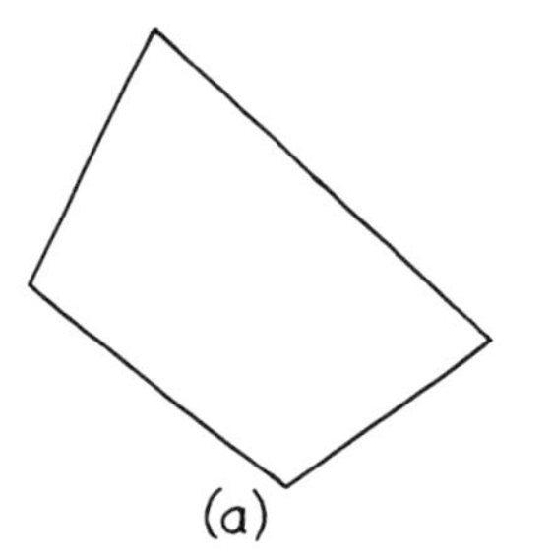

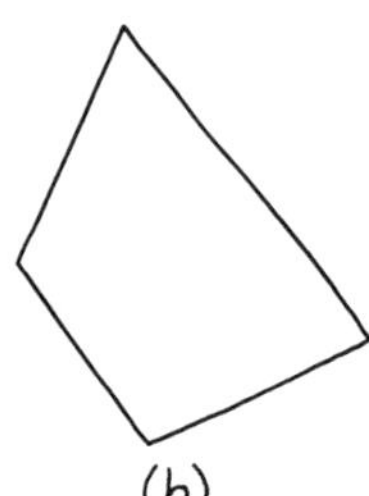

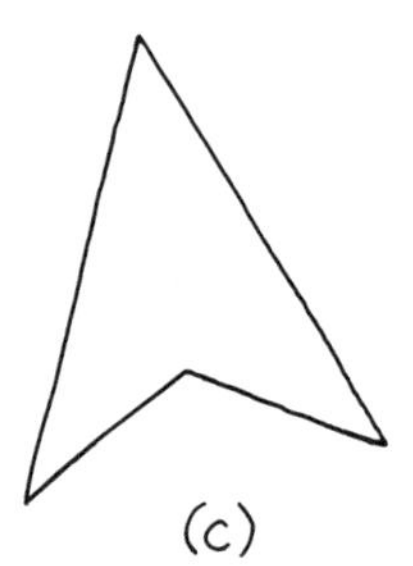

Figure 3-4

(a) (b)

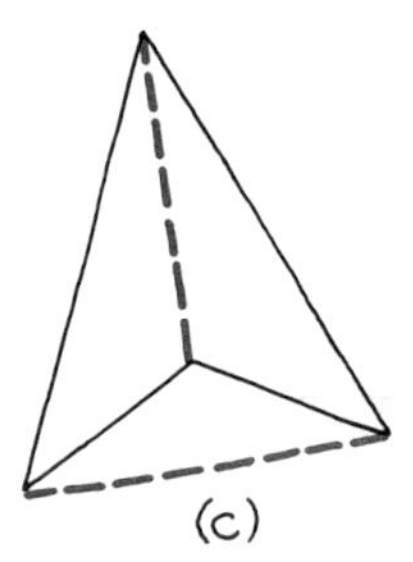

Figure 3-5

Convex Quadrilaterals

In Figure 3-4, quadrilateral (c) is different from quadrilaterals (a) and (b). How is it different? As we saw in Chapter 1, polygons like (a) and (b) are convex. If we draw all the diagonals in each polygon (Figure 3-5), we see that in (a) and (b) the diagonals are always entirely inside the polygon, while in (c) there is a diagonal located outside the polygon. The quadrilateral in (c) is nonconvex. You might say that the sides of one angle in (c) seem to "cave in."

The quadrilaterals in (a) and (b) of Figure 3-4 are called **convex quadrilaterals.** In the following sections we will examine different types of convex quadrilaterals.

Trapezoids

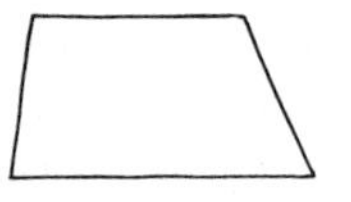

Figure 3-6

The convex quadrilateral pictured in Figure 3-6 has a special characteristic or property. One pair of opposite sides is parallel. More precisely, this means that if you imagine a pair of opposite

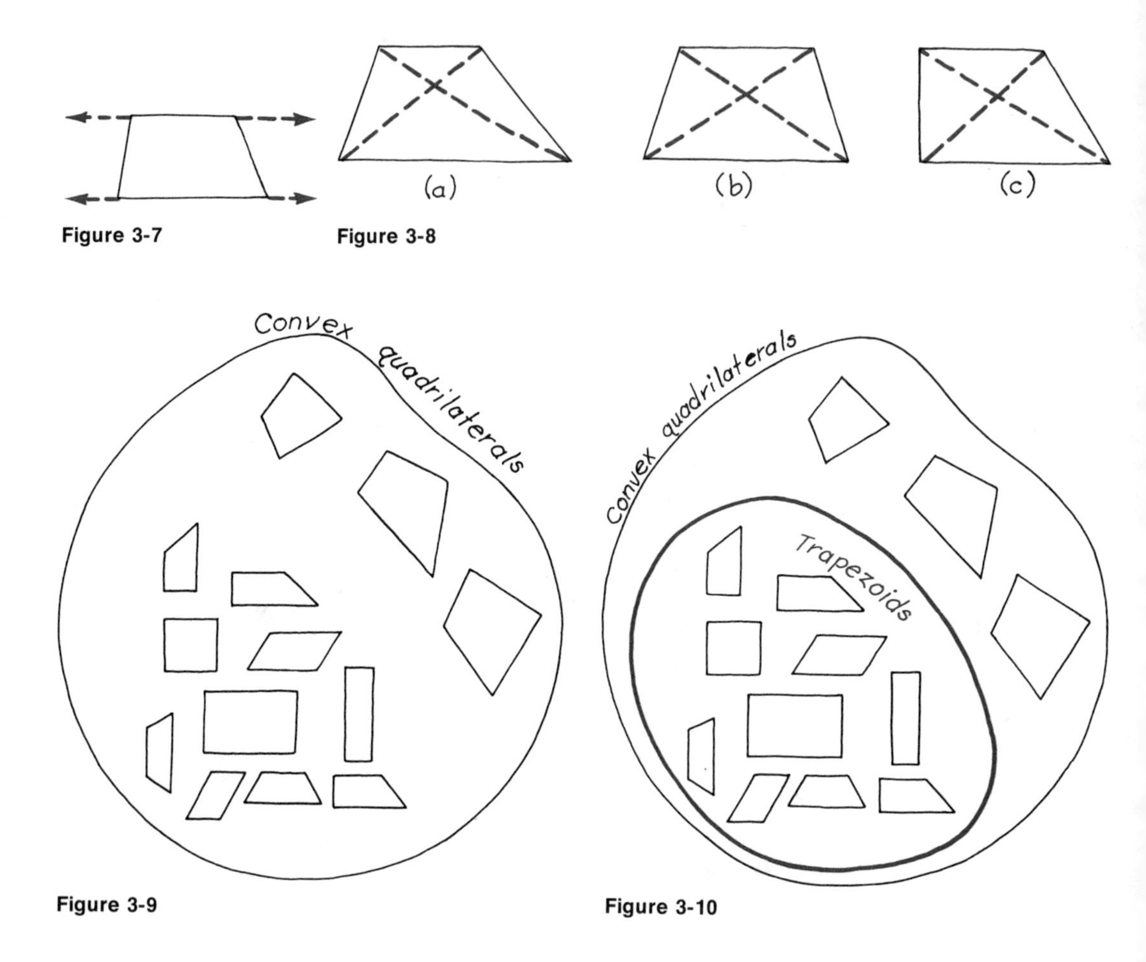

Figure 3-7

Figure 3-8

Figure 3-9

Figure 3-10

sides to be extended to form lines, those lines would be parallel (Figure 3-7).

Any quadrilateral with one pair of opposite sides parallel is called a trapezoid.* Several trapezoids are pictured in Figure 3-8. In each case one pair of opposite sides is parallel. These parallel sides are often referred to as the **bases** of the trapezoid.

The trapezoids in (b) and (c) of Figure 3-8 have special names. If you examine the trapezoid in (b), you will see that the nonparallel pair of sides is equal in measure. This is called an **isosceles trapezoid.** The trapezoid in (c) has two right angles and thus is called a **right trapezoid.**

Let us take a look at the diagonals of the trapezoids in Figure 3-8. Are the diagonals of a trapezoid congruent? Measure them and find out. The two diagonals in trapezoids (a) and (c) are *not* equal in measure. Only the two diagonals in (b), the isosceles trapezoid, are congruent.

**Sometimes a trapezoid is defined as a quadrilateral with only one pair of opposite sides parallel and the other pair not parallel. In this book we will consider a quadrilateral a trapezoid if one pair of opposite sides is parallel. This means that a quadrilateral with both pairs of opposite sides parallel qualifies as a trapezoid.*

If you can imagine all the convex quadrilaterals as a set represented in the simple closed curve in Figure 3-9, then trapezoids would be a subset. This can be shown by a diagram as in Figure 3-10. The trapezoids form a subset because they are a special type of convex quadrilateral; they have one pair of opposite sides parallel. This type of diagram is called a **Venn diagram.** It illustrates that the set of trapezoids is a subset of the set of convex quadrilaterals. *All* trapezoids are convex quadrilaterals (that is, trapezoids belong to the set of convex quadrilaterals). *Some* convex quadrilaterals are trapezoids.

Trapezoids and Parallelograms

We have defined a trapezoid as a convex quadrilateral with one pair of opposite sides parallel. Using the strips, can you make a convex quadrilateral with both pairs of opposite sides parallel? The model in Figure 3-11 is a convex quadrilateral with both pairs of opposite sides parallel. It qualifies as a trapezoid since it does have *one* pair of opposite sides parallel. It is a special trapezoid called a **parallelogram.**

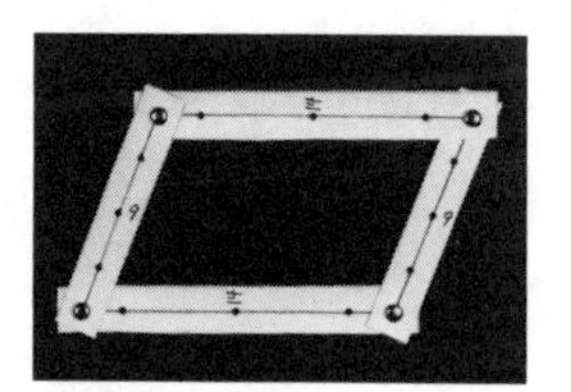

Figure 3-11

The name *parallelogram* suggests the unique property of this convex quadrilateral: every side is parallel to some other side. We can think of parallelograms as being a subset of the set of trapezoids since every parallelogram is a trapezoid. Again, we can use a Venn diagram to express this relationship (Figure 3-12).

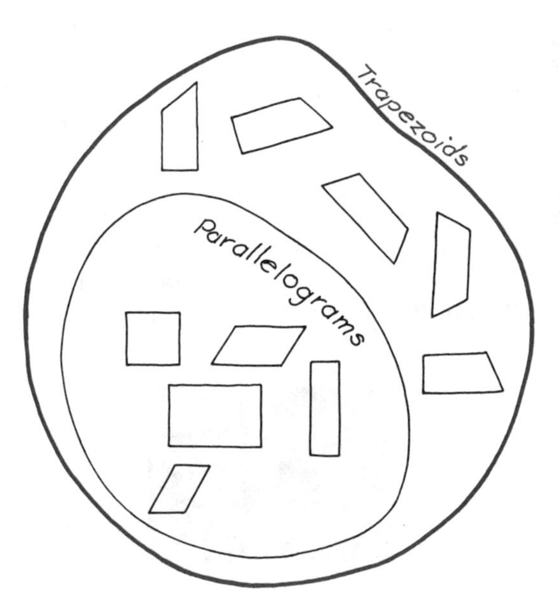

Figure 3-12

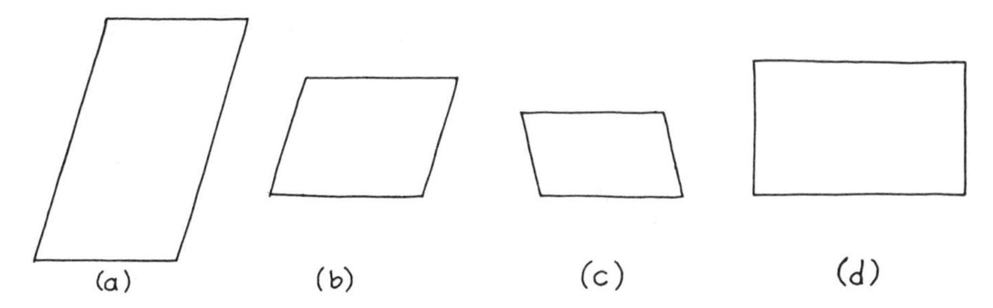

Figure 3-13

Make some other parallelograms using the strips. In each case be certain that both pairs of opposite sides are parallel. What do you notice about the lengths of the sides of your parallelograms? In any parallelogram you make, the opposite sides are equal in measure. Several parallelograms are shown in Figure 3-13. What is true about the opposite sides of each parallelogram? Is this true for all parallelograms? Why?

Parallelograms and Rectangles

The parallelogram in Figure 3-13(d) has a special property. What is it? Note that its angles are all right angles. A parallelogram with right angles is called a **rectangle.** A rectangle, then, is a special kind of parallelogram.

Using the strips, make a rectangle like the one in Figure 3-13(d). With just a little push inward at one vertex or one of the sides, you can transform the rectangle into a parallelogram (Figure 3-14). In fact, by pushing slowly inward you can transform the rectangle into many different parallelograms. This transformation shows that the rectangle belongs to the set of parallelograms. The rectangle is a special member of the parallelogram family. How is it special? It is the only parallelogram with right angles.

As we transform the rectangle into different parallelograms (Figure 3-14), what is happening to the *sides* of the rectangle, the

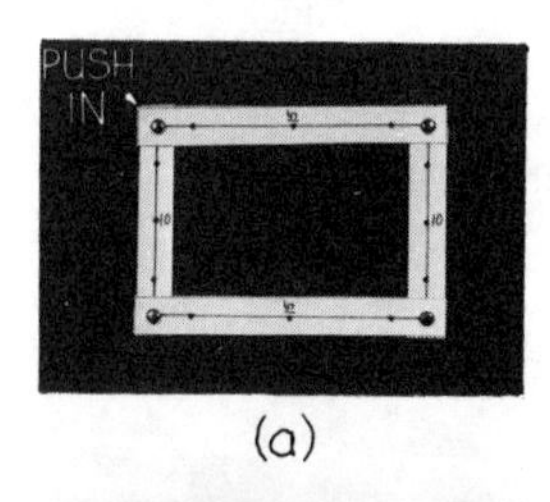

(a)

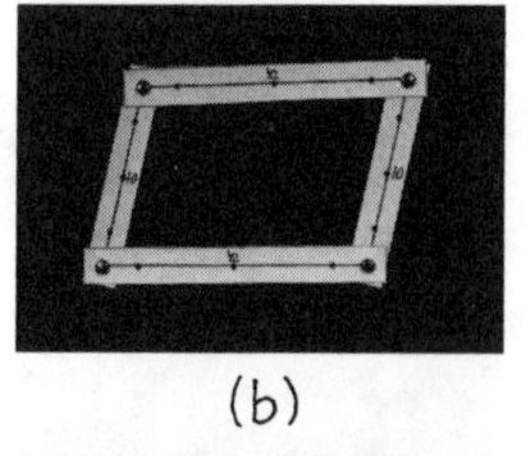

(b)

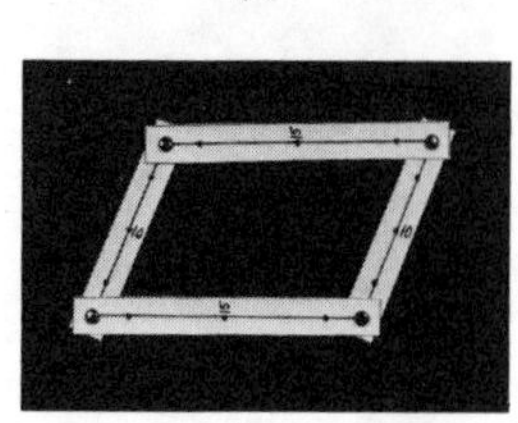

(c)

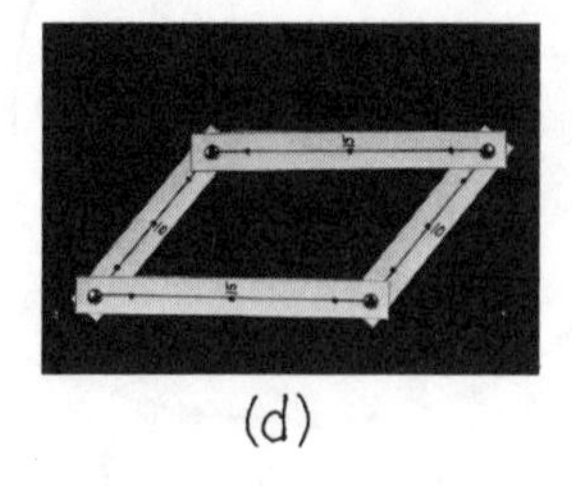

(d)

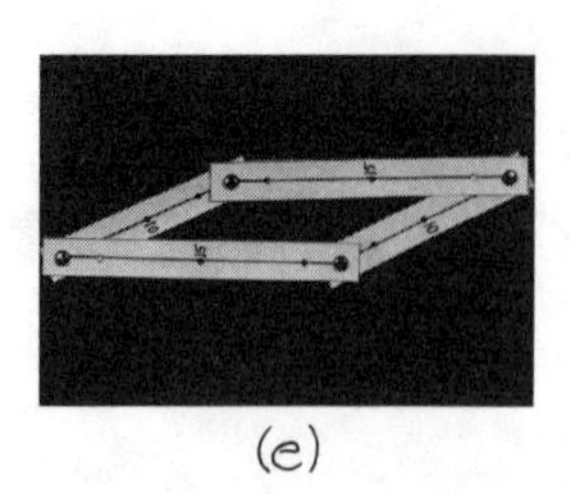

(e)

Figure 3-14

angles, and the *diagonals?* It is easy to see that the sides do *not* change in length and that opposite sides remain parallel and congruent.

In transforming the rectangle to parallelograms, the angles do *not* remain right angles. As you push inward slowly at one of the vertices, you find that two opposite angles become *acute* angles and the other two opposite angles become *obtuse* angles. Yet the two angles in each pair of opposite angles appear to be the same size.

As you continue, slowly transforming the parallelograms into other parallelograms, you realize that one pair of opposite angles is getting smaller and smaller while the other pair is increasing. Each of the decreasing angles seems to approach an **extreme** or **limit** of 0° [Figure 3-14(e)] while the increasing angles approach the size of straight angles (180°). The rectangle itself is approaching a limit: it is getting flatter and flatter until the extreme case when two strips overlap the other two strips and the rectangle becomes a line segment.

As you examine the angles more carefully and see that two angles are increasing in measure while the other two decrease, you might wonder whether the sum of the number of degrees in all four angles remains the same, with two angles gaining what the other two have lost. If you examine the case when the parallelogram becomes a line segment, this guess seems reasonable. In the case of the rectangle, there are four right angles of 90° each for a total of 360°. As you transform the rectangle to parallelograms (Figure 3-14), two angles increase in measure while the other two decrease, until the extreme case when two angles approach the size of a straight line (180°) and the other two decrease toward 0°. Again the sum of all four angles in degrees approaches 360 (180 + 180 + 0 + 0). It becomes convincing, intuitively, that **the sum of the measures of all four angles remains 360° for all parallelograms.**

Let us take another look at this transformation. In Figure 3-15 rectangle *ABCD* is transformed into parallelogram *A'B'CD*. As we effect this transformation, $\angle A$ increases in measure as it becomes $\angle A'$ and $\angle B$ decreases in measure as it becomes $\angle B'$. Is one angle gaining what the other is losing?

The angle-fixers can help us gain some insight into this problem. When the parallelogram is a rectangle, a 90° angle-fixer can be

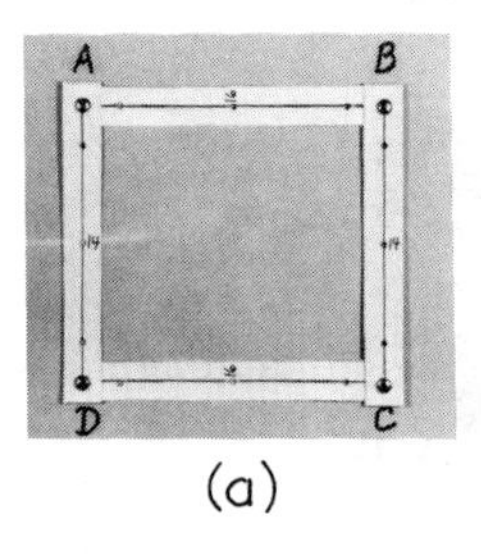

(a)

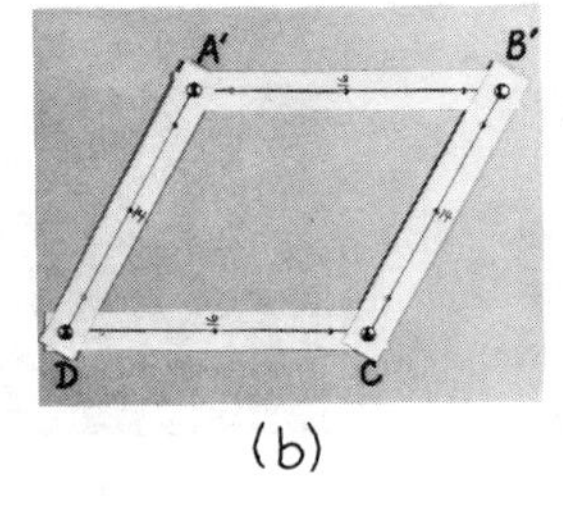

(b)

Figure 3-15

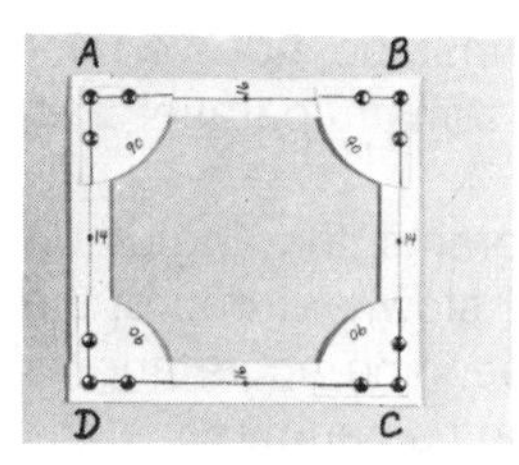

Figure 3-16

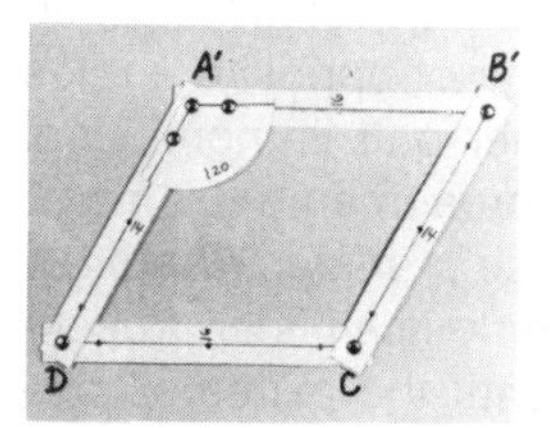

Figure 3-17

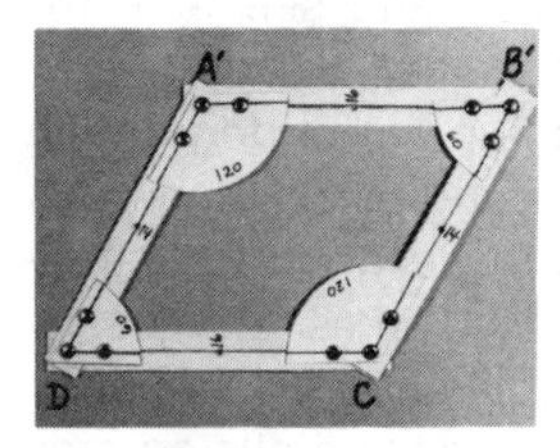

Figure 3-18

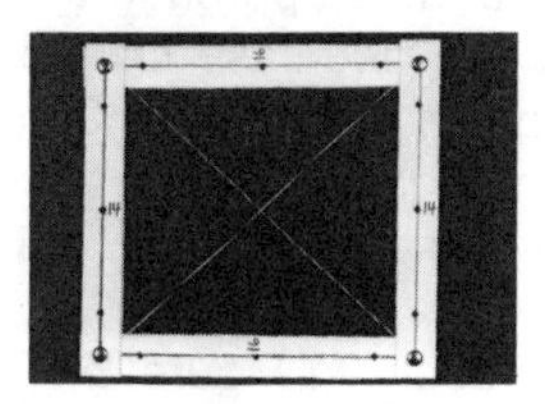

Figure 3-19

attached at each vertex (Figure 3-16). Angles A and B are adjacent angles of figure $ABCD$. The sum of the measures of $\angle A$ and $\angle B$ is 180°. Angles A and B are supplementary. When $ABCD$ is transformed into $A'B'CD$, the 120° angle-fixer fits at A' (Figure 3-17). Which angle-fixer will fit at B'? How about at C? As Figure 3-18 shows, a 60° angle-fixer fits at B' and another 120° angle-fixer fits at C. Also, another 60° angle-fixer fits at D. We find that opposite angles have the same measure and adjacent angles are supplementary.

Will this always be true? Transform $ABCD$ into other parallelograms so that angle-fixers with the following degree measures fit at vertex A': 45, 75, 105, 150. What are the measures of $\angle B'$, $\angle C$, and $\angle D$ in each case? What relationships are always true?

This simple transformation of a rectangle into parallelograms can also give us information about the diagonals. Attach two pieces of elastic thread to the model of a rectangle made from strips to form two diagonals as shown in Figure 3-19. You can wind the thread around the brass fasteners to make it taut. What do you notice about these two diagonals? By measuring the diagonals, you can verify that they are *equal in measure* (congruent) and they *bisect each other.* That is, they intersect at a point which partitions each diagonal into two congruent segments, or, more briefly, they cut each other in half.

Now transform the rectangle into parallelograms as we did before. What happens to the diagonals? One diagonal is stretched while the other contracts (Figure 3-20). You can tell without even measuring that one diagonal increases while the other decreases. Are the diagonals still congruent? Do they still bisect each other? By this transformation we see that the diagonals do *not* remain congruent, but they always bisect each other. The diagonals are congruent only when the parallelogram is a rectangle. We can conclude that the diagonals of a parallelogram are not always congruent but always bisect each other.

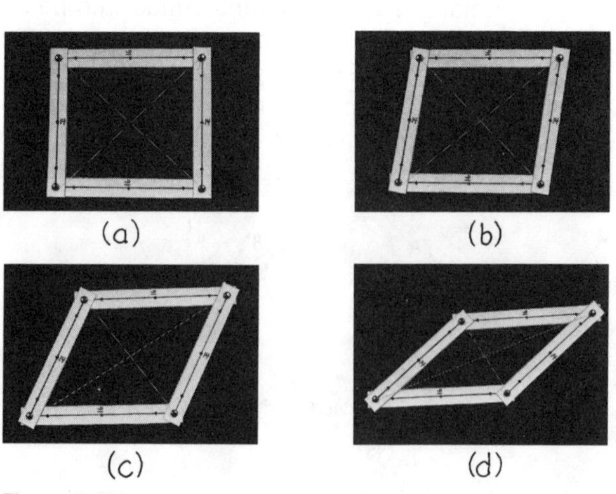

Figure 3-20

Let us summarize our observations about parallelograms. Every parallelogram has the following properties:

1. Opposite sides are parallel.
2. Opposite sides are congruent.
3. The sum of the measures of all four angles is 360°.
4. Opposite angles are congruent.
5. Adjacent angles are supplementary.
6. Diagonals bisect each other.

These are the properties that remained unchanged or **invariant** as we transformed the rectangle into many different parallelograms. We will be discussing other transformations like this. When we transform a figure into a different figure and whatever was parallel in the original figure remains parallel in the resulting figure, we have performed an **affine transformation.** The original and the resulting figures are called **affine figures.** As we transformed the rectangle into different parallelograms, both pairs of opposite sides remained parallel. This was an affine transformation, and the rectangle and resulting parallelograms were affine figures.

Since a rectangle is a parallelogram, it has all the properties of a parallelogram. In addition, all the angles of the rectangle are right angles. We also saw that the diagonals of a rectangle are congruent. So the rectangle is a special kind of parallelogram. Again we can use a Venn diagram to show this (Figure 3-21). In the set of all possible parallelograms, the set of rectangles is a subset of the parallelograms. To belong to that subset of rectangles, a parallelogram must have right angles.

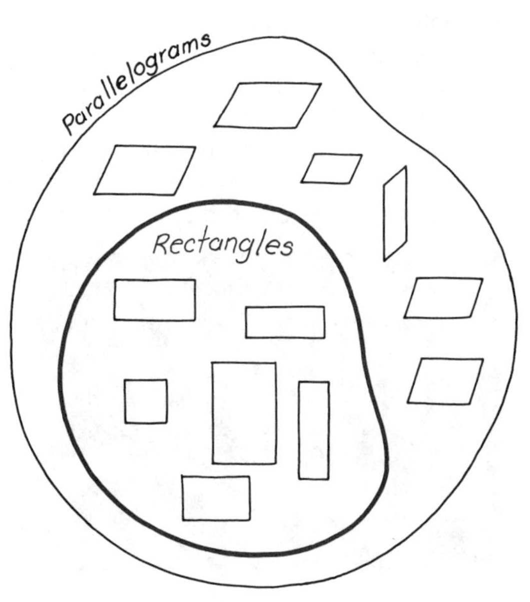

Figure 3-21

Parallelograms and Rhombuses

Figure 3-22

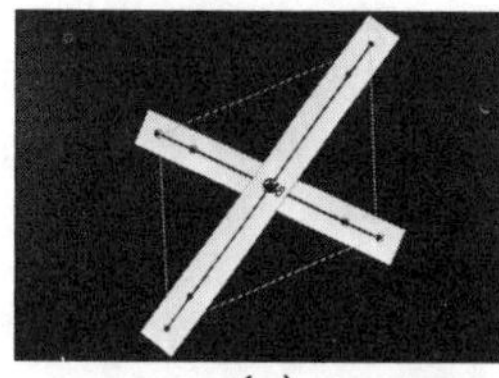

(a)

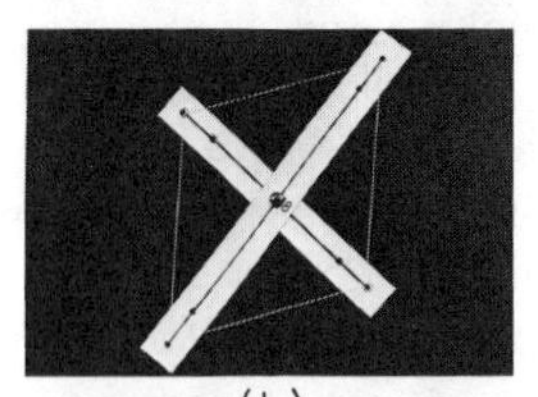

(b)

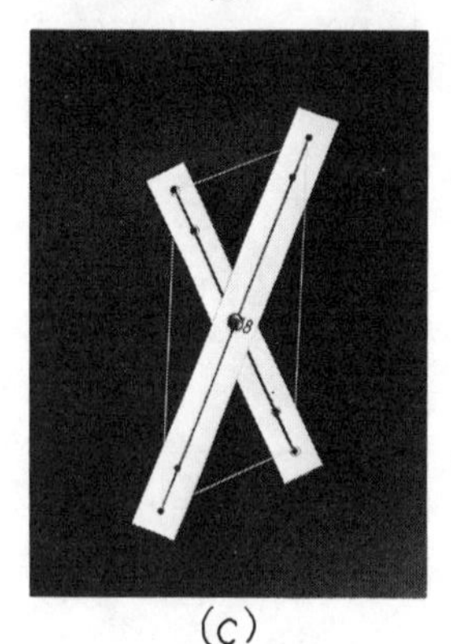

(c)

Figure 3-23

The fact that the diagonals of a parallelogram always bisect each other suggests another way to make a parallelogram. Take any two strips (not congruent) and join them at the hole in the middle of the strip (the midpoint of the line segments represented by the strips). Then pass a long piece of elastic thread through the end holes of the strips, one hole at a time, and tie the ends of the thread together to form a quadrilateral with the thread as shown in Figure 3-22. What kind of quadrilateral is formed by the thread? A parallelogram! As you move the strips, the elastic thread *always* forms a parallelogram (Figure 3-23).

At one point in this transformation the diagonals form right angles; they are perpendicular to each other [Figure 3-23(b)]. What do you notice about the sides of the parallelogram formed by the elastic thread? The figure formed is a special kind of parallelogram. **All four sides of this parallelogram are congruent and the diagonals are perpendicular.** This special parallelogram is called a **rhombus.**

Suppose you take four strips the same length and join them to form a polygon as in Figure 3-24. What kind of polygon is formed? It is a parallelogram with all four sides congruent. The polygon is a rhombus. Whenever you take four strips the same length and join them to form a polygon, you will have a model of a rhombus. Every rhombus is a parallelogram and has all the properties of a parallelogram. But the rhombuses have the additional property that all four sides are always congruent. We can think of a rhombus as an equilateral parallelogram.

Make several rhombuses with strips and then make their diagonals with elastic thread as in Figure 3-25. Examine the diagonals in each rhombus. What do you notice? **In every rhombus the diagonals bisect each other and are perpendicular.**

Now let us take one of the rhombuses made from the strips and transform it into other rhombuses the way we transformed the

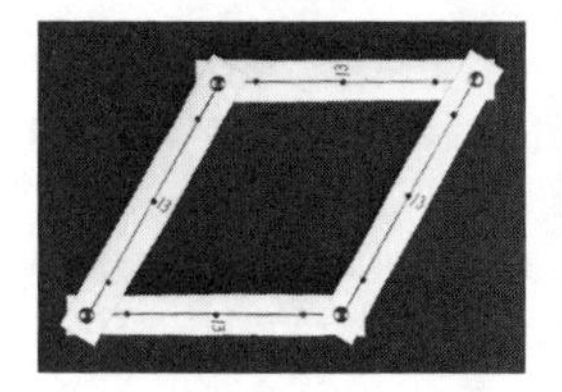

Figure 3-24

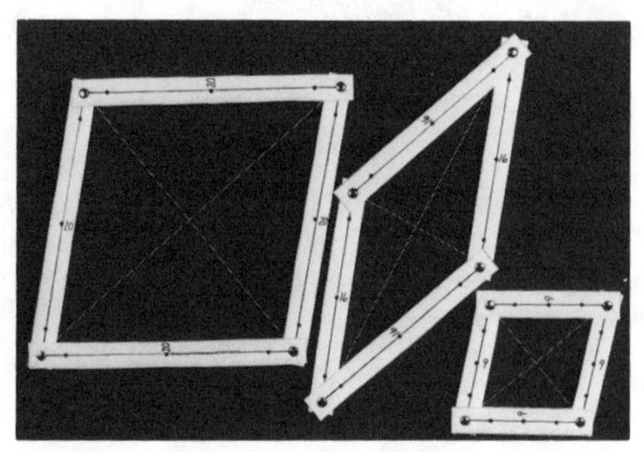

Figure 3-25

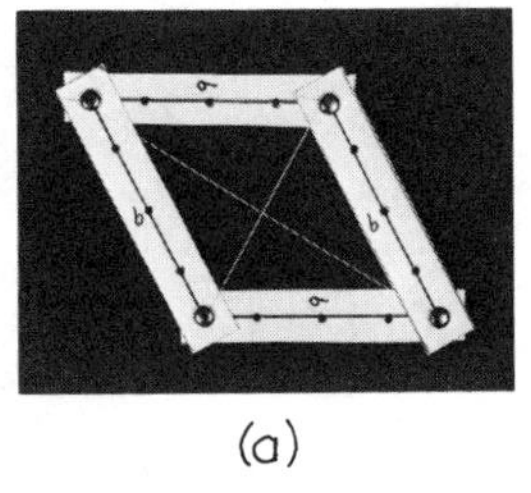
(a)

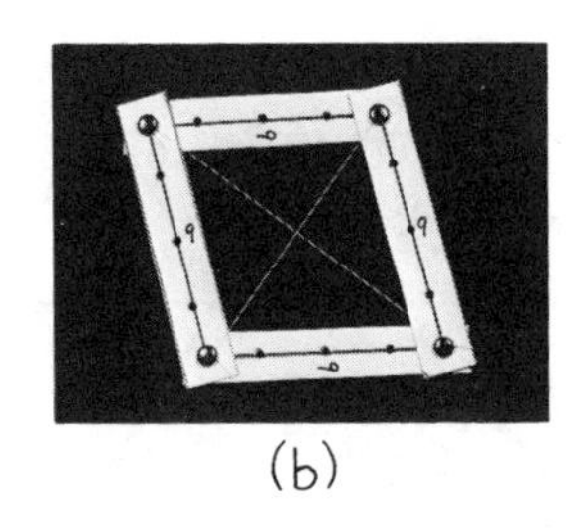
(b)

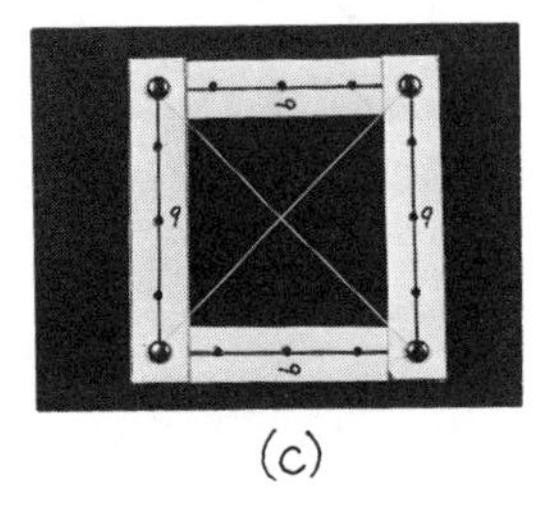
(c)

Figure 3-26

rectangle into parallelograms. As you can see (Figure 3-26), the rhombus can be transformed into many different rhombuses. At one point, the angles formed by the sides of the rhombus all become right angles [Figure 3-26(c)]. We now have a special kind of rhombus called a **square.** We can think of a square as a **rhombus whose sides form right angles.**

Starting with the model of the square made from strips, push in one side to transform the square into different rhombuses (Figure 3-27). *What remains the same and what changes as we transform the square into other rhombuses?* If you carefully examine the rhombuses formed, you will note that the following statements are true:

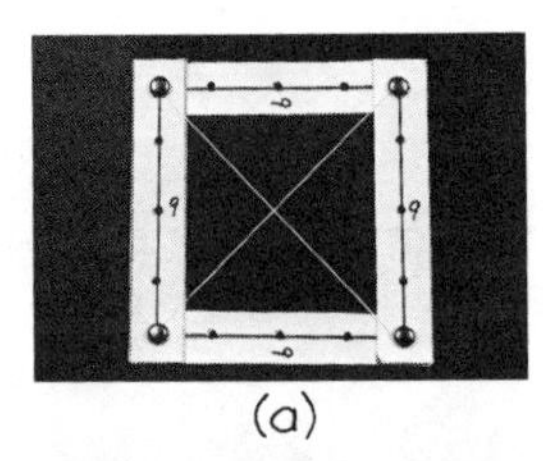
(a)

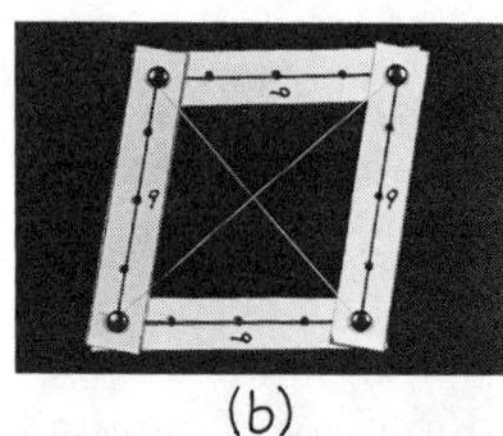
(b)

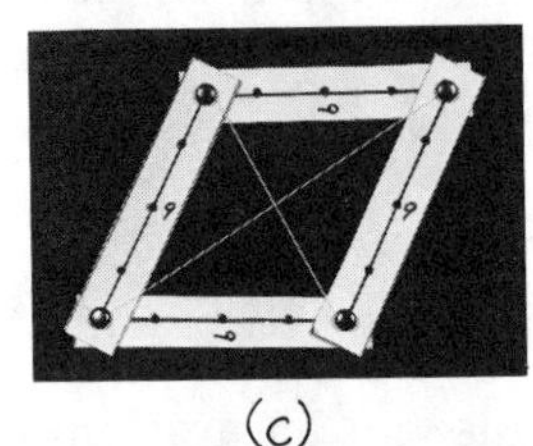
(c)

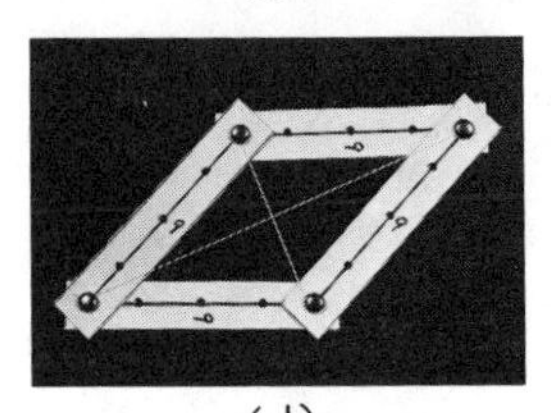
(d)

Figure 3-27

1. Opposite sides remain parallel.
2. The measure of each angle changes; one pair of opposite angles is increasing while the other pair is decreasing.
3. Opposite angles are always congruent.
4. Adjacent angles are always supplementary.
5. The sum of the measures of the angles remains 360°.

These statements are not surprising, since rhombuses are parallelograms and must have all the properties of a parallelogram. Note also that the transformation from square to rhombuses is like the transformation from rectangle to parallelograms. This is another affine transformation, and the resulting rhombuses are affine figures. (You will learn more about affine transformations in Chapter 9.)

Let us transform this model of a square once more. This time examine the diagonals as you transform the square into rhombuses. What is always true? For every rhombus formed, the diagonals *always* bisect each other and are perpendicular. **When the rhombus is a square, the diagonals are also congruent** (Figure 3-27).

This information about the diagonals of a rhombus can be used to develop an easy method for drawing a rhombus. All you need do is draw two perpendicular line segments that bisect each other and use the end points of these line segments as vertices for the

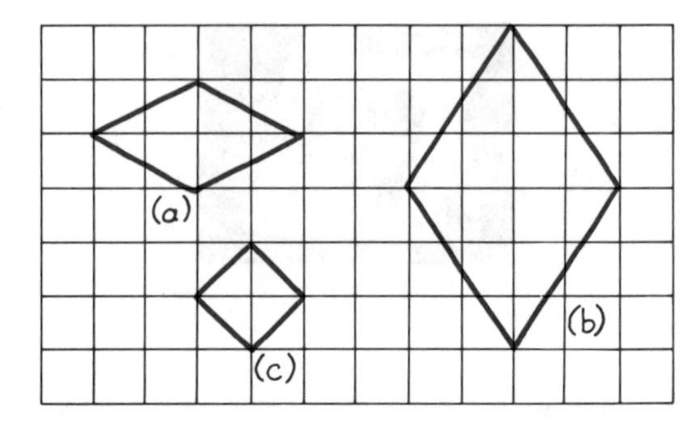

Figure 3-28

rhombus. Squared paper might be used, as in Figure 3-28. If the two perpendicular line segments (diagonals) are congruent, you will have a special rhombus [Figure 3-28(c)]. What is that special rhombus?

Rectangles and Squares

Earlier in this chapter we studied another parallelogram with congruent diagonals, the rectangle. Both the rectangle and the square have that same property: the diagonals are always congruent. Actually, *a square qualifies as a rectangle.*

We defined a rectangle as a parallelogram with right angles. Since a square is a parallelogram with right angles, you can call it a rectangle. The square has an extra feature: all its sides are congruent. A square is a special rectangle. It is an equilateral rectangle.

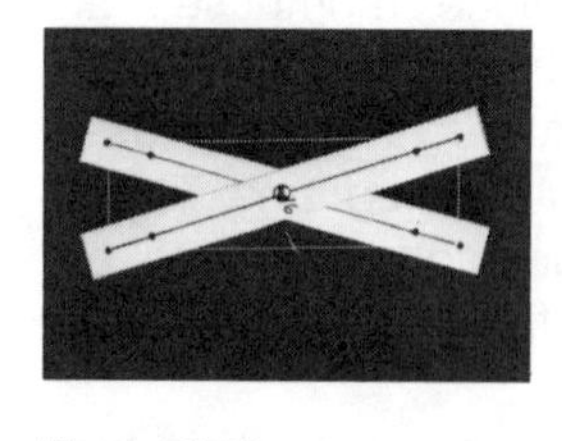

Figure 3-29

We can see the relationship between the square and the rectangle by making a model of a rectangle and examining its transformation. Take two congruent strips and join them at their midpoints. Since the diagonals of a rectangle are always congruent and bisect each other, these strips can be used as the diagonals for the model of a rectangle. Next pass a long piece of elastic thread through the end holes, one hole at a time, and tie the ends of the thread together as in Figure 3-29. The thread will form a rectangle.

You can transform this model of a rectangle and form other rectangles by moving the diagonal strips as shown in Figure 3-30. As other rectangles are formed, the sides of the rectangles change in length: one pair of opposite sides gets larger while the other pair gets smaller. How about the angles formed by the diagonals? How do they change?

At one point in this transformation all four sides of the rectangle are congruent [Figure 3-30(c)]. The rectangle becomes a square.

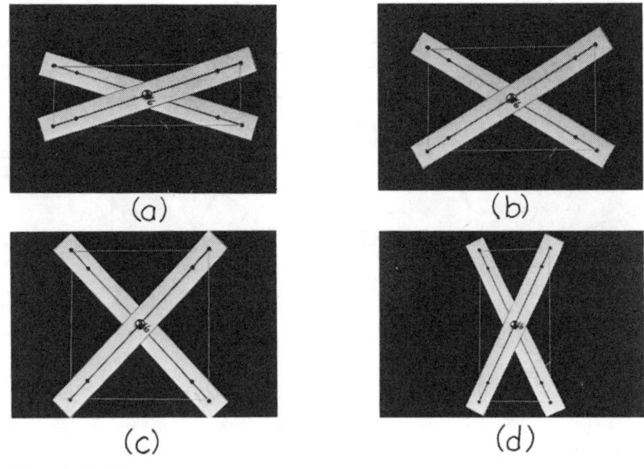

Figure 3-30

What kinds of angles are formed by the diagonals of that square? Right angles! The other rectangles did *not* have diagonals that were perpendicular (formed right angles). It is only when the rectangle is a square that its diagonals are perpendicular. You might have expected this since you know that a square is a rhombus.

Squares, Rhombuses, and Rectangles

We have seen that:

1. The square belongs to the set of rhombuses; it is a special kind of rhombus.
2. The square belongs to the set of rectangles; it is a special kind of rectangle.

Squares, then, belong to two sets at the same time. We can say that the **intersection** of the set of rhombuses and the set of rectangles is the set of squares. That is, squares belong to both sets at the same time.

We can express this idea by using the Venn diagram in Figure 3-31. Suppose we were to put every rhombus inside the boundary marked "Rhombuses" and every rectangle inside the brown boundary marked "Rectangles." Every square would qualify both as a rhombus (parallelogram with all sides congruent) and as a rectangle (parallelogram with right angles). To place the squares so that they are inside the rhombus boundary *and* inside the brown rectangle boundary at the same time, we must put them in the region marked "Squares."

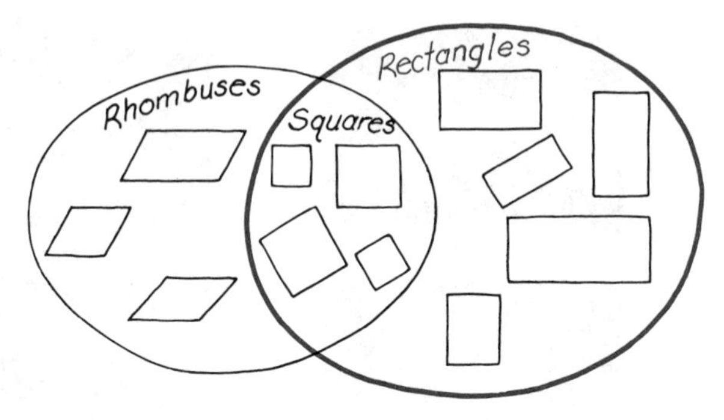

Figure 3-31

Classification of Quadrilaterals

We began our study of quadrilaterals by examining convex quadrilaterals. We discussed trapezoids, parallelograms, rectangles, rhombuses, and squares, and we studied the relationships among these figures. One way of indicating these relationships is by using another Venn diagram, as in Figure 3-32. The diagram shows the following:

1. If we start with the set of all convex quadrilaterals, then the set of trapezoids is a subset of that set. The trape-

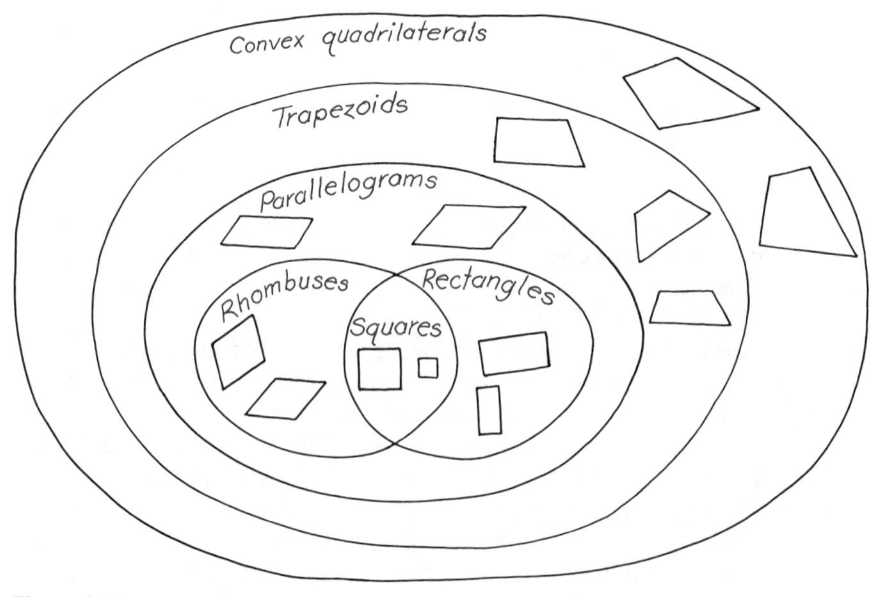

Figure 3-32

zoids are the quadrilaterals with one pair of opposite sides parallel.

2. A subset of the set of trapezoids is the set of parallelograms. Parallelograms are trapezoids with *both* pairs of opposite sides parallel.
3. The set of rhombuses and the set of rectangles are both subsets of the set of parallelograms. Rhombuses are special parallelograms with all sides congruent. Rectangles are special parallelograms whose angles are right angles.
4. The intersection of the set of rhombuses and the set of rectangles is the set of squares.

The squares are located within every boundary because they qualify for every set. A square is a rhombus, a rectangle, a parallelogram, a trapezoid, and a convex quadrilateral. A square is a very special figure.

Other Polygons

We have been considering polygons with exactly four sides—quadrilaterals. With the strips you can make other types of polygons having as many sides as you wish. And, as you might imagine, these polygons can be given special names according to the number of sides they have. The prefixes of these special names give a clue as to how many sides the polygon has. Table 3-1 gives some of these names.

Table 3-1

Sides	Special name
5	Pentagon
6	Hexagon
7	Heptagon
8	Octagon
9	Nonagon
10	Decagon
12	Dodecagon

When all the sides of a polygon are congruent, that polygon can be called **equilateral.** If no sides of a polygon have the same measure, the polygon is **scalene.** As you might expect, when all the angles of a polygon are congruent, it is called **equiangular.**

Which polygons in Figure 3-33 are equilateral? Scalene? Equiangular? Which are both equilateral and equiangular? As you will recall from Chapter 1, polygons that are both equilateral and equiangular are called regular polygons.

A polygon can be equilateral and *not* equiangular, or vice versa. All the models of polygons in Figure 3-34 were made from six congruent strips. Each polygon is equilateral. Note that the polygons in (a) and (c) are *not* equiangular. In fact, the polygon in (c) is not even convex. Only (b) is a regular polygon.

In the set of quadrilaterals, only the square is a regular polygon. The rhombus is equilateral, but not necessarily equiangular. Similarly, the rectangle is equiangular, but not necessarily equilateral.

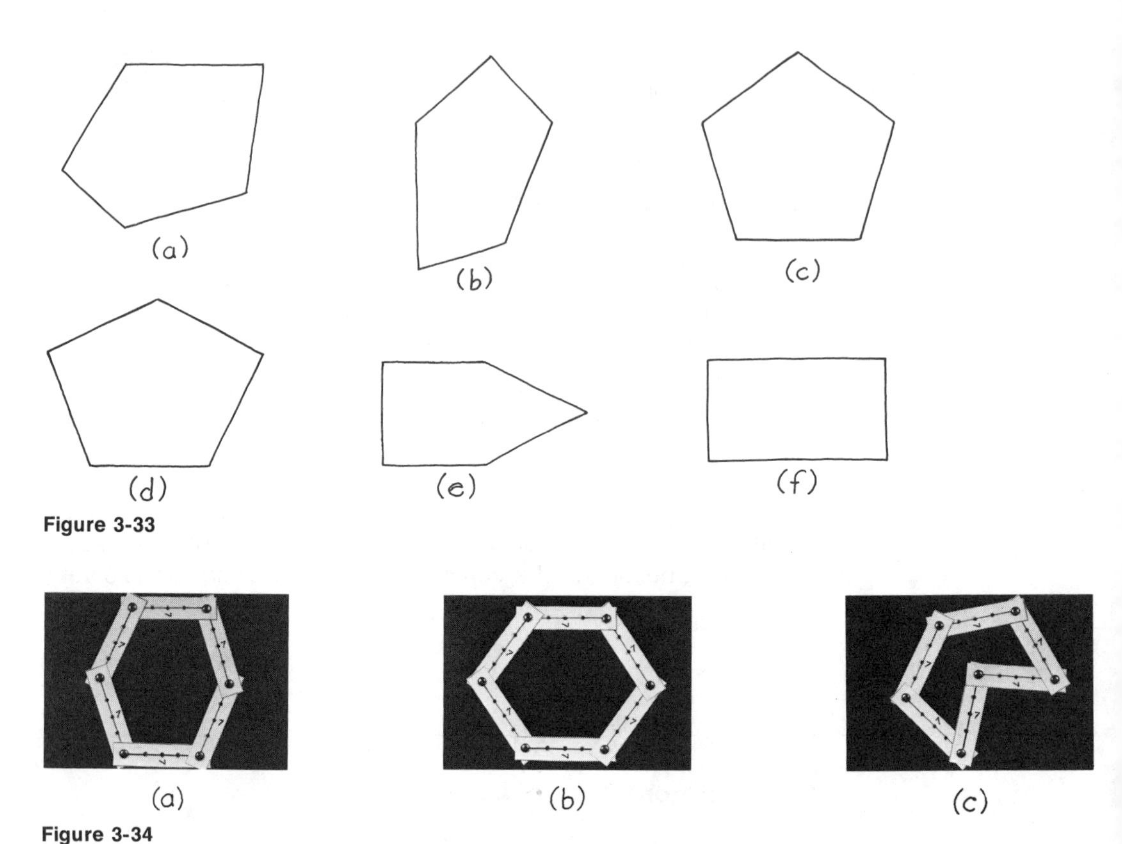

Figure 3-33

Figure 3-34

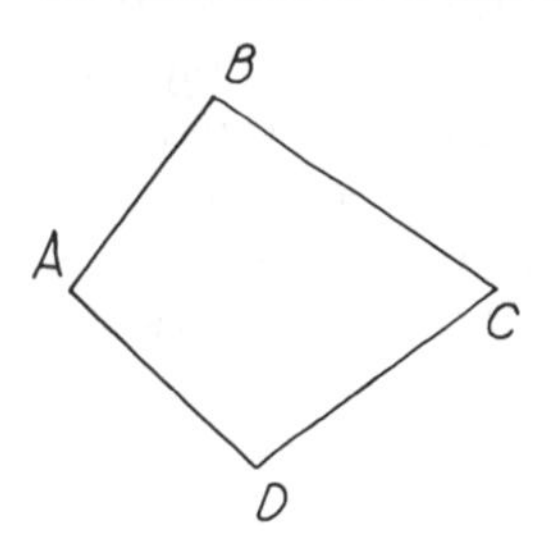

Figure 3-35

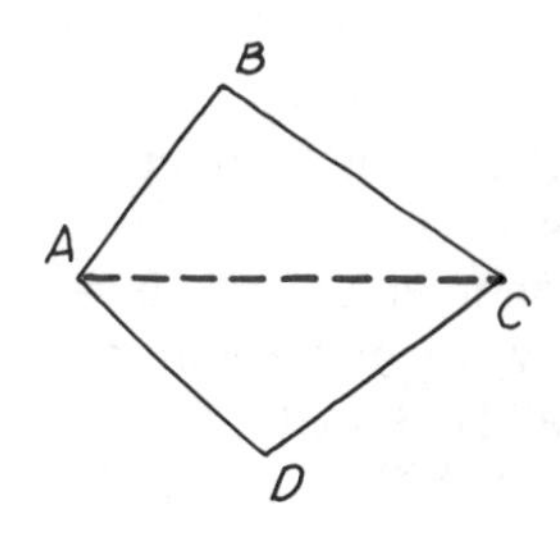

Figure 3-36

Sum of the Measures of the Angles of a Polygon

When studying quadrilaterals we found that the sum of the measures of the angles of *any* parallelogram is always 360°. As we transformed the rectangle to other parallelograms and the square to other rhombuses, that sum remained constant. Do you think the sum of the measures of the angles of *any* quadrilateral is 360°?

Let us take a look at this possibility. Draw *any* quadrilateral, such as *ABCD* in Figure 3-35. Since *ABCD* is a quadrilateral, it has four sides and four angles. Points *A, B, C,* and *D* are the vertices for these angles. Suppose we choose a vertex, such as *A,* and draw the diagonals of the quadrilateral that have *A* as an end point. There is only one such diagonal: $\overline{AC}$. As you can see in Figure 3-36, we have formed two triangles, *ABC* and *ACD.*

You will remember that the sum of the measures of the angles of *any* triangle is 180°. So the sum of the measures of the angles

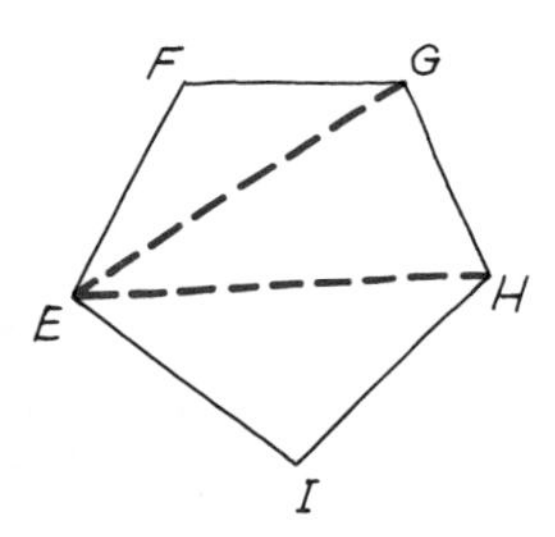

Figure 3-37

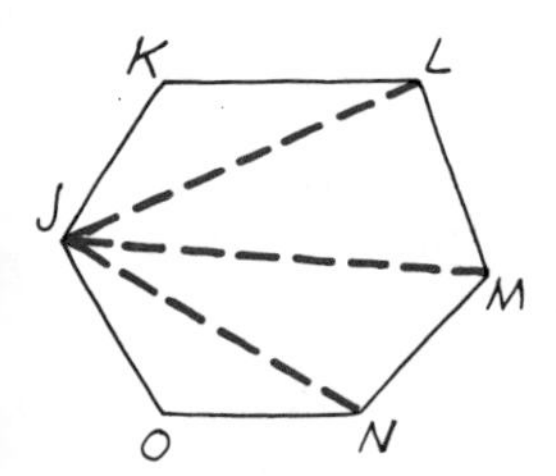

Figure 3-38

of *both* triangles is 360°. But the measure of the angles of the original quadrilateral is the same as the total measure of the angles of both triangles. What can you conclude about the sum of the measures of the angles of the quadrilateral?

Let us consider other polygons. In pentagon *EFGHI* (Figure 3-37), all the diagonals that have one of the vertices as an end point have been drawn, and three triangles have been formed. Similarly, in hexagon *JKLMNO* (Figure 3-38) four triangles have been formed. Can you see a pattern?

1. Quadrilateral (four sides) into two triangles
2. Pentagon (five sides) into three triangles
3. Hexagon (six sides) into four triangles

In this way a 10-sided polygon would form 8 triangles, a 100-sided polygon would form 98 triangles, and, in general, a polygon with "n" number of sides, or an n-sided polygon, would form $(n - 2)$ triangles.

In each case the sum of the measures of the angles of the original polygon is equal to the sum of the measures of the angles in all the triangles formed. If we know how many triangles are formed, we need only multiply that number by 180. Table 3-2 shows these relationships.

Table 3-2

Polygon	Sides	Triangles formed	Sum of measures of angles of polygon
Quadrilateral	4	2	360°
Pentagon	5	3	540°
Hexagon	6	4	720°

How can we find the sum of the measures in degrees of the angles of a polygon with 50 sides? Following the same pattern, we could make 48 triangles and the total number of degrees would be 48×180 or 8640°.

In general, for a polygon with n sides we can say that the sum of the measures in degrees of the angles of that polygon can be found as follows:

$$(n - 2) \times 180$$

In this way you can determine the total number of degrees of the angles of any polygon.

More Experiences

3-1. Make models of the following shapes on a geoboard. Include rubber bands to represent the diagonals of each figure. Then describe the relationship of the diagonals for each figure.

(*a*) Parallelogram that is not a rectangle or rhombus
(*b*) Rectangle
(*c*) Rhombus
(*d*) Square
(*e*) Trapezoid that is not a parallelogram
(*f*) Quadrilateral that is none of the above

3-2. Suppose each pair of line segments on the geoboards below represents diagonals of a quadrilateral. What kinds of quadrilaterals are formed in each case?

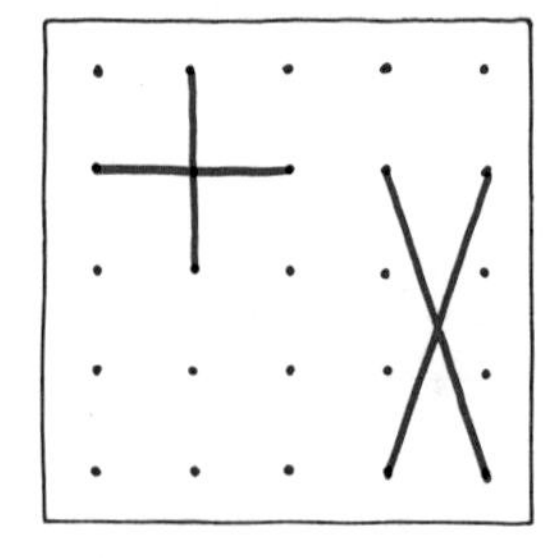

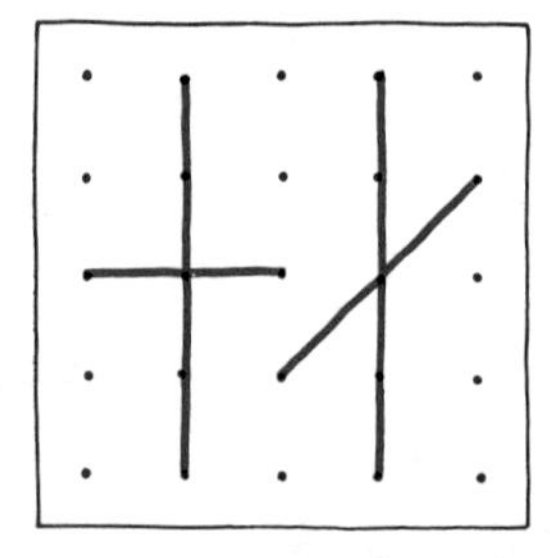

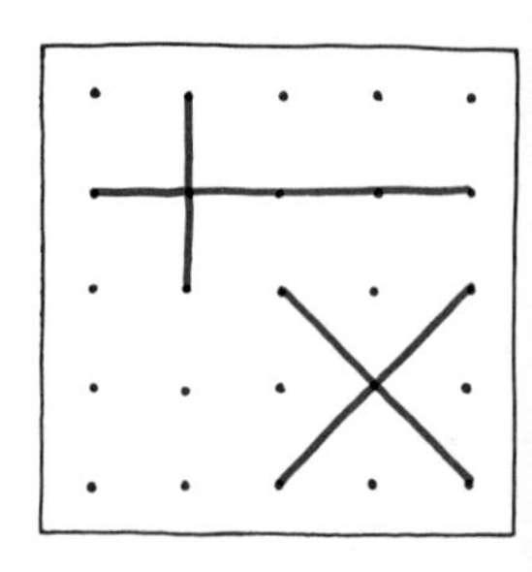

3-3. A square with two nails along each side (2 × 2) is the smallest square that can be made on a geoboard. How many 2 × 2 squares can be made on a 5 × 5 geoboard? How many 3 × 3 squares can be made? 4 × 4? 5 × 5?

It is possible to make other squares on the geoboard, as shown at the left. Can you discover still others? How many different kinds of squares is it possible to make on a 5 × 5 geoboard?

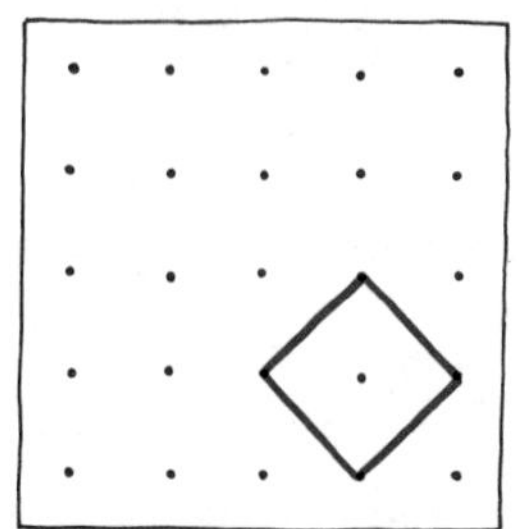

3-4. Construct a quadrilateral by using two congruent strips of one length and two congruent strips of another length so that strips with the same length are adjacent to each other. What does the resulting quadrilateral look like? This shape is sometimes called an antiparallelogram or kite.

Use elastic thread to make the diagonals of this model. What do you notice about the diagonals? Push in at two opposite vertices of the model and transform it into other kites. What happens to the diagonals?

3-5. Make (with strips and elastic thread) or draw models of quadrilaterals with the following diagonals. What kind of quadrilateral is formed in each case?

(*a*) Two noncongruent diagonals that bisect each other
(*b*) Two noncongruent diagonals that bisect each other and are perpendicular
(*c*) Two congruent diagonals that bisect each other
(*d*) Two noncongruent diagonals with only one diagonal bisected and the diagonals perpendicular
(*e*) Two noncongruent diagonals that do *not* bisect each other and are *not* perpendicular
(*f*) Two congruent diagonals that bisect each other and are perpendicular

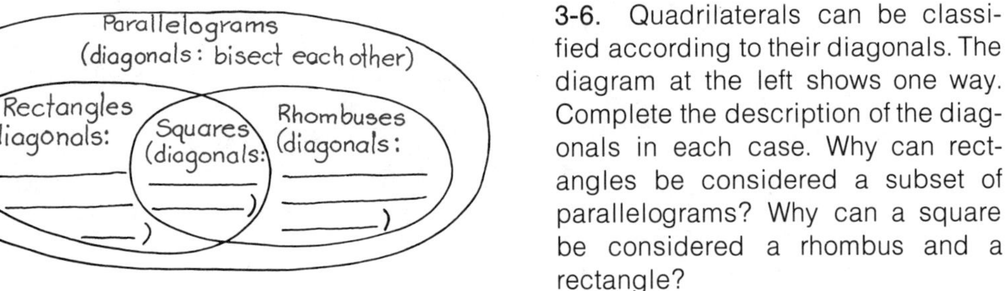

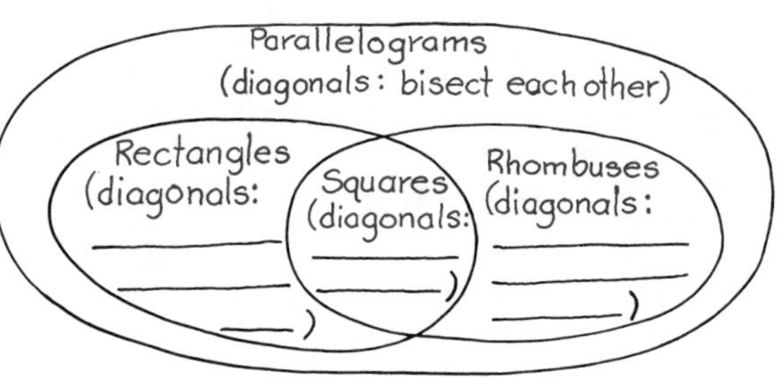

3-6. Quadrilaterals can be classified according to their diagonals. The diagram at the left shows one way. Complete the description of the diagonals in each case. Why can rectangles be considered a subset of parallelograms? Why can a square be considered a rhombus and a rectangle?

3-7. If you join five congruent strips to form the model of a polygon, this polygon will be an *equilateral pentagon.* Is this a model of a *regular* pentagon? Explain your response.

3-8. Starting with four strips of different lengths, construct a quadrilateral. Draw a diagram of that quadrilateral. Now take the model apart and put it together another way to form a *different* quadrilateral. Draw a diagram of that quadrilateral. How many different quadrilaterals can you make with these four strips?

3-9. Construct a quadrilateral by using three congruent strips of one length and one strip twice that length. Can you transform that model into a trapezoid? What kind of trapezoid results?

Suppose you include strips that extend from each end hole of the shorter base of the trapezoid to the midpoint of the longer base. What kinds of triangles have been formed?

3-10. Using a ruler and protractor, can you draw a regular pentagon? How about a regular hexagon?

How many degrees are there in each angle of a regular polygon with 20 sides? How many degrees are there in each angle of a regular polygon with n sides (where n represents any number of sides)?

3-11. A triangle has no diagonals; a quadrilateral has two diagonals. How many diagonals does a pentagon have?

Can you complete the chart at the left? Do you see a pattern? What is it?

Sides	Diagonals
3	0
4	2
5	
6	
7	
8	
.	
.	
.	

4 Symmetry

A Kind of Balance

The shapes and designs pictured in Figure 4-1 have a very special property. We see this property often in many things in our environment. It is a kind of balance. This special kind of balance is called **symmetry.** If a shape or figure has symmetry, we call it symmetrical. But what is symmetry? Are there different kinds of symmetry? How do we know if a shape has symmetry? These are some of the questions we will explore in this chapter.

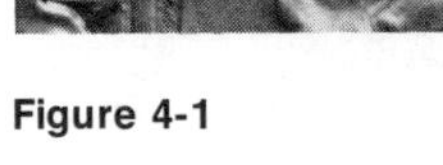

Figure 4-1

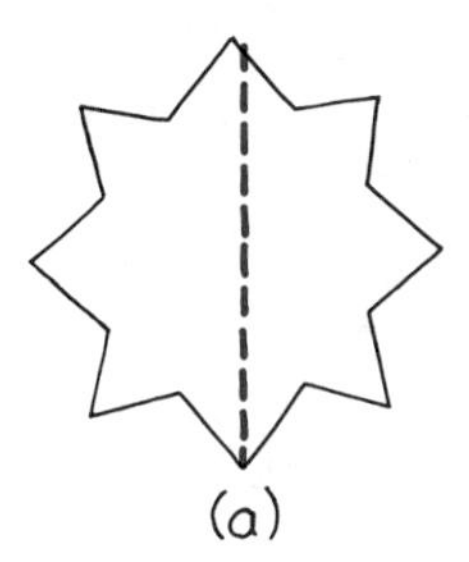

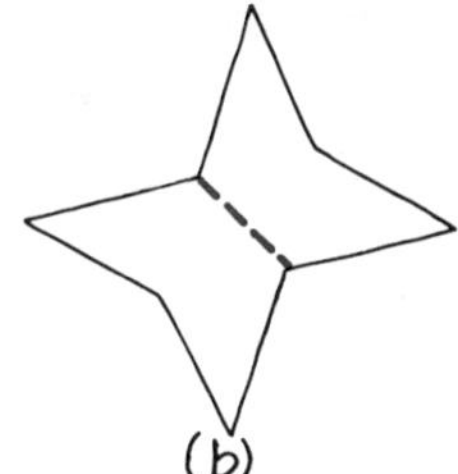

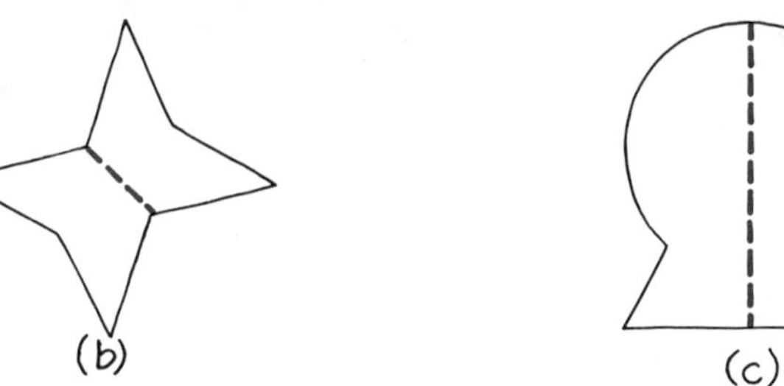

Figure 4-2

Bilateral Symmetry

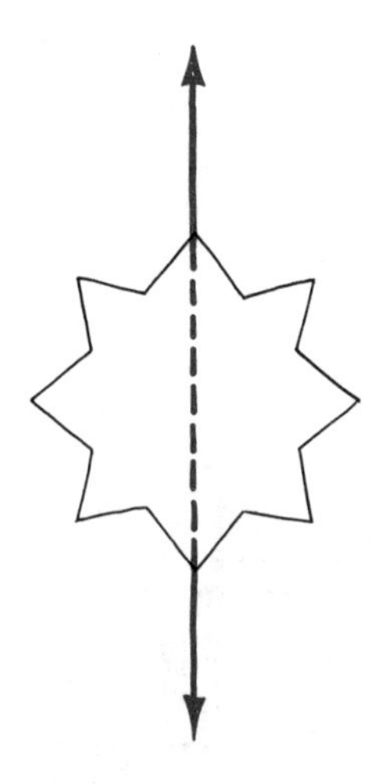

Figure 4-3

Use some tracing paper to copy the shapes in Figure 4-2. Cut out the shapes along the solid curve. Now fold each one along the dotted line segment. What do you notice? In each case the two parts of the shape coincide or match completely. These shapes have what is called **bilateral symmetry.** They are symmetrical with respect to the line segment along the crease. That line segment belongs to the **line of symmetry** (also called **axis of symmetry**) for the shape (Figure 4-3).

Can you fold the shapes in Figure 4-2 in any *other* way so that the two parts that result coincide or match completely? Shape (a) can be folded as in Figure 4-4 so that the two parts coincide. It has two lines of symmetry. How many lines of symmetry do shapes (b) and (c) have?

It is easy to make a shape that has bilateral symmetry. Here is one way. Fold a piece of paper in half. Draw a design as in Figure 4-5 so that the end points of the design are on the crease. Cut out your design and open the paper. The resulting shape will have bilateral symmetry. The crease lies along the line of symmetry (Figure 4-6). You may remember this method as the long-established procedure for making the shapes of symmetrical pumpkins, Christmas trees, and valentine hearts as in Figure 4-7.

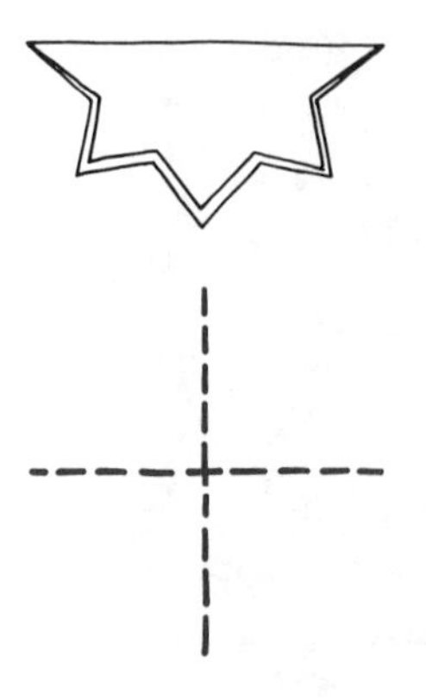

Figure 4-4

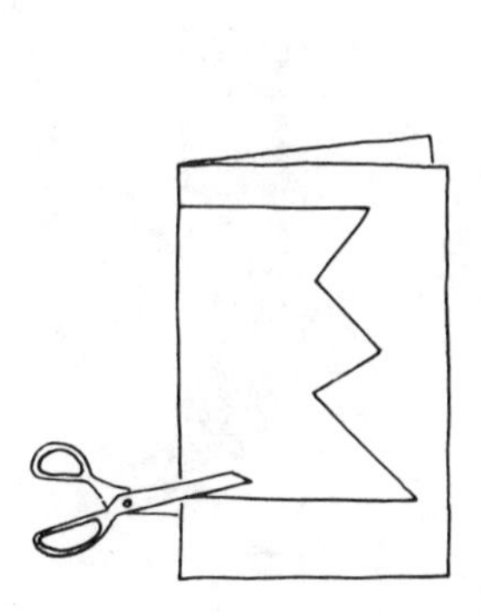

Figure 4-5

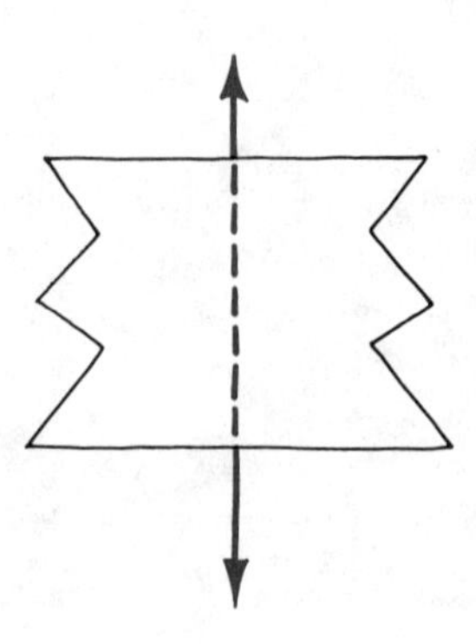

Figure 4-6

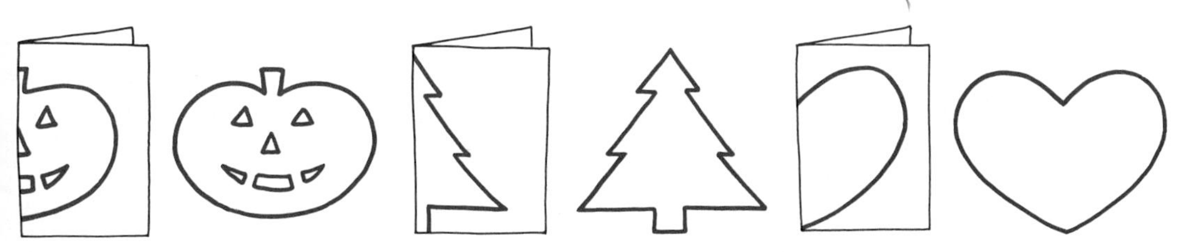

Figure 4-7

Figure 4-8

The paper folding technique is also useful in making figures with more than one line of symmetry. Fold a piece of paper in half. Then fold it in half again, and draw a design with one end point on one crease and the other end point on the second crease, as in Figure 4-8. When you cut out your design you will find it has two lines of symmetry represented by the two creases (Figure 4-9). How were the dancing people in Figure 4-10 made? Try making some interesting symmetrical designs by folding paper.

A mirror can be very useful for locating lines of symmetry. Let us experiment by using a small mirror and the shape depicted in Figure 4-11. Place the mirror perpendicular to the shape so that its edge coincides with one of the dotted line segments (Figure 4-12). What do you see in the mirror? The mirror reflects the part of the shape on the mirror side. Notice that the shape outside the mirror together with its reflection form a different, symmetrical shape. In fact, whenever you position the mirror on *any* of the dotted line segments, the result is another symmetrical shape, half outside

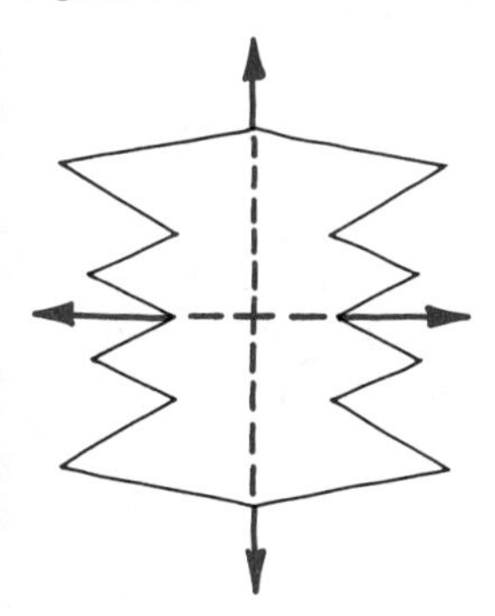

Figure 4-9

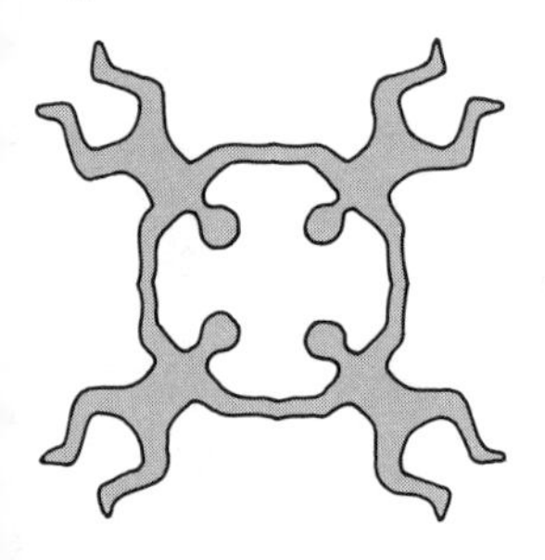

Figure 4-10

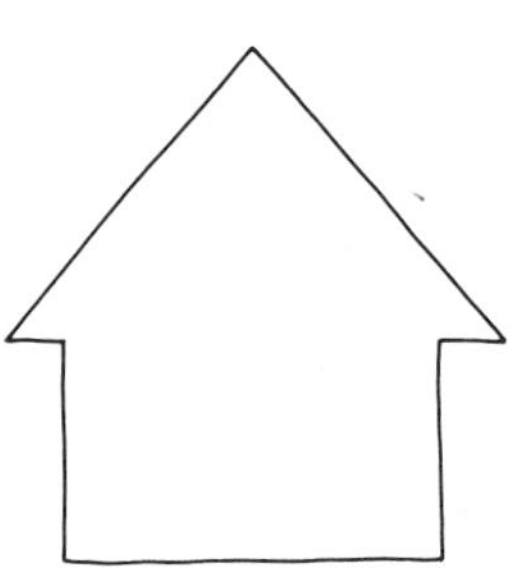

Figure 4-11

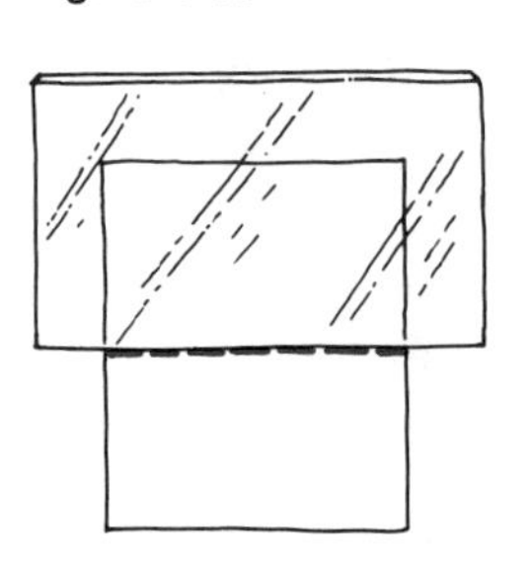

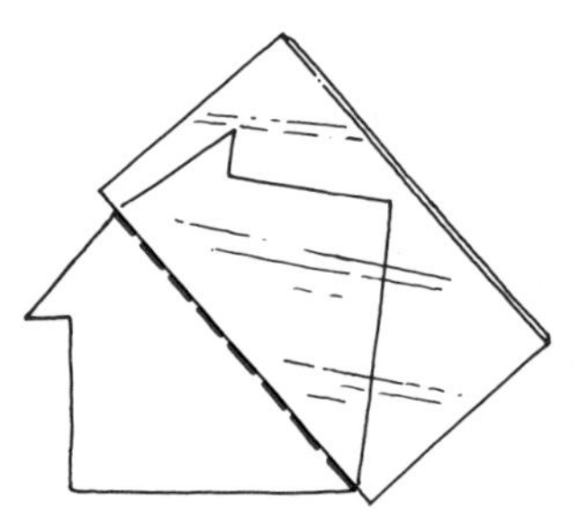

Figure 4-12

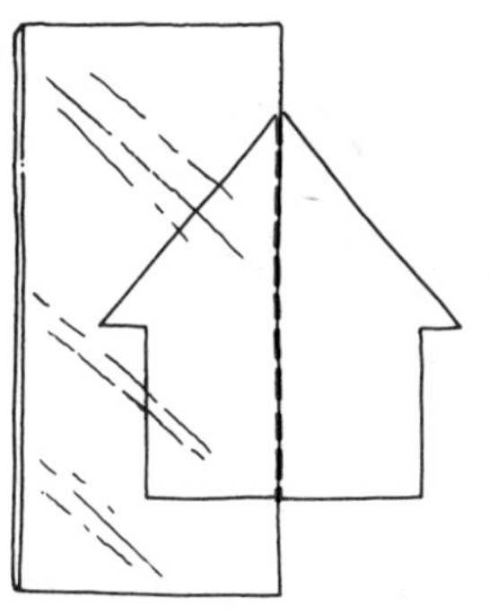

Figure 4-13

the mirror and half "inside" the mirror. When you place the mirror on a certain dotted line segment (Figure 4-13), however, something even more special happens! The symmetrical shape looks just like the original shape. Note that this dotted line segment is on a line of symmetry.

This experiment shows us how a mirror can help locate lines of symmetry of a shape. Suppose we have a shape such as the one shown in Figure 4-14. Does it have any lines of symmetry? Let us try to position the mirror in different ways to see if we can find a place where the entire original shape appears—half outside the mirror and half as a reflection. There are three such places. The edge of the mirror lies on a line of symmetry in each case (Figure 4-15). We can locate three lines of symmetry.

Using a mirror, locate the lines of symmetry of shapes in Figure 4-16. Does a shape always have a line of symmetry? As you will find, shape (d) has no line of symmetry. There is no place you could put the mirror so that the part of the shape outside the mirror and its reflection will form the original shape.

Let us try something else with a mirror. Take a penny, a nickel, and a quarter. Place them in front of a mirror as in Figure 4-17. The penny is closest to the edge of the mirror and the nickel is the next closest, while the quarter is furthest from that edge. Look at the reflection of these coins. What do you notice? Again, in the reflection the penny is closest to the edge, the nickel is the next closest, and the quarter is the furthest away.

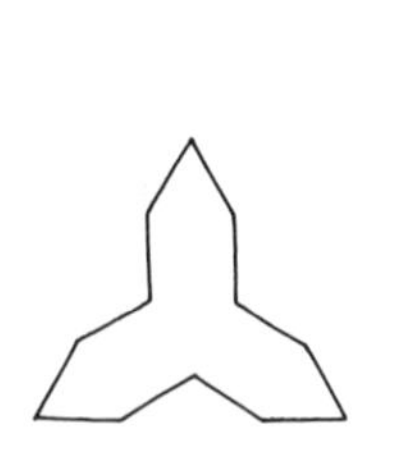

Figure 4-14

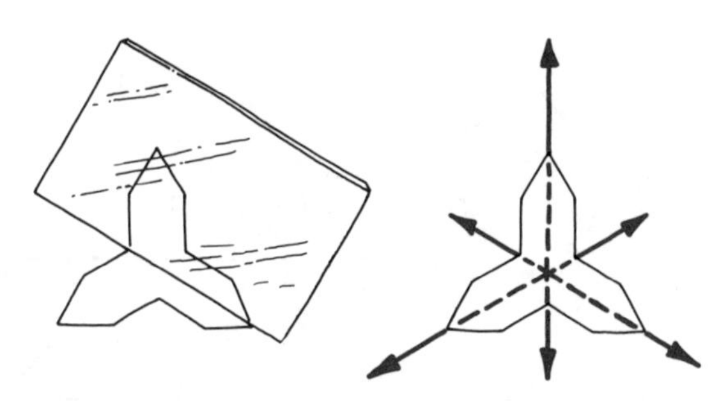

Figure 4-15

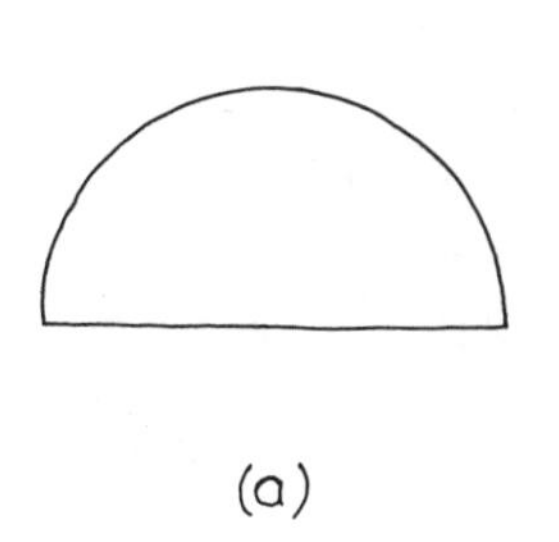

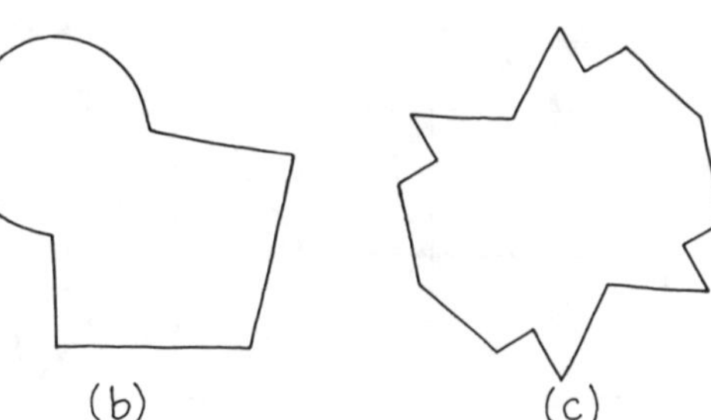

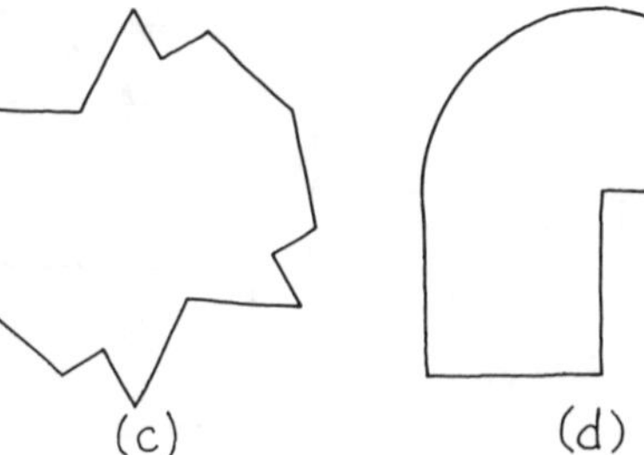

Figure 4-16

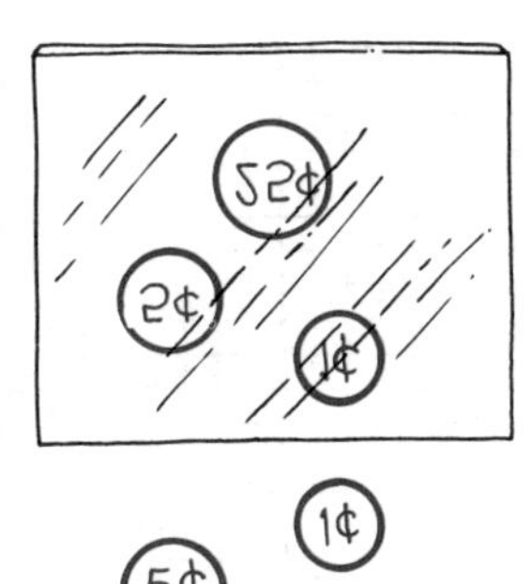

Figure 4-17

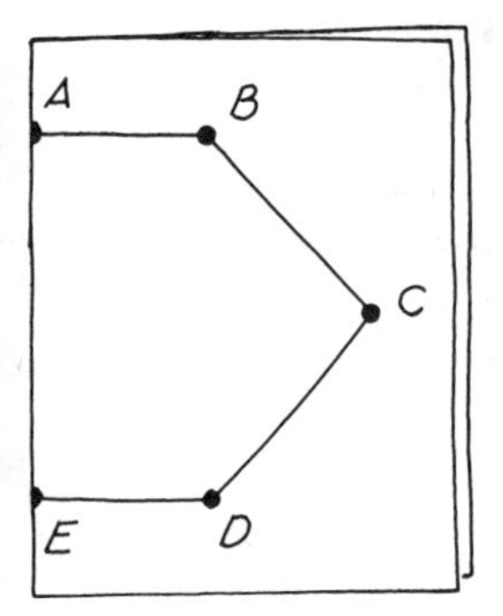

Figure 4-18

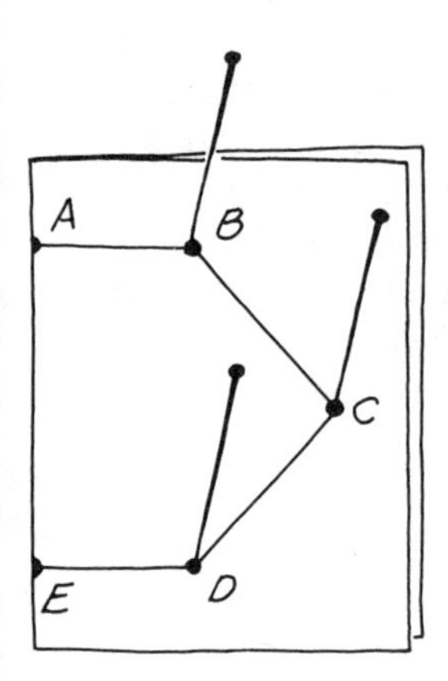

Figure 4-19

Fold a piece of paper and draw the design shown in Figure 4-18. Using a pin or the point of a compass, make a hole through both halves of the paper at points *B*, *C*, and *D* (Figure 4-19). Now open the paper. The holes on the other part of the crease correspond to points *B*, *C*, and *D*. Label them *B′*, *C′*, and *D′* (Figure 4-20). Draw the line segments $\overline{AB'}$, $\overline{B'C'}$, $\overline{C'D'}$, and $\overline{D'E}$. The entire shape is symmetrical, with the crease through *A* and *E* as the line of symmetry (Figure 4-21). Notice that the distance from point *B* to the line of symmetry (represented by $\overline{BA}$) is the same as the distance from the corresponding point *B′* to the line of symmetry (represented by $\overline{B'A}$). Similarly $\overline{DE}$ is as long as $\overline{D'E}$, and if we draw the distance of *C* from the line of symmetry as $\overline{CF}$, then $\overline{CF}$ is as long as $\overline{C'F}$. The crease that represents the line of symmetry is like the edge of the mirror. The distance of each point on one side to the crease is the same as the distance of the corresponding point on the other side to the crease. In other words, the corresponding points of a design that has bilateral symmetry are the same distance from the line of symmetry.

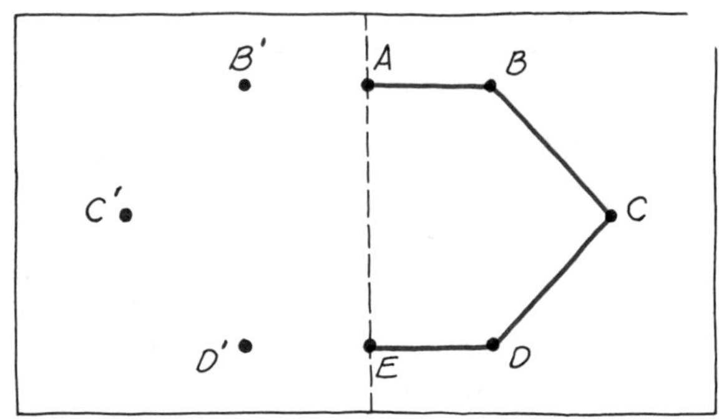

Figure 4-20

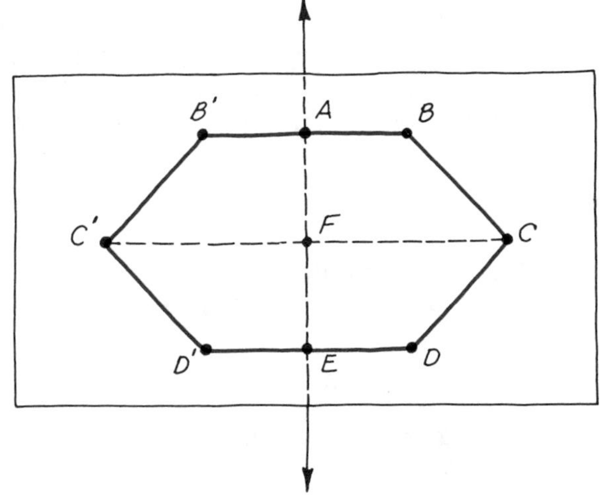

Figure 4-21

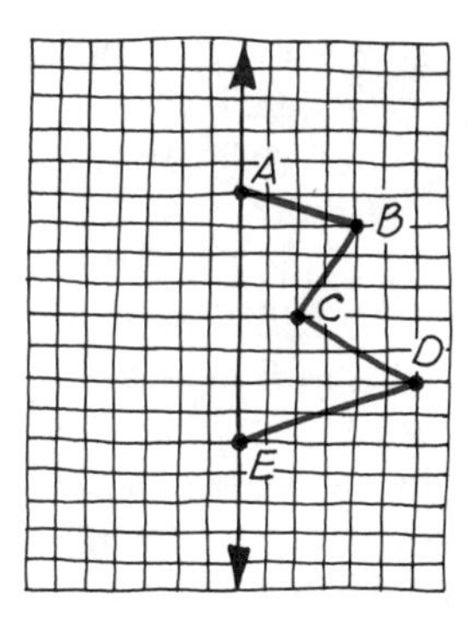

Figure 4-22

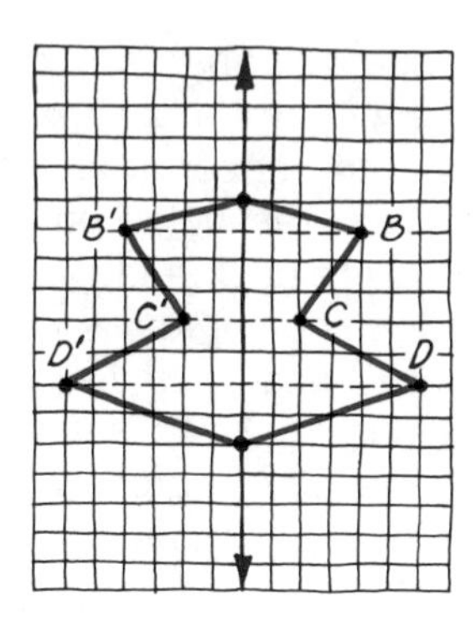

Figure 4-23

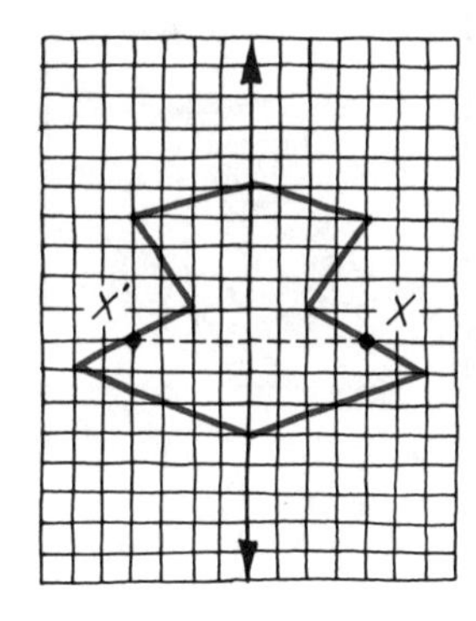

Figure 4-24

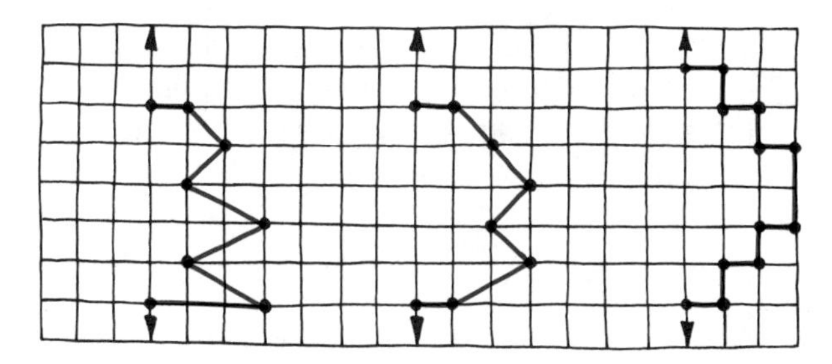

Figure 4-25

This suggests another way to make figures with bilateral symmetry. Look at the portion of a shape drawn on squared paper in Figure 4-22. Line $\overleftrightarrow{AE}$ is indicated as a line of symmetry. Can you locate the points on the other side of $\overleftrightarrow{AE}$ that correspond to *B, C,* and *D?* Since corresponding points are the same distance from the line of symmetry, the resulting shape looks like the one in Figure 4-23. Notice that if we choose any point of the figure on one side of the line of symmetry, say point *X,* we can find the corresponding point of the figure (*X'*) on the other side of the line of symmetry by finding the distance between *X* and the line of symmetry (Figure 4-24). Can you draw the other parts of the symmetrical figures suggested in Figure 4-25?

Bilateral Symmetry in Quadrilaterals and Triangles

You are familiar with the quadrilateral shapes shown in Figure 4-26. Make a paper model of each and, by folding, find how many lines of symmetry each shape has. You can see from Figure 4-27 that the square can be folded in four different ways so that both

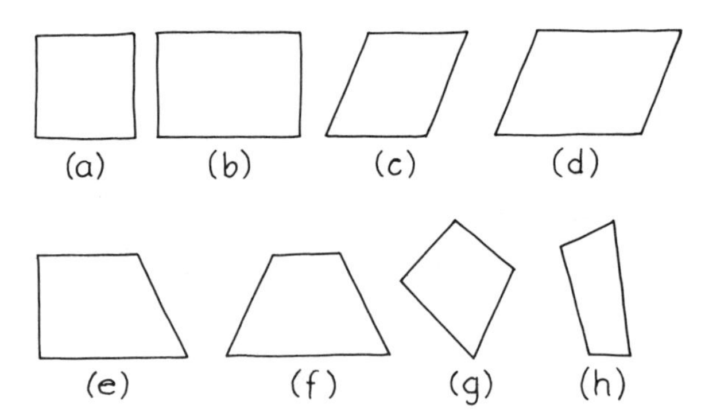

Figure 4-26

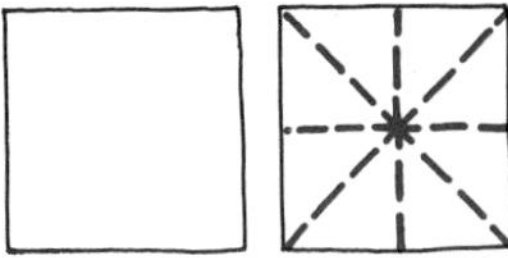

Figure 4-27

parts match completely and the crease represents a line of symmetry. Table 4-1 indicates the number of lines of symmetry each shape in Figure 4-26 has. Can you find all the lines of symmetry?

Table 4-1

Shape	Lines of symmetry
(a) Square	4
(b) Rectangle	2
(c) Rhombus	2
(d) Parallelogram	0
(e) Trapezoid	0
(f) Isosceles trapezoid	1
(g) Kite	1
(h) Scalene quadrilateral	0

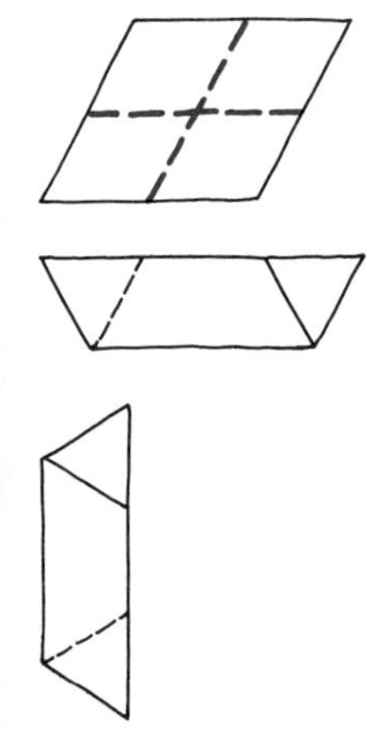

Figure 4-28

You might imagine that the parallelogram* has two lines of symmetry as indicated in Figure 4-28. But when you fold along these lines, you realize that the two parts do not coincide. These creases are *not* along lines of symmetry. Why don't rectangles or rhombuses have four lines of symmetry? Why do some quadrilateral shapes have more lines of symmetry than others?

Look at the triangular shapes shown in Figure 4-29. Do any of these shapes have lines of symmetry? How many lines of symmetry does each shape have? Make paper shapes like these and test for lines of symmetry by folding. Keep a list of your results like the one begun in Table 4-2, and classify each triangular shape as equilateral, isosceles, or scalene.

**Actually, each shape represents a region bounded by the quadrilateral. For convenience, we will refer to each shape by the type of quadrilateral or triangle that forms its boundary.*

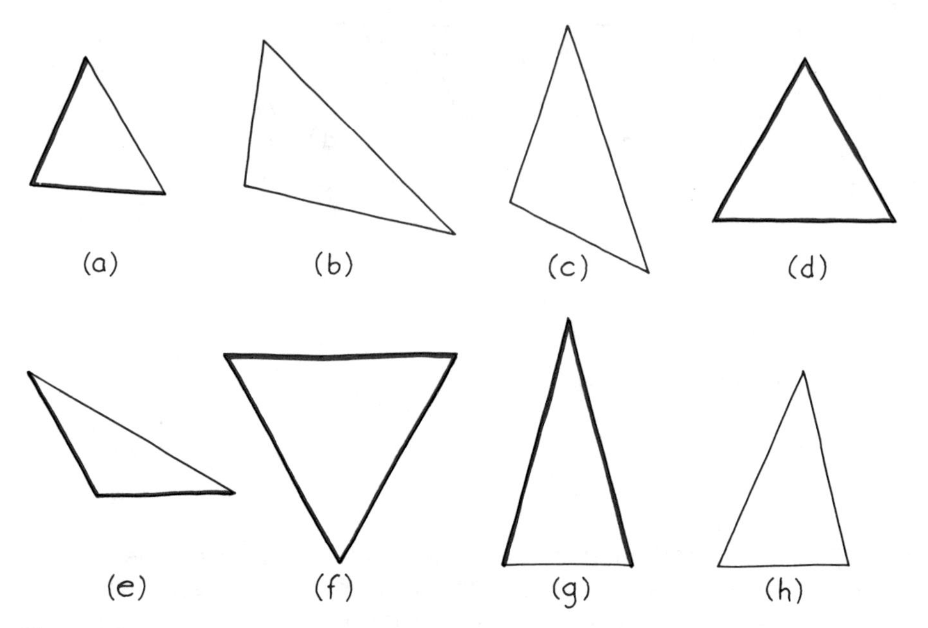

Figure 4-29

Table 4-2

Triangular shape		Lines of symmetry
(a)	Isosceles	1
(b)	?	?
(c)	?	?
(d)	?	?
.	.	.
.	.	.
.	.	.

What did you discover? How many lines of symmetry does every equilateral triangle have? How many does every isosceles triangle have? How about scalene triangles? This shows us another way to classify triangles.

Rotational Symmetry

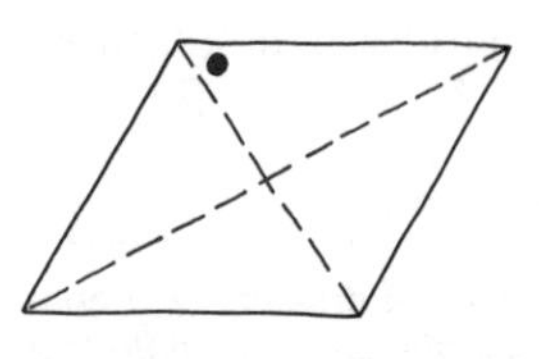

Figure 4-30

We found that a parallelogram (one that is *not* a square, rhombus, or rectangle) has no lines of symmetry. But it does seem to have some sort of balance. On a piece of paper, draw a parallelogram like the one shown in Figure 4-30 and then draw another just like

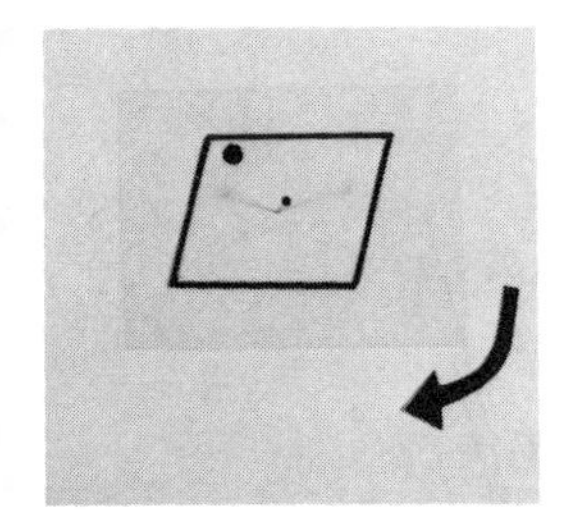

Figure 4-31

it on a piece of plastic (a section of clear acetate used for overhead transparencies will do). Locate the center as the intersection of the diagonals. Also draw a dot in the upper left-hand corners. Now place the plastic shape over the paper one (so that the dots coincide) and stick a pin into the shapes at the point of intersection of the diagonals as in Figure 4-31. Turn the plastic shape clockwise. When do the two shapes coincide? How many times do they coincide as you rotate the plastic shape all the way around through 360°? Notice that the shapes coincide twice in Figure 4-32, at positions (b) and (d). The parallelogram has what is called **rotational symmetry.** The point where we have the pin is the **center of rotation.** Since the design coincides with itself two times as we rotate through 360°, we can say that the **order** of rotational symmetry is 2.

Try the same experiment with a regular pentagon as in Figure 4-33. The center of the pentagon is indicated. Does the figure have rotational symmetry? What is the order of rotational symmetry?

Suppose the regular pentagon has the shape of a boxtop as in Figure 4-34. Each vertex is numbered. How many ways can we place the lid on the box? By turning the lid, we can place it so that each numbered vertex corresponds to the front right-hand edge of the box (Figure 4-35). There are five ways. We might have expected

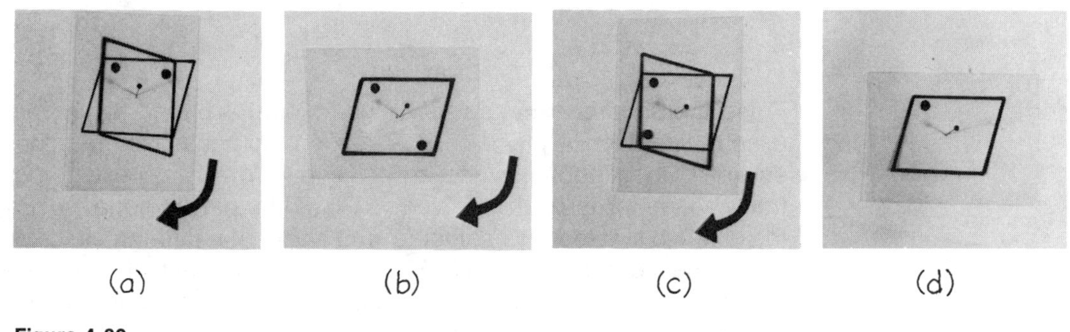

(a) (b) (c) (d)

Figure 4-32

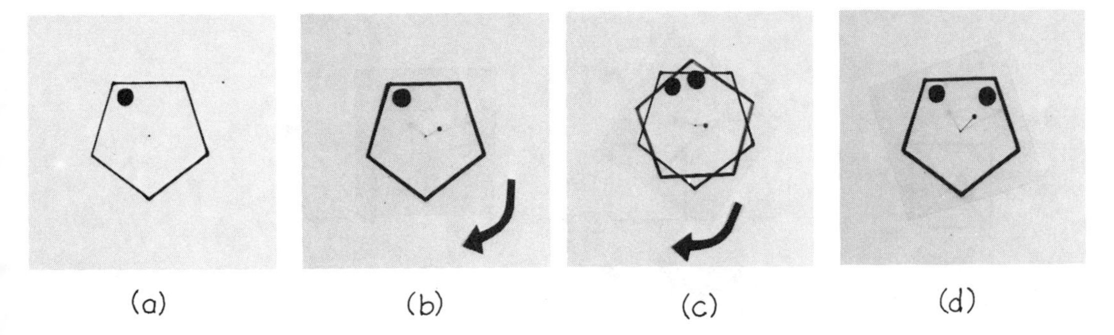

(a) (b) (c) (d)

Figure 4-33

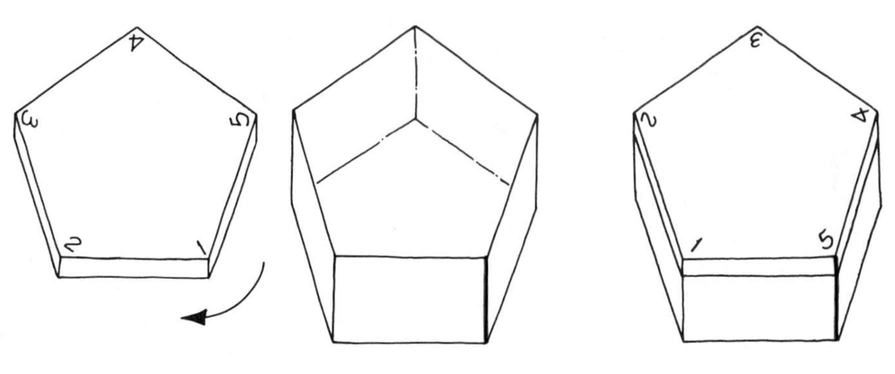

Figure 4-34

Figure 4-35

this, since the regular pentagon has an order of rotational symmetry of 5. How many ways could the lid be replaced if it were shaped like a square? A rectangle? What would be the order of rotational symmetry in each case?

Angles of Rotation

Trace the square in Figure 4-36 on a piece of paper. Make one exactly like it on a piece of plastic. Find the center as the intersection of the diagonals. Also, in each square, draw a ray extending from the center through one vertex. Place the plastic square over the other so that the rays coincide, and stick a pin through the center of the squares (Figure 4-37). Now slowly rotate the plastic square until the two squares coincide again. Where is the ray of

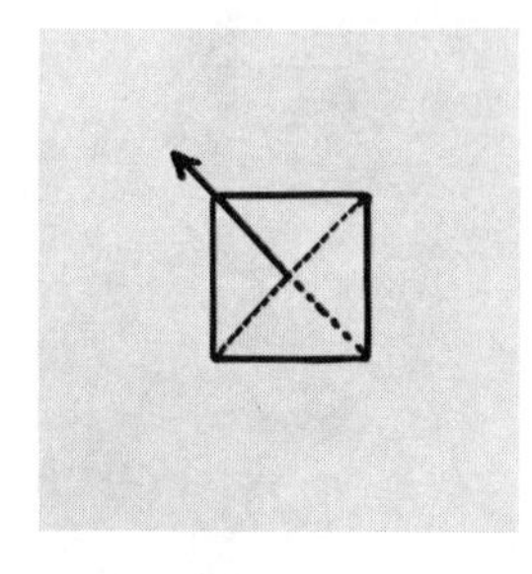

Figure 4-36

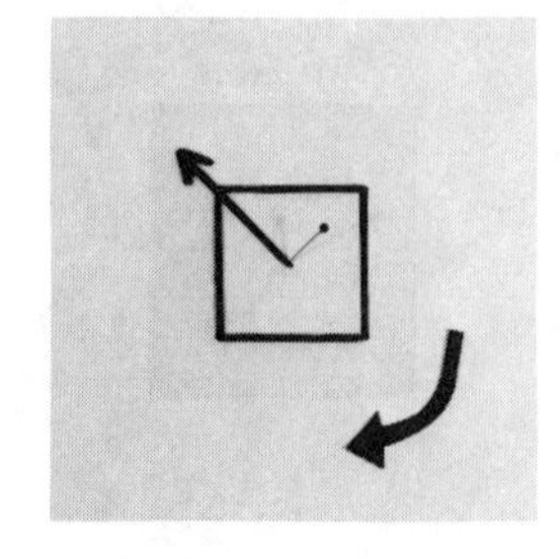

Figure 4-37

Figure 4-38

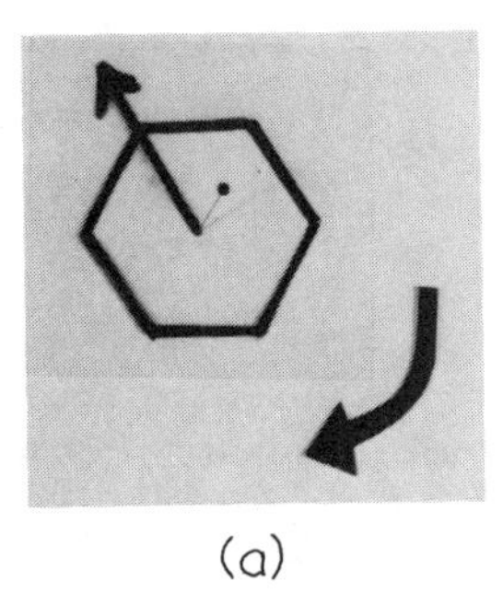

(a) (b)

Figure 4-39

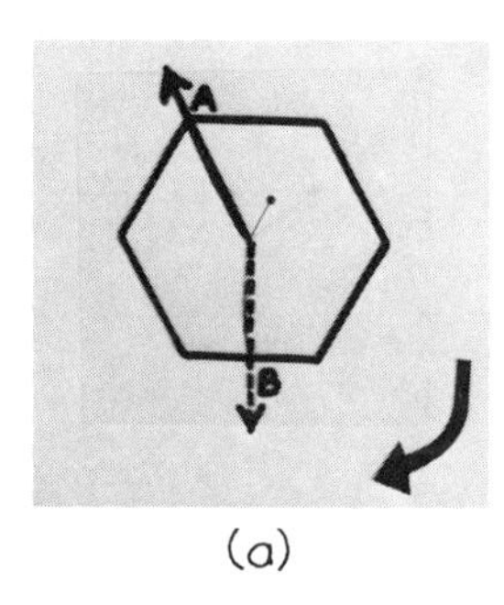

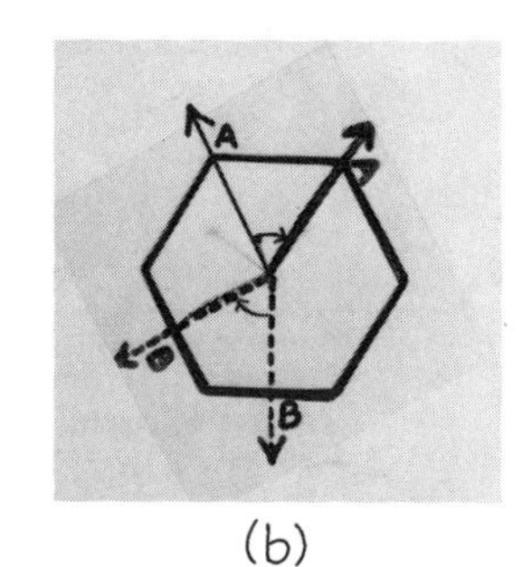

(a) (b)

Figure 4-40

the plastic square? What kind of angle has been formed by the two rays? As you can see from Figure 4-38, a right angle has been formed. This angle is called the **angle of rotation** for the square. It tells you how much you should rotate the squares to make them coincide again.

Let us do the same thing with a regular hexagon, as in Figure 4-39. Notice that the angle of rotation is smaller. How many degrees are in the angle of rotation? If you measure with a protractor, you will find that the angle of rotation contains 60°. But you might figure it out another way. As we rotate the hexagon, the ray moves through 360°. The hexagon coincides with itself six times as we revolve the plastic copy (the order of rotational symmetry is 6). Each time we rotate the hexagon we move it the same amount—$360 \div 6 = 60$.

Note that as we rotate this shape through 60°, each point of the shape is rotated. A dotted ray has been drawn from the center through another point on the hexagons in Figure 4-40. As we rotate the plastic shape through 60°, we can see that the ray with point *B* has been rotated through 60° just as the ray with point *A* has been rotated by that same amount. In fact, every ray emanated from the center through any point of the hexagon is rotated through 60°. Every point of the hexagon moves along a circular path.

Symmetry Everywhere

In this chapter we have discussed two kinds of symmetry on a plane surface—bilateral (line) symmetry and rotational symmetry. We will be referring to concepts of symmetry again as we study other plane shapes, learn about transformations, and experiment with three-dimensional shapes. Meanwhile, be on the lookout for symmetry. You can find it everywhere—in plants, animals, art, architecture, music, dance, poetry—the list is endless. By the way, are *you* symmetrical?

More Experiences

4-1. Which letters of the alphabet have bilateral symmetry? Which have an order of rotational symmetry greater than 1?

Which numerals have bilateral symmetry? Which have an order of rotational symmetry greater than 1?

ABCDEFGHIJKLMNOPQRSTUVWXYZ

1234567890

4-2. Print the word HAD on a piece of paper and place a mirror perpendicular and to the right of it. The reflection should look like the diagram. Why are the letters reversed? Why do the A and the H look the same in the reflection but the D does not?

Now place the mirror perpendicular to the paper but above the word HAD, as in the diagram. Why do the D and the H look the same in this reflection but the A does not? Why did the H look the same in both reflections?

Print the word ICE on a piece of paper. Place a mirror perpendicular to the paper and above the word as in the diagram. What do you notice abouts its reflection? Why does this happen? Can you find other words like this?

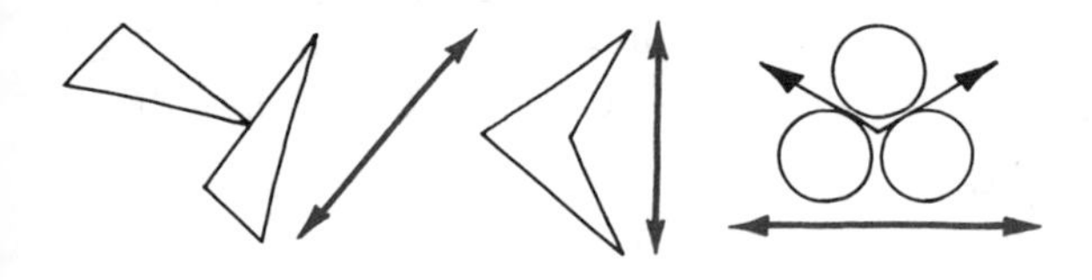

4-3. Draw each of the figures depicted at the left on a piece of tracing paper along with the axes of symmetry shown. Now try to draw the reflection of each figure as it should appear on the other side of this axis of symmetry.

Fold the paper along each axis of symmetry. Through the paper you will be able to see the original figure and the exact location of its reflection. How close were your drawings to the reflected figures?

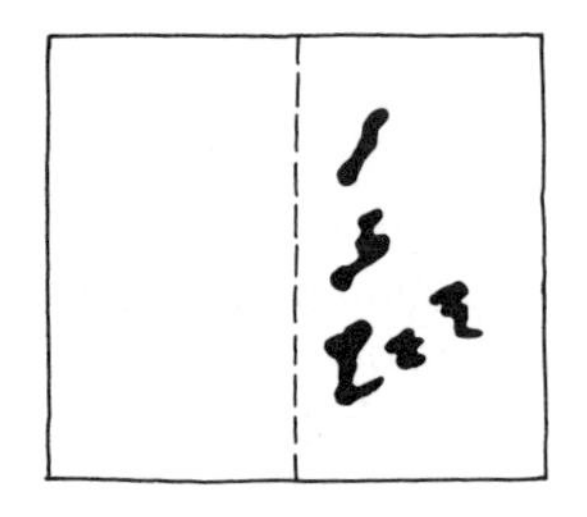

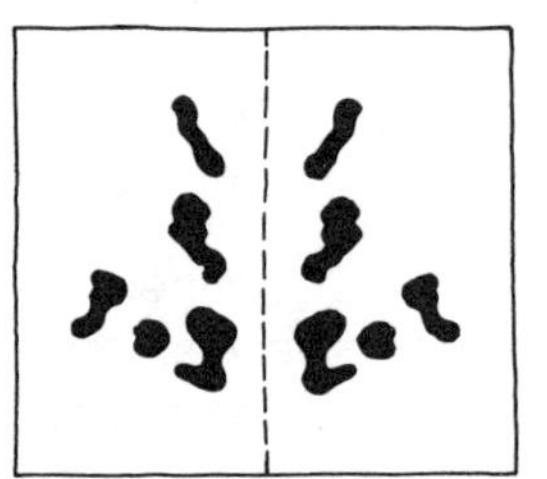

4-4. Bilateral symmetry is vividly demonstrated in making ink blots. Fold a piece of paper. Then place some drops of ink or paint on one side of the crease and refold the paper along that crease. When you open the paper, you will have a design that exhibits bilateral symmetry.

Make some unusual designs from ink blots. Can you make an ink blot with two lines of symmetry?

4-5. Can you draw a figure (not a polygon) with exactly:

(*a*) One line of symmetry
(*b*) Two lines of symmetry
(*c*) Three lines of symmetry
(*d*) Four lines of symmetry

How can paper folding help you draw these figures?

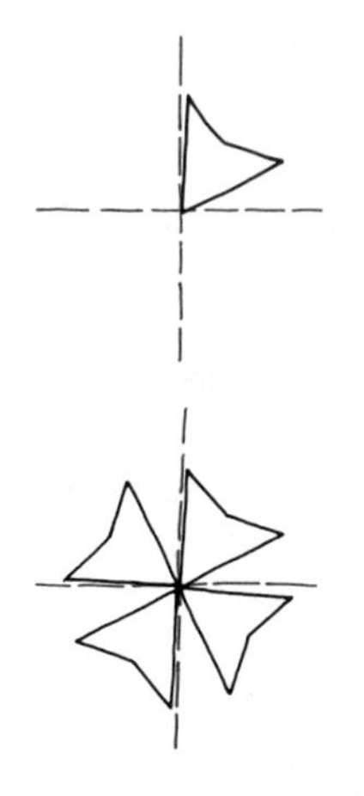

4-6. Take a piece of paper and fold it to make two perpendicular creases. Cut out a design form and place it on one of the four sections of the paper as shown at the left. Draw the design (outline of the cardboard shape). Now place the cardboard shape in the same way on the other sections and draw it. The resulting design has a rotational symmetry of order 4. Does it have bilateral symmetry?

Can you make other designs with rotational symmetry of order 4? Of order 6? Of order 8? Do they have bilateral symmetry?

4-7. Can you draw figures with the following kinds of symmetry?

	Lines of symmetry	Order of rotational symmetry
Figure *A*	2	2
Figure *B*	0	3
Figure *C*	2	1
Figure *D*	4	4

Can the order of rotational symmetry be less than the number of lines of symmetry for a figure? Explain.

4-8. Examine the figures at the left and complete the chart:

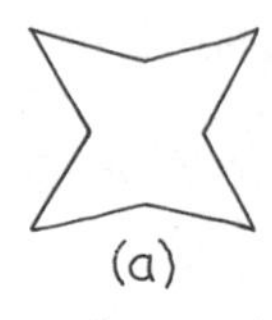
(a)

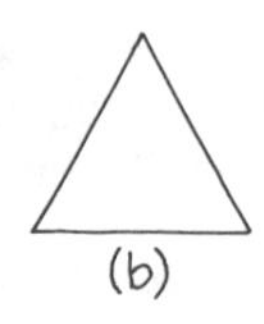
(b)

(c)

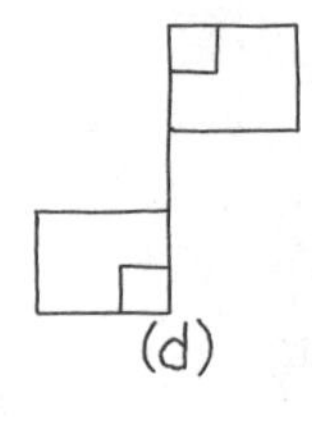
(d)

(e)

(f)

Figure	Axes of symmetry	Order of rotational symmetry
(a)		
(b)		
(c)		
(d)		
(e)		
(f)		

What can you conclude from the chart? If a figure has rotational symmetry of order 2 or more, does it have to have bilateral symmetry? If a figure has bilateral symmetry, does it have to have rotational symmetry of order 2 or more?

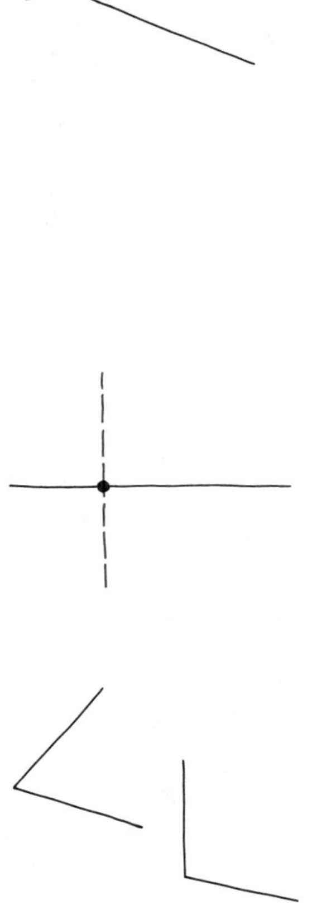

4-9. Draw a line segment on a piece of paper. Place a mirror perpendicular to the paper and intersecting the line segment at an end point, as shown at the left. The line segment and its reflection look like an angle.

Move the mirror (keeping it perpendicular to the paper and intersecting the line segment at an end point) and form different angles. In each case half the angle appears in front of the mirror and the other half appears as a reflection. What is the relationship between the edge of the mirror and this angle? When do the segment and its reflection form a right angle? An obtuse angle? A straight angle?

Draw a point on the line segment. How can you use the mirror to locate another line segment that is perpendicular to this segment at that point?

4-10. Draw several angles like those at the left on a piece of tracing paper. For each angle, fold the paper so that the sides of the angle coincide and the crease goes through the vertex of the angle. What does the crease do to the angle?

Make several triangular shapes out of paper (equilateral, isosceles, scalene). By paper folding, find the three creases that can be formed for each triangle as was done above. How are the three creases in each triangle related to one another? What are they called?

4-11. Paper folding can be very useful for finding perpendiculars and

WITHDRAWN

ITHACA COLLEGE LIBRARY

for bisecting line segments and angles. Try the following with tracing paper.

(*a*) Draw a line segment. Indicate a point on the segment. Can you fold the paper so that the crease you form is perpendicular to the line segment at that point?

(*b*) Draw another line segment. How can you fold the paper so that the crease formed is the perpendicular bisector of the line segment?

(*c*) Draw another line segment. Indicate a point on the paper that is *not* on the line segment. How can you fold the paper so that the crease is perpendicular to the line segment and the point is on the crease?

(*d*) Draw a line segment. Indicate a point on the paper that is *not* on the line segment. Can you locate, by paper folding, a line segment that is parallel to the original line segment and includes the point? (*Hint:* You will need more than one fold.)

4-12. Cut out several large triangular paper shapes (equilateral, isosceles, scalene). By paper folding, try to find the three altitudes of each triangle. When can you find all three altitudes in this way? When is it not possible? Explain.

4-13. How many axes of symmetry does an equilateral triangle have? A square? A regular pentagon? A regular polygon with n sides?

How can you draw the axes of symmetry for a regular polygon with an *even* number of sides? How can you draw the axes of symmetry for a regular polygon with an *odd* number of sides?

4-14. Where would each of the shapes below be located on the geoboard if they were reflected about the axis indicated? How did you find the vertices of the transformed shape?

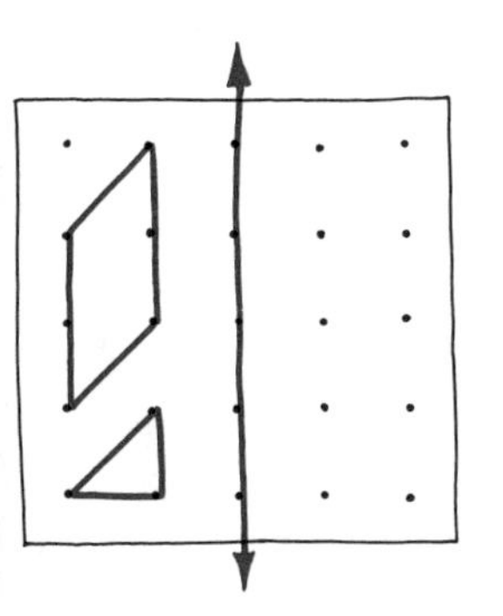

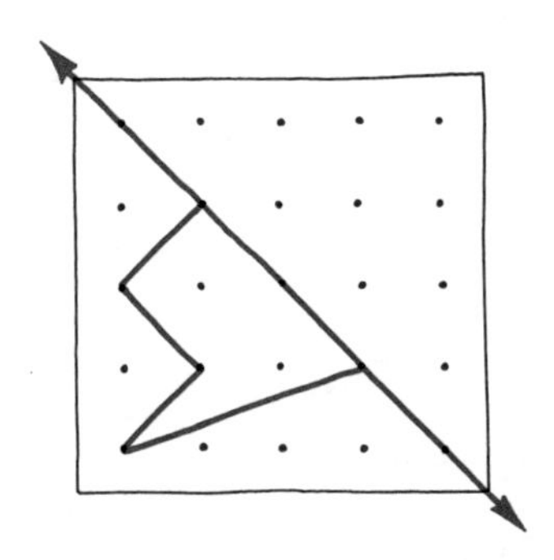

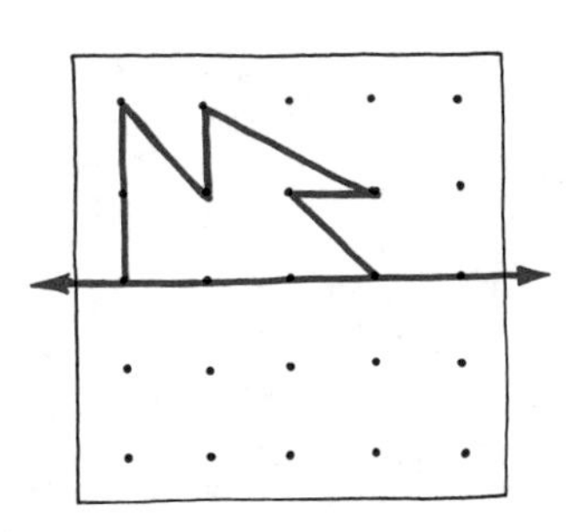

4-15. How many axes of symmetry do each of the shapes on the geoboards below have? What is the order of rotational symmetry for each shape?

Make other shapes on the geoboard, and find their axes of symmetry and order of rotational symmetry.

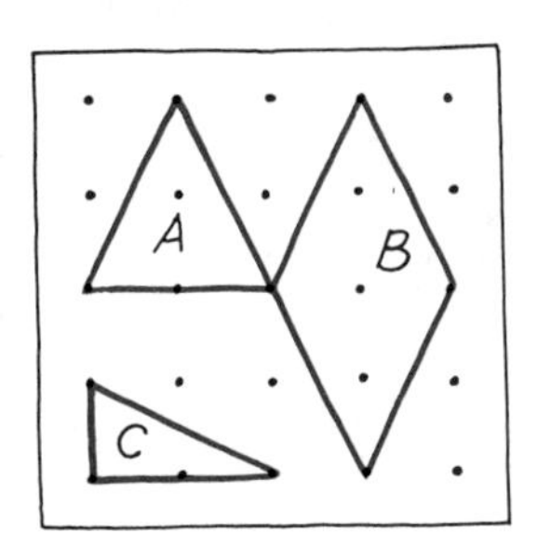

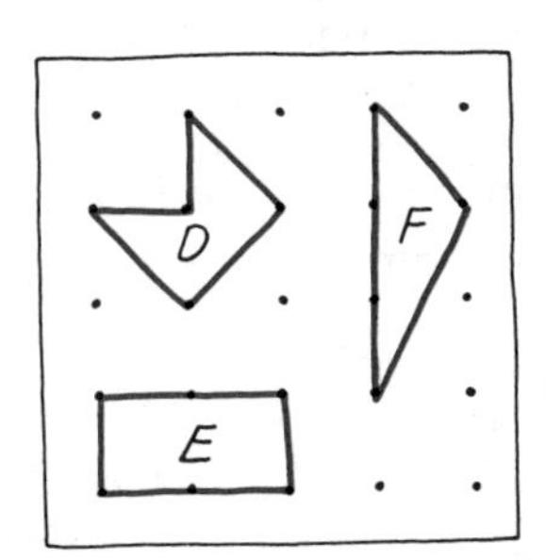

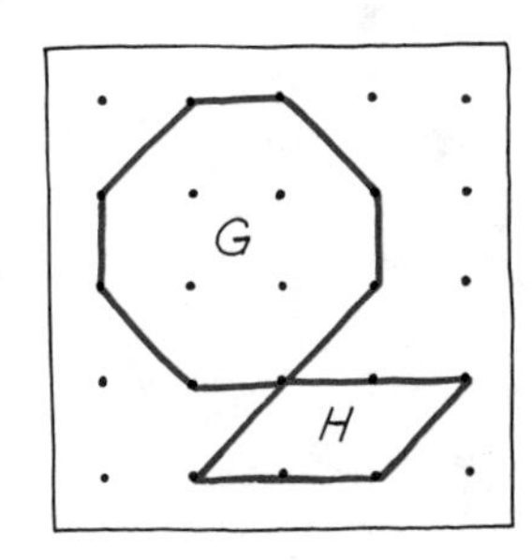

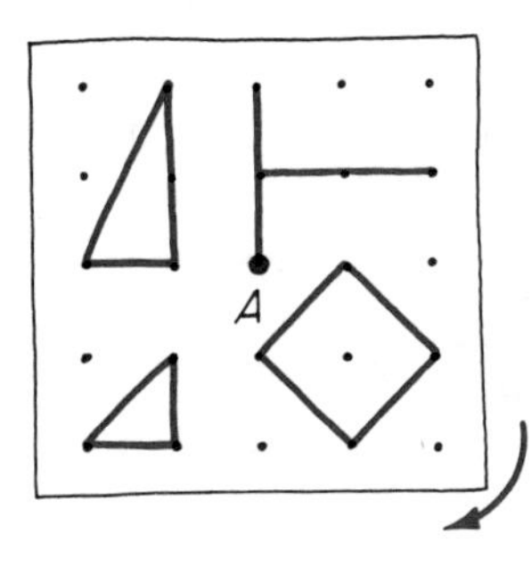

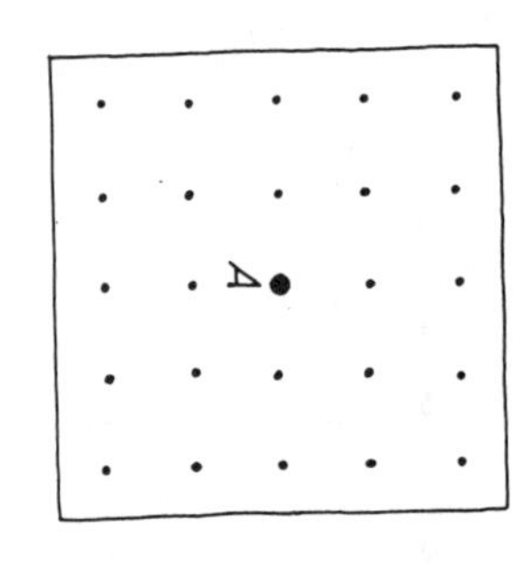

4-16. Suppose the geoboard on the far left were rotated 90° about point *A*. Where would each shape be located on the transformed board (left)? Check your answer by using a geoboard.

Make other shapes on a geoboard and try to predict where they will be when the geoboard is rotated 90, 180, or 270°.

5 Perimeter and Area

Which Is Larger?

Two rectangles are pictured in Figure 5-1. Which is larger, (a) or (b)? We can see that (a) is taller while (b) is wider. But which one is *larger?* Before we can answer this question, we have to decide what we mean by larger. How do we compare two rectangles? How can we measure each rectangle? Let us consider some practical examples.

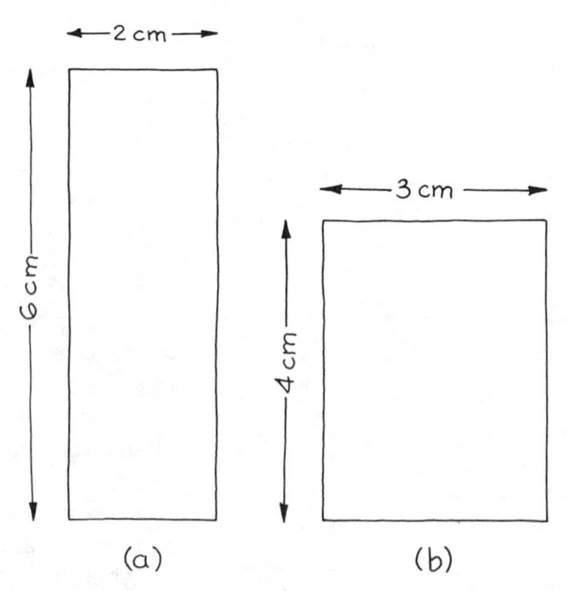

Figure 5-1

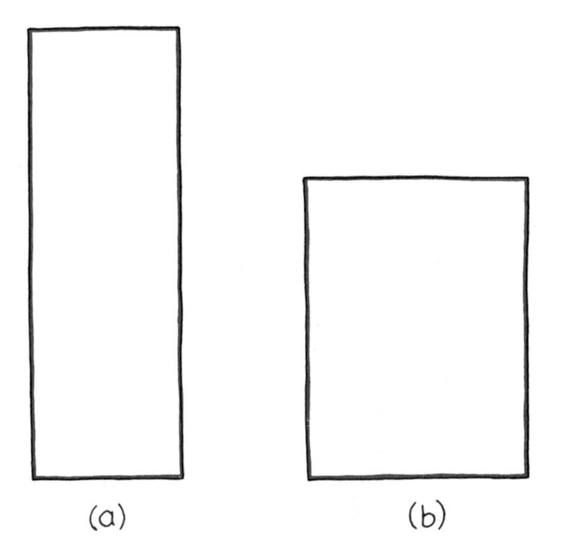

Figure 5-2

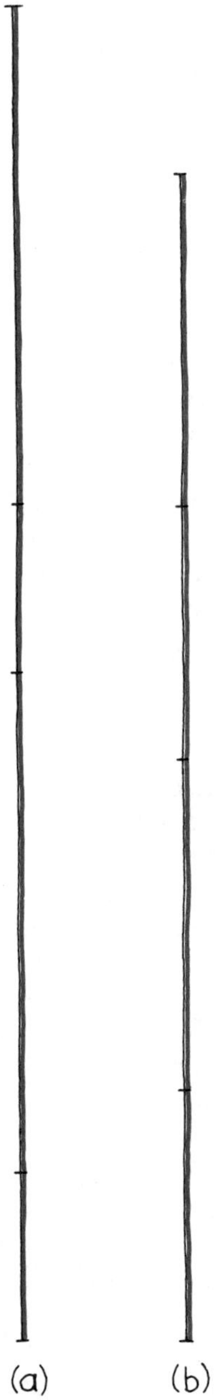

Suppose that (a) and (b) represent two pictures, and we want to frame the pictures with tape and attach them to a wall or a page of a book. If we frame each picture, as in Figure 5-2, would we need more tape for (a) or (b)? This is an easy type of measurement because we are trying to determine the *length* of the tape used for each figure. If we imagine each side of the picture to be a line segment and join the line segments end to end, we form line segments as in Figure 5-3. We can now compare the two line segments and find that rectangle (a) requires more tape than rectangle (b).

But how much more tape is used for (a) than for (b)? This is like measuring two line segments and comparing lengths. In Chapter 1 we found we could use a pencil, a paper clip, or some other unit to measure length. Using a standard unit like a centimeter (cm), inch (in.), or foot (ft) makes it possible for everyone to be familiar with the unit being used. We find that the length of the tape for (a) is 16 cm, while the length of the tape for (b) is 14 cm. It takes 2 cm of tape more to frame (a) than to frame (b).

The measure of the boundary of a figure is called the **perimeter.*** We can think of the boundary as a rim and its measure as peRIMeter. If the figure is considered a simple closed curve, then its perimeter would be the length of that simple closed curve. One way to measure the perimeter is to use string to frame the figure and then measure the length of the string used. Since finding the perimeter involves measuring the length of a line segment (as we did for the pictures), this is considered a *linear measurement.*

Of course we could have simplified the problem by simply measuring the height and width of each picture. For example, in Figure 5-2 rectangle (a) has a height of 6 cm and a width of 2 cm. Since

(a) (b)

Figure 5-3

**How does the derivation of the word perimeter indicate its meaning?*

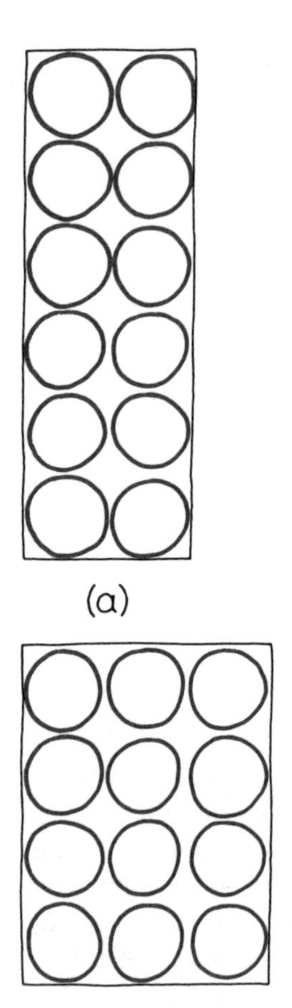

Figure 5-4

the opposite sides of a rectangle are equal in measure, we have two sides with a length of 6 cm each and two sides with a length of 2 cm each. The total perimeter or sum of the measures of all sides in centimeters would be 16 cm (6 + 2 + 6 + 2).

But we may not be concerned with the sides of the rectangles. Sometimes we need to examine the interior of a rectangle. Consider this very practical problem: Imagine that the two rectangles in Figure 5-1 represent trays. If we put as many cookies as possible on each tray (covering each tray with a layer of cookies and not having the cookies overlap), which tray would hold more cookies?

Figure 5-4 illustrates each tray covered with cookies. What do we discover? Each tray holds just as many cookies! Whichever tray we use, we will be able to serve just as many cookies.

In this problem the trays were rectangular in shape, and we were concerned with how much **surface** we had on each tray. We measured this surface by counting how many cookies would fit on each tray. We used a round cookie as a unit for this measurement. We might have used some other shape of cookie to get the same information, possibly triangular or square cookies.

Whenever we are concerned with the amount of surface enclosed by a geometric figure, we are concerned with the **area** within that figure. Suppose the rectangle in Figure 5-5 represents a set of points on a plane. Since the rectangle represents a simple closed curve, the plane is partitioned into three sets of points: (1) points forming the rectangle, (2) points outside the rectangle, and (3) points inside the rectangle. The rectangle, along with the points inside it, forms a **rectangular region.** In determining the area within each rectangle of Figure 5-4, we were attempting to measure the size of that region.

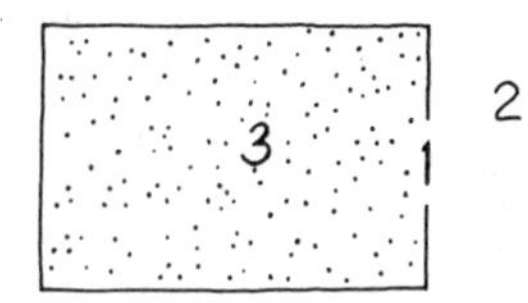

Figure 5-5

Now let us return to the original question: Which is larger, (a) or (b) in Figure 5-1? We found that, in considering the length of the curves forming each rectangle, (a) has a greater perimeter than (b). In this sense, (a) is larger. But usually this question means: Which rectangle contains the most surface area? Now we are interested in the size of the rectangular regions or portions of the plane bounded by the rectangles. This is *area* measurement. We found that each region contains the *same area.* When two regions have the same area they are called **equivalent.** The regions bounded by (a) and (b) are equivalent.

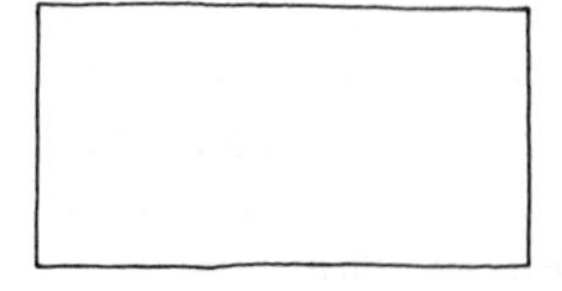
Figure 5-6

The concepts of perimeter and area are helpful in real-life situations. Suppose Figure 5-6 represents a bird's-eye view of a piece of land. We can use both linear measurement and area measurement to give us information about the land. Linear measurement can help us find the length of the sides of this property. If we find the total length of all the sides, we will know the perimeter of the land. That information will tell us how many feet of fence we would need to fence in the property.

How can we measure the area of this rectangular-shaped piece of land? Using cookies is rather awkward. Besides, if we find that 10,500 cookies cover the land, the information is meaningless to someone who does not know how big our cookie unit is! As with linear measurement, we need a *standard unit* for area measurement. We need some unit that is familiar to everyone.

It is convenient to use a square-shaped unit. We can imagine that we cover the land completely with squares that have 1-ft sides. Each square can be called a **square foot.** The square foot would then be our unit of measurement for finding the area. If we find that it takes 15,000 squares to cover this property, we can say that its area is 15,000 sq ft.

In fact, the area of property is usually measured by using square feet as a unit. You have no doubt heard the word *acre* used to describe the size of property. An acre represents 43,560 sq ft. In an urban area, pieces of property are usually smaller than an acre. Such a plot of ground is often called a *lot.* Although the size of a lot varies from city to city, a common size is 5000 sq ft.

In the remainder of this chapter we will discuss ways to find the area of different-shaped regions without having to actually cover the region with a particular unit. We will refer to the area within a rectangle or within some other polygon. What we really mean is the area of the region or portion of the plane bounded by the polygon. Keep in mind that the *polygon itself* is a simple closed curve made up entirely of line segments and does not have area.

Area within a Rectangle

Imagine that the rectangle pictured in Figure 5-7 has a length of 4 cm and a height of 3 cm. How can we find the area bounded by this rectangle? We have seen that finding area involves measuring the amount of surface enclosed by the figure. Let us use a square tile with sides 1 cm in length as our unit of measurement.

If we fill the interior of the rectangle with the tiles, we find that we need 12 tiles (Figure 5-8). We can say that the area within the rectangle (meaning the area of the region bounded by the rectangle) is 12 tiles. Since each tile represents a square with a 1-cm

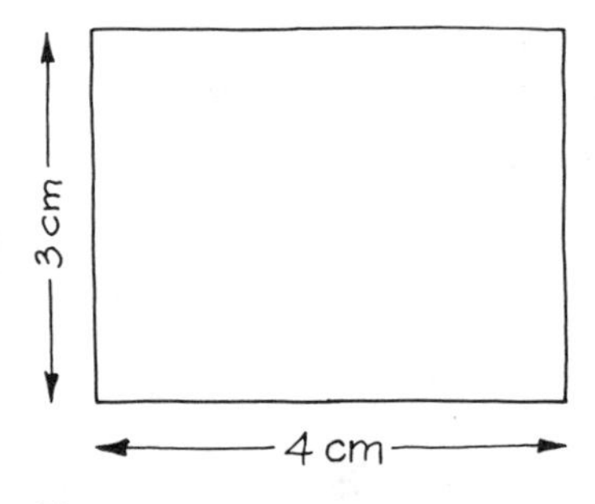

Figure 5-7

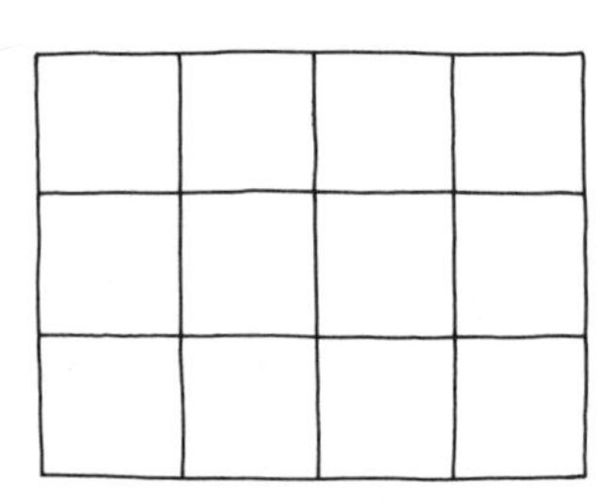

Figure 5-8

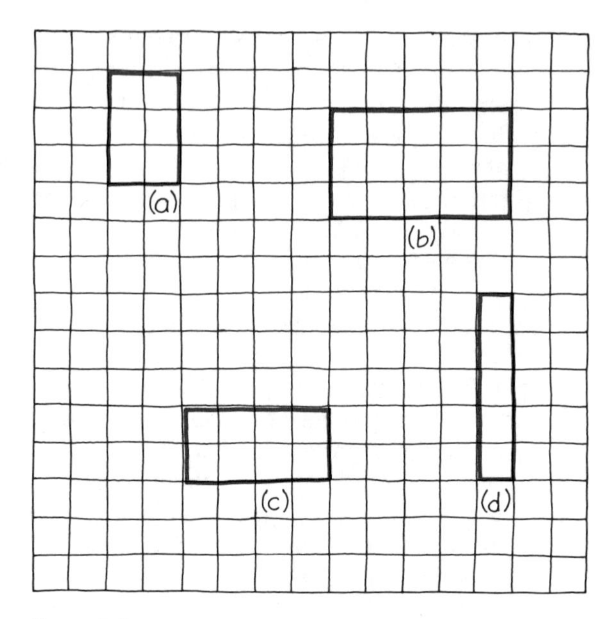

Figure 5-9

side, each tile might be called a **square centimeter.** The area within the rectangle, then, would be 12 sq cm.

Rather than filling the interior of rectangular regions with tiles, we could draw rectangles on squared paper and easily discover their areas. Several rectangles are drawn in this way in Figure 5-9. If each square represents a square centimeter, we can find the area within each rectangle by counting the number of squares within each rectangle.

We can call one side of the rectangle the **base** of the rectangle. Then an adjacent side would be the **height** or **altitude** of the rectangle. The measure of the base and the measure of the height are called the **dimensions** of a rectangle. The word *dimension* suggests its meaning: *di* meaning "two" and *mension* meaning "measure." Let us make a chart showing the dimensions of each rectangle and the resulting areas (Table 5-1).

Table 5-1

Rectangle	Base (in cm)	Height (in cm)	Area (in sq cm)
(a)	2	3	6
(b)	5	3	15
(c)	4	2	8
(d)	1	5	5

Examine the table carefully. Do you see a short way to find the area of a rectangle if you know its dimensions?

Figure 5-10

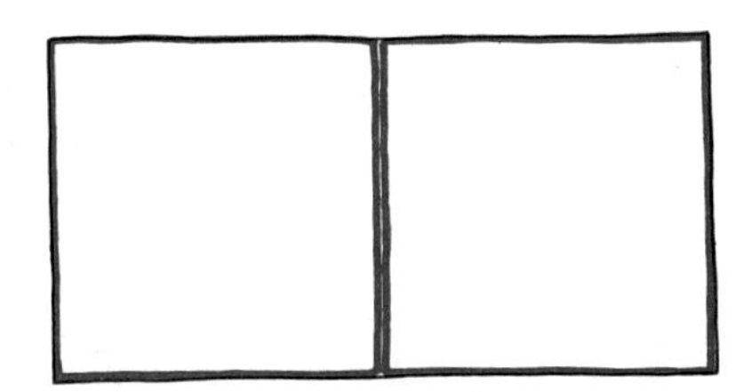

Figure 5-11

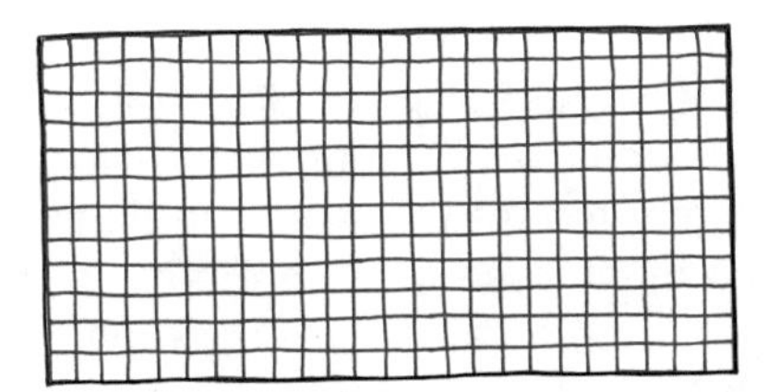

Figure 5-12

Explain the formula below, which summarizes this discovery:*

$$A_{\text{rectangle}} = b \times h \qquad \text{or} \qquad A_{\text{rectangle}} = b \cdot h$$

What would be the area, *A,* within a rectangle with base, *b,* 5 in. and height, *h,* 6 in.? Following the same pattern as before, the area would be 5×6 or 30 sq in.

Suppose the rectangle pictured in Figure 5-10 has dimensions 2 ft and 1 ft. What is its area? In this case it is more convenient to use squares that have 1-ft sides to cover the interior of the rectangle —that is, square feet. If a square foot is the unit of area measurement, then the area within the rectangle would be 2 sq ft (Figure 5-11).

Suppose we need to know the area within the rectangle in Figure 5-10 in square inches. We would then find the dimensions in *inches* (24 in. and 12 in.) and find that it takes $24 \times 12 = 288$ sq in. to cover the interior of the rectangle (Figure 5-12).

In each case we are expressing the measures of the base and the height of the rectangle in the same linear units (centimeter, meter, inch, foot) and the measure of the area in the corresponding area units (square centimeter, square meter, square inch, square foot). We can now say that the area within the rectangle in Figure 5-10 is 2 sq ft or 288 sq in. This means we could completely cover the region bounded by the rectangle by using either two large squares with 1-ft sides or 288 small squares with 1-in. sides.

**In this book both the symbols × and · will be used to indicate multiplication.*

Area within a Square

Figure 5-13

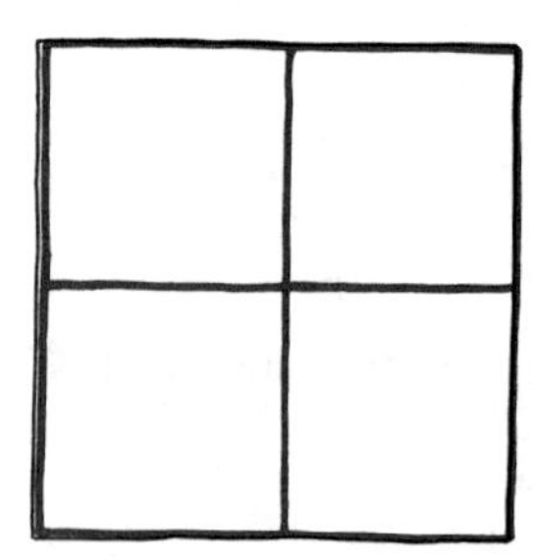

Figure 5-14

We know that a square is a special kind of rectangle; it is a rectangle with all sides congruent. What is the area of the region bounded by the square in Figure 5-13? Each side of the square has a measure of 2 m (meters). The area of the square region is 4 sq m (Figure 5-14).

Do you see a short way to find the area? Since the square is a rectangle, we can use the same method to find this area as we used before: multiply the measures of the dimensions in meters to find the measure of the area in square meters.

Since a square has all sides congruent, the base and the height are the same measure. We find that it is only necessary to know the measure of one side of the square. If s represents the measure of the side of a square, then the area can be found by using the following formula:

$$A_{\text{square}} = s \cdot s$$

In words, the measure of the area in square units equals the measure of a side in units times the measure of a side in units.

But there is a more concise way to write that formula. In the equation

$$24 = 3 \cdot 4 \cdot 2$$

24 is called the **product** and 3, 4, and 2 are called **factors.** Similarly, in $8 = 2 \times 2 \times 2$, the 8 is the product and 2, 2, and 2 are factors. In the second example we have a factor of 2 repeated three times. We can use an abbreviated form called **exponent form** to express this product as follows:

$$8 = 2^3$$

In the exponent form, the number 2 represents the **base** and the small ^{3} represents the **exponent.** The exponent tells you how many times the base is used as a factor. Therefore, 2^3 means that 2 is used as a factor three times, or $2^3 = 2 \cdot 2 \cdot 2$.

In the formula for the area within a square, $A = s \cdot s$, the measure of the side of the square is used as a factor two times. We can write $A = s \cdot s$ as

$$A_{\text{square}} = s^2$$

When a number is used as a factor two times, it is common to say that the number is **squared** or multiplied by itself. Thus, $16 = 4^2$ can be read "16 equals 4 squared." The formula $A = s^2$ can be read "the measure of the area, A, within a square equals the measure of a side, *s*, squared" (with the side expressed in linear units and the area expressed in the corresponding square units).

Suppose we know the area within a square and want to determine the length of a side of that square. If the area within a square is 16 sq m, what is the length of a side of the square? We are looking for a number that, when squared (used as a factor twice), equals 16. That number is 4.

The number 4 is called the **square root** of 16. This is written as follows:

$$\sqrt{16} = 4$$

The symbol $\sqrt{}$ represents "square root" and the statement reads, "the square root of 16 equals 4." If the area within the square is 16 sq m, the side of the square has a length of 4 m. Similarly, if the area within a square is 36 sq cm, a side of that square will have a length of 6 cm, since 6 is the square root of 36. We know that $36 = 6 \times 6$, and it follows that $\sqrt{36} = 6$.

Graphing the Area within a Square

The area within a square increases very rapidly as the length of a side of the square is increased. If a side of a square is 1 unit (centimeter, meter, inch, foot), the area is 1 square unit (square centimeter, square meter, square inch, square foot). By *doubling* the length of the side to 2 units, the area is quadrupled to 4 square units. Tripling the length of the original side to 3 units increases the area to nine times the original square, or 9 square units.

Let us make a table listing the length of the sides of different squares and their corresponding areas (Table 5-2).

Table 5-2

Measure of side, x (in units)	Measure of area, y (in square units)
1	1
2	4
3	9
4	16
5	25
.	.
.	.
.	.

In this table, x represents the measure of a side of a square and y represents the measure of the corresponding area of that square. The table shows that if a side of a square measures 3 units, the area is 9 square units. In this case 3 and 9 can be thought of as a **pair** of numbers. Other pairs are 2 and 4, 5 and 25, and so forth.

To understand the relationship between these pairs of numbers better, we can construct a special picture called a **graph.** A graph is composed of a **set of points.** So first we need to devise a way to relate the points in a plane to pairs of numbers. The following method is a common one for giving number names to points on the plane. Draw two perpendicular lines on a piece of squared paper, as indicated in Figure 5-15. We will call the horizontal line the **x axis** and the vertical line the **y axis.** Each line represents a set of points. Instead of labeling the points on the lines as *A, B, C,* and so on, we give the points number labels. We can name the intersection of the two lines 0 and call it the **origin.** On the horizontal or *x* axis note that certain points are labeled 1, 2, 3, and so on. They are labeled in such a way that the distance between 0 and 1 is the same as the distance between 1 and 2, 2 and 3, and so forth. The numbers are increasing from 0 on the horizontal axis in the right direction. Similarly, points on the vertical or *y* axis are labeled with numbers increasing as you read upward. Notice that the distance between 0 and 1 on the *x* axis is the same as the distance between 0 and 1 on the *y* axis and the numbers 0, 1, 2, 3, 4, and so on are equally spaced on the *y* axis also. Each axis looks very much like a ruler. The distance from 0 to 1 is the **unit.** The distance from 0 to 2 is 2 units, from 0 to 3 is 3 units, and so forth.

Now we devise a rule for naming any number on the plane. Every point will have a first name and a last name. We will use numbers

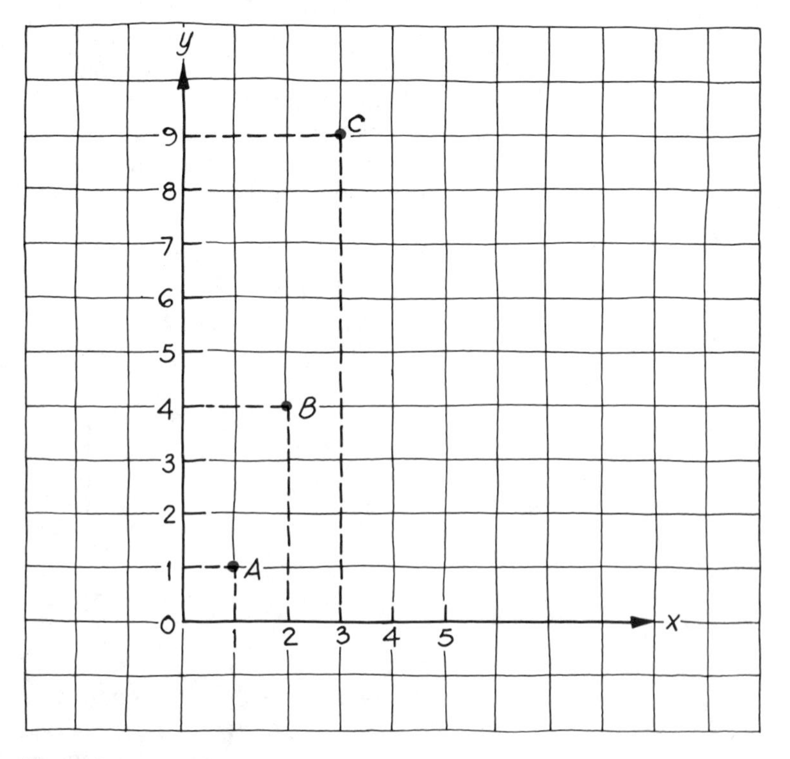

Figure 5-15

as these names. Point A in Figure 5-15 will be assigned the name (1, 1) because to travel from the origin to point A we can move 1 unit to the right on the x axis and 1 unit upward parallel to the y axis. Similarly, point B is assigned the name (2, 4) since to travel to B from the origin we can move 2 units along the x axis and 4 units parallel to the y axis.

In this way we can assign number names to every point. Each number name will be a pair of numbers like (3, 5), where the first number (3) represents the distance traveled from the origin along the x axis and the second number (5) represents the distance traveled parallel to the y axis to reach the point. The two numbers naming the point are called the **coordinates** of the point. Each number is a coordinate. The first number in the pair is called the x coordinate and the second number is called the y coordinate.

Earlier we constructed a table listing the measures of the sides of different squares and their corresponding areas. Suppose each pair of numbers represents the coordinates of a point as given in Table 5-3. The first three points are represented in the graph (Figure 5-15).

Table 5-3

Measure of side, x (in units)	Measure of area, y (in square units)	Coordinates of corresponding point
1	1	(1, 1)
2	4	(2, 4)
3	9	(3, 9)
.	.	.
.	.	.
.	.	.

But the table we used gave only a few values for the measure of a side of a square and its corresponding area. What is the area within a square if a side measures 0.5 units? The area would be 0.25 square units. Similarly, when a side measures 1.5 units, the area within the square will be 2.25 square units. We can extend the table to include as many different pairs of numbers as we want. The expanded table might look like Table 5-4.

Table 5-4

Measure of side, x (in units)	Measure of area, y (in square units)	Coordinates of corresponding point
0.5	0.25	(0.5, 0.25)
1	1	(1, 1)
1.5	2.25	(1.5, 2.25)
2	4	(2, 4)
2.5	6.25	(2.5, 6.25)
3	9	(3, 9)

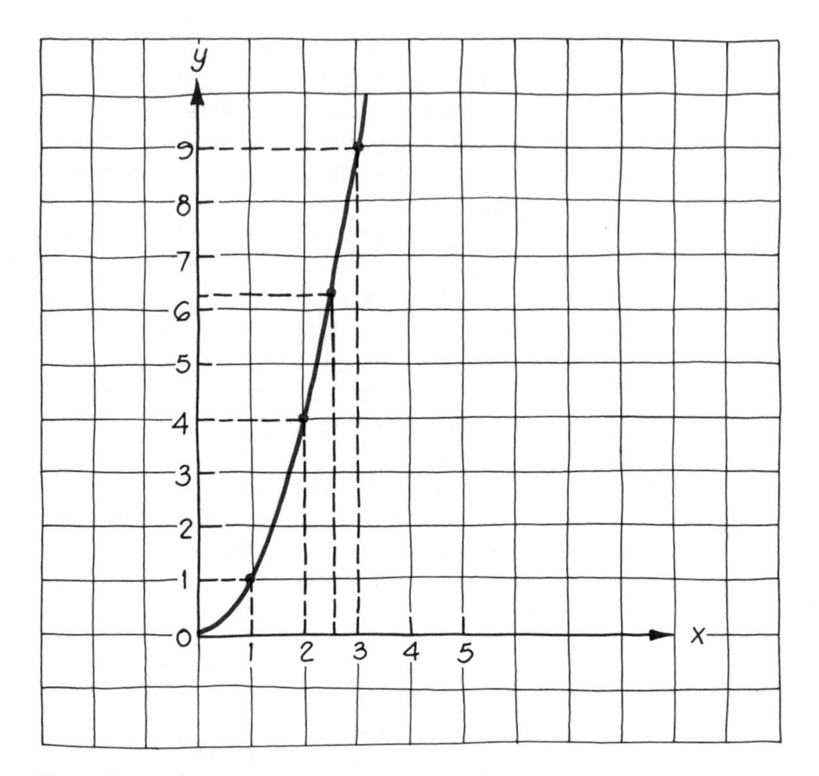

Figure 5-16

How do we locate the point (0.5, 0.25) on the graph in Figure 5-15? If you think of the axes as rulers, 0.5 on the axis would be halfway between 0 and 1. Similarly, 0.25 on the y axis would be one-quarter of the distance from 0 to 1. Can you locate all the points described by Table 5-4?

We can also draw a curve passing through these points, as in Figure 5-16. This curve is called a **parabola.** Examine the points that form this curve. If you consider the coordinates of the points of this curve, for every point the y coordinate is equal to the x coordinate squared. This can be expressed as follows:

$$y = x^2$$

This is called the **equation** of the parabola. This parabola is a picture of the relationship between the sides of different squares and their areas. For example, the point on the parabola with x coordinate 2.5 has a y coordinate of 6.25 square units.

Actually, the curve we have drawn is only one **branch** of a parabola. A parabola has a form as indicated in Figure 5-17. Two parabolas are shown; you can imagine that their branches extend indefinitely.

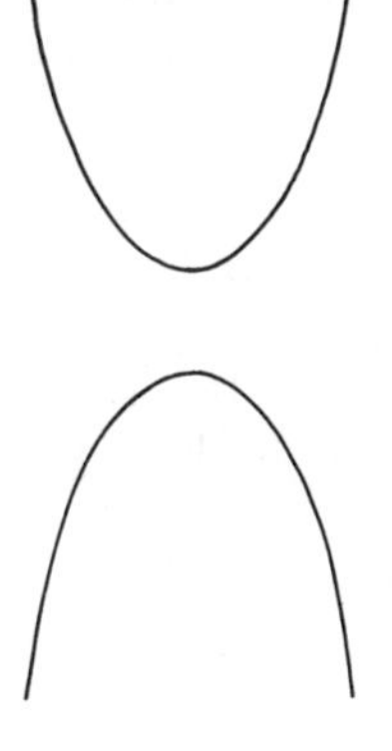

Figure 5-17

We often see shapes in real life that look like parabolas. The path of a bouncing ball forms a parabola (Figure 5-18). The supporting structure for many bridges looks like a parabola (Figure

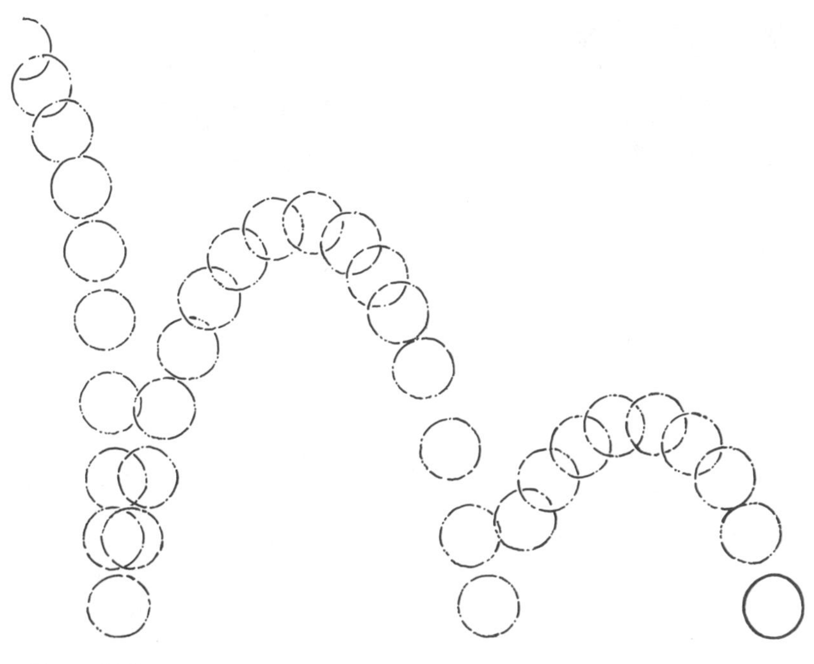

Figure 5-18

Figure 5-19

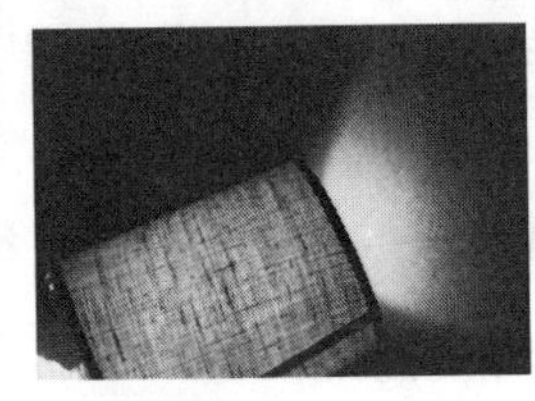

Figure 5-20

5-19). The path of the water descending in a waterfall is part of a parabola. In Figure 5-20 you see a parabolic shape formed on a wall by the light from a table lamp (when the lamp is not held parallel to the wall).

More About Rectangles

Let us make some models of rectangular regions in a different way. Take a piece of string and tie its ends together. Using the thumb

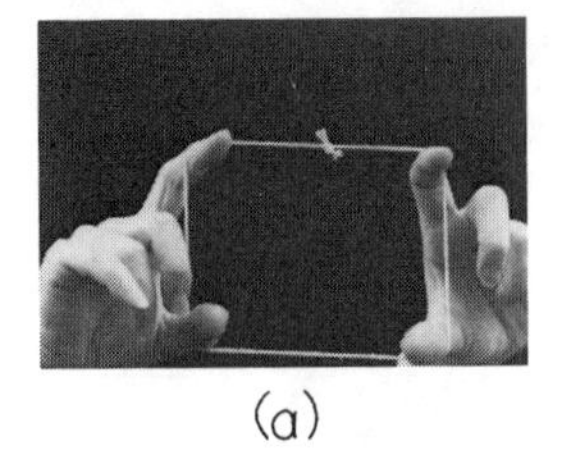

(a)

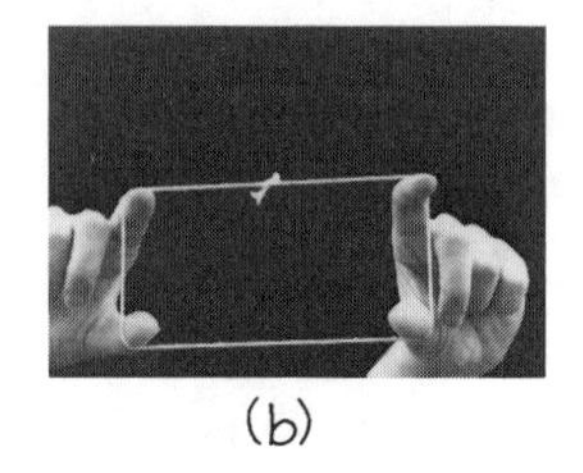

(b)

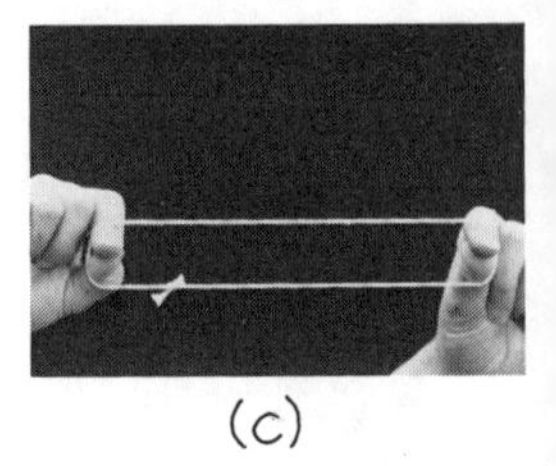

(c)

Figure 5-21

and forefinger of both hands, form a rectangle with the string as in Figure 5-21(a). As you bring the two fingers of each hand closer together and your hands get further apart, you are creating different rectangles with longer and longer bases and shorter and shorter altitudes. But each rectangle has the same perimeter since the sides of each are formed by the same piece of string.

As you form these rectangles, does the area within each rectangle remain the same? You will realize that as your two fingers come closer together, the altitude decreases and the area within the resulting rectangles decreases. You have the maximum area when the rectangle becomes a square. You will form the square when the altitude and the base have the same measure.

In using the string you have formed models of rectangles with the same perimeter. But the regions bounded by these rectangles have different areas. Rectangles with the same perimeter are called **isoperimetric rectangles.**

Graphing Isoperimetric Rectangles

Let us consider a set of rectangles with the same perimeter (isoperimetric rectangles). Imagine a rectangle with a perimeter of 24 cm. It might have a base of 8 cm and a height of 4 cm. Can you think of other rectangles with a perimeter of 24 cm?

Table 5-5 lists some other possibilities. In each case x represents the measure of the base of the rectangle and y represents the measure of the height.

These rectangles represent only a few possibilities. Clearly, there are any number of different rectangles with a perimeter of 24 cm. You will note that if the perimeter is 24 cm, the sum of x and y is always 12. The sum of the measures of two adjacent sides like x and y is called the **semiperimeter** of the rectangle.

If we think of the dimensions of these rectangles as coordinates for points on a plane, we can graph them as in Figure 5-22. As you can see, if you draw a curve passing through all these points, the curve will be part of a line. Choose different points on this line and find their coordinates. You will see that for every point on the line,

Table 5-5

Base, x (in cm)	Height, y (in cm)	Coordinates of corresponding point
0.5	11.5	(0.5, 11.5)
1	11	(1, 11)
2	10	(2, 10)
3.5	8.5	(3.5, 8.5)
4	8	(4, 8)
.	.	.
.	.	.
.	.	.
6	6	(6, 6)
.	.	.
.	.	.
.	.	.
10	2	(10, 2)
11	1	(11, 1)
.	.	.
.	.	.
.	.	.

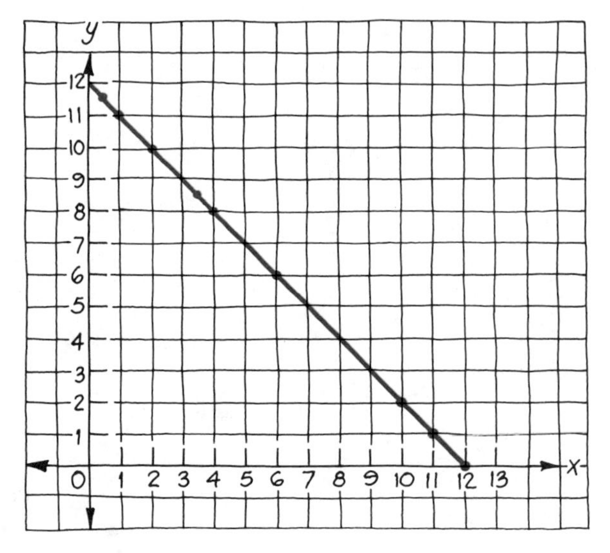

Figure 5-22

the sum of the x coordinate and the y coordinate is 12. We can state that finding as follows:

$$x + y = 12$$

The coordinates of every point on this line correspond to the dimensions of a rectangle with a perimeter of 24 cm. For example, in Figure 5-23 point A corresponds to a rectangle with a base of 1 cm and a height of 11 cm; points B and C correspond to rectangles as indicated. You can imagine many different rectangular

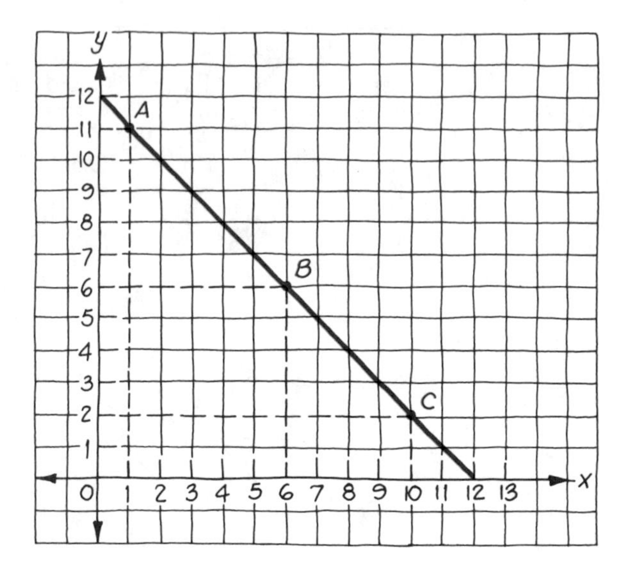

Figure 5-23

Table 5-6

Base, x (in cm)	Height, y (in cm)	Area (in sq cm)
0.5	11.5	5.75
1	11	11
2	10	20
3.5	8.5	29.75
4	8	32
.	.	.
.	.	.
.	.	.
6	6	36
.	.	.
.	.	.
.	.	.
10	2	20
11	1	11
.	.	.
.	.	.
.	.	.

regions drawn "under" the line in this way. All these rectangular regions have the same perimeter, but their area is *not* the same. We can list their dimensions and areas as in Table 5-6.

Which region will have the greatest area? Examine Table 5-6 and explain why the following statement seems reasonable: **In a set of regions bounded by isoperimetric rectangles, the region bounded by a square has the greatest area.**

Graphing Equivalent Rectangular Regions

We have seen that the area within a rectangle with a length of 6 cm and a height of 2 cm is 12 sq cm. Suppose we take 12 small squares with 1-cm sides and form a rectangular region like Figure 5-24(a). Can you form different rectangular regions by using the 12 squares? Regions (b) and (c) in Figure 5-24 are other possibilities.

Notice that the area of regions (a), (b), and (c) is the same—12 sq cm. These are equivalent regions. We can say that these regions contain the same area. Is the perimeter the same for each region? Their perimeters are as follows: (a) 16 cm, (b) 14 cm, and (c) 26 cm. We have formed three *rectangular regions that contain the same area but have different perimeters.*

Suppose we consider all rectangular regions with an area of 36 sq cm. The rectangle might have a base of 9 cm and a height of 4 cm. Can you think of other rectangles with an area of 36 sq cm?

There are any number of possibilities. In each case the measure of the base in centimeters multiplied by the measure of the height in centimeters must equal 36. Table 5-7 lists the dimensions of some rectangles that would qualify, with x again representing the measure of the base and y the measure of the height of each rectangle.

If we consider the dimensions of these rectangles as the coordinates of points on a plane, we can graph them as in Figure 5-25. The curve drawn through these points is called a **hyperbola.** Choose any point on the hyperbola and find its coordinates. You will see that the product of the x and the y coordinates for any point on the hyperbola is 36. We can write this finding as follows:

$$x \cdot y = 36$$

This is called the equation of the hyperbola.

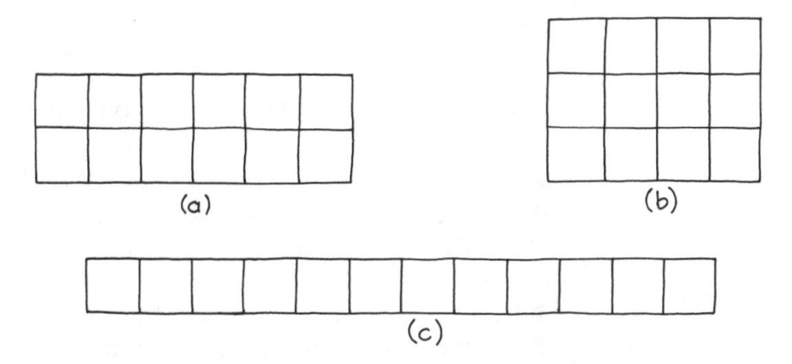

Figure 5-24

Table 5-7

Base, x (in cm)	Height, y (in cm)	Coordinates of corresponding point
1	36	(1, 36)
2	18	(2, 18)
3	12	(3, 12)
4	9	(4, 9)
5	7.2	(5, 7.2)
6	6	(6, 6)
.	.	.
.	.	.
.	.	.
8	4.5	(8, 4.5)
.	.	.
.	.	.
.	.	.
12	3	(12, 3)
.	.	.
.	.	.
.	.	.
36	1	(36, 1)

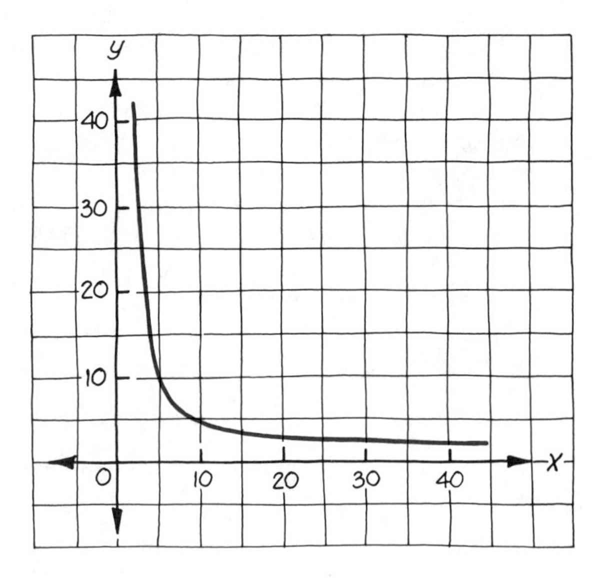

Figure 5-25

The coordinates of every point on that hyperbola correspond to the dimensions of a rectangle with an area of 36 sq cm. For example, in Figure 5-26 point D corresponds to a rectangle with a base of 1 cm and a height of 36 cm; points E and F correspond to rectangles as indicated. Again, you can imagine different rectangular regions drawn under the curve in this way.

All these rectangular regions have the *same area* but *different perimeters.* Table 5-8 lists the dimensions and perimeters of several of these rectangles.

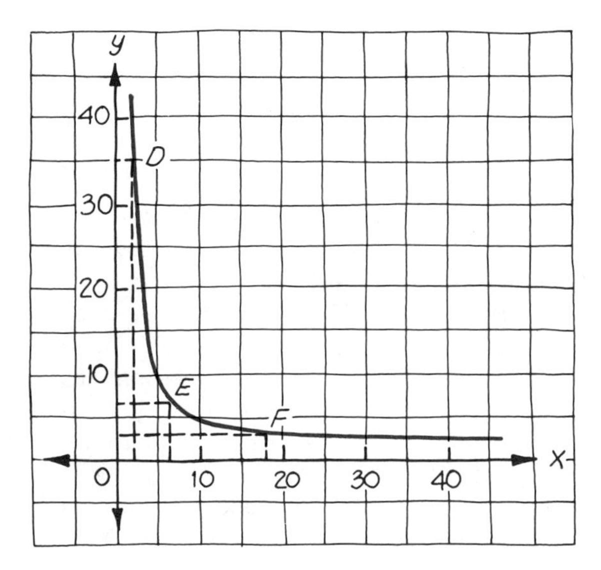

Figure 5-26

Table 5-8

Base, x (in cm)	Height, y (in cm)	Perimeter (in cm)
1	36	74
2	18	40
3	12	30
4	9	26
5	7.2	24.4
6	6	24
.	.	.
.	.	.
.	.	.
8	4.5	25
.	.	.
.	.	.
.	.	.
12	3	30
36	1	74

Which of these equivalent rectangular regions has the least or smallest perimeter? Examine Table 5-8 and explain why the following statement seems reasonable: **In a set of equivalent rectangular regions, the region bounded by the square has the least perimeter.**

In practical terms, if you consider the rectangular regions determined by *D, E,* and *F* in Figure 5-26 as pieces of land, each piece has the same area. But the square piece requires the least amount of fencing since its boundary has the least perimeter.

Area within a Parallelogram

We know how to find the area within a rectangle. A rectangle is a special type of parallelogram. How can we find the area within *any* parallelogram? We could draw a parallelogram on squared paper, like *CDEF* in Figure 5-27, and count the total number of squares contained in the region bounded by the parallelogram. We would have to count all the complete squares and include all the squares that can be formed by the parts of squares within that region. This is quite tedious. Is there a shortcut?

Suppose we make a cardboard model of a region bounded by a parallelogram like the one pictured in Figure 5-28. The base is indicated by *b* and the height by *h* (the heavy line segment). Now imagine that we cut this model along *h* and rearrange the pieces as shown in Figure 5-29. The resulting region is bounded by a *rectangle.* The rectangular region has the same area as the original region because the same amount of cardboard is used for each model.

But we *know* how to find the area of a rectangular region. We measure the base and the height of the rectangle, using the same

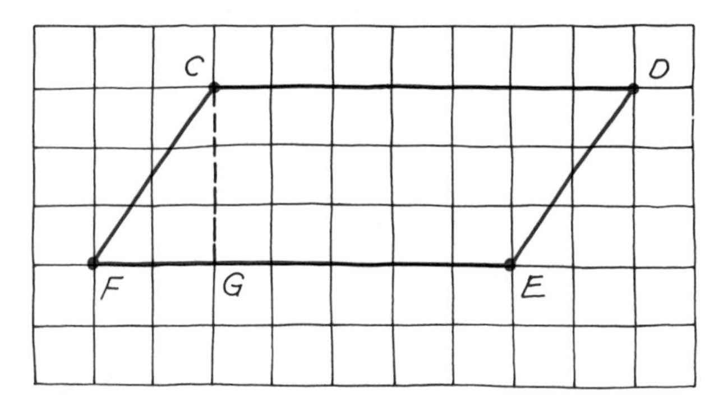

Figure 5-27

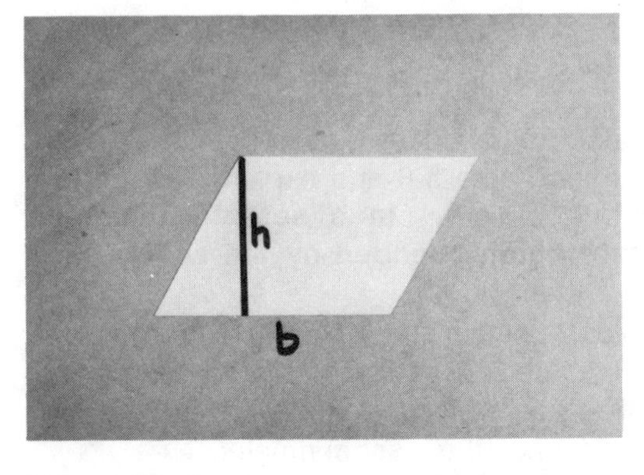

Figure 5-28

Figure 5-29

unit, and multiply those two measures. If the base, *b*, of our model has a length of 6 cm and the height, *h*, has a length of 4 cm, then the area of the rectangular region must be 24 sq cm. Since the original region bounded by the parallelogram and the rectangular region have the same area, the area of the original region bounded by a parallelogram must also be 24 sq cm.

Do you see an easy way to find the area within *any* parallelogram? A formula for the area within any parallelogram would be the following:

$$A_{\text{parallelogram}} = b \cdot h$$

We can use this formula to find the area of parallelogram *CDEF*. If we choose $\overline{FE}$ as the base of parallelogram *CDEF* (Figure 5-27), then the measure of the base is 7 units. With $\overline{FE}$ as the base, the height of the parallelogram is the perpendicular line segment $\overline{CG}$ from vertex *C* to side $\overline{FE}$. The measure of the height is 3 units. We know, then, that the area within parallelogram *CDEF* is 21 square units. We do not need to *count* the square units.

(*Caution:* Do not confuse side $\overline{FC}$ with the height of the parallelogram. It is not perpendicular to the base. Only in the case of a rectangle do you have a side that can be considered the height, since all the angles of a rectangle are right angles.)

Area within a Rhombus

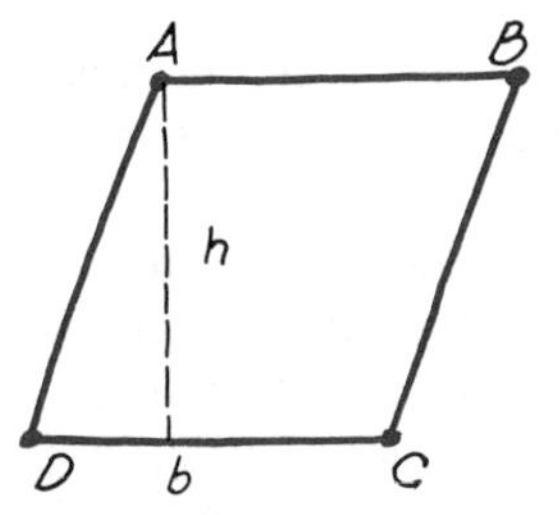

Figure 5-30

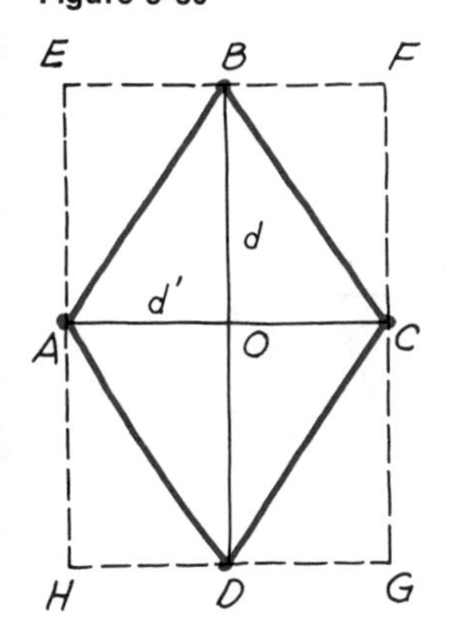

Figure 5-31

How can you find the area of the region bounded by the rhombus *ABCD* of Figure 5-30? If the measure of side $\overline{DC}$ is called *b* and the measure of the altitude from vertex *A* to side $\overline{DC}$ is called *h*, then, *since the rhombus is a parallelogram,* the measure of the area within the rhombus will equal the product of these two measures:

$$A_{\text{rhombus}} = b \cdot h$$

Keep in mind that *b* and *h* are measured with the same unit (centimeter, meter, inch, foot) and the area is expressed with the corresponding square unit (square centimeter, square meter, square inch, square foot).

Sometimes we do not know the measures of the base and the height of a rhombus but are given the measures of the diagonals. There is an easy way to find the area within a rhombus when you know the measures of its diagonals. Suppose we draw rhombus *ABCD* in a different position, as in Figure 5-31. We will include the diagonals of the rhombus and call the measure of one diagonal *d* and the measure of the other diagonal d'. Now draw a rectangle whose sides intersect the vertices of the rhombus as shown.

You can see that the region bounded by rectangle *EFGH* has been partitioned into eight triangular regions that are just alike. There are *four* triangular regions within the rhombus *ABCD* and four outside it. *So the area within rhombus ABCD must be one-half the area within rectangle EFGH.*

We know how to find the area of a rectangular region: $b \cdot h$. But in this case the measure of the base of the rectangle *HG* is the same as d' and the measure of the height of the rectangle is the same as d. Then the area within rectangle *EFGH* is seen to be equal to $d \cdot d'$. Since the area within the rhombus *ABCD* is half that area, we have:

$$A_{\text{rhombus}} = \frac{1}{2} \cdot (d \cdot d') \qquad \text{or} \qquad A_{\text{rhombus}} = \frac{d \cdot d'}{2}$$

This formula tells us that to find the measure of the area within a rhombus we can simply multiply the measures of the diagonals and divide that product by 2.

Area within a Triangle

In the previous section we could have calculated the area within rhombus *ABCD* by finding the total area of the four triangular regions within the rhombus. But how do you find the measure of the area of a triangular region?

Make a model of a triangular region out of cardboard. This region is bounded by a triangle; label the measure of a base of that triangle b and the measure of the corresponding altitude h, as indicated in Figure 5-32. Now make another triangular region just like the first one. Combine the two models, as indicated in Figure 5-33, to form a region bounded by a parallelogram.

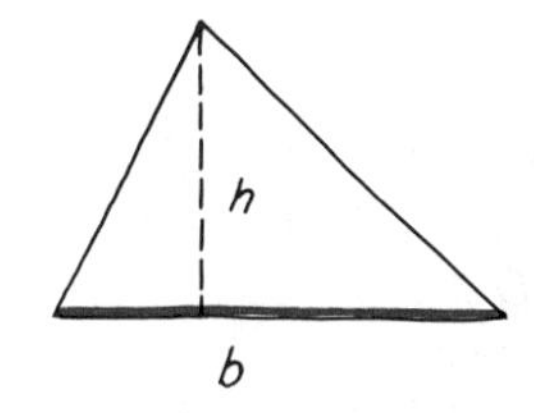

Figure 5-32

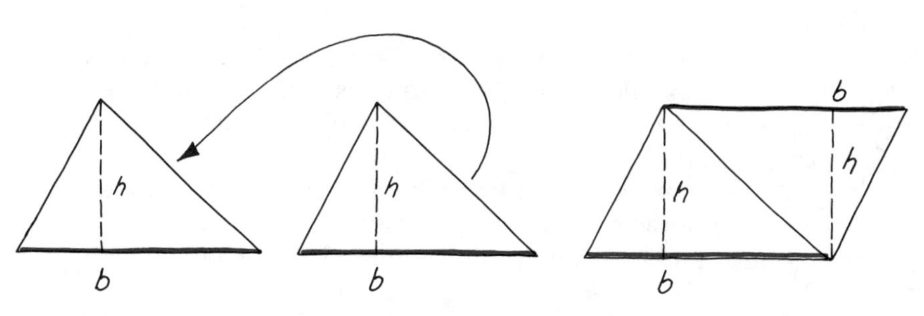

Figure 5-33

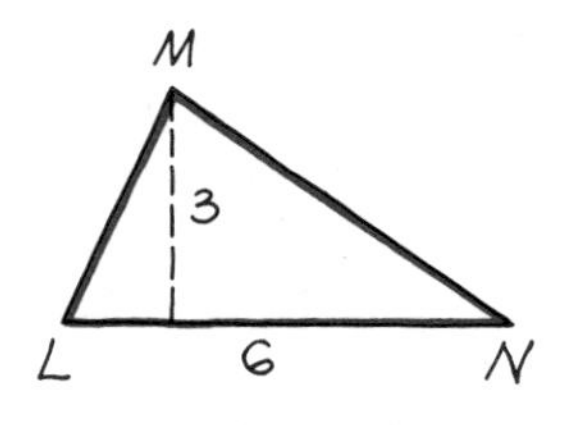

Figure 5-34

Clearly, the measure of the area of the region bounded by the parallelogram is the product of b and h. How can we express the measure of the area of each triangular region? Each triangular region can be considered one-half the region bounded by the parallelogram. It follows that:

$$A_{\text{triangle}} = \frac{1}{2} \cdot (b \cdot h) \qquad \text{or} \qquad A_{\text{triangle}} = \frac{b \cdot h}{2}$$

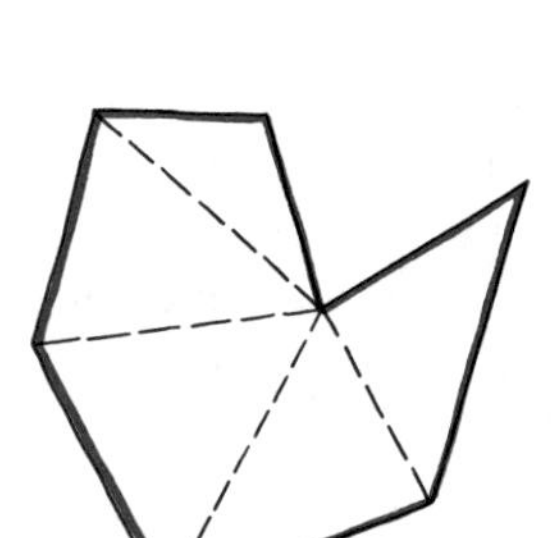
Figure 5-35

Now we can find the measure of the area within any triangle when we know the measures of a base and corresponding altitude. In Figure 5-34 the length of a base of triangle *LMN* is 6 cm and the length of its altitude is 3 cm. This indicates that the area within the triangle must be 9 sq cm, $(6 \cdot 3) \div 2$.

Sometimes you can find the area of a polygonal region more easily by partitioning it into triangular regions and finding their areas. For example, the area within the polygon in Figure 5-35 can be determined by finding the sum of the measures of the areas of the triangular regions formed.

Some Equivalent Triangular Regions

The isosceles triangle pictured in Figure 5-36 has a base with a measure of 6 units and a corresponding height of 4 units. Consequently, the measure of the area bounded by this triangle is 12 square units.

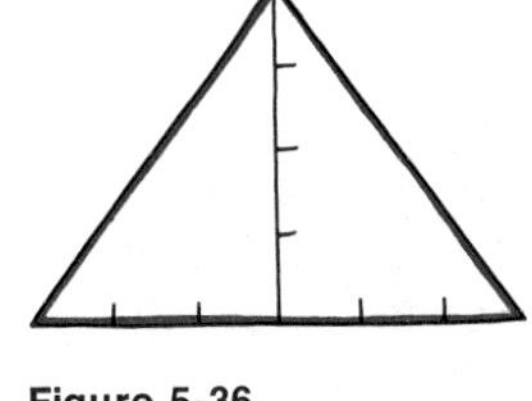
Figure 5-36

Can you draw other triangles that have a base of 6 units and a height of 4 units? As you will realize, there are any number of triangles possible. A few such triangles are indicated in Figure 5-37. Since each of these triangles has a base that measures 6 units and a height that measures 4 units, the area within each triangle is 12 square units. In other words, the triangular regions bounded by these triangles are *equivalent*. In Figure 5-37 you will notice that the base for each triangle is $\overline{AC}$, a line segment of $\overleftrightarrow{AC}$, and that the

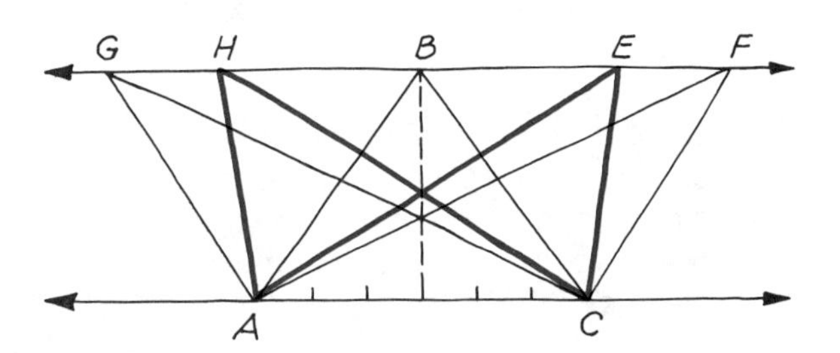

Figure 5-37

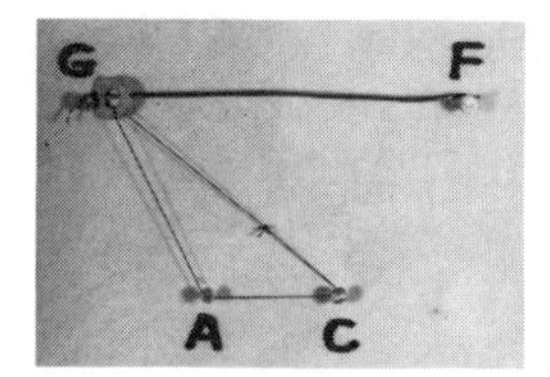

Figure 5-38

vertices *A* and *C* are 6 units apart. The third vertex of each triangle is located on a line parallel to $\overleftrightarrow{AC}$.

You can make a more dynamic model for these equivalent triangular regions by using a piece of wood, nails, wire, and an elastic band. The model is pictured in Figure 5-38. A nail is driven into the wood at points *G*, *F*, *A*, and *C*. These points are chosen so that the lines $\overleftrightarrow{GF}$ and $\overleftrightarrow{AC}$ are parallel. A piece of wire extends from *G* to *F*. The triangle is formed by an elastic band wrapped around the nails at *A* and *C* (which form vertices for the triangle) and around the wire—or, better yet, it is put through a metal ring that can be moved along the wire. As the metal ring is moved in a horizontal direction along the wire, many different triangles are formed by the elastic band. The area within each triangle is the same since all the triangles have bases the same size and altitudes the same size. For example, if $\overline{AC}$ has a measure of 6 units and the distance between $\overleftrightarrow{GF}$ and $\overleftrightarrow{AC}$ is 4 units, then the area within every triangle formed is 12 square units.

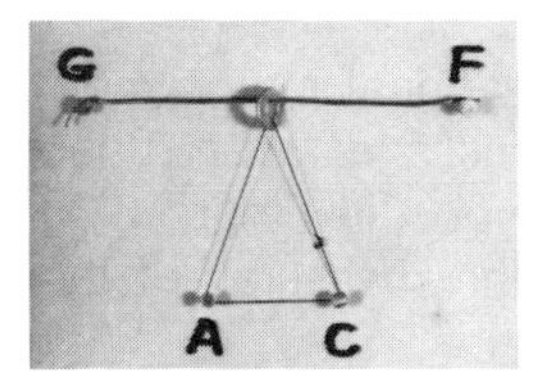

Figure 5-39

If you let go of the ring and allow it to move freely along the wire, when the ring stops moving you will find that an isosceles triangle has been formed by the elastic band (Figure 5-39). This indicates that forming an isosceles triangle causes the least tension in the elastic band, meaning that the isosceles triangle has the least perimeter. **Consequently, in this set of equivalent triangular regions, the region bounded by an isosceles triangle has the least perimeter.**

Some Isoperimetric Triangles

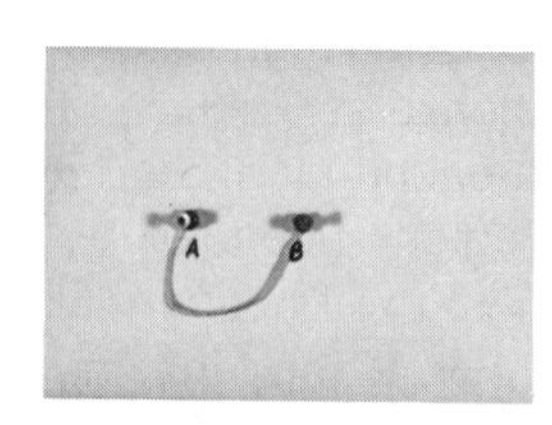

Figure 5-40

Let us consider a set of triangles whose perimeters are equal in measure. Suppose these triangles also have bases with the same measure. We can visualize this set of triangles better by constructing a model. All we need is a piece of cardboard or wood, two thumbtacks or nails, and a piece of string. Put the two tacks into the cardboard, say at points *A* and *B* as in Figure 5-40, and fasten the end points of the string to the tacks.

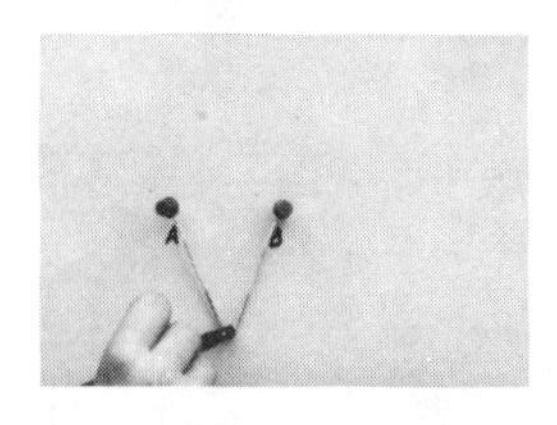

Figure 5-41

Imagine the line segment from *A* to *B* as the fixed base for the triangles. The string will be used to form the other two sides. In this way the perimeters of the resulting triangles will always be the same measure. Points *A* and *B* will be two vertices of the triangles and $\overline{AB}$ will be one side. Use the tip of a pencil to stretch the string, forming two more sides with the tip as a third vertex (Figure 5-41). To form another triangle with base $\overline{AB}$ and the same perimeter, keep the string taut and move the point of the pencil sideways (Figure 5-42). Your new triangle has the same perimeter. In this way you can make any number of isoperimetric triangles with

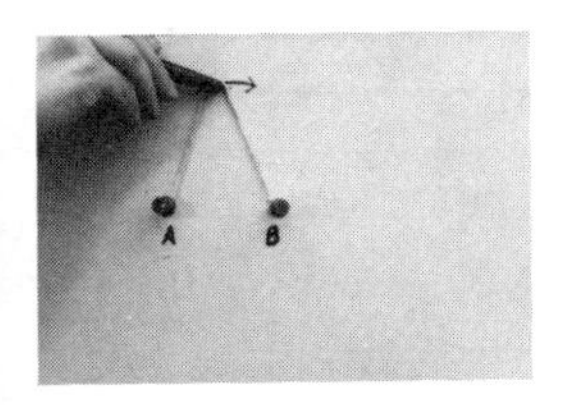

Figure 5-42

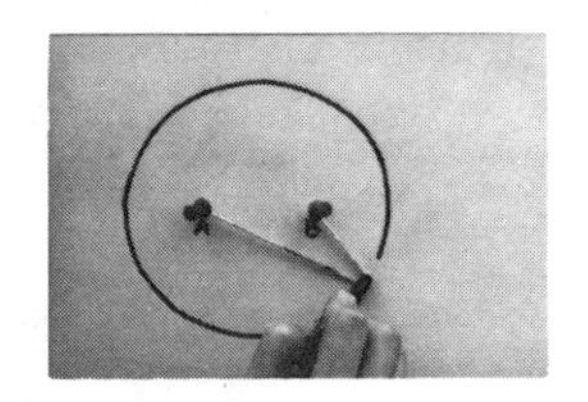

Figure 5-43

Figure 5-44

base $\overline{AB}$. If you draw the path of the vertices formed by the tip of the pencil, that path will look like an **ellipse** (Figure 5-43).

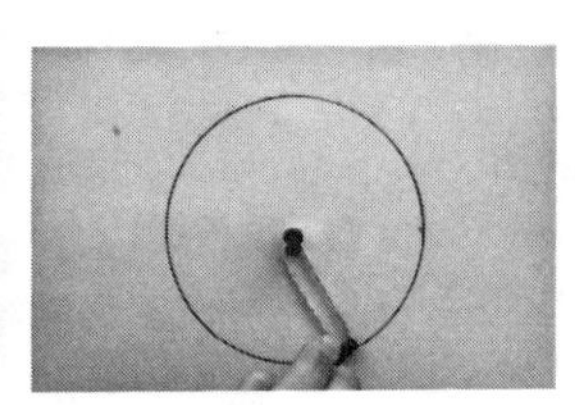

Figure 5-45

Is the measure of the area within each triangle the same? As you form different triangular regions you can judge that they are *not* equivalent. These regions are bounded by triangles that have the same base but different altitudes. Of all the triangles formed, the isosceles triangle has an altitude of maximum size. **Consequently, the measure of the area within the isosceles triangle will be the greatest.**

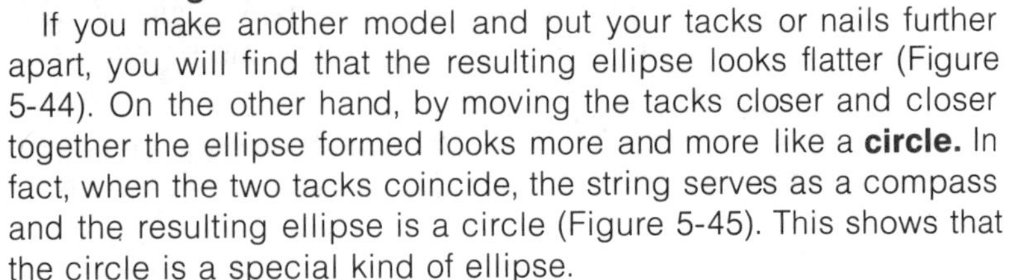

If you make another model and put your tacks or nails further apart, you will find that the resulting ellipse looks flatter (Figure 5-44). On the other hand, by moving the tacks closer and closer together the ellipse formed looks more and more like a **circle.** In fact, when the two tacks coincide, the string serves as a compass and the resulting ellipse is a circle (Figure 5-45). This shows that the circle is a special kind of ellipse.

Figure 5-46

The method described above is a handy way to draw an ellipse. We see many examples of ellipses in our daily lives. When we see a circular object—like a dish or the rim of a glass or vase—from a distance, the shape of the object often looks like an ellipse (Figure 5-46).

Area within a Trapezoid

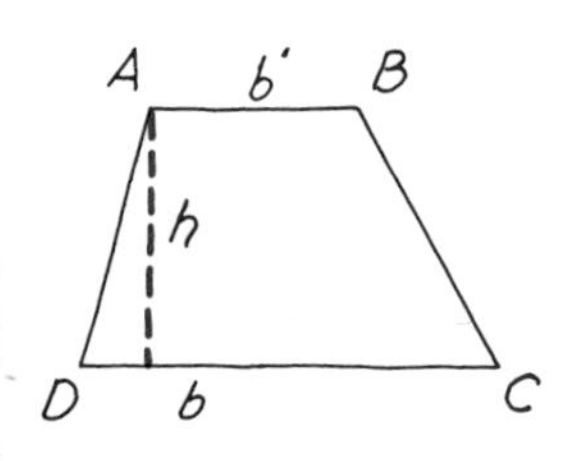

Figure 5-47

In Chapter 3 we examined a quadrilateral with one pair of sides parallel—the trapezoid. It is sometimes necessary to calculate the area of a region bounded by a trapezoid. Suppose trapezoid *ABCD* (Figure 5-47) represents the shape of the floor of an entry hall. How can we find the area within the trapezoid?

In trapezoid *ABCD* the parallel sides $\overline{AB}$ and $\overline{DC}$ are called the **bases** and are not congruent. We can call the measure of one of these bases b and the measure of the other base b'. The measure of the altitude is labeled h.

We can find an easy way to calculate the area by following a procedure like that used to discover the formula for the area within

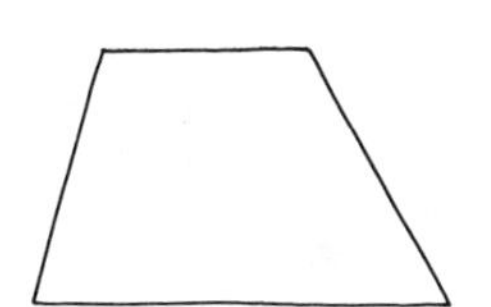

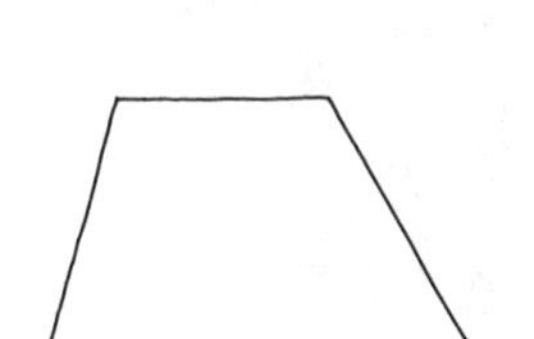

Figure 5-48

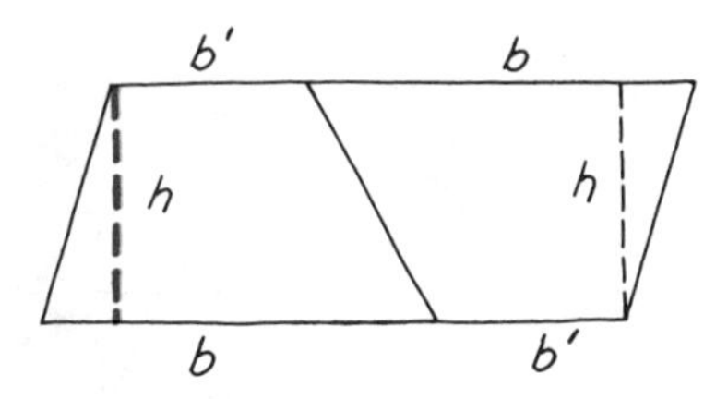

Figure 5-49

a triangle. Let us make two identical cardboard models of the trapezoid $ABCD$ as in Figure 5-48. If you combine the two trapezoids as in Figure 5-49, the resulting model is a parallelogram.

We know how to calculate the area of a parallelogram. We multiply the measure of the base by the measure of the height. This parallelogram has a base that measures $b + b'$ (the sum of the measures of the bases of the original trapezoid) and a height of h (the same as the height of the trapezoid). So the area of the region bounded by the parallelogram would be expressed as $(b + b') \cdot h$.

Each of the two regions bounded by the trapezoids represents one-half that area. So the measure of the area of a region bounded by one of the trapezoids can be expressed as follows:

$$\underset{\text{trapezoid}}{A} = \frac{1}{2} \cdot [(b + b') \cdot h]$$

$$= \frac{(b + b') \cdot h}{2}$$

If the bases of the trapezoid have measures of 3 and 4 cm and the height has a measure of 2 cm, the measure of the area within the trapezoid is 7 sq cm.

When the bases of a trapezoid are not only parallel but also congruent, the trapezoid is called a parallelogram. Notice that if the bases of the trapezoid are congruent, then $b = b'$ (Figure 5-50).

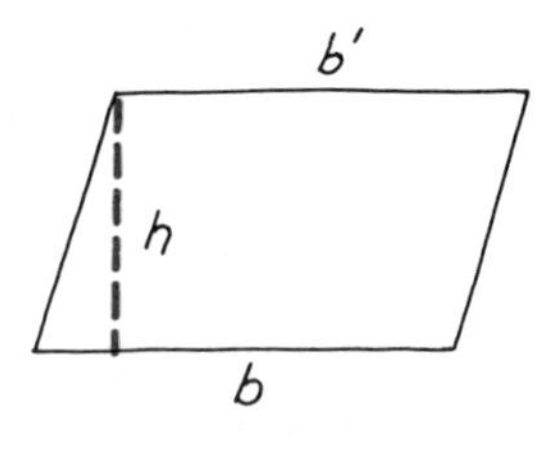

Figure 5-50

The formula for finding the measure of the area within a trapezoid,

$$\underset{\text{trapezoid}}{A} = \frac{(b + b') \cdot h}{2}$$

becomes

$$\underset{\text{trapezoid}}{A} = \frac{(b + b) \cdot h}{2}$$

$$= \frac{(2b) \cdot h}{2}$$

$$= b \cdot h$$

The resulting formula is the one used to find the area within a parallelogram. This makes sense, since the trapezoid is a parallelogram when the bases have the same measure.

Area within a Regular Hexagon

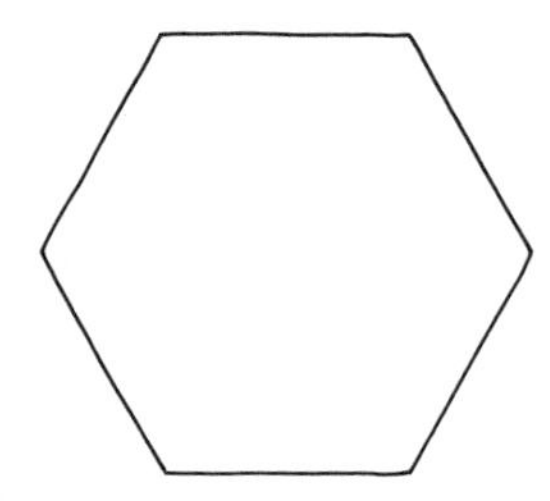

Figure 5-51

The shape in Figure 5-51 is a regular hexagon. How can we find its area? Again, we can discover a way to find the area within a new shape by using area formulas we already know.

Suppose we draw three diagonals of the regular hexagon as indicated in Figure 5-52. These diagonals intersect at a point labeled *O*. Into what shapes do the diagonals partition the interior of the regular hexagon? The interior has been partitioned into six regions, each bounded by equilateral triangles.

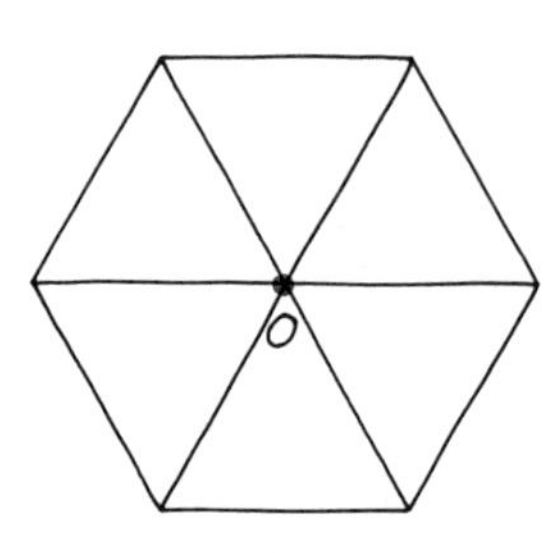

Figure 5-52

Now we can find the area within the regular hexagon by finding the area within the six equilateral triangles. Suppose we call the measure of the length of a side of the regular hexagon *s*. If we draw a perpendicular from *O* to a side *s*, that line segment represents the altitude or height of one of the equilateral triangles. Let us call the measure of the length of the altitude *h* (Figure 5-53).

We know that the area within a triangle is given by the formula $A = 1/2 \cdot (b \cdot h)$, where *b* is the measure of the length of the base and *h* is the measure of the height. The area within the equilateral triangle with base *s* and height *h* can be expressed as $A = 1/2 \cdot (s \cdot h)$. Similarly, since each of the six triangles has a height *h* and a base *s*, the area within each triangle can be expressed as $A = 1/2 \cdot (s \cdot h)$. See Figure 5-54.

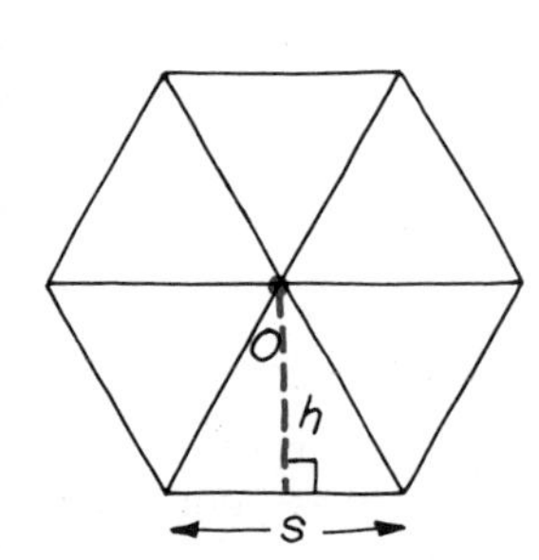

Figure 5-53

The area within the regular hexagon is the same as the sum of the areas within the six triangles. Why? We can represent the area within the regular hexagon as

$$A_{\text{hexagon}} = \frac{1}{2}\cdot(s\cdot h) + \frac{1}{2}\cdot(s\cdot h) + \frac{1}{2}\cdot(s\cdot h) + \frac{1}{2}\cdot(s\cdot h) + \frac{1}{2}\cdot(s\cdot h) + \frac{1}{2}\cdot(s\cdot h)$$

or six times the area within one of the triangles:

$$A_{\text{hexagon}} = 6\cdot\left[\frac{1}{2}\cdot(s\cdot h)\right]$$

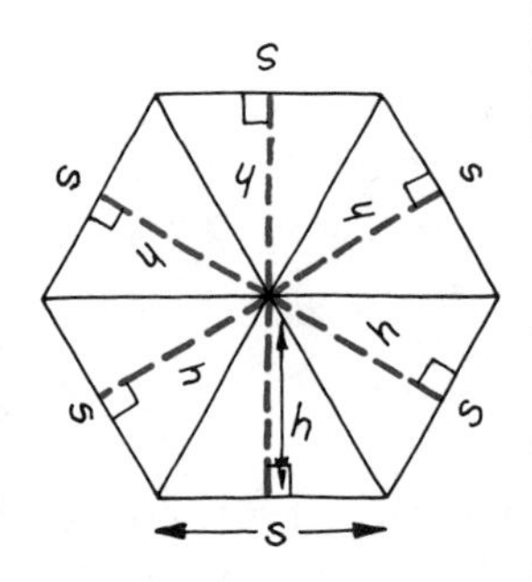

Figure 5-54

This tells us that one way to find the area within a regular hexagon is to partition its interior into equilateral triangles and find the measure of the length of a side and the measure of the length of the altitude of a triangle. We can then find the area of one triangle by using the formula for the area within a triangle. The area within the hexagon will be six times that area.

In Figure 5-55 the length of one side of the regular hexagonal shape is 6 m, and the length of *h* is 5.2 m. What is the area of this shape? Since the area within one triangular region is $1/2 \cdot (b \cdot h)$ or

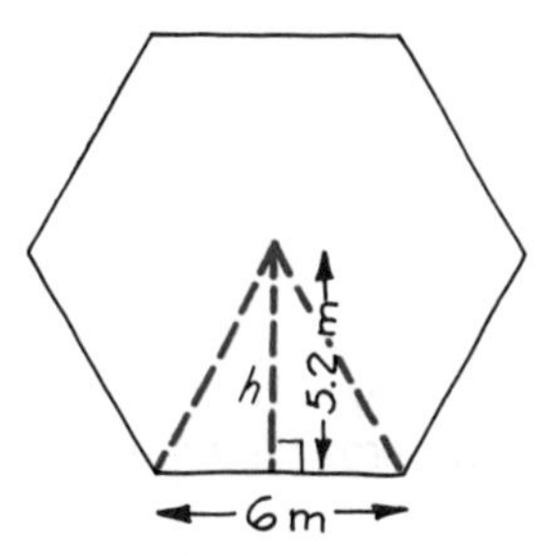

Figure 5-55

$\frac{1}{2} \cdot (6 \cdot 5.2)$, the area within the hexagonal shape is $6 \cdot \frac{1}{2} \cdot (6 \cdot 5.2) = 6 \cdot 15.6 = 93.6$ sq m.

You can also find the area within a regular hexagon by considering the measure of its perimeter. By partitioning the interior of the hexagon into six equilateral triangles, we know that the area within the hexagon can be represented as the sum of the areas of those six triangles:

$$A_{\text{hexagon}} = \frac{1}{2} \cdot (s \cdot h) + \frac{1}{2} \cdot (s \cdot h) + \frac{1}{2} \cdot (s \cdot h) + \frac{1}{2} \cdot (s \cdot h) + \frac{1}{2} \cdot (s \cdot h) + \frac{1}{2} \cdot (s \cdot h)$$

By what is called the **commutative principle of multiplication** $(5 \cdot 3 = 3 \cdot 5)$, we know that $(s \cdot h) = (h \cdot s)$. So we can write the preceding expression as follows:

$$A_{\text{hexagon}} = \frac{1}{2} \cdot (h \cdot s) + \frac{1}{2} \cdot (h \cdot s) + \frac{1}{2} \cdot (h \cdot s) + \frac{1}{2} \cdot (h \cdot s) + \frac{1}{2} \cdot (h \cdot s) + \frac{1}{2} \cdot (h \cdot s)$$

By what is called the **associative principle of multiplication** $[5 \cdot (3 \cdot 2) = (5 \cdot 3) \cdot 2]$, we know that $\frac{1}{2} \cdot (h \cdot s) = (\frac{1}{2} \cdot h) \cdot s$. So we can write the preceding expression as follows:

$$A_{\text{hexagon}} = \left(\frac{1}{2} \cdot h\right) \cdot s + \left(\frac{1}{2} \cdot h\right) \cdot s + \left(\frac{1}{2} \cdot h\right) \cdot s + \left(\frac{1}{2} \cdot h\right) \cdot s + \left(\frac{1}{2} \cdot h\right) \cdot s + \left(\frac{1}{2} \cdot h\right) \cdot s$$

Finally, by what is called the **distributive principle of multiplication over addition** $[(5 \cdot 3) + (5 \cdot 3) = 5 \cdot (3 + 3)]$, we can remove the $(\frac{1}{2} \cdot h)$ from each term of the equation above and write the expression as follows:

$$A_{\text{hexagon}} = \left(\frac{1}{2} \cdot h\right) \cdot (s + s + s + s + s + s)$$

But $(s + s + s + s + s + s)$ represents the perimeter of the hexagon. If we call the perimeter of the hexagon p, we can now say that the area within the hexagon can be found in the following way:

$$A_{\text{hexagon}} = \left(\frac{1}{2} \cdot h\right) \cdot p \quad \text{or} \quad \frac{1}{2} \cdot (h \cdot p) \quad \text{or} \quad \frac{h \cdot p}{2}$$

This shows us another way to find the area within a regular hexagon. Consider the regular hexagon in Figure 5-55. Since the length of one side is 6 m, the perimeter is $6 \cdot 6 = 36$ m. We know that h

is 5.2 m. We can easily find the area within the hexagon as follows:

$$A_{\text{hexagon}} = \frac{h \cdot p}{2} = \frac{(5.2) \cdot (36)}{2} = 93.6 \text{ sq m}$$

Area within Polygonal Regions

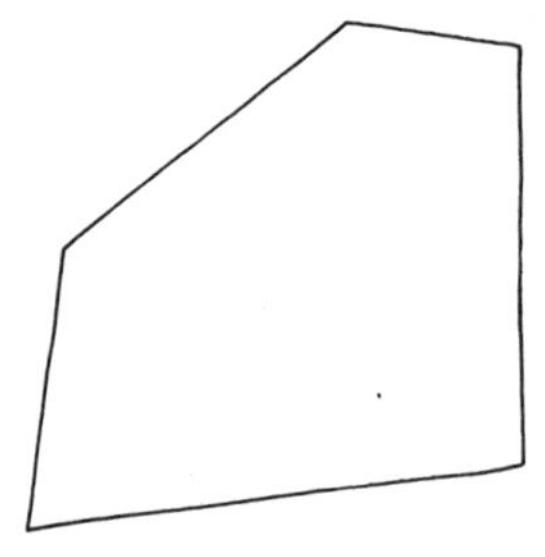

Figure 5-56

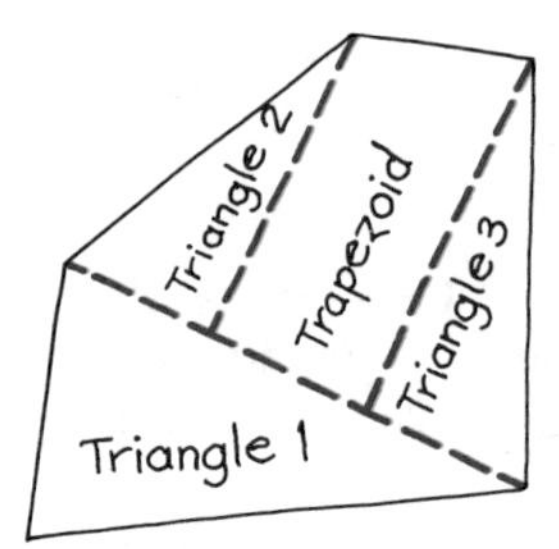

Figure 5-57

In this chapter we have discovered how to find the area within such polygonal shapes as rectangles, squares, parallelograms, rhombuses, triangles, trapezoids, and regular hexagons. Knowing how to find the area within these shapes can help us calculate the area within other shapes.

Suppose the diagram in Figure 5-56 represents a plot of land, and we want to know its area. How can we do it? What measurements do we need? Figure 5-57 suggests one way: the interior of this shape is partitioned into three triangles and a trapezoid. By measuring the altitudes and bases of these triangles and the trapezoid, we can proceed to use the formulas for finding the areas within these shapes. The area within the original shape will be the sum of these areas. Similarly, shapes (a) to (c) in Figure 5-58 have been partitioned into triangles, rectangles, and trapezoids. How might shapes (d) to (f) be partitioned so that the areas within them can be more easily calculated?

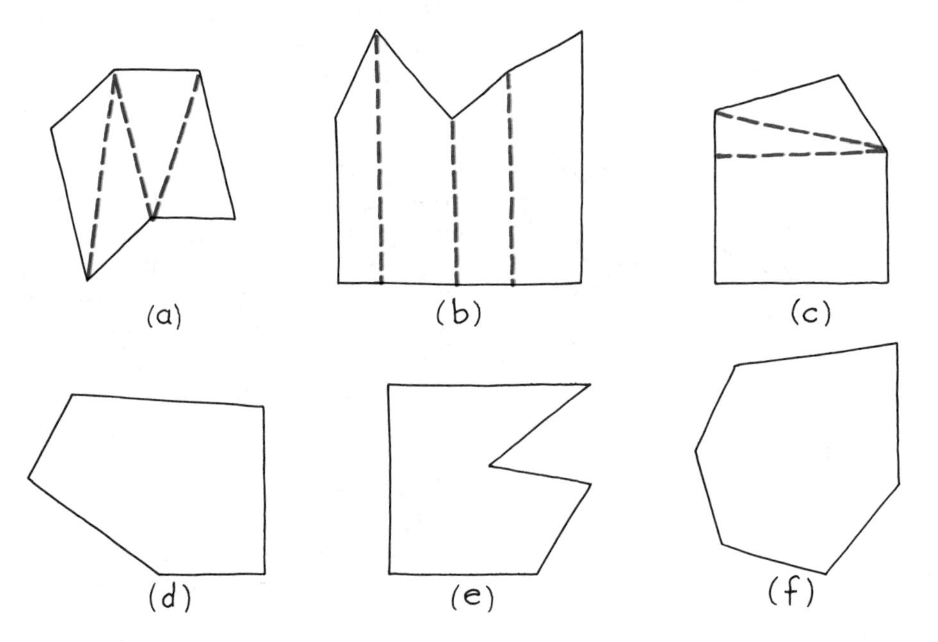

Figure 5-58

More Experiences

5-1. If a rectangle has a perimeter of 32 cm, what might its dimensions be? How many different possibilities are there? Keep track of the dimensions of the rectangles you find by using a chart like this:

Rectangle	Length of one side	Length of other side

Can you graph the pairs of dimensions for these rectangles with perimeter 32 cm? What kind of graph is formed?

Do all the possible rectangles with perimeter 32 cm have the same area? What would be the dimensions of the rectangle with perimeter 32 cm that has the greatest area?

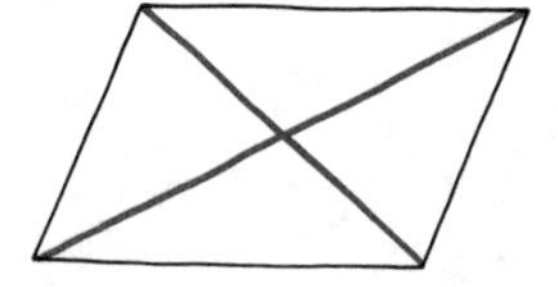

5-2. The diagonals of a parallelogram partition its interior into four triangular regions. While these four regions do not necessarily have the same shape, they always have the same area—they are equivalent. Can you explain why this must be true? (*Hint:* How do you find the area of a triangle?)

5-3. A **tangram** puzzle is a square shape (cardboard, wood, etc.) that has been cut into seven pieces in a

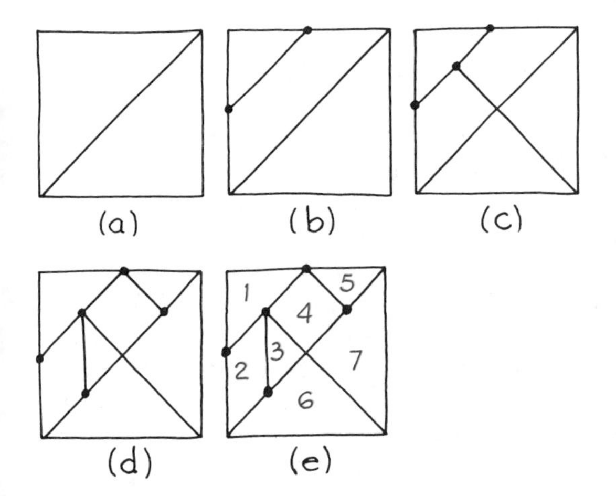

special way, as indicated at the left. (The dots represent the midpoints of the segments in each case.)

The resulting seven shapes are related to each other in many interesting ways. Make a tangram puzzle out of oaktag or cardboard and label the pieces 1 to 7 as indicated.

(*a*) Suppose shape 3 has an area of 1 square unit. What are the areas of the other shapes? What shapes have the same area?

(*b*) Which shapes have the same area as shape 4? Do all these shapes have the same perimeter? How can you explain your findings?

(*c*) Suppose shape 4 has an area of 1 square unit. What are the areas of the other shapes?

(*d*) The larger shapes can be covered with smaller shapes in different ways. For example, shape 6 can be covered by shapes 3, 4, and 5. How? Can you find other ways any one of the tangram shapes can be covered by other tangram pieces? How many different ways can you find?

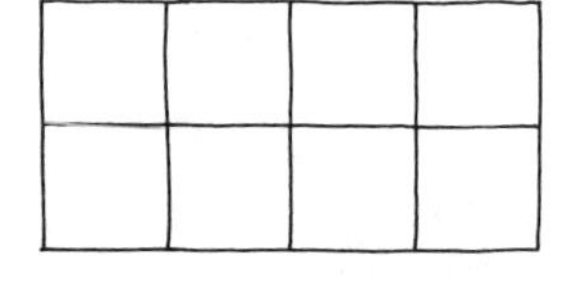

5-4. The square is a convenient shape to use as a unit of area since you can easily cover a plane surface with square regions.

Other polygonal-shaped regions can be used to cover a plane surface like those shown at the left.

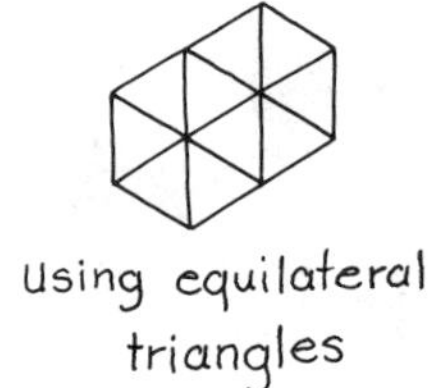

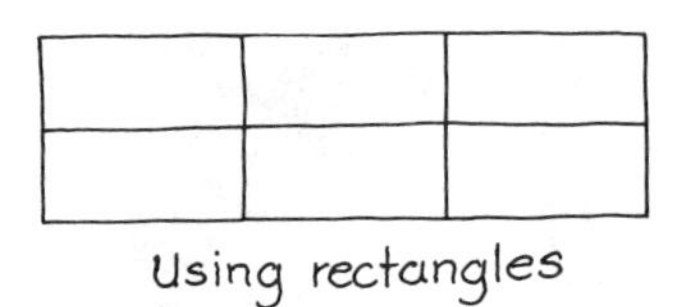

A design used to cover a plane surface is often called a **tessellation** or a tiling pattern. Which of the following shapes can be used to make a tessellation? (Cut out oaktag models of these shapes and test them.)

(*a*) Isosceles trapezoid
(*b*) Rhombus
(*c*) Regular hexagon
(*d*) Parallelogram

(*e*) Regular octagon
(*f*) Scalene triangle

Is there a way to know (without having to make the shapes and test them) whether a shape can be used to form a tessellation? Can a regular pentagon be used to make a tessellation? Explain.

The diagrams show some combinations of different-shaped regions that form tessellations. Can you explain why the shapes fit together in each case? Try making some tessellations of your own.

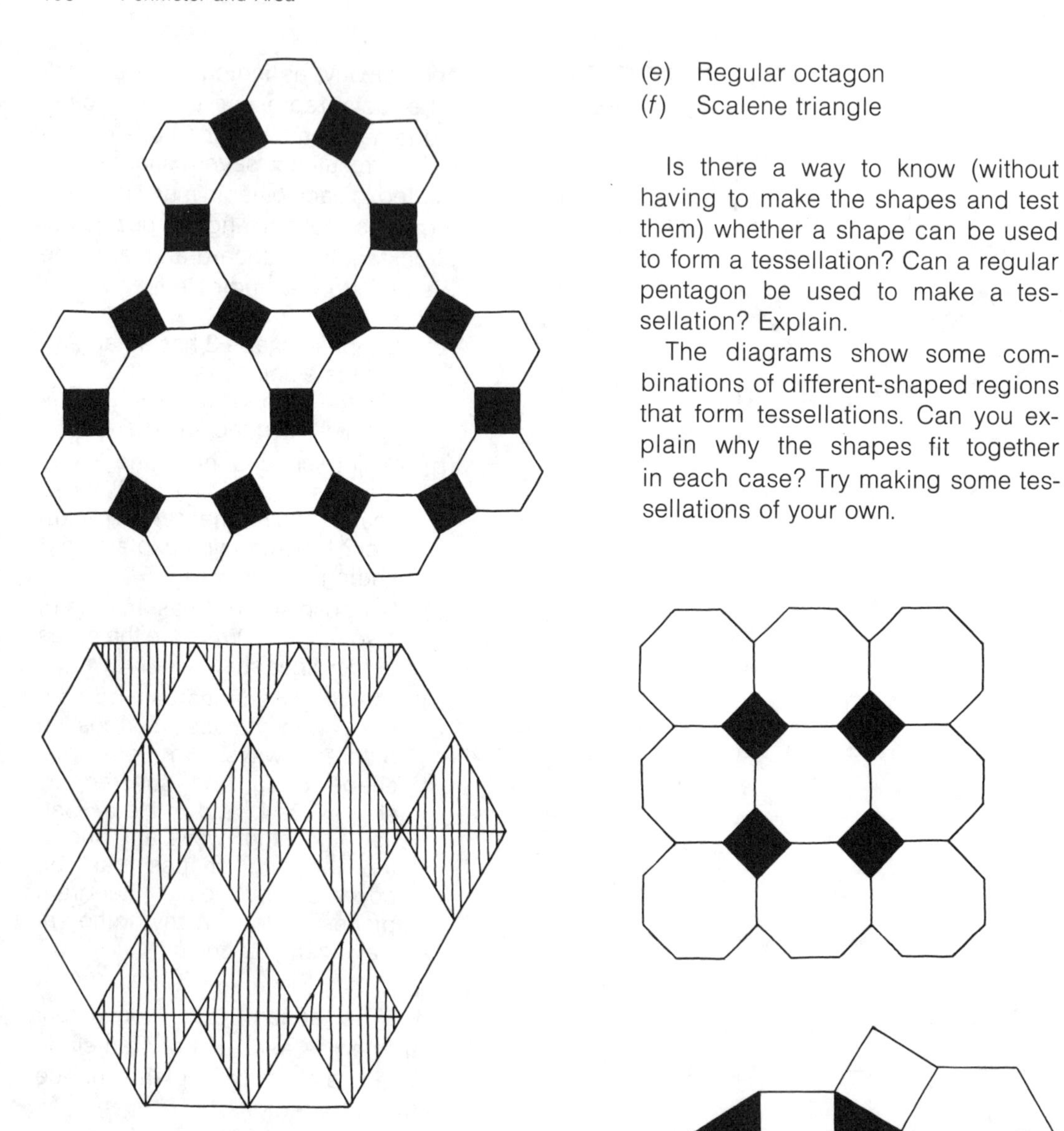

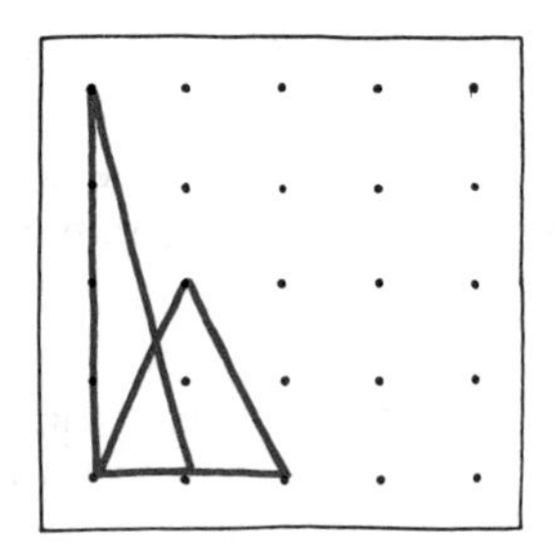

5-5. The triangles on the geoboard at the left are equivalent. How can you show this? How many different triangles can you make on the geoboard that are equivalent to these triangles? Draw them on dot paper.

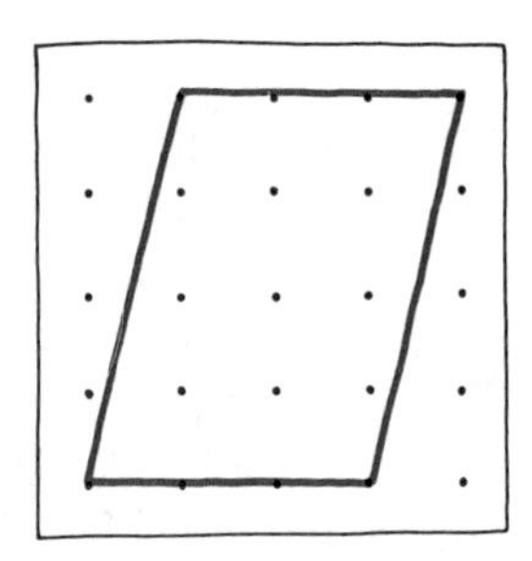

5-6. Suppose that on a geoboard the vertical or horizontal distance between two nails represents a unit of linear measurement and the smallest square that can be made represents a unit of area measurement. Then the figure depicted at the left would have an area of ______.

You can find the area by partitioning the interior of the shape into small squares and triangles and finding the sum of the areas of these shapes. Or you can apply the appropriate formula.

Find the area of the shapes below in two ways.

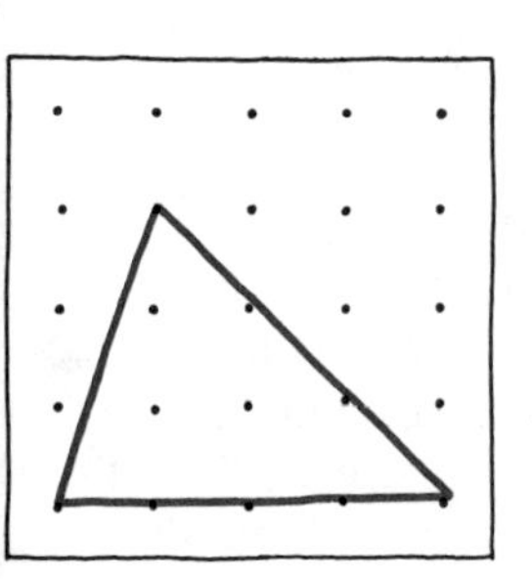

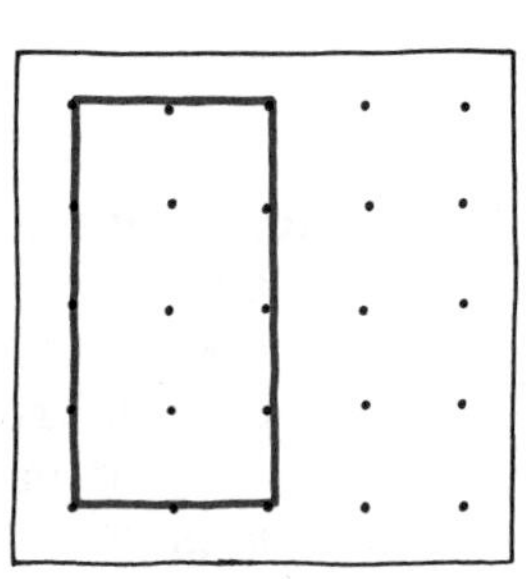

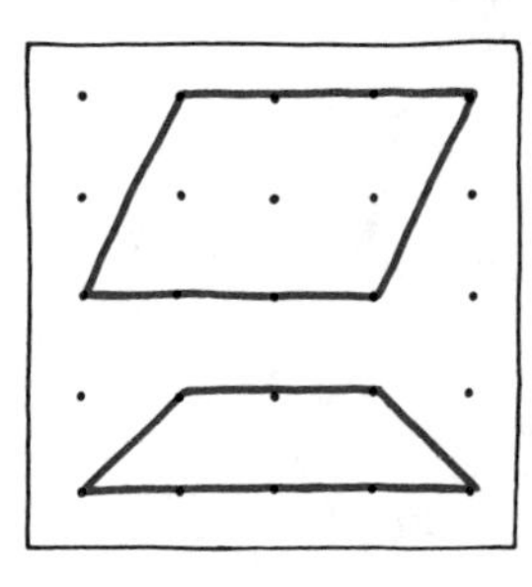

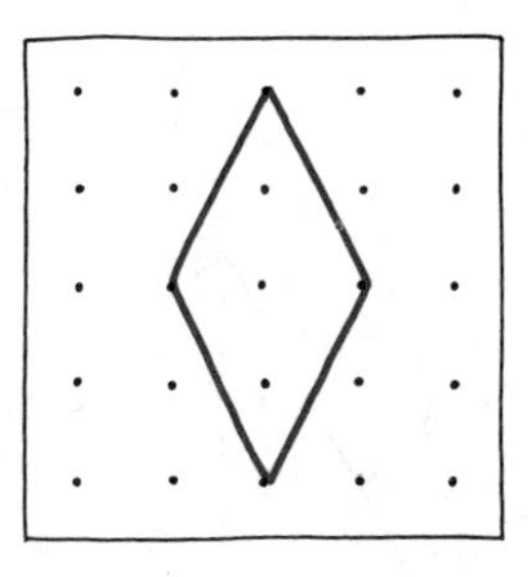

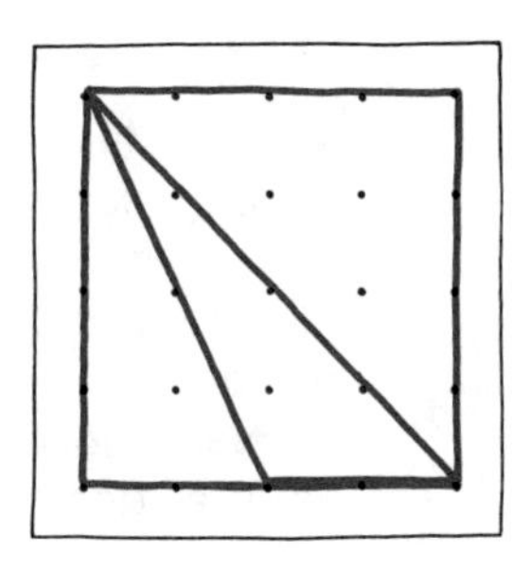

5-7. A square and a triangle are shown on the geoboard at the left. Compare the area of the triangle to the area of the square. How does this relationship suggest a way to partition a square into four parts so that the resulting parts are equivalent although not the same shape?

5-8. Draw a square whose side is 8 cm. Join the midpoints of the sides of the square with line segments to form another square. How does the

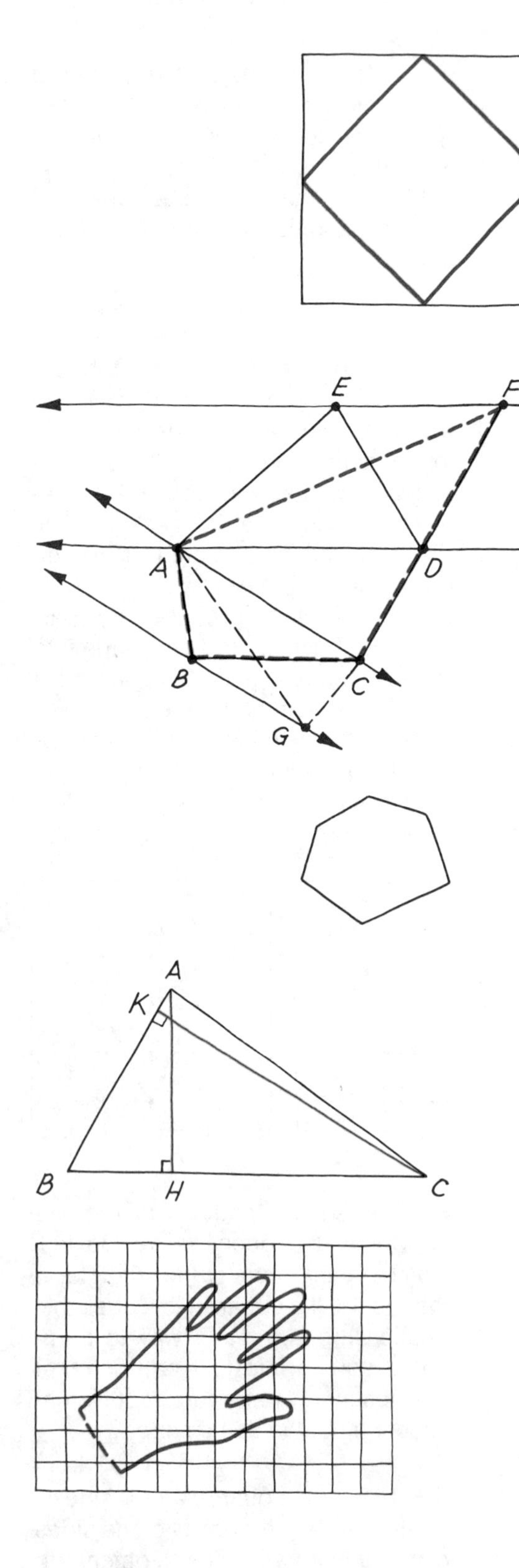

area of the second square compare to the area of the first?

Continue the procedure by joining the midpoints of the second square with line segments and forming a third square. How does the area of the third square compare to that of the second and that of the first? Can you keep doing this? What happens each time?

5-9. In the figure at the left, $\overleftrightarrow{EF}$ and $\overleftrightarrow{AD}$ are parallel. How can you show that pentagon *AEDCB* has the same area as quadrilateral *ABCF*? (They are equivalent.)

Similarly, quadrilateral *ABCF* has the same area as triangle *AGF*. Why?

Can you draw a pentagon that has the same area as the hexagon shown at the left? Can you draw an equivalent quadrilateral? An equivalent triangle?

Is the following statement true? "Any polygon can be transformed into a triangle that is equivalent to the original polygon." Explain.

5-10. In the scalene triangle shown at the left, side $\overline{BC}$ is longer than side $\overline{AB}$. How do you know that altitude $\overline{CK}$ must be longer than altitude $\overline{AH}$? What would have to be true for the altitudes to be the same length?

5-11. Many things have irregular shapes so that you cannot find their areas by using a formula. Trace the outline of your hand on a piece of squared paper and close the curve formed. How might you proceed to find a good approximation of the area within this curve? Find some other irregular shapes and use your approach to find an approximation of their areas.

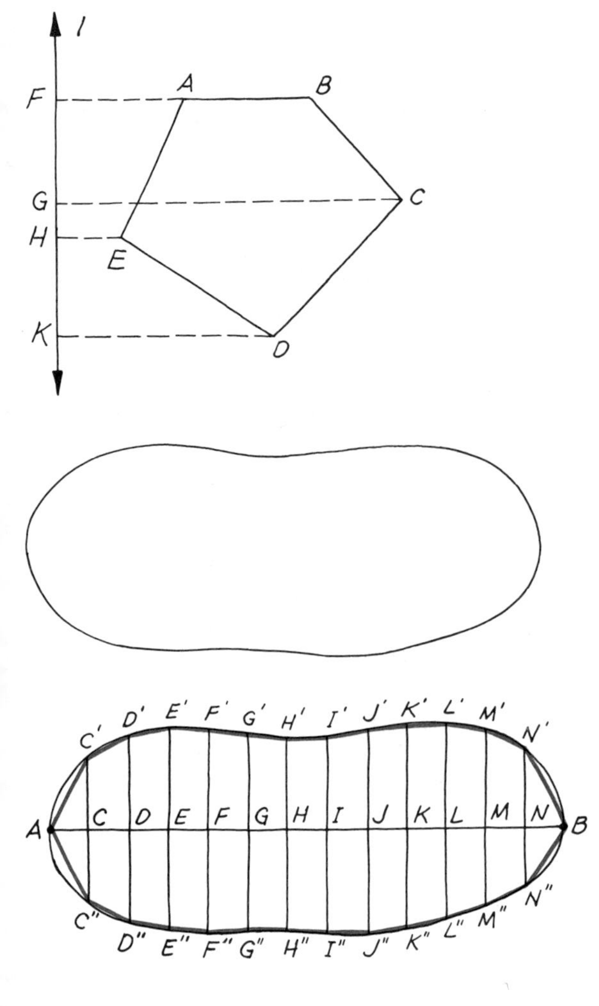

5-12. One way to find the area of a polygonal piece of land like *ABCDE* at the left is to make a reference line *l* and then find the measures $\overline{FB}$, $\overline{GC}$, $\overline{HE}$, $\overline{KD}$, $\overline{FG}$, $\overline{GH}$, and $\overline{HK}$. How can these measures help you calculate the area of *ABCDE*?

5-13. How can you find the area of a piece of land with a curved boundary? One way is to proceed as follows:

1. Choose two points such as *A* and *B* and draw the line segment *AB*.
2. Draw points *C, D, E, F, G, . . . , N* to partition *AB* into congruent parts.
3. Draw perpendicular line segments at each of the points *C, D, . . . , N* to intersect the curved boundary at *C′, C″, D′, D″, . . . , N′, N″*.
4. Connect point *A* with *C′*, point *C′* with *D′*, point *D′* with *E′*, and so on, as shown.
5. The original land has been partitioned into many trapezoids with the same height ($\overline{CD}$, $\overline{DE}$, and so on) and triangles at the ends.

What measures will you have to find for an approximate area of this curved piece of land?

Suppose that *AB* had been partitioned into more parts. Would the approximate area calculated be *more* or *less* accurate?

5-14. The floor of a rectangular room has dimensions 14 ft by 20 ft. A rug is on the floor and is 18 in. from each wall. What is the perimeter and area of the rug? What is the area of the part of the floor not covered by the rug?

5-15. The top of a rectangular counter is 18 in. wide and 5 ft long. What is the area of the counter top in square inches? In square feet? In square yards?

5-16. Estimate the area of the walls of a room in square feet. Convert your estimate to square yards. Then measure the walls of the room. How close was your estimate?

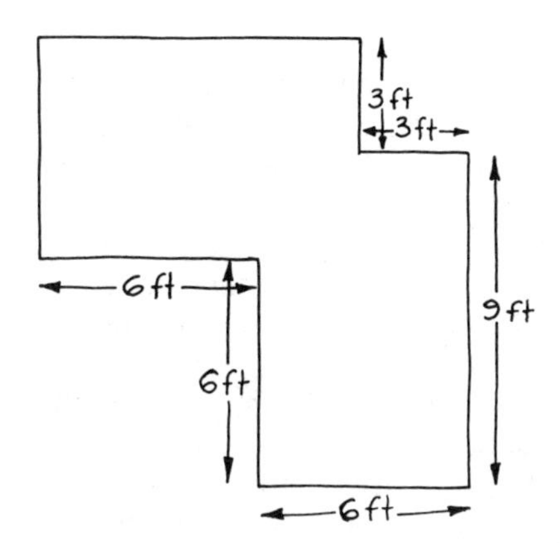

5-17. The diagram at the left is the floor plan of the entry hall of a house.

(*a*) Suppose you want to cover the floor with 1-ft-square tiles. How many will you need?

(*b*) How many 1-in.-square mosaic tiles would you need?

(*c*) How many square yards of carpeting would cover it?

The Pythagorean Relationship

Making a Square: The Cord Method

In the last chapter we found a way to trace the path of a circle by using a piece of cord and a nail as a pivot. When we used two nails with the cord and formed many triangles with the same base and the same perimeter, we located the path of an ellipse.

Suppose we want to form a square shape on the ground. Imagine we have no ruler or compass. We have only some spikes and some cord. How can we do it?

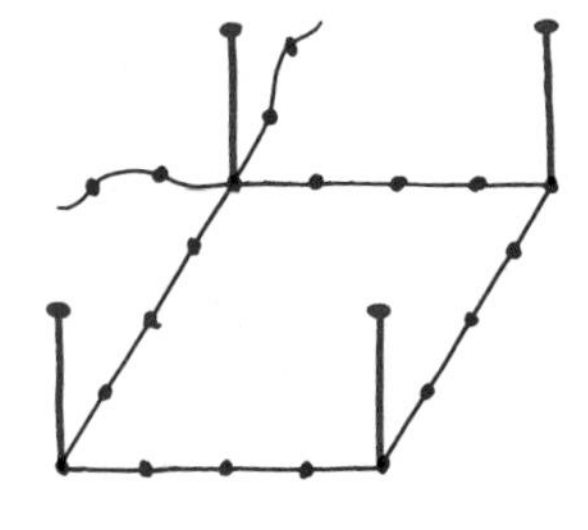

Figure 6-1

Forming a quadrilateral with four congruent sides is no great problem. We simply make any kind of unit out of the string by making equally spaced knots in it. The space between any two knots will be one unit. We can then form this special quadrilateral out of the string as in Figure 6-1. The spikes form the vertices, and the sides of the quadrilateral are congruent since each side is the same number of units in measure.

You can make another model of this quadrilateral by using some string, four thumbtacks, and a piece of corkboard. By folding the string in half over and over again you can easily make units. Rather than knotting the string you may simply want to mark the units with a felt pen. You then place the four thumbtacks as vertices for your quadrilateral so that there are an equal number of units between each two tacks (Figure 6-2). That is easy enough. But we want to construct the model of a square, and a square is not only equilateral but also equiangular. Each angle must be a right angle. How can we be certain that each angle is a right angle without using a protractor or any other instrument?

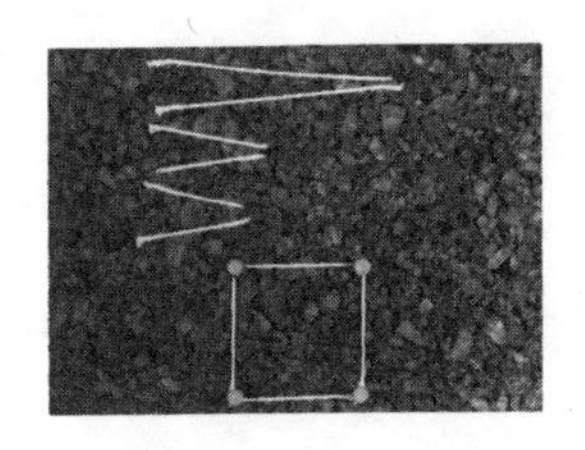

Figure 6-2

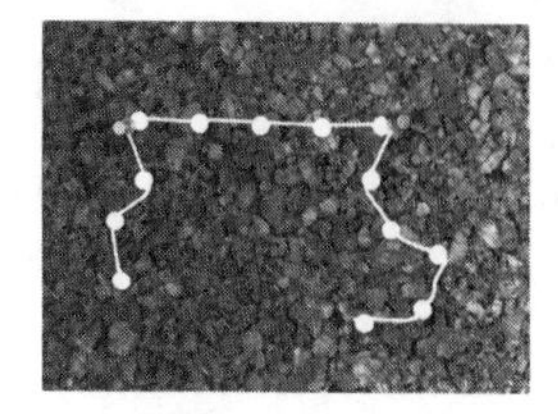

Figure 6-3

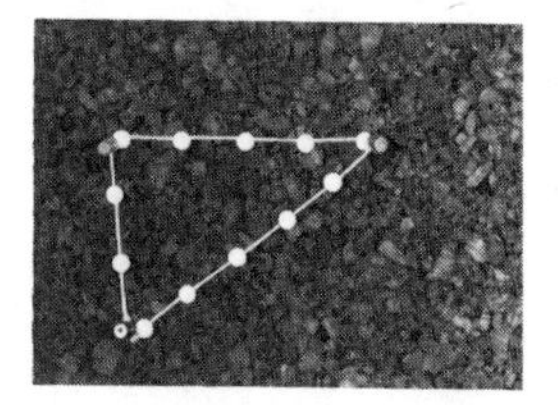

Figure 6-4

This is a very old problem. In fact, at least as early as 2900 B.C. the ancient Egyptians needed to solve this same problem in order to form a square base for their pyramids. How did they form the right angles?

Here is how they did it: They took a piece of cord that was knotted to form 12 units. They formed a special triangle on the ground by using three spikes as vertices and the cord as the sides. This special triangle was to have sides of 3 units, 4 units, and 5 units. They then drove two spikes into the ground 4 units apart. Next, the middle portion of the cord was extended between the spikes so that 3 units of cord remained at one end and 5 units remained at the other (Figure 6-3). The two ends of the cord were then extended and brought together so that the end points met (Figure 6-4).

What did the resulting shape look like? A *right* triangle! With this cord method, three other right angles could be formed and a square of any size could be developed with the aid of only the cord and the spikes (Figure 6-5).

Why do you think 3, 4, and 5 were chosen as the numbers of units for the sides of the triangle formed by the knotted cord? This trio of numbers has a special property:

$$3^2 + 4^2 = 5^2$$
$$9 + 16 = 25$$

By doubling each of the three original numbers, we can get another trio or triple of numbers having that property—6, 8, 10:

$$6^2 + 8^2 = 10^2$$
$$36 + 64 = 100$$

In fact, if we choose any number and multiply each number of the triple (3, 4, 5) by that number, we will find other triples with this property. For example, multiplying by 3 we get 9, 12, 15:

$$9^2 + 12^2 = 15^2$$
$$81 + 144 = 225$$

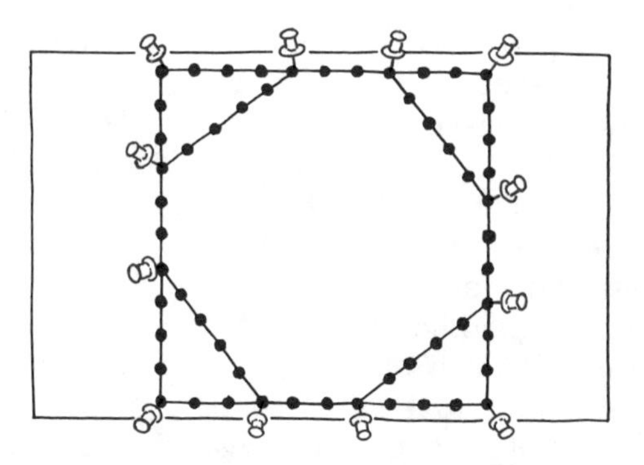

Figure 6-5

Multiplying by $\frac{1}{5}$ we get $\frac{3}{5}$, $\frac{4}{5}$, 1:

$$\left(\frac{3}{5}\right)^2 + \left(\frac{4}{5}\right)^2 = (1)^2$$
$$\left(\frac{9}{25}\right) + \left(\frac{16}{25}\right) = 1$$

Can you find others? In each case, if you use a triple of numbers with this special property to form a triangle whose sides have those numbers as their measures, you always form a right triangle.

Triples formed from the basic 3, 4, 5 trio are not the only ones with that special property. There are many others. In early times the Indians and the Chinese formed right angles by the cord method and divided the cord into sections of 5, 12, and 13 or 8, 15, and 17 units. These triples have the special property:

$$5^2 + 12^2 = 13^2 \qquad 8^2 + 15^2 = 17^2$$
$$25 + 144 = 169 \qquad 64 + 225 = 289$$

In this chapter we will explore this strange relationship between right triangles and these special triples of numbers.

Pythagoras and the Pythagorean Relationship

A very useful discovery about right triangles is credited to Pythagoras, a Greek who lived in the sixth century B.C. Not much is known about Pythagoras. He was born on the Aegean island of Samos and spent 20 years in Egypt. Because he met with political difficulties upon his return to Samos, he migrated to Crotona, a Greek seaport colony in southern Italy.

It was in Crotona that he founded his famous Pythagorean school. Preliminary instruction at the school included moral and religious preparation along with the first elements of mathematics and music, which Pythagoras considered to be of the utmost importance. After this initial instruction, the best students had the opportunity to study directly under Pythagoras and formed a kind of fraternity known as the Pythagorean Society. The symbol for this brotherhood was the five-pointed star.

It was considered a great honor to belong to this society whose members shared everything—their material goods, their feelings, and their ideas. Because of their communal efforts, however, it is difficult to know which discoveries were made by Pythagoras and which were made by others of his society.

The most famous discovery attributed to Pythagoras is known as the **Pythagorean relationship** among the sides of a right

Figure 6-6

triangle. This relationship was illustrated on a commemorative stamp issued in Greece in 1955 to celebrate the twenty-five hundredth anniversary of the founding of the Pythagorean school.

Essentially, Pythagoras stated that if you start with a right triangle and construct three squares by using each side of the triangle as a side for one of the squares, then the area within the largest square (formed by the hypotenuse) is equal to the sum of the areas within the two smaller squares (formed by the legs). The stamp illustrates this relationship (Figure 6-6): the area within a large square (25 square units) is equal to the sum of the areas within two smaller squares ($9 + 16 = 25$ square units). In this case the original triangle had a hypotenuse with a length of 5 units and legs of 3 and 4 units. But this relationship is valid for *any* right triangle. How can we show this?

Figure 6-7

Figure 6-8

Before we make a model to demonstrate that this relationship holds for all right triangles, let us take a look at a preliminary idea we will need. Suppose we draw the outlines of three congruent rectangles like the one shown in Figure 6-7 and make six oaktag triangular shapes like the one in Figure 6-8. Now let us place two triangular shapes within each rectangle in different ways as shown in Figure 6-9. In each case we are covering part of the area within the rectangle. The uncovered regions within the rectangles are outlined in brown. Each uncovered region in (a), (b), and (c) encloses the same amount of area. The regions are equivalent. Why? We started out with three congruent rectangles. We then covered the same amount of area within each rectangle (using two

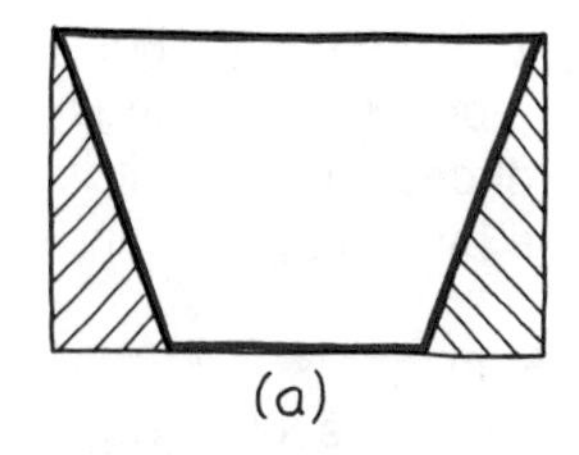

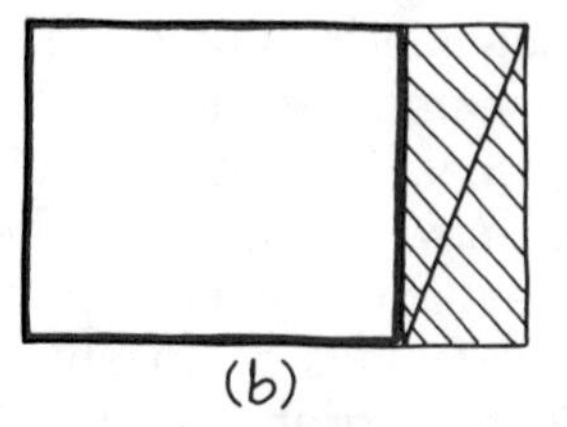

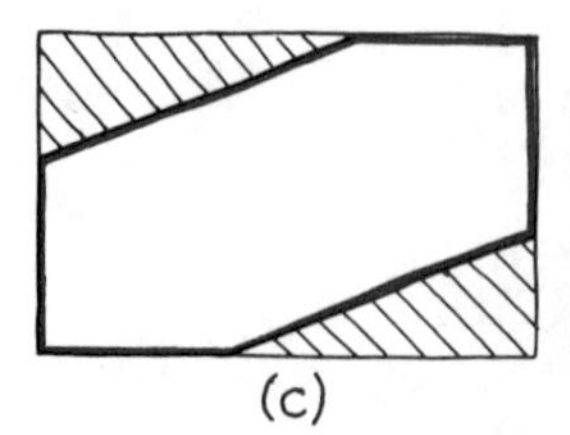

Figure 6-9

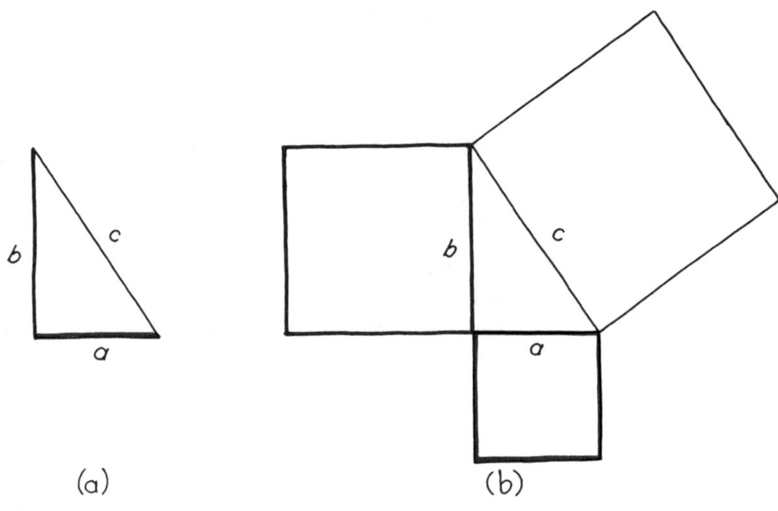

Figure 6-10

of the six congruent triangles in each case). The remaining uncovered areas must be the same (subtracting equals from equals gives results that are equal).

We can now use that very basic idea to demonstrate the Pythagorean relationship. We want to show that for *any* right triangle the area within the largest square formed by using the hypotenuse as a side is equal to the sum of the areas within the two smaller squares formed by using one leg as a side. If we designate the lengths of the sides of a right triangle as a, b, and c, then we are saying that the area within the square with side c is equal to the sum of the areas within the squares with sides a and b respectively (Figure 6-10).

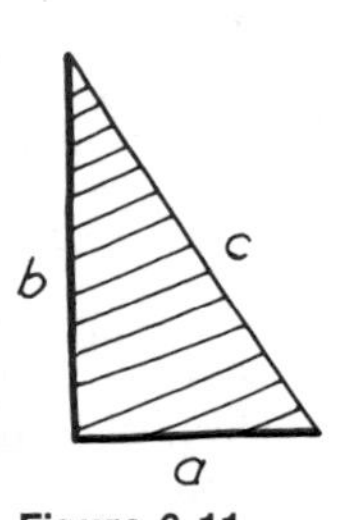

Figure 6-11

Let us make eight congruent right triangles (heavy paper or oaktag will do). Make the sides of the right triangle any size you want. Label the shorter leg of each triangle a; label the other leg b. We will label the hypotenuse c (Figure 6-11). Now draw two congruent squares on a piece of paper with sides equal in length to the lengths of $a + b$ like the one in Figure 6-12.

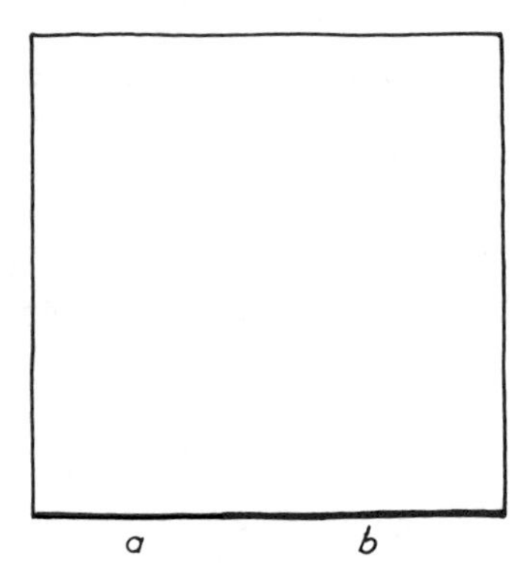

Figure 6-12

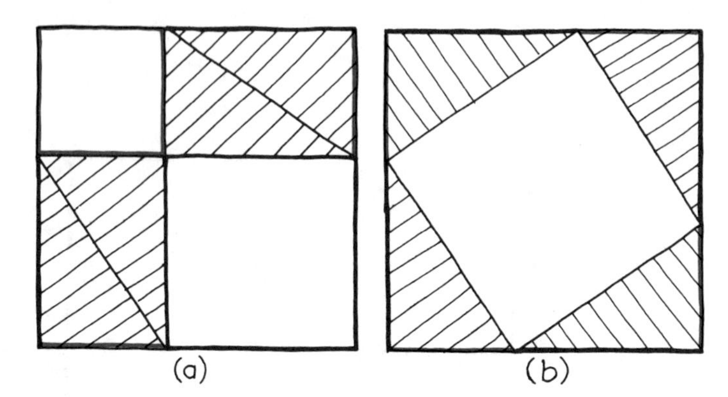

Figure 6-13

Place four triangles in each of the two squares as in Figure 6-13. Look at what is formed by the uncovered regions within the squares. In part (a) the uncovered portions form two regions, the smaller enclosed by a square with a side equal to a and the larger enclosed by a square with a side equal to b. In part (b) the uncovered portion forms a region enclosed by a square with a side c. Now since we started with two congruent squares and covered just as much of the area within each square (using four congruent triangles in each case), the remaining (uncovered) areas within each square must be the same. In other words, the area within the square with side c must be equal to the area within the square with side a plus the area within the square with side b. That is just what we wanted to show!

This special relationship belonging to right triangles is usually expressed algebraically as follows:

> If a and b represent the lengths of the legs of a right triangle, and c represents the length of the hypotenuse of that triangle, then
>
> $$a^2 + b^2 = c^2$$

This statement corresponds to our geometric interpretation, since a^2 is the value of the area of the square with side a, b^2 is the value of the area of the square with side b, and c^2 is the value of the area of the square with side c.

This is the same relationship we found for those special triples of numbers at the beginning of this chapter. Remember?

$$3, 4, 5 \longrightarrow 3^2 + 4^2 = 5^2$$
$$5, 12, 13 \longrightarrow 5^2 + 12^2 = 13^2$$

Since all right triangles have this special property, these triples can be used as the measures of the sides of a right triangle. These triples of numbers are often called Pythagorean triples.

Some Applications

Aside from being an interesting phenomenon, the Pythagorean relationship is very useful. Whenever we have a problem involving a right triangle, if we know the measures of any two sides of that triangle, we can always find the measure of the third side. Some practical examples are given below.

Problem 1

Imagine a young tree supported by wires that have been tied to the tree and staked in the ground (Figure 6-14). We can see that the tree, the ground, and a wire form the shape of a triangle—that wonderfully rigid figure. If we assume the tree is vertical and the ground is reasonably level, we can think of that triangle as a right triangle (indicated with dashes). If we know that the stake is 7 ft from the tree and the wire has been attached to the tree 6 ft from the ground, how long is the wire? What we want to know is the length of the hypotenuse of this right triangle.

Solution

By the Pythagorean relationship, we know that if a and b represent the measures of the legs of the right triangle and c represents the measure of the hypotenuse, then

$$a^2 + b^2 = c^2$$

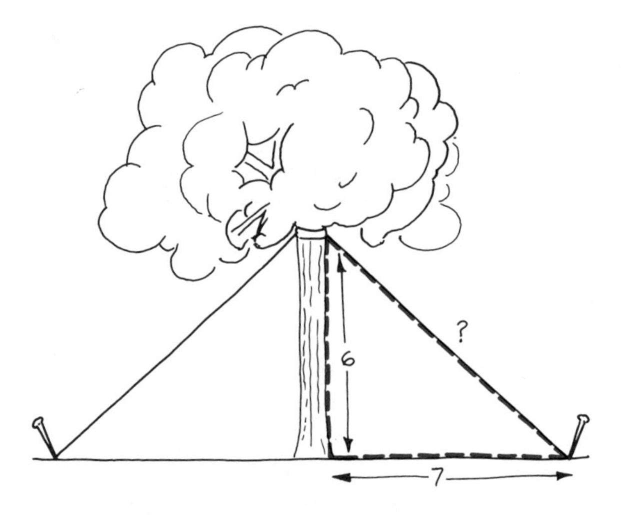

Figure 6-14

In this case we have $a = 6$ and $b = 7$, and we want to find c. Therefore,

$$\begin{aligned} 6^2 + 7^2 &= c^2 \\ 36 + 49 &= c^2 \\ 85 &= c^2 \end{aligned}$$

Here c is a number whose square is 85; that is, $c = \sqrt{85}$. Since $9 \times 9 = 81$ and $10 \times 10 = 100$, c must be between 9 and 10:

$$9 < c < 10$$

We can try squaring the numbers 9.1, 9.2, 9.3, and so forth to get a better approximation of the value of c^2. Since

$$\begin{aligned} (9.2)^2 &= 84.64 \\ (9.3)^2 &= 86.49 \end{aligned}$$

c must be between 9.2 and 9.3:

$$9.2 < c < 9.3$$

We can say that the length of the wire $c \approx 9.2$. The symbol $\approx$ means "approximately equals." If we need a more precise value for c, we can square the numbers 9.21, 9.22, 9.23, and so on to see which square is closest to the square of c (which is 85). In this way we can obtain as precise a value for c as we need.

Problem 2

Let us look at another situation. Suppose we have a kite that has been staked to the ground. At a certain time the kite is flying directly above a bush 70 ft from the stake If the string on the kite is 100 ft long, how high is the kite? We can imagine that the line segment representing the height of the kite forms one leg of a right triangle, the line segment representing the distance from the bush to the stake forms the other leg, and the line segment representing the string* on the kite forms the hypotenuse (Figure 6-15).

Solution

Again we have formed a right triangle and can use the Pythagorean relationship, $a^2 + b^2 = c^2$. This time we can call one leg $a = 70$ and the hypotenuse $c = 100$. We want to find the other leg, b. Substituting, we get

$$\begin{aligned} a^2 + b^2 &= c^2 \\ (70)^2 + b^2 &= (100)^2 \\ 4900 + b^2 &= 10{,}000 \end{aligned}$$

**We imagine the kite string to be a line segment although in reality it would not be perfectly straight*

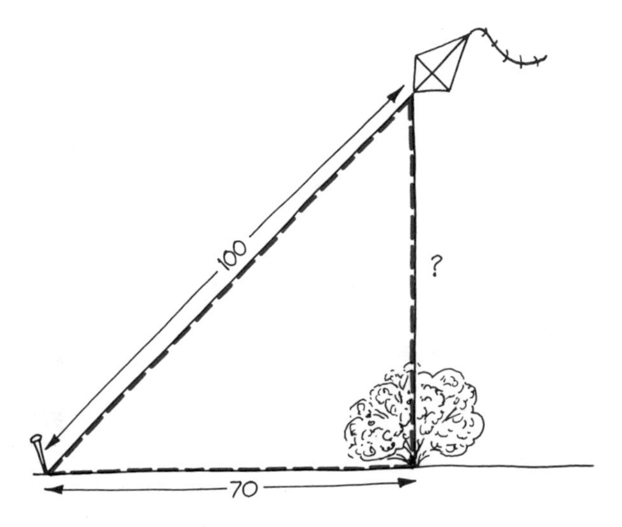

Figure 6-15

or

$$b^2 = 5100$$

We are looking for a number, b, which when squared gives us 5100. Since $(70)^2 = 4900$, we can try $(71)^2$, $(72)^2$, and so on to find out which is nearest to 5100:

$$(71)^2 = 5041$$
$$(72)^2 = 5184$$

Therefore,

$$(71)^2 < b^2 < (72)^2$$
$$71 < b < 72$$

The height of the kite is between 71 and 72 ft. If we need to know its height more precisely, we can square the numbers 71.1, 71.2, 71.3, and so on to find out which square comes closest to 5100. Can you find its height to the nearest tenth of a foot?

Other Considerations

Does this special property of right triangles apply to any other type of triangle? Try making some acute and some obtuse triangles and construct the squares corresponding to each side as in Figures 6-16 and 6-17. You will see that for the obtuse triangles the sum of the measures of the areas within the squares formed by using the

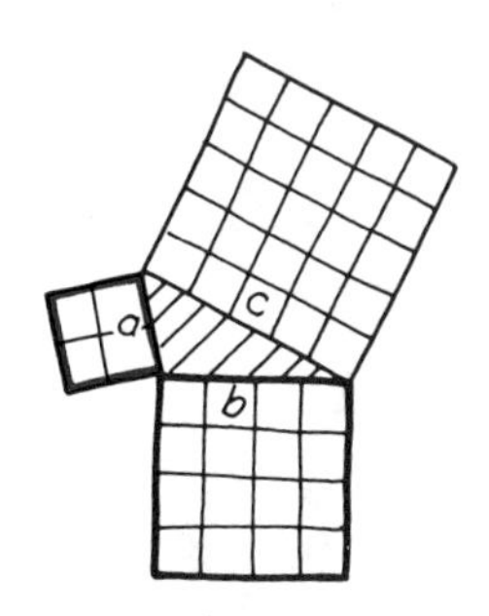

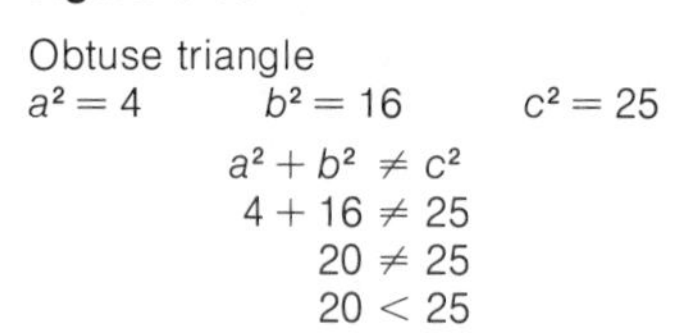

Figure 6-16

Obtuse triangle
$a^2 = 4$ $b^2 = 16$ $c^2 = 25$

$$a^2 + b^2 \neq c^2$$
$$4 + 16 \neq 25$$
$$20 \neq 25$$
$$20 < 25$$

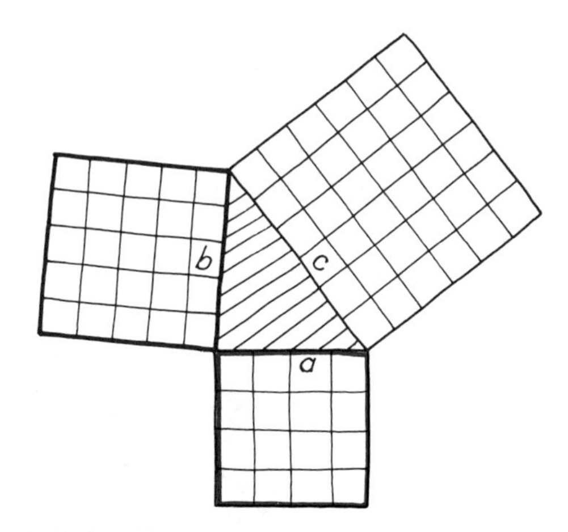

Figure 6-17

Acute triangle
$a^2 = 16$ $b^2 = 25$ $c^2 = 36$

$$a^2 + b^2 \neq c^2$$
$$16 + 25 \neq 36$$
$$41 \neq 36$$
$$41 > 36$$

two smaller sides, a and b, is *less than* the measure of the area within the square formed by the largest side c (that is, $a^2 + b^2 < c^2$). For the acute triangles, the sum of the measures of the areas within the two squares formed by the smaller sides, a and b, is *greater than* the measure of the area within the square formed by the largest side c (that is, $a^2 + b^2 > c^2$).

Let us investigate this finding a little further. Suppose we construct an isosceles triangle ABC as in Figure 6-18. Let point F be the midpoint of the base $\overline{BC}$ and let $\overline{EF}$ be perpendicular to $\overline{BC}$ as shown. Imagine that the base $\overline{BC}$ remains fixed and that we move vertex A along $\overline{EF}$. As the illustration indicates, many isosceles triangles can be formed, such as ABC, DBC, and EBC. Consider the squares that can be formed by using the corresponding sides of these triangles.

In the case of triangle DBC, $\angle D$ is obtuse and the triangle is an obtuse triangle. The sum of the measures of the areas within the square formed by side $\overline{BD}$ and the square formed by side $\overline{DC}$ is *less than* the measure of the area within the square formed by side $\overline{BC}$. You can see this intuitively if you imagine these squares to be made of cardboard. The two smaller squares could not be made to cover the large square with side $\overline{BC}$.

As the vertex A is moved upward toward E, what happens to the size of the squares corresponding to the two sides that are changing in length? Those sides are increasing in length and the corresponding squares formed are increasing in size. At point A,

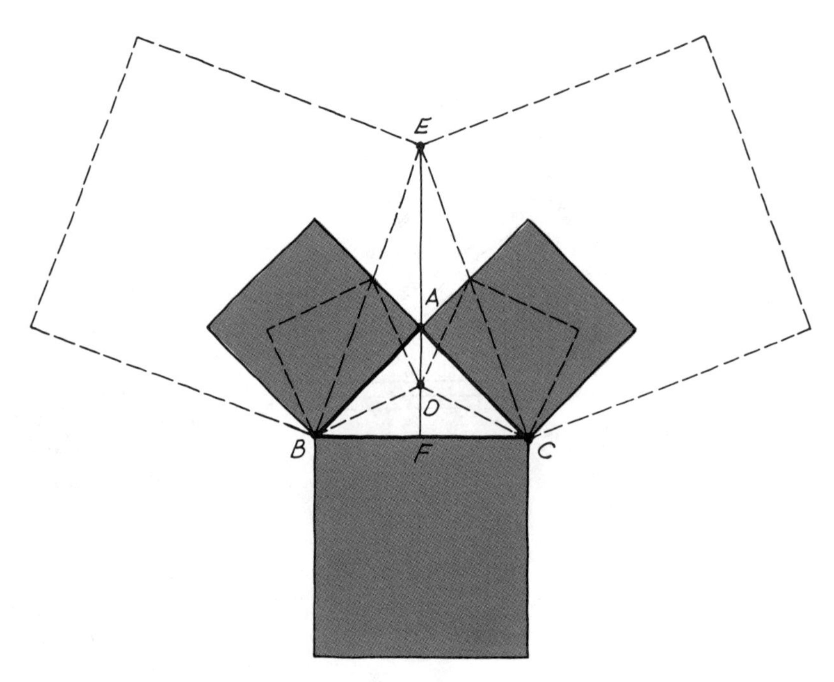

Figure 6-18

$\angle A$ is a right angle and triangle ABC is a right triangle. By the Pythagorean relationship, we know that the sum of the measures of the areas within the squares formed by sides $\overline{AB}$ and $\overline{AC}$ *equals* the measure of the area within the square formed by the hypotenuse $\overline{BC}$ (brown shaded regions).

As we continue to move the vertex A along $\overline{EF}$ toward E, side $\overline{BC}$ remains the same while the other two sides increase in length. The angle at A decreases in size and becomes acute. At point E, $\angle E$ is acute and triangle EBC is acute. The squares formed by sides $\overline{EB}$ and $\overline{EC}$ are very large, and you can see that the sum of the measures of the areas within these two squares is greater than the measure of the area within the square formed from base $\overline{BC}$.

What can we conclude from this experience? We formed many isosceles triangles by keeping base $\overline{BC}$ fixed and moving the opposite vertex along $\overline{EF}$. When that vertex is between points A and F, obtuse triangles are formed and the Pythagorean relationship is *not* true. Similarly, when that vertex is between points A and E, acute triangles are formed and again the Pythagorean relationship does *not* hold. In fact, the only time this special relationship exists is when that vertex is at point A and a right triangle ABC is formed. The Pythagorean relationship belongs *exclusively* to *right* triangles.

More Experiences

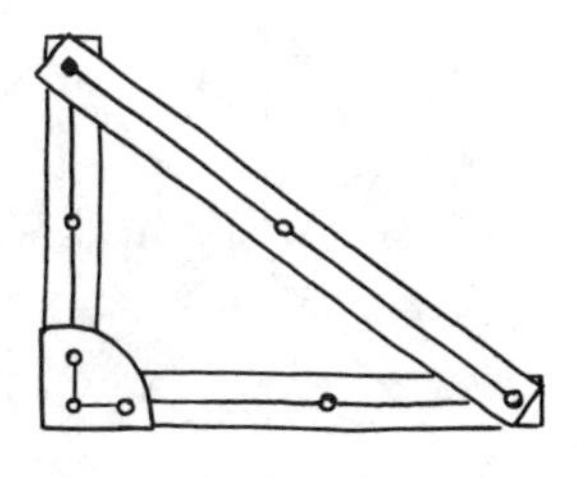

6-1. Using the oaktag strips and angle-fixers you made for the activities in previous chapters, make some right triangles. Use a right angle-fixer and any three strips that fit together to make a right triangle. How many different right triangles can you make? What are the lengths of the sides of these right triangles?

6-2. If the vertical or horizontal distance between two adjacent nails on a geoboard is 1 in., can you find the length of each line segment represented on the geoboard at the left?

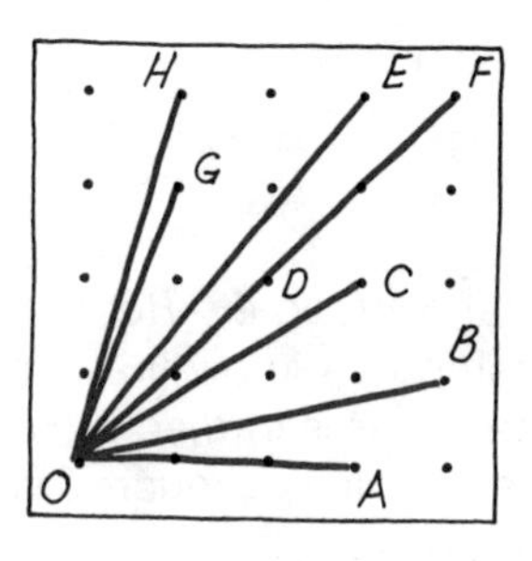

$m(\overline{OA}) =$
$m(\overline{OB}) =$
$m(\overline{OC}) =$
$m(\overline{OD}) =$
$m(\overline{OE}) =$
$m(\overline{OF}) =$
$m(\overline{OG}) =$
$m(\overline{OH}) =$

What is the length of the sides of each square shown below? What is the area of each square?

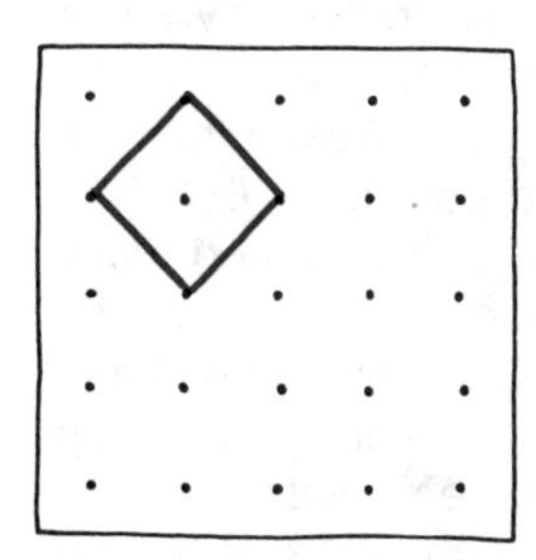

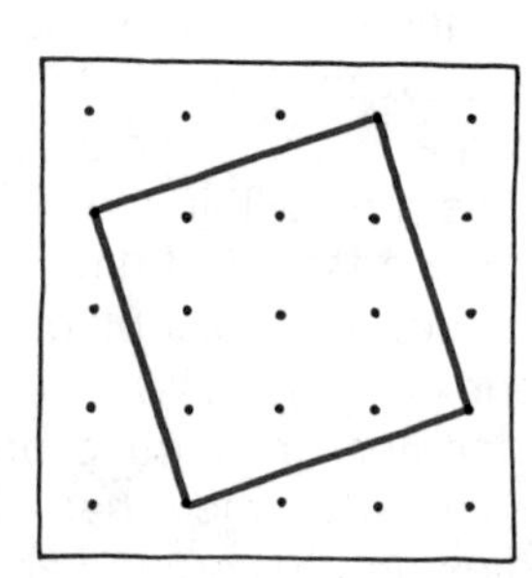

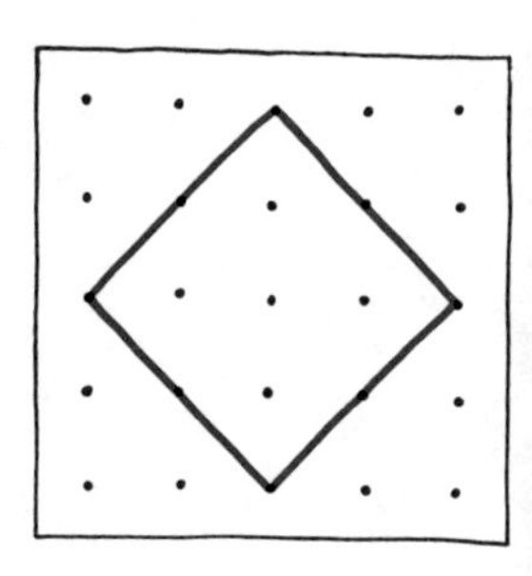

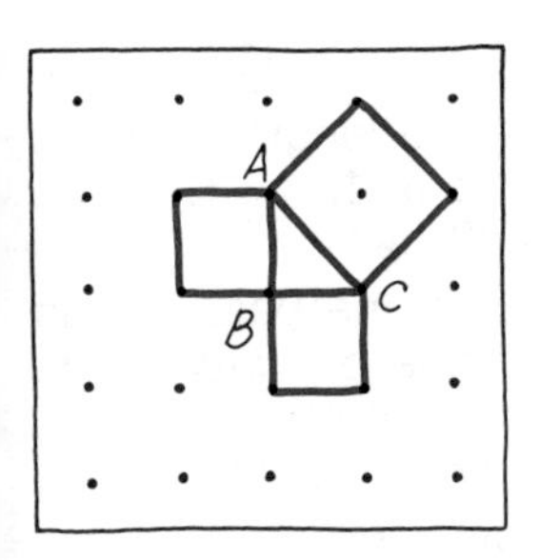

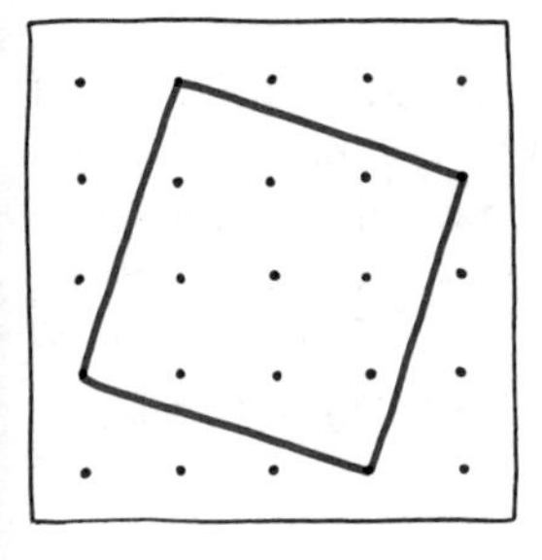

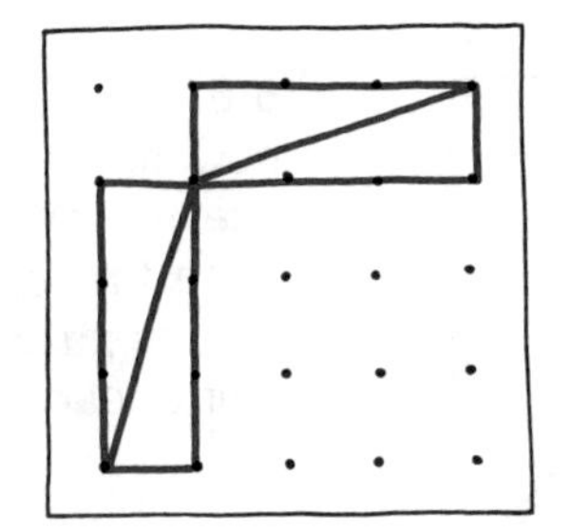

6-3. Triangle *ABC* on the geoboard at the left is a right triangle. How does this diagram demonstrate the Pythagorean relationship?

How do the shapes made on the geoboards at the left demonstrate the Pythagorean relationship?

6-4. Can the following shapes be made on a 5×5 geoboard?

(*a*) Rhombus (one that is not a square)
(*b*) Equilateral triangle
(*c*) Regular hexagon
(*d*) Isosceles trapezoid
(*e*) Octagon

Draw the shapes you can make on dot paper. Explain why certain shapes cannot be made.

6-5. The Pythagorean relationship can help you solve these problems:

(*a*) A square, an equilateral triangle, and a regular hexagon have the same perimeter of 24 cm. What are the areas of these shapes? Which has the greatest area?
(*b*) Construct a square on each side of a regular hexagon as in the diagram at the left. By joining the adjacent outer vertices of these squares, form a regular dodecagon. If the length of one side of the original hexagon is 10 cm, what is the area of the dodecagon?

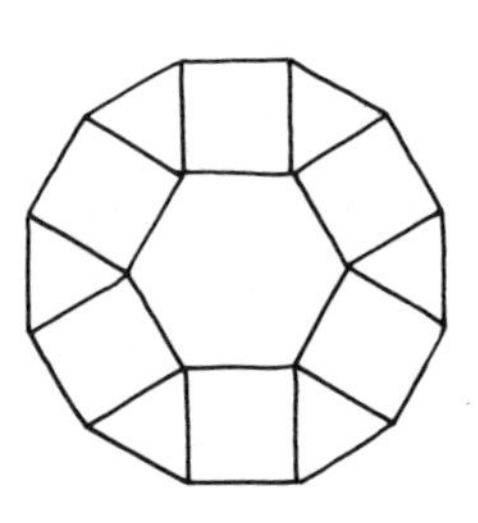

(*c*) A rhombus has a perimeter of 100 in. and the length of one of its diagonals is 12 in. Can you find its area?

(*d*) The lengths of the bases of an isosceles trapezoid are 12 in. and 4 in. and its height is 3 in. What is its perimeter? How long are its diagonals?

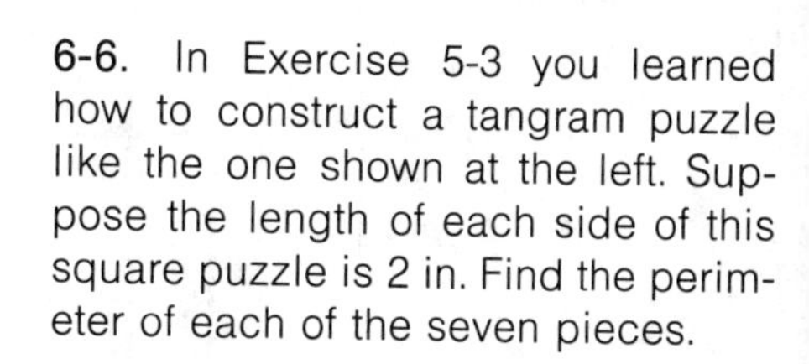

6-6. In Exercise 5-3 you learned how to construct a tangram puzzle like the one shown at the left. Suppose the length of each side of this square puzzle is 2 in. Find the perimeter of each of the seven pieces.

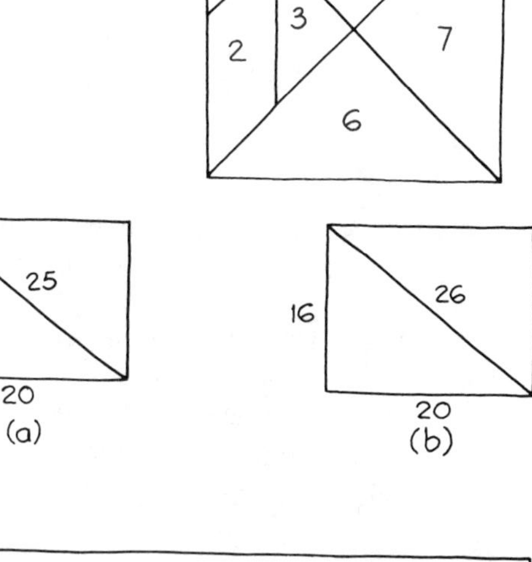

6-7. The Pythagorean relationship is useful in surveying. For example, which of the diagrams at the left represents a rectangular piece of land? How do you know this?

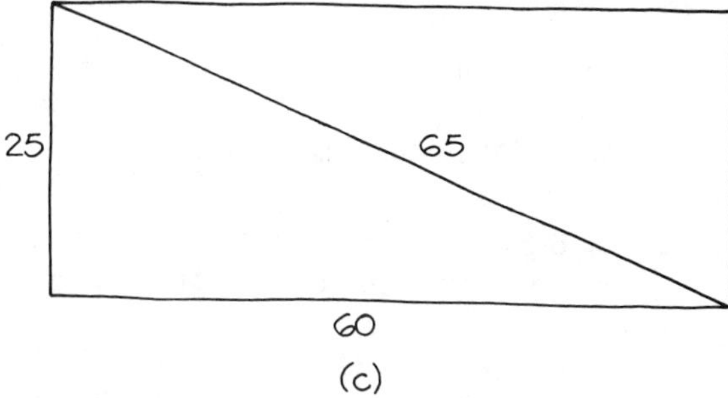

7 Congruence and Similarity

Matching Shapes

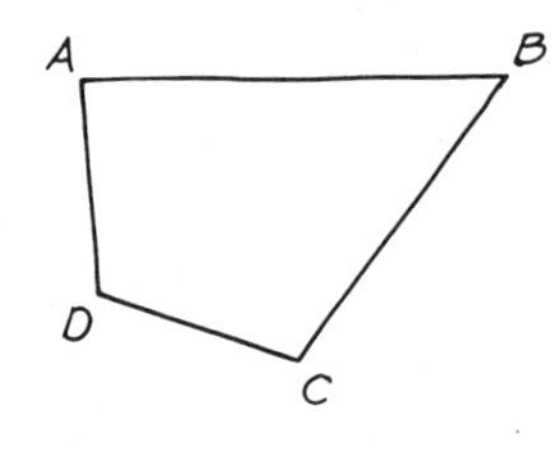

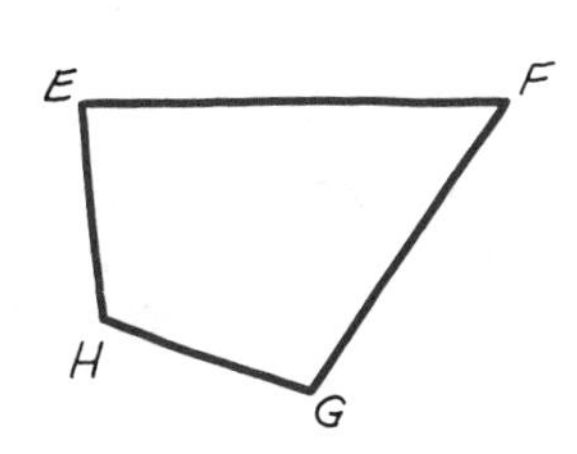

Figure 7-1

In the preceding chapters we came across certain figures we described as being just alike or exactly alike, such as quadrilaterals *ABCD* and *EFGH* in Figure 7-1. What we meant was that the figures have the *same shape and size*. You can check this by drawing quadrilateral *EFGH* on a piece of tracing paper and then placing the traced figure over *ABCD* (Figure 7-2) so that vertex *E* matches with *A*, vertex *F* with *B*, and so on. The two quadrilaterals coincide or match completely. Notice that if you attempt to match vertex *A* with *F*, vertex *B* with *G*, or some other matching, you cannot make the quadrilaterals coincide. Why not?

You might respond that by just looking at the shape of the quadrilaterals you can tell you have to match *E* with *A*, *F* with *B*, and so on to make them coincide. Vertices *A*, *B*, *C*, and *D* are in the same order in the first quadrilateral as are *E*, *F*, *G*, and *H* in the

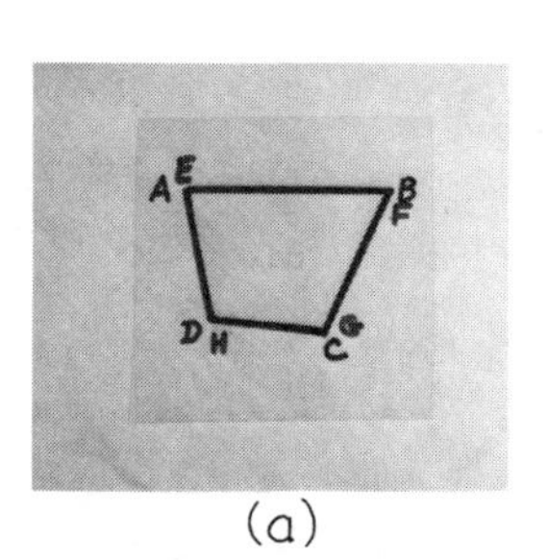

(a)

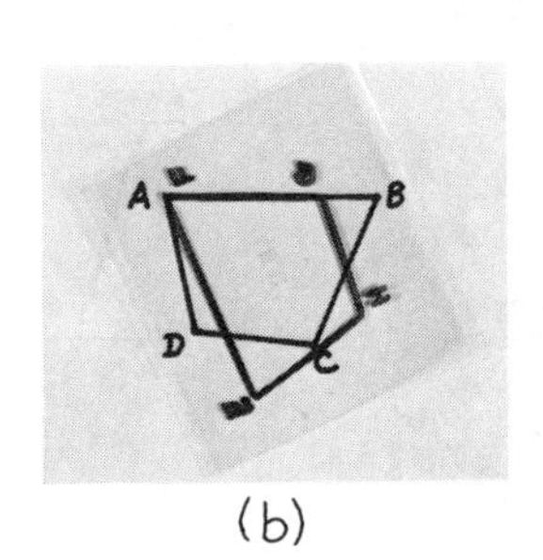

(b)

Figure 7-2

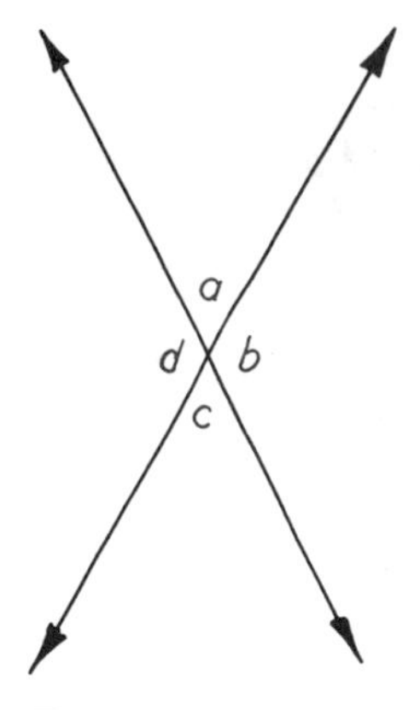

Figure 7-3

second. These pairs of vertices are called **corresponding vertices.** Similarly, the sides in the same order are called **corresponding sides.** Sides $\overline{AB}$ and $\overline{EF}$ are corresponding sides, as are $\overline{BC}$ and $\overline{FG}$, $\overline{CD}$ and $\overline{GH}$, and $\overline{DA}$ and $\overline{HE}$.

When two figures have the same shape and size, they are called **congruent figures.** As we just saw, if two figures are congruent, you can show that they coincide or match completely. This word *congruent* is not a new one for us. We have been using it with line segments and angles. We found that two line segments are congruent when they have the same measure. **Similarly, two angles are congruent when they have the same measure.** But since the sides of an angle are rays, the measure of an angle refers to the size of its opening. We found that in a parallelogram the opposite angles are congruent. This also applies to all rectangles, rhombuses, and squares since they are parallelograms.

If you draw two intersecting lines in a plane, you will form four angles. In Figure 7-3 the angles are labeled *a, b, c,* and *d.* Note that *a* and *c* are congruent and *b* and *d* are congruent. Angles *a* and *c* are called a **pair of vertical angles.** So *b* and *d* are also a pair of vertical angles. Since the sum of the measures of $\angle a$ and $\angle b$ equals the measure of a straight angle, *a* and *b* are supplementary angles. Can you find other pairs of supplementary angles?

Two sticks can be used as a model for intersecting lines and vertical angles. As you can see in Figure 7-4, in each case **the angles in a pair of vertical angles are congruent.** When the intersecting lines are perpendicular, all four angles are congruent; each of them is a right angle [Figure 7-4(c)].

If two lines in a plane do not meet, no angles are formed. The two lines are parallel. Let us draw two parallel lines and a third line (in brown) that intersects them (Figure 7-5). This third line is called a **transversal.** It is interesting to consider the angles that are formed.

As you can see in Figure 7-5, eight angles are formed. Besides the vertical angles being congruent, there are other pairs of congruent angles. Suppose you were to slide line *m* down the transversal toward line *n* so that *m* is always parallel to *n.* When *m* reaches *n* and the two lines match or coincide, $\angle a$ will coincide

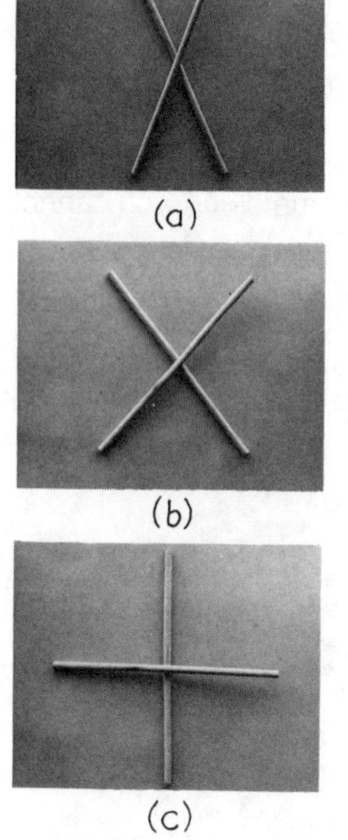

Figure 7-4

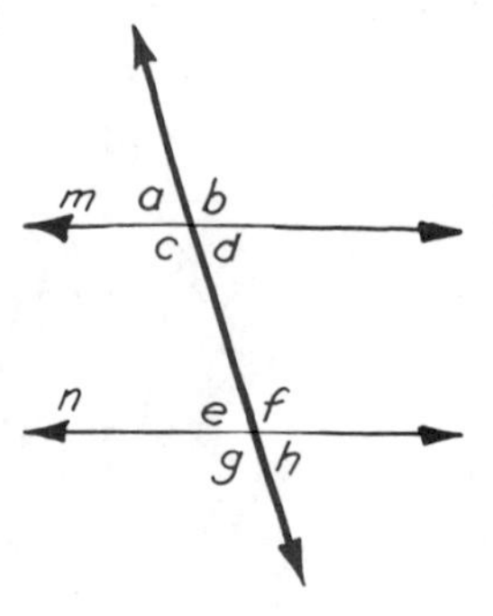

Figure 7-5

with $\angle e$, $\angle d$ will coincide with $\angle h$, and so on. You can use tracing paper to verify this. *Angles a and e are found to be congruent; they are called* **corresponding angles.** Similarly, the other pairs of angles forming corresponding angles are congruent—*d* and *h*, *b* and *f*, and *c* and *g*.

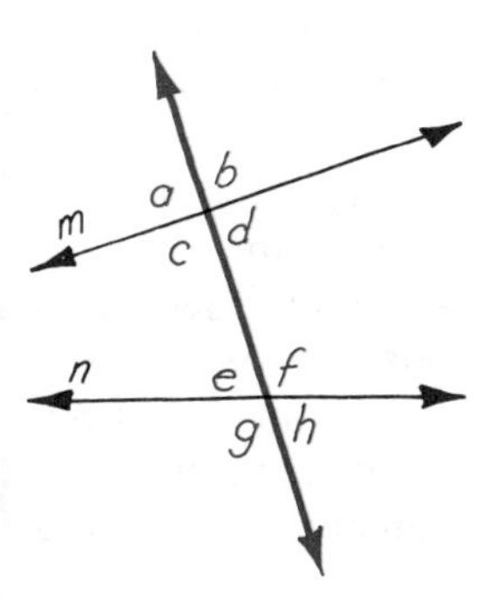

Figure 7-6

But we can find still other pairs of angles that are congruent. For example, *c* and *f* are congruent. How do we know this? Well, we know that *c* and *b* are congruent (vertical angles) and *b* and *f* are congruent (corresponding angles), so *c* and *f* must also be congruent. Since both *c* and *f* are congruent to *b*, they must be congruent to each other. Angles *c* and *f* are called **alternate interior angles.** Another pair of alternate interior angles is *d* and *e*. Can you see why *d* and *e* must be congruent?

If we begin with two lines that are not parallel and a third intersecting line (transversal), then neither corresponding angles nor alternate interior angles are congruent. In Figure 7-6, $\angle a$ and $\angle e$ are not congruent nor are $\angle d$ and $\angle e$. The two original lines must be parallel if we want corresponding angles and alternate interior angles to be congruent.

Congruent Polygons

It is easy to construct an angle congruent to another angle. All we have to do is be certain the two angles have the same measure. But suppose we have a polygon and want to form another polygon just like it (that is, a *congruent polygon*). How can we do it?

Your first response might be to measure the sides of the polygon and form a new polygon so that its corresponding sides are congruent. This seems reasonable, but it does not guarantee that we will form a congruent polygon.

Consider rectangle *ABCD* formed by strips in Figure 7-7. When the rectangle is transformed into the parallelogram *A′B′CD*, we find that the corresponding sides of the original rectangle and parallelogram are equal in measure (that is, congruent), but the figures are *not* congruent. Why not? You can see that the angles are different in the two figures. Angles *A* and *A′* are *not* congruent.

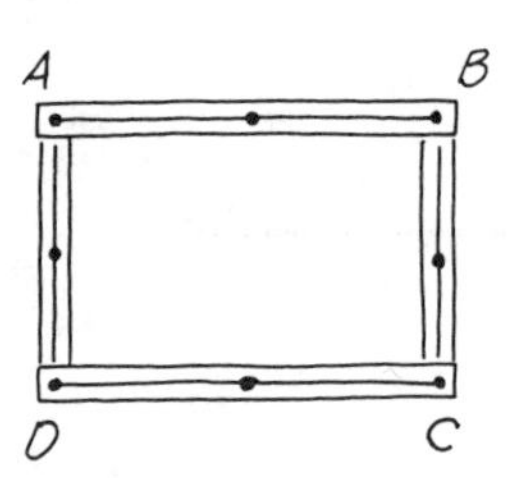

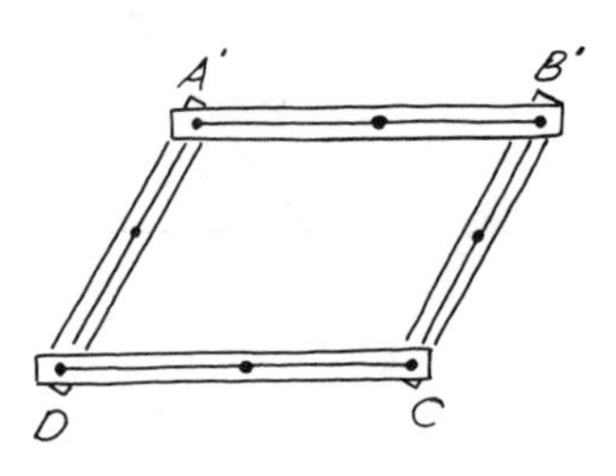

Figure 7-7

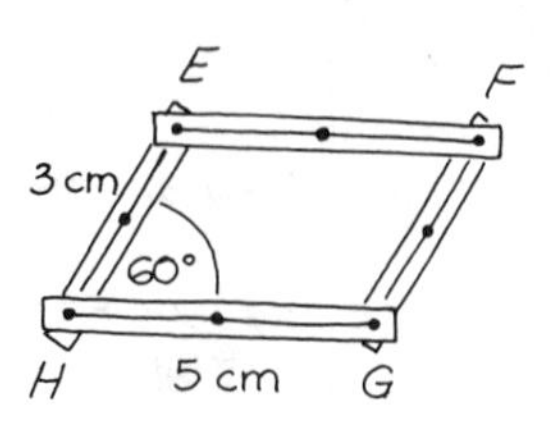

Figure 7-8

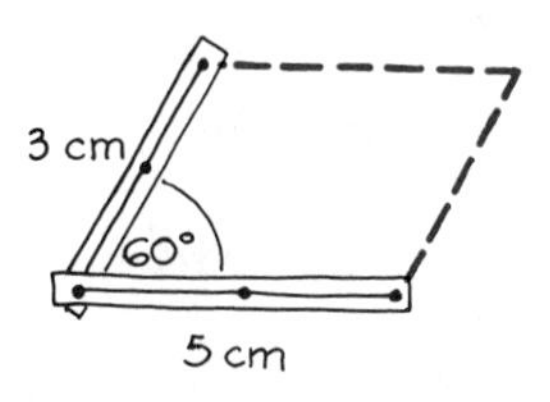

Figure 7-9

This indicates that to form a polygon congruent to a given polygon, we need to know the measures of the sides *and* the measures of the angles of the original polygon.

Actually, in the case of a parallelogram you do not have to measure *all* the angles and sides to form a congruent parallelogram. If we want to make a parallelogram like *EFGH* in Figure 7-8, we really only have to measure two adjacent sides and the angle they form. If you take a strip congruent to $\overline{HG}$ and another strip congruent to $\overline{EH}$ and join them so that they form an angle congruent to $\angle H$ (Figure 7-9), the remainder of the new parallelogram is fixed since the opposite sides must be parallel and congruent. So to construct a parallelogram congruent to a given parallelogram, you must know *at least* the measure of two adjacent sides of the original parallelogram and the measure of the angle formed by those sides.

Congruent Triangles

What do we have to know about a triangle in order to construct a congruent triangle? If we know the measures of the sides and the angles, we can certainly form a congruent triangle. But what is the *least* we need to know about the original triangle?

Suppose you draw a triangle like the one in Figure 7-10 and other people want to draw a triangle exactly like yours, that is, congruent to yours. Pretend you are describing your triangle to these people on the telephone. You give each of them different bits of information about your triangle. Let us see which ones would be able to construct a triangle definitely congruent to yours (measurements are approximate).

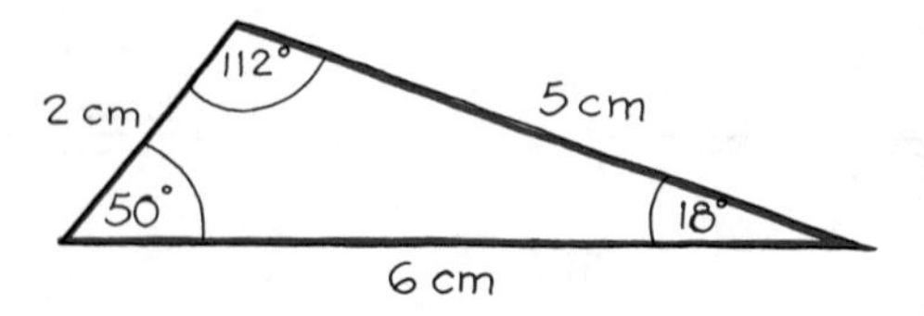

Figure 7-10

Each person is going to attempt to make a triangle congruent to this one

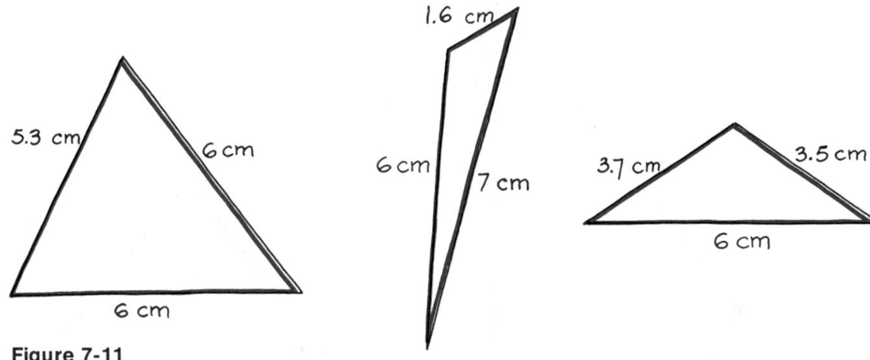

Figure 7-11

Given: S

Person 1

Information given—S: The length of one side is 6 cm.
Result: This person can make any number of triangles with a 6-cm side (Figure 7-11). More information is needed.

Person 2

Information given—SS: The length of one side is 6 cm. Another side is 5 cm long.
Result: This person can make several triangles with two sides having these lengths (Figure 7-12). More information is needed.

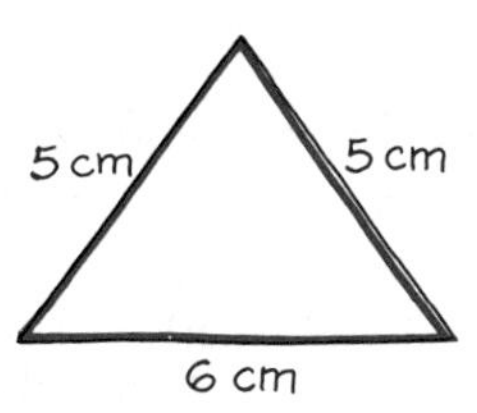

Figure 7-12

Given: SS

Person 3

Information given—SSS: The lengths of the three sides are 6, 5, and 2 cm.
Result: This person can construct a triangle exactly like yours. In fact, Figure 7-13 shows that this person can form only one possible triangle (as we discovered in Chapter 2 when we saw that three strips may form only one possible triangle).

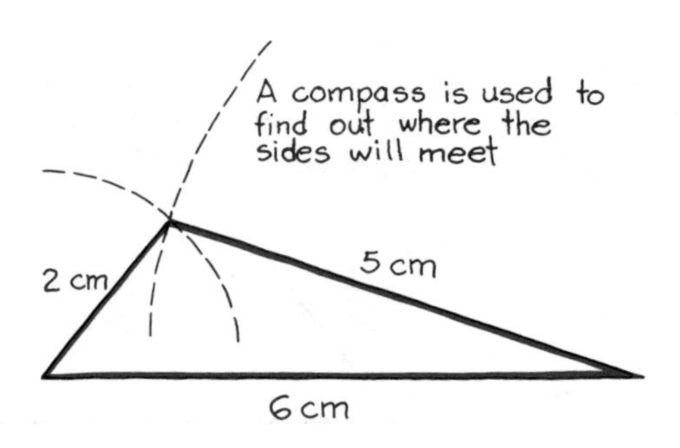

Figure 7-13

Given: SSS

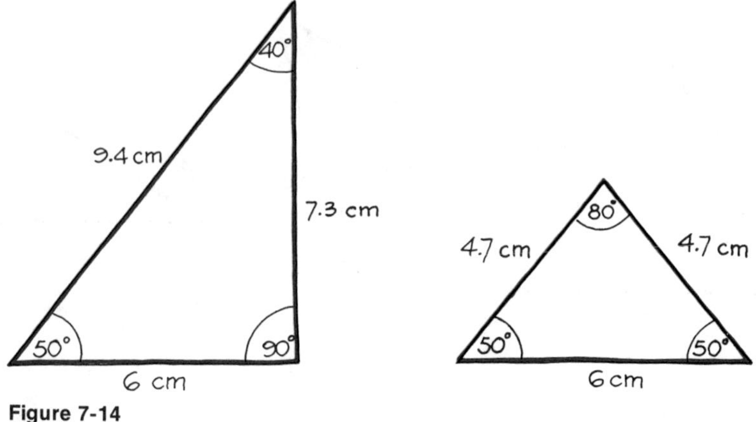

Figure 7-14

Given: SA

Person 4

Information given—SA: The size of one angle is 50°; the length of an adjacent side is 6 cm.
Result: This person can construct several triangles with a 6-cm side and a 50° angle (Figure 7-14). More information is needed.

Person 5

Information given—SAS: The lengths of two sides are 6 and 2 cm; the size of the included angle (angle formed by those two sides) is 50°.
Result: This person can construct only one triangle (Figure 7-15). Once the two sides are joined and the angle is fixed, the third side is fixed. So this person will be able to construct a triangle that is definitely congruent to yours.

Person 6

Information given—ASA: The sizes of two angles are 50 and 112°; the length of the side common to both angles (included side) is 2 cm.
Result: Only one triangle can be formed (Figure 7-16). Once the two angles and the side are drawn, the remainder of the triangle is determined. This person's triangle will be congruent to yours.

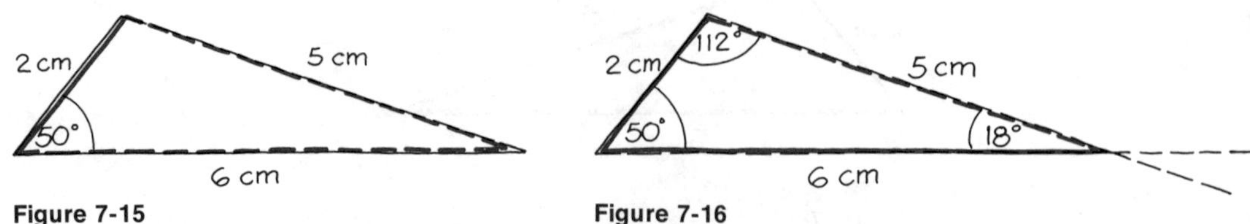

Figure 7-15

Given: SAS

Figure 7-16

Given: ASA

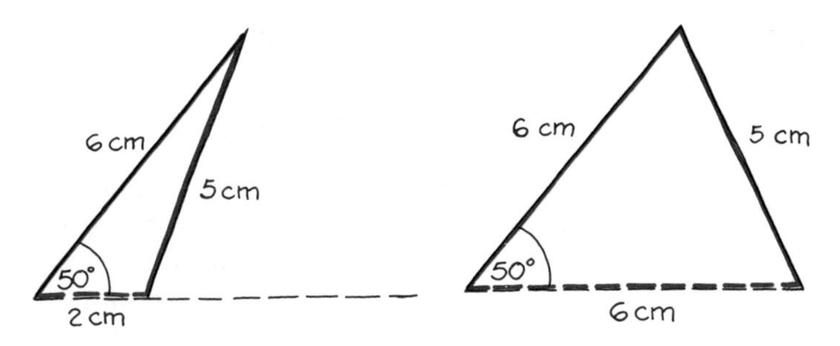

Figure 7-17

Given: SSA

Person 7

Information given—SSA: The lengths of two sides are 5 and 6 cm; one angle (not formed by these sides) is 50°.
Result: This person can form two different triangles (Figure 7-17) and cannot be certain which is congruent to yours. More information is needed.

Person 8

Information given—AAA: The sizes of the three angles of the triangle are 50, 112, and 18°.
Result: Any number of triangles can be formed with angles of these sizes (Figure 7-18). Notice that all the triangles have the same shape. This person needs more information to determine which of these triangles is congruent to yours.

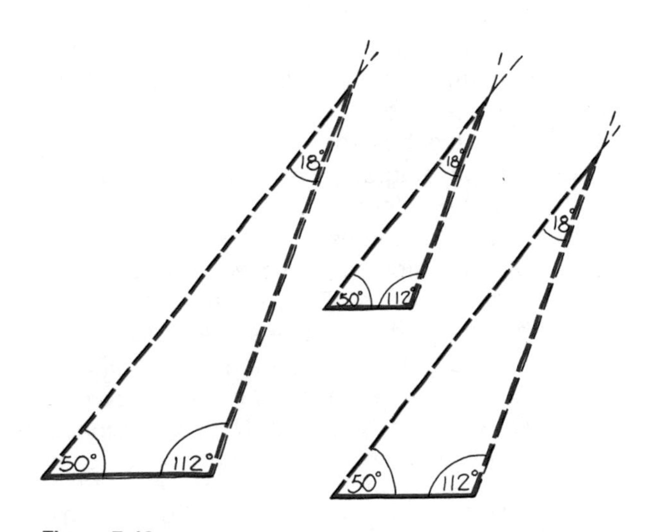

Figure 7-18

Given: AAA

As you can see from these telephone exchanges, not everyone was able to construct a triangle definitely congruent to yours. Sometimes your information led to several different triangles. At other times the information led to only one possible triangle, which was congruent to yours. What does this tell us? When we are comparing two triangles to find out whether they are congruent, it is not necessary to compare *every* pair of corresponding sides and *every* pair of corresponding angles of the two triangles. We can be *certain* the triangles are congruent:

1. If each side of one triangle is congruent to the corresponding side of the other triangle (SSS)
2. Or if two sides and the included angle of one triangle are congruent to the corresponding parts of the other triangle (SAS)
3. Or if two angles and the included side of one triangle are congruent to the corresponding parts of the other triangle (ASA)

Similarity and Ratios

So far we have been discussing figures which are congruent, that is, have the same shape *and* the same size. Sometimes figures have the same shape but not necessarily the same size. These figures are called **similar** figures. We see many examples of similar figures in everyday life—the different sizes of an item in the supermarket, the enlargement of a photograph, or a slide and its projection on the screen. When you attempt to copy a diagram from the blackboard onto your paper, you are trying to make a similar diagram; that is, the shape is the same although the size may be different.

Similar figures are interesting because they exhibit important relationships. Suppose we have two maps of the United States that are different sizes (one larger than the other). Let us focus our attention on three cities—New York, Cleveland, and Atlanta—and imagine that on each map these cities form the vertices of a triangle like the ones pictured in Figure 7-19, with vertices labeled *C, N, A* and *C′, N′, A′*. Notice that these triangles have the same shape. Let us compare corresponding measurements:

$m\angle A = 36$	$m\angle A' = 36$	$m\,\overline{CA} = 3$	$m\,\overline{C'A'} = 6$
$m\angle C = 102$	$m\angle C' = 102$	$m\,\overline{CN} = 2$	$m\,\overline{C'N'} = 4$
$m\angle N = 42$	$m\angle N' = 42$	$m\,\overline{AN} = 4$	$m\,\overline{A'N'} = 8$

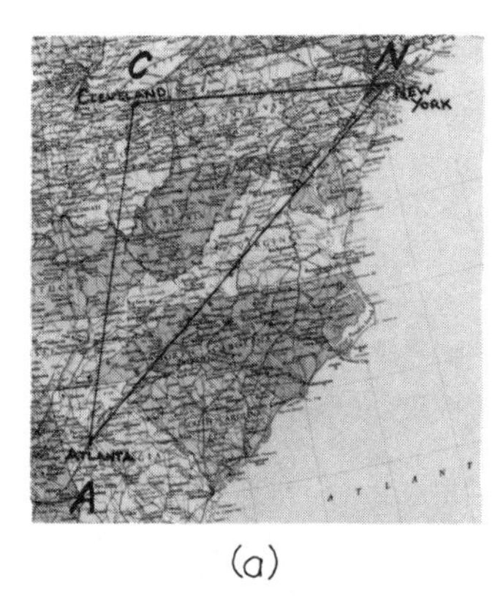

(a)

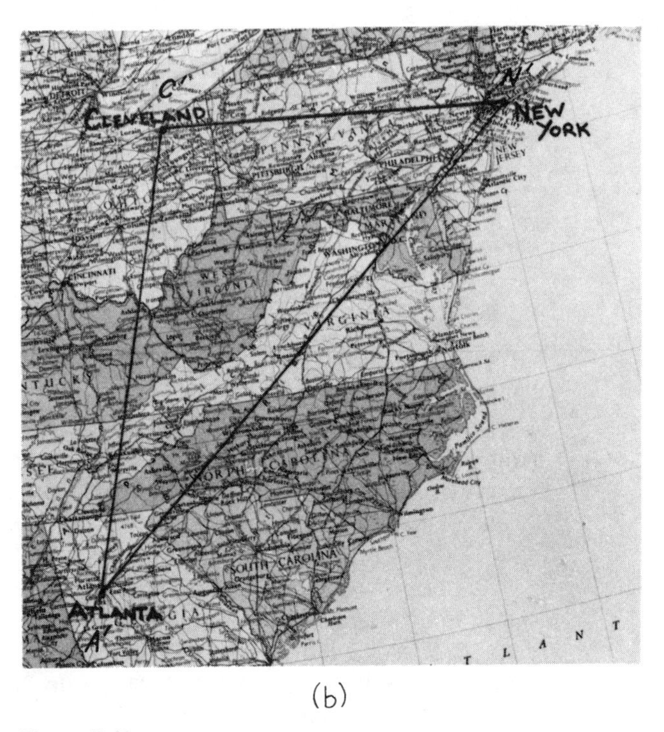

(b)

Figure 7-19

What do we find? The corresponding angles of these two similar triangles are equal in measure. What about the corresponding sides? Each side of triangle $C'N'A'$ is twice as long as the corresponding side in triangle CNA.

Mathematically, we can say that the **ratio** of the measure of a side of triangle CNA to the measure of a corresponding side of triangle $C'N'A'$ is 1 to 2. The term *ratio* simply indicates **two numbers in a definite order.** We might also have written the ratio like this: 1:2 or $1/2$. The ratios are as follows:

$$\frac{\text{Measure of side of triangle } CNA}{\text{Measure of corresponding side of triangle } C'N'A'} = \frac{1}{2}$$

$$\frac{m\,\overline{CA}}{m\,\overline{C'A'}} = \frac{3}{6} = \frac{1}{2}$$

$$\frac{m\,\overline{CN}}{m\,\overline{C'N'}} = \frac{2}{4} = \frac{1}{2}$$

$$\frac{m\,\overline{AN}}{m\,\overline{A'N'}} = \frac{4}{8} = \frac{1}{2}$$

If we change the order of the measures, we can form the ratio 2:1 or $^2/_1$, indicating the ratio of the measure of a side of triangle *C'N'A'* to the measure of a corresponding side of triangle *CNA:*

$$\frac{\text{Measure of side of triangle } C'N'A'}{\text{Measure of corresponding side of triangle } CNA} = \frac{2}{1}$$

$$\frac{m\,\overline{C'A'}}{m\,\overline{CA}} = \frac{6}{3} = \frac{2}{1}$$

$$\frac{m\,\overline{C'N'}}{m\,\overline{CN}} = \frac{4}{2} = \frac{2}{1}$$

$$\frac{m\,\overline{A'N'}}{m\,\overline{AN}} = \frac{8}{4} = \frac{2}{1}$$

As you see, a ratio can be represented as a fraction and, like a fraction, can be reduced to lowest terms (for example, $^2/_4 = {}^1/_2$). The critical thing is to know the *order* of the terms for the ratio. The ratio of the measure of a side of triangle *CNA* to the measure of a corresponding side of triangle *C'N'A'* ($^1/_2$) is *different* from the ratio of the measure of a side of triangle *C'N'A'* to the measure of a corresponding side of triangle *CNA* ($^2/_1$).

A special relationship among similar figures is that if you take pairs of measures of corresponding sides in the same order, you get the same ratio all the time! In the first instance, when we made pairs of measures of the sides of triangle *CNA* to the corresponding measures of sides of triangle *C'N'A'*, we always got the ratio 1:2 or $^1/_2$. In the second instance, reversing the order and making pairs of measures of the sides of triangle *C'N'A'* to the corresponding sides of triangle *CNA*, we always got the ratio 2:1 or $^2/_1$.

This will prove to be a very useful bit of information. Whenever we have two similar triangles and we know the length of the sides of one triangle and the ratio of the measures of corresponding sides of these triangles, we can find the length of the sides of the second triangle. Suppose we make two similar triangles like *ABC* and *A'B'C'* so that the ratio of the measure of the sides of triangle *ABC* to the measure of the corresponding sides of *A'B'C'* is 3:1 (Figure 7-20). If we know that $m\,\overline{AB} = 6$, then $m\,\overline{A'B'}$ must be 2. If $m\,\overline{BC} = 9$, then $m\,\overline{B'C'} = 3$. And if $m\,\overline{AC} = 5.1$, then $m\,\overline{A'C'} = 1.7$.

In each case the ratio of the measure of a side of triangle *ABC* to the measure of a corresponding side of triangle *A'B'C'* is preserved:

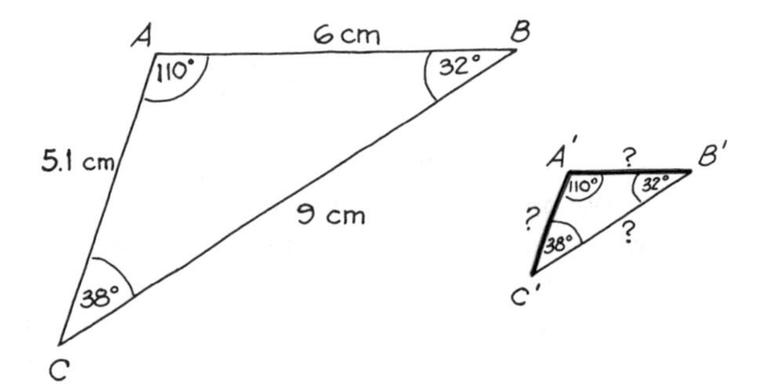

Figure 7-20

$$\frac{m\,\overline{AB}}{m\,\overline{A'B'}} = \frac{6}{2} = \frac{3}{1}$$

$$\frac{m\,\overline{BC}}{m\,\overline{B'C'}} = \frac{9}{3} = \frac{3}{1}$$

$$\frac{m\,\overline{AC}}{m\,\overline{A'C'}} = \frac{5.1}{1.7} = \frac{3}{1}$$

Each ratio has been expressed in fraction form

$$\frac{\text{Numerator}}{\text{Denominator}}$$

and is seen to be 3:1 when reduced to lowest terms.

Forming Other Ratios

Let us take another look at our similar triangles *CNA* and *C′N′A′* (Figure 7-19). We might also form ratios *within* each triangle by forming the ratio of the measure of one side to the measure of another side of that same triangle. For example, the ratio of the measure of $\overline{CA}$ to the measure of $\overline{AN}$ is 3:4 or $3/4$. What is the ratio of the measures of the corresponding sides of triangle *C′N′A′*, that is, m $\overline{C'A'}$ to m $\overline{A'N'}$? That ratio is 6:8 or $6/8$. But we have two ratios that are equivalent:

$$\frac{3}{4} = \frac{6}{8}$$

This is called a **proportion.** Two equivalent ratios form a proportion. Can we find other proportions? Certainly! How about the following:

$$\frac{m\,\overline{CA}}{m\,\overline{CN}} = \frac{3}{2} \qquad \frac{m\,\overline{C'A'}}{m\,\overline{C'N'}} = \frac{6}{4} \qquad \text{and} \qquad \frac{3}{2} = \frac{6}{4}$$

or

$$\frac{m\,\overline{AN}}{m\,\overline{CN}} = \frac{4}{2} \qquad \frac{m\,\overline{A'N'}}{m\,\overline{C'N'}} = \frac{8}{4} \qquad \text{and} \qquad \frac{4}{2} = \frac{8}{4}$$

Can you find other pairs of equivalent ratios?

We have seen that when two triangles are similar, the measures of their corresponding sides are in proportion. That is, if we form a ratio of the measures of two sides of one triangle, that ratio will be equivalent to the ratio formed by the measures of the corresponding two sides of the other triangle in the same order. Suppose we have two similar triangles like triangles *DEF* and *D′E′F′* in Figure 7-21. If we know the measures of the three sides of one triangle and the measure of just one side of the other, we can find the measures of all the sides of that other triangle. Since triangle *DEF* is similar to triangle *D′E′F′*,

$$\frac{m\,\overline{DE}}{m\,\overline{DF}} = \frac{m\,\overline{D'E'}}{m\,\overline{D'F'}}$$

$$\frac{4}{5} = \frac{m\,\overline{D'E'}}{10}$$

$$m\,\overline{D'E'} = 8 \qquad \text{since } \frac{4}{5} = \frac{8}{10}$$

We might have solved this problem another way. Previously, we discovered that the ratio of the measures of corresponding sides of similar figures remains the same. In this case we find that the ratio

$$\frac{m\,\overline{DF}}{m\,\overline{D'F'}} = \frac{5}{10} = \frac{1}{2}$$

This means that the ratio

$$\frac{m\,\overline{DE}}{m\,\overline{D'E'}} = \frac{1}{2}$$

If $m\,\overline{DE} = 4$, then $m\,\overline{D'E'} = 8$ since

$$\frac{m\,\overline{DE}}{m\,\overline{D'E'}} = \frac{4}{8} = \frac{1}{2}$$

Can you find the measure of $\overline{E'F'}$?

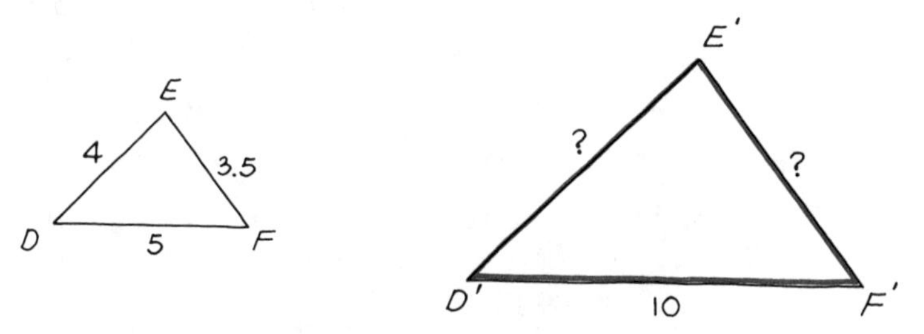

Figure 7-21

Other Similar Polygons

So far we have been considering the relationships that exist among similar triangles. Let us take a look at similar polygons with more than three sides. Going back to our different-sized maps of the United States, we can choose five cities to be vertices of a polygon. Using New York (N), Detroit (D), Omaha (O), Dallas (S), and Atlanta (A) as vertices of polygons, we get the two pentagons (polygons with five sides) shown in Figure 7-22. If we measure the corresponding angles we find that they are equal in measure:

$$\begin{aligned} m\angle O &= m\angle O' = 105 \\ m\angle D &= m\angle D' = 164 \\ m\angle N &= m\angle N' = 55 \\ m\angle A &= m\angle A' = 137 \\ m\angle S &= m\angle S' = 79 \end{aligned}$$

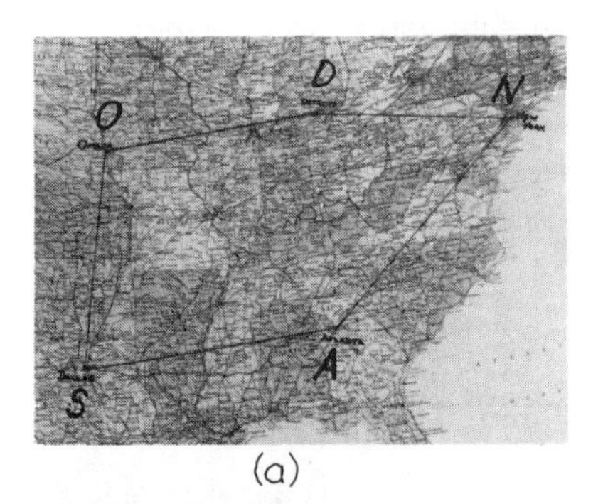

(a)

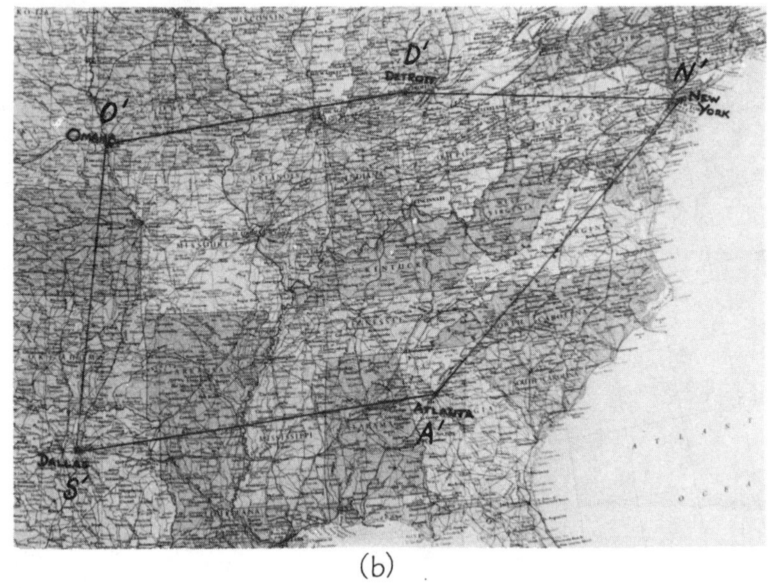

(b)

Figure 7-22

What about corresponding sides?

$$\begin{array}{ll} m\,\overline{OD} = 7 & m\,\overline{O'D'} = 14 \\ m\,\overline{DN} = 5 & m\,\overline{D'N'} = 10 \\ m\,\overline{NA} = 8 & m\,\overline{N'A'} = 16 \\ m\,\overline{AS} = 7.5 & m\,\overline{A'S'} = 15 \\ m\,\overline{SO} = 6 & m\,\overline{S'O'} = 12 \end{array}$$

Again the ratio formed by the measures of corresponding sides of the two figures remains the same:

$$\frac{m\,\overline{OD}}{m\,\overline{O'D'}} = \frac{7}{14} = \frac{1}{2} \qquad \frac{m\,\overline{DN}}{m\,\overline{D'N'}} = \frac{5}{10} = \frac{1}{2} \qquad \text{and so on}$$

We can conclude that the two pentagons are similar, since corresponding angles have the same measure and the measures of corresponding sides form the same ratio.

Suppose we have two polygons with corresponding angles equal in measure *and* corresponding sides equal in measure. Then the two polygons are congruent. Are they similar? Yes! The requirements for similarity are satisfied: corresponding angles have the same measure, and the measures of corresponding sides form the same ratio, in this case 1:1. We can say that if two polygons are congruent, they must be similar. Yet if two polygons are similar, they are not necessarily congruent (since their corresponding sides are not necessarily equal in measure).

Forming Similar Triangles

In the previous sections we saw that to show two polygons are similar we must demonstrate that corresponding angles have the same measure *and* the measures of corresponding sides form the same ratio. Suppose we know only that the corresponding angles of two polygons are equal in measure. Can we be certain the polygons are similar? Look at the rectangle and square in Figure 7-23. Corresponding angles are equal in measure (90°)

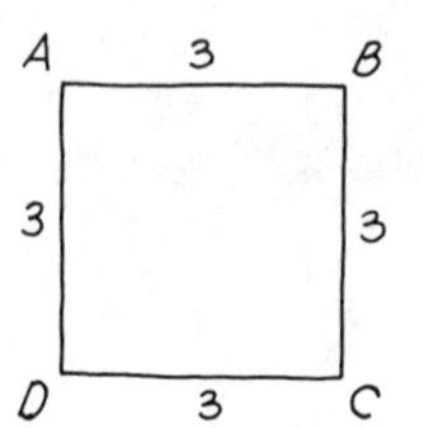

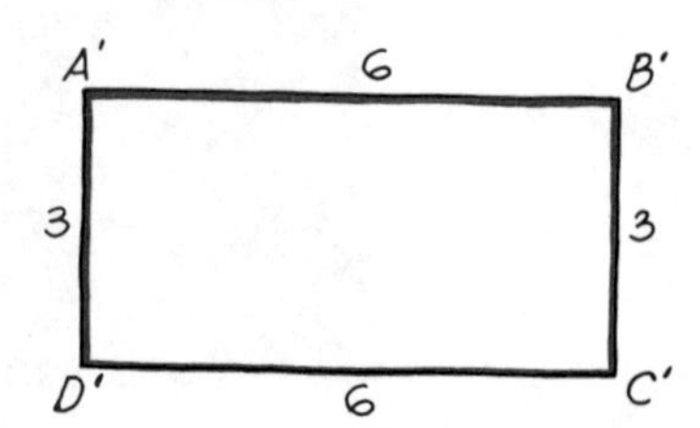

Figure 7-23

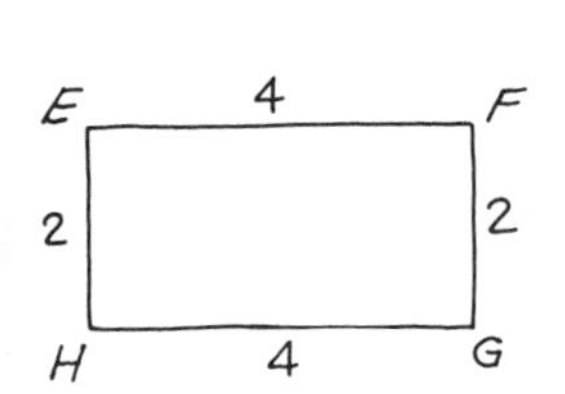

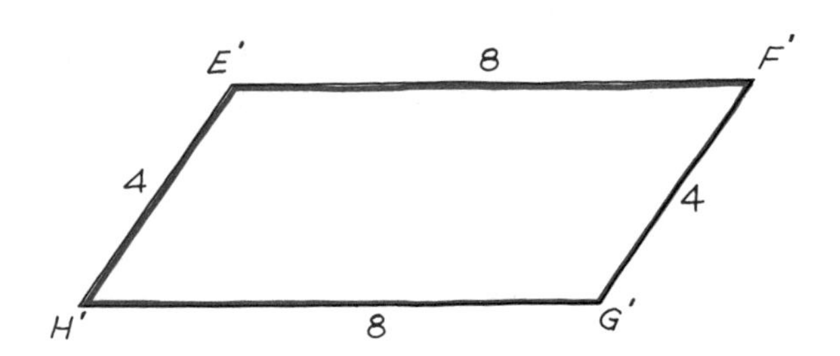

Figure 7-24

but the measures of corresponding sides do not form the same ratio:

$$\frac{m\,\overline{AD}}{m\,\overline{A'D'}} = \frac{3}{3} \qquad \text{or } \frac{1}{1}$$

while

$$\frac{m\,\overline{DC}}{m\,\overline{D'C'}} = \frac{3}{6} \qquad \text{or } \frac{1}{2}$$

Two polygons can have corresponding angles equal in measure and *not* be similar.

Now let us imagine we have two polygons and the measures of corresponding sides form the same ratio. Are the two polygons similar? Not always. Look at the parallelogram and rectangle in Figure 7-24. The measures of corresponding sides form the same ratio, 1:2 or ½:

$$\frac{m\,\overline{EH}}{m\,\overline{E'H'}} = \frac{2}{4} = \frac{1}{2} \qquad \frac{m\,\overline{EF}}{m\,\overline{E'F'}} = \frac{4}{8} = \frac{1}{2} \qquad \text{and so on}$$

But the corresponding angles do not have the same measure ($\angle E$ is smaller than $\angle E'$). The polygons do not have the same shape. They are not similar.

In general, if we want to show that two polygons are similar we must satisfy both requirements. The corresponding angles must be equal in measure *and* the measures of corresponding sides must form the same ratio.

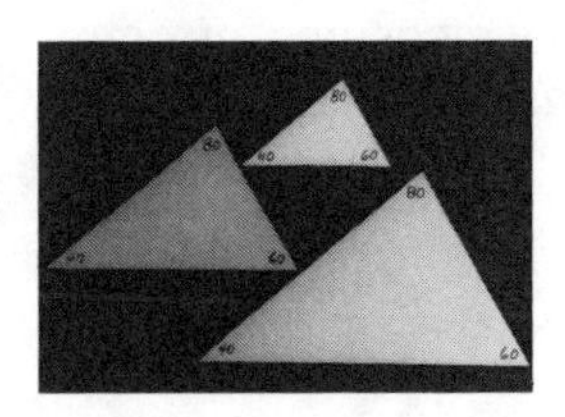

Figure 7-25

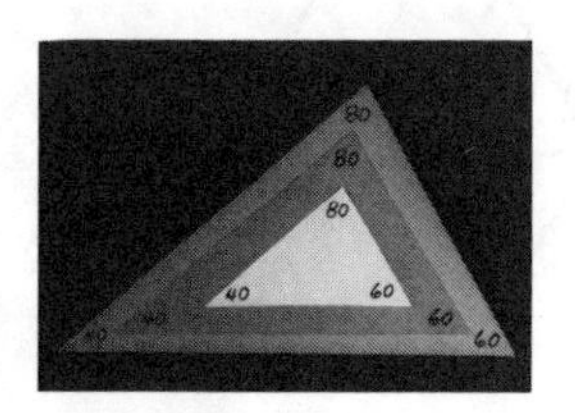

Figure 7-26

It is only when the polygons are triangles that we do not have to show that *both* requirements have been met. Let us discover why. Make several cardboard triangles with corresponding angles equal in measure like those in Figure 7-25. You will find that every triangle in your set has the same shape. The triangles are similar. You can see this more clearly if you stack the triangular shapes one on top of another as in Figure 7-26. This shows us that if triangles have corresponding angles equal in measure, the triangles are similar. In a like manner, if we make a set of triangles so that the measures of corresponding sides have the same ratio, we will find that the corresponding angles have the same measure and the triangles are similar. This tells us that to show two triangles are similar it is only necessary to show *either* that corresponding angles are equal in measure *or* that the measures of corresponding sides form the same ratio. In the case of triangles, whenever one of these requirements is satisfied, the other requirement is also satisfied.

Ratio of the Measures of the Areas within Similar Polygons

Before you read on, make many congruent square shapes and equilateral triangle shapes of cardboard. Form the figures pictured in Figure 7-27 with your cardboard shapes. These shapes can represent pairs of similar polygons (brown outline).

In (a) we have two squares. The ratio of the measure of corresponding sides is 1:2 or $1/2$. The perimeter of the smaller square is 4 units; the perimeter of the larger is 8 units. If we form a ratio of the measures of the perimeters, we obtain the same ratio, $1/2$. Let us take a look at the areas within the squares. The smaller square region has an area of 1 square unit; the larger square region has an area of 4 square units. The ratio of the measures of the corresponding areas within these squares is 1:4.

Now let us look at the rectangles in (b) and form ratios in the same way. The ratio of the measures of corresponding sides is $4/6$ or $2/3$. The ratio of the measures of corresponding perimeters is $12/18$ or $2/3$. The ratio of the measures of corresponding areas is $8/18$ or $4/9$. In the same way, we can form a set of ratios for (c). A summary of these ratios appears in Table 7-1.

As the table indicates, in each case **the ratio of the measures of corresponding perimeters of similar polygons is the same as the ratio of the measures of corresponding sides.** Do you see a relationship between the ratio of the measures of corresponding sides of a pair of similar polygons and the ratio of the

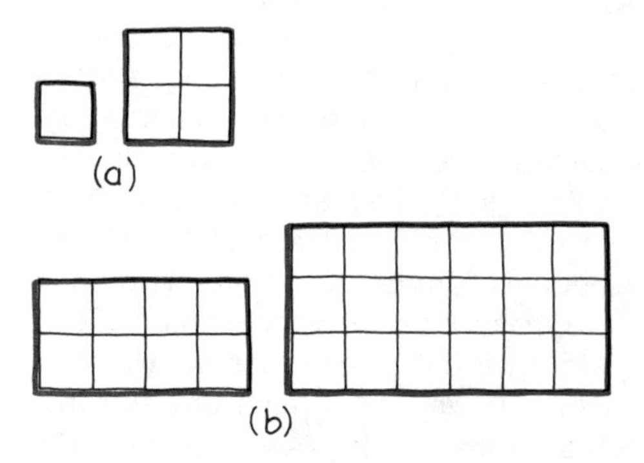

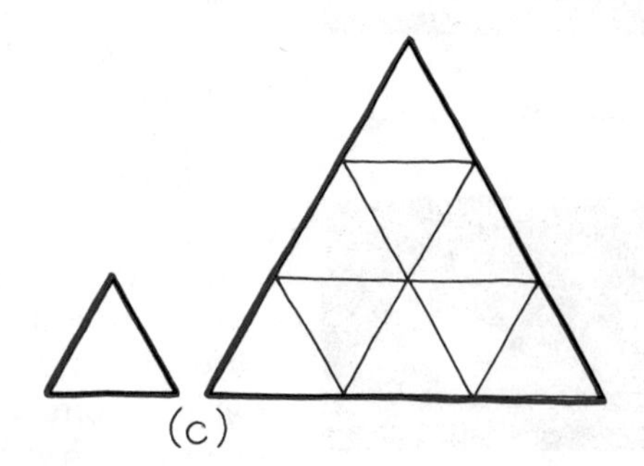

Figure 7-27

Table 7-1

Pair of similar polygons	Ratio of measures of corresponding sides	Ratio of measures of corresponding perimeters	Ratio of measures of corresponding areas
(a)	$\frac{1}{2}$	$\frac{1}{2}$	$\frac{1}{4}$
(b)	$\frac{2}{3}$	$\frac{2}{3}$	$\frac{4}{9}$
(c)	$\frac{1}{3}$	$\frac{1}{3}$	$\frac{1}{9}$

measures of the areas within those polygons? In each case the ratio of the measures of the areas can be formed by squaring the numbers that represent the measures of corresponding sides:

$$\frac{1^2}{2^2}=\frac{1}{4} \qquad \text{for pair (a)}$$

$$\frac{2^2}{3^2}=\frac{4}{9} \qquad \text{for pair (b)}$$

$$\frac{1^2}{3^2}=\frac{1}{9} \qquad \text{for pair (c)}$$

This tells us that whenever we have two similar polygons, if we know the ratio of the measures of corresponding sides to be *a/b*, then the ratio of the corresponding measures of area within these figures will be a^2/b^2.

Try making other pairs of similar polygons and finding these ratios.

More Experiences

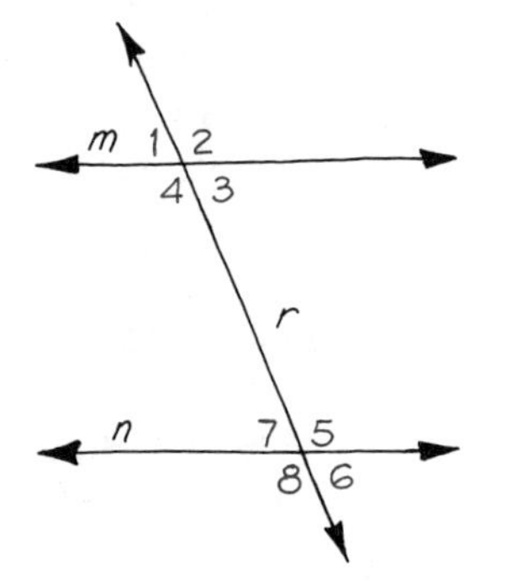

7-1. Lines *m* and *n* are parallel. Line *r* intersects these two lines to form the angles numbered 1, 2, . . . , 8 as at the left. If the measure in degrees of $\angle 1 = 69$, what are the measures of the other angles?

$m\angle 1 = 69$
$m\angle 2 =$
$m\angle 3 =$

$m\angle 4 =$
$m\angle 5 =$
$m\angle 6 =$
$m\angle 7 =$
$m\angle 8 =$

Draw two other parallel lines and a third line that intersects them. Measure each angle formed as done previously.

What angles are congruent when two parallel lines are drawn and a third line intersects them?

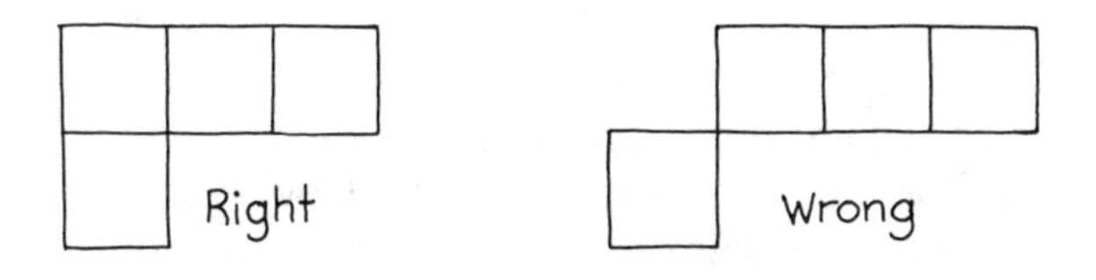

7-2. Make some cardboard squares that have sides 2 cm in length. Arrange four of these squares so that they form a polygonal region, making certain that *at least* one entire side always touches an entire side of another square.

(*a*) How many *different* ways can you arrange the four squares? Draw your arrangements. Are they all different? Or are some congruent? How might you test to see if any are congruent?

(*b*) Each of the polygonal regions formed has the same area. How do you know this must be true? Do they have the same perimeter?

(*c*) Try to make polygonal regions in the same way by using *five* cardboard squares. How many *different* regions can you find? Check to be certain that none of your regions are congruent. Can you find 12 different regions? Which has the least perimeter? Why?

(*d*) What would your results be if you used six cardboard squares?

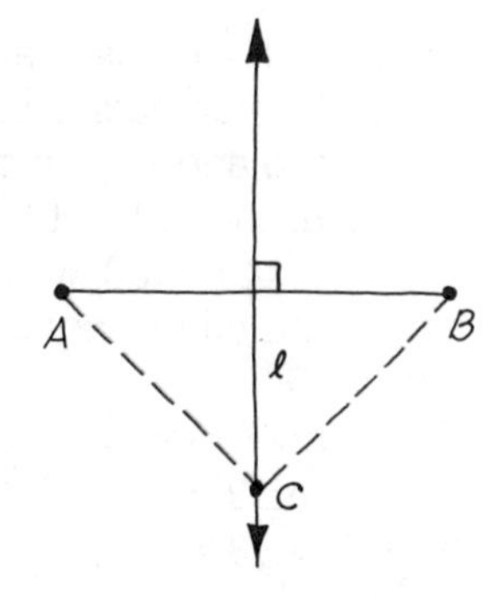

7-3. Suppose that line ℓ is the perpendicular bisector of $\overline{AB}$ and that C is a point on ℓ. Can you show that C must be equidistant from A and B [that is, $m(\overline{AC}) = m(\overline{CB})$]?

Is *every* point on this perpendicular bisector equidistant from *A* and *B?* Explain.

7-4. Many triangles are shown on the geoboards below. Find as many pairs of similar triangles as you can. Which triangles are congruent?

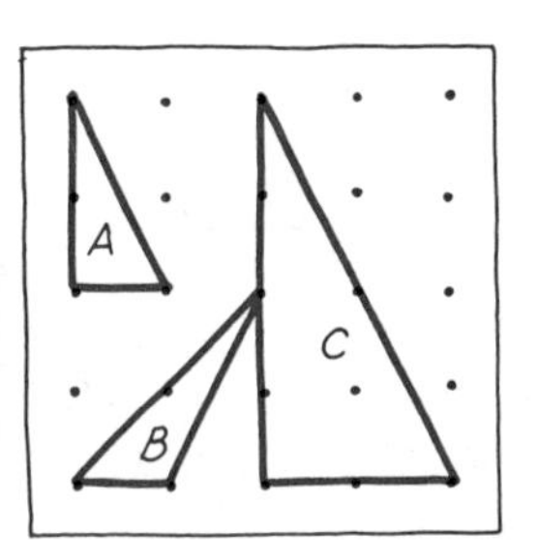

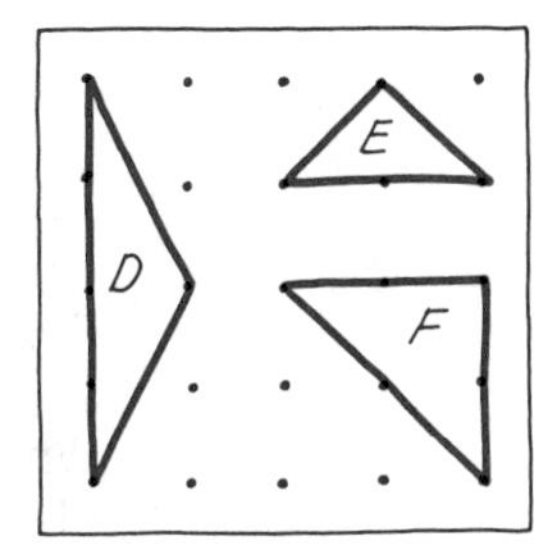

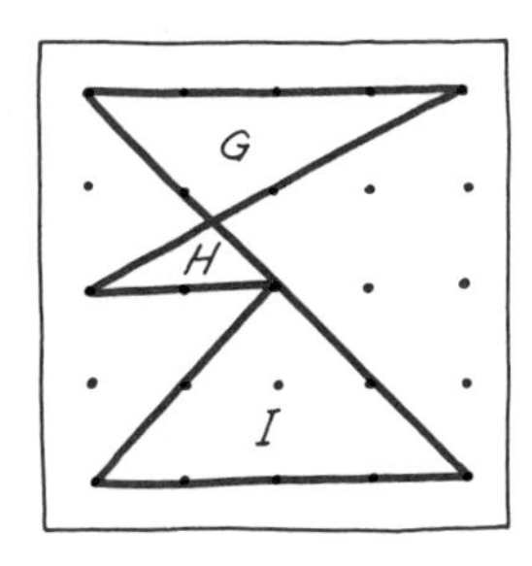

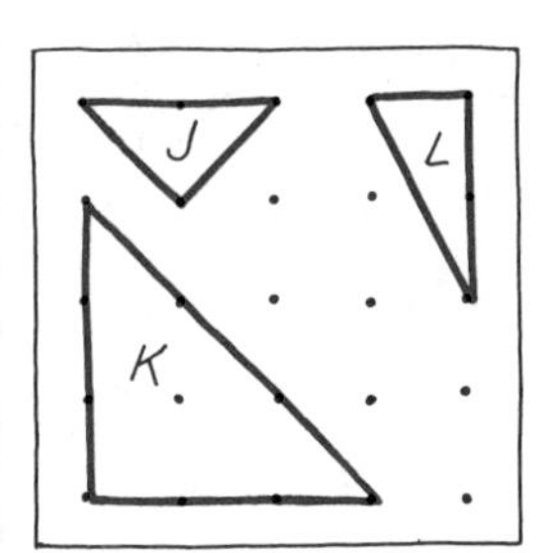

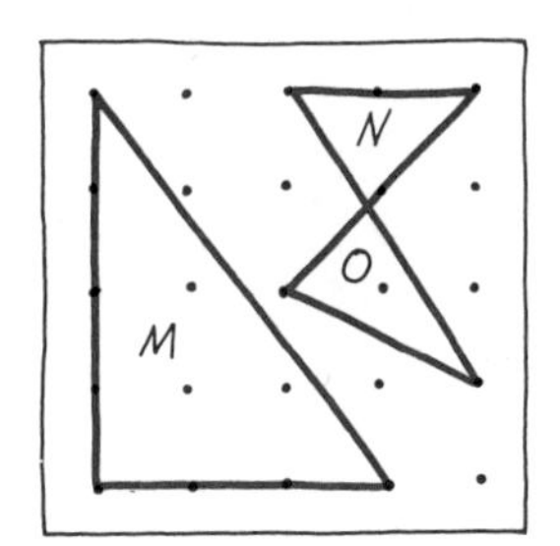

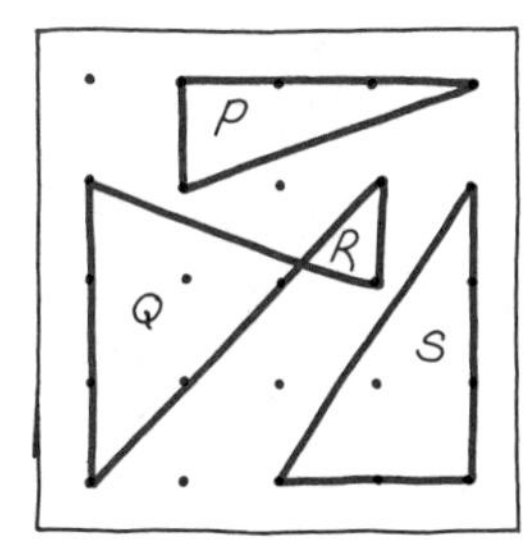

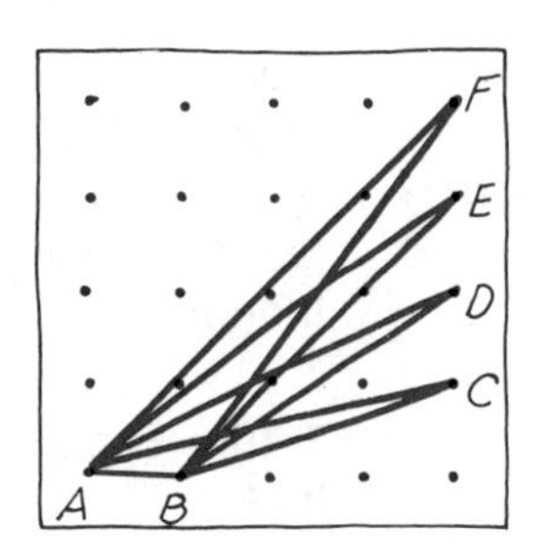

7-5. Four triangles are shown on the geoboard at the left. Each has *A* and *B* as vertices and a different third vertex (*C, D, E,* or *F*). What is the ratio of the areas of the four triangles?

7-6. Suppose two triangles are similar, with a ratio of similitude of 3:4 or 3/4. The lengths of the sides of the smaller triangle are 12, 21, and 30 cm. What are the lengths of the corresponding sides of the larger triangle?

7-7. The corresponding sides of the two parallelograms below form the same ratio. Are the two parallelograms similar? Explain.

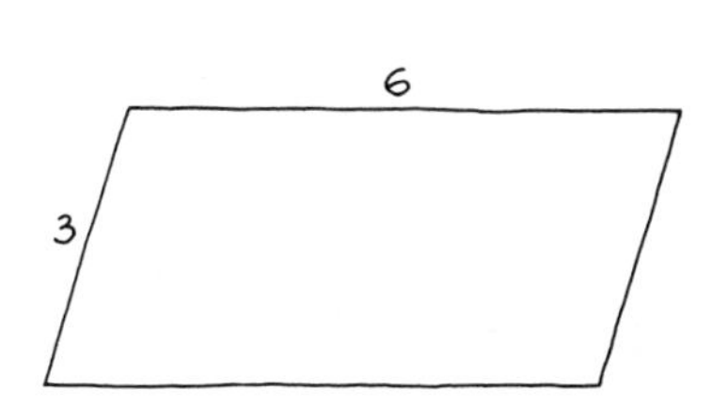

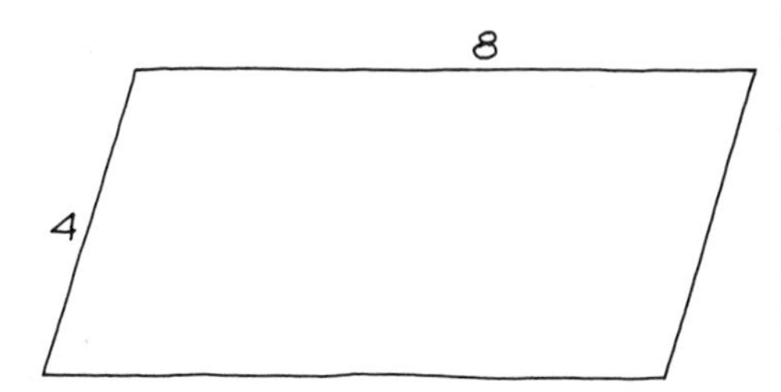

7-8. The two polygons below are similar. What is the length of $\overline{NO}$? What is the ratio of the measures of corresponding sides?

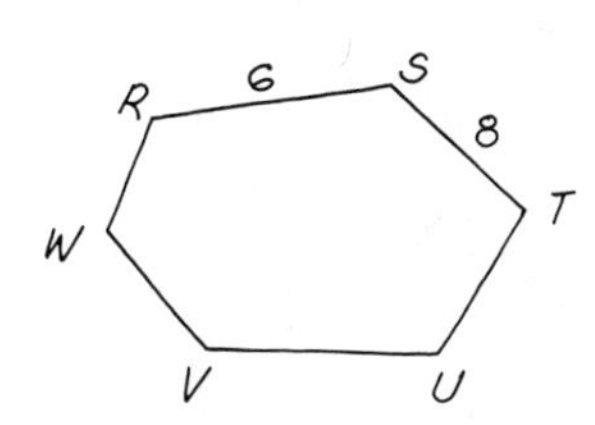

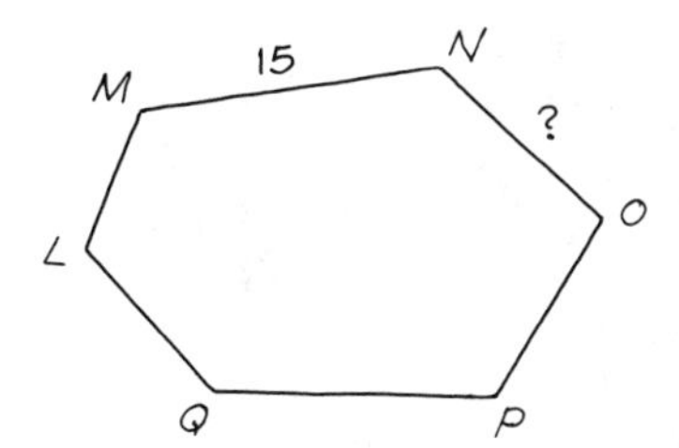

7-9. The measure of the area of one equilateral triangle is four times the measure of the area of another equilateral triangle. The measure of a side of the smaller triangle is 6 cm. What is the measure of a side of the larger triangle?

7-10. Explain why the following are true:

(*a*) If two right triangles each have one acute angle of 28°, the two triangles must be similar.
(*b*) If the vertex angles of two isosceles triangles are each 44°, the triangles must be similar.
(*c*) If two regular polygons have the same number of sides, they must be similar.

7-11. Suppose you are examining a road map that has no scale. You know, however, that the actual distance between two cities is 15 mi and they are $1\frac{1}{2}$ in. apart on the map. Can you determine the scale of the map? If the distance between two other cities is 4 in. on the map, what would be the actual distance?

7-12. A square-shaped photograph is enlarged by doubling the measure of each side. What is the area of the enlarged photograph? How does it compare to the area of the original photograph?

How might a 4×6 in. photograph be enlarged? Explain why the enlargement cannot have a square shape.

Suppose you have a square-shaped photograph you want to enlarge to twice the size of the original (in area). How can you do it? Explain.

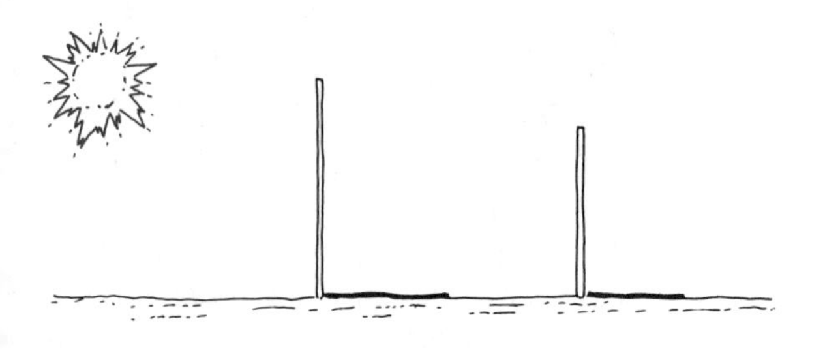

7-13. Take two dowels (of different lengths) outdoors on a sunny day and stick them into the ground near each other. Make sure the dowels form a right angle with the ground. Measure the length of the part of each dowel above the ground. Then measure the shadow formed by each dowel. Come back at different times during the day and measure the shadows again. Keep a record of your results like this:

Time	Dowel length 1	Shadow length 1	Dowel length 2	Shadow length 2	Ratio of dowel length to shadow length 1	Ratio of dowel length to shadow length 2

Examine your results. Do you see any relationships between the ratios of the length of the dowel to the shadow? Can you explain your results?

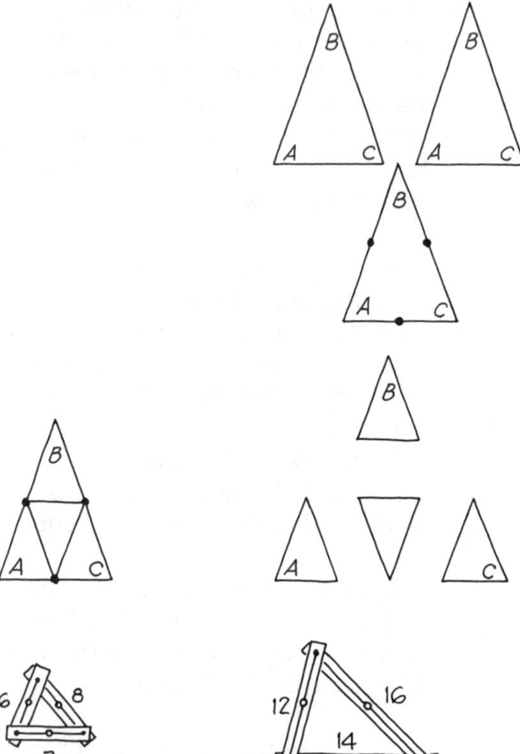

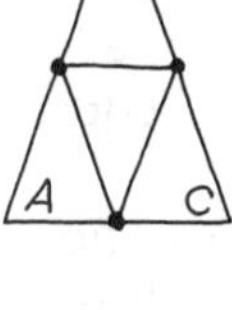

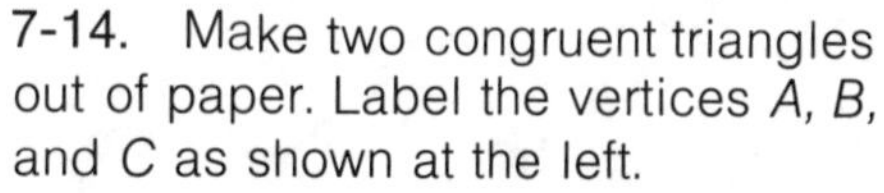

7-14. Make two congruent triangles out of paper. Label the vertices *A, B,* and *C* as shown at the left.

Take *one* paper triangle and fold each side to locate its midpoint. Draw line segments from one midpoint to the next. The interior of this triangle has been partitioned into four triangles.

Cut out these four triangles. How are they related to each other? How is any one of them related to the other paper triangle *ABC?*

Try this procedure with other pairs of congruent paper triangles. Do you always get the same results?

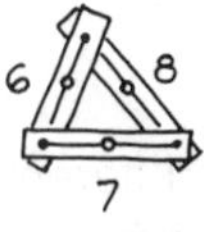

7-15. Make the pairs of triangles shown at the left by using strips and brass fasteners. Measure the angles of these triangles with a protractor. Compare your results. What do you notice about the angles of these triangles? How are the lengths of the sides related?

Make the following pairs of triangles (numbers represent lengths of sides). Which are similar triangles?

(*a*) 6, 8, 10
12, 16, 20

(*b*) 6, 6, 8
10, 10, 11

(*c*) 6, 6, 8
18, 18, 24

(*d*) 6, 7, 8
9, 10, 11

What relationships do you find between the measures of the corresponding sides?

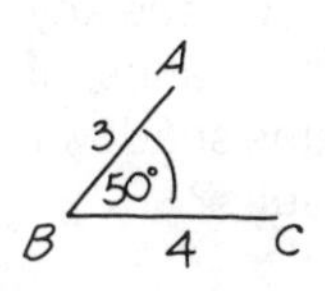

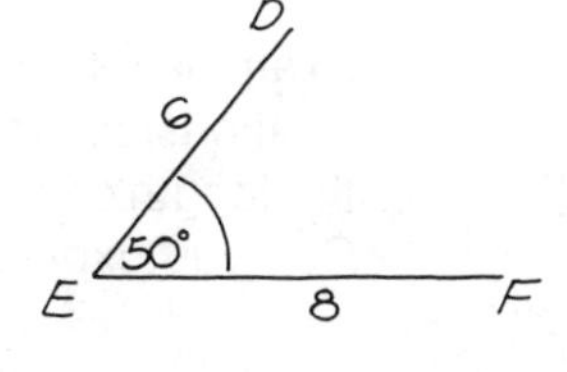

7-16. Using the strips and angle-fixers or using a protractor and ruler, make or draw the angles at the left with measures as indicated.

Angles *ABC* and *DEF* have the same measure, and the line segments along their corresponding sides have the same ratio. What is that ratio?

What are the measures of $\overline{AC}$ and $\overline{DF}$?

What is the ratio between m($\overline{AC}$) and m($\overline{DF}$)?

What is true about triangles *ABC* and *DEF*? Explain.

7-17. Make or draw the following pairs of congruent angles with lengths of line segments along their sides in proportion as indicated below:

(*a*) 3, 60°, 7
6, 60°, 14

(*b*) 4, 45°, 10
6, 45°, 15

(*c*) 4, 120°, 8
10, 120°, 20

Suppose a triangle were formed in each case. What would be the lengths of the corresponding third sides? How would the measures of those corresponding third sides be related? What is true about the triangles formed in each case?

What might you conclude about two triangles with two pairs of corresponding sides in proportion and the angles included between those sides congruent?

7-18. If each of the four sides of one quadrilateral is congruent to a corresponding side of another quadrilateral (SSSS = SSSS), will the two quadrilaterals be congruent? If SASS of one quadrilateral are congruent to the corresponding SASS of another quadrilateral, will the two quadrilaterals be congruent? Draw diagrams to illustrate your answer.

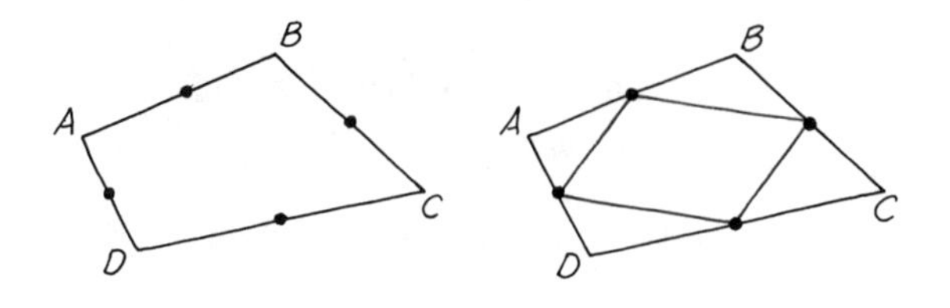

7-19. Draw a quadrilateral *ABCD* like the one at the left. Locate the midpoints of each side. Draw line segments to connect the midpoints as shown. What figure is formed by these line segments?

Draw other quadrilaterals and repeat this procedure. What figure is formed in each case? Can you explain the results?

What figure is formed when the quadrilateral is a trapezoid? A rectangle? A parallelogram? A rhombus? A square?

Circles

Introduction

If we look at the different shapes in our environment, we can find many things with a circular shape—the base of a glass, a wheel on a car, or the brim of a lampshade. Have you ever wondered *why* something has a circular shape?

People have always found the circle to be a very important shape. No one knows exactly when we began to make circular things, but one possibility is that when our ancestors noticed how a tree trunk could roll down a hill, they realized it would be easier to roll a heavy weight than to drag it. Voilà! The wheel was born!

Some people believe that the idea of rolling something is instinctive. Isn't it instinct that guides the golden beetle to round off the edges and corners of objects so it can transport them more easily from one place to another? The ancient Egyptians used to worship this insect, which in mythology is credited with the ability to make the sun move.

Figure 8-1

People found they could make clay vases by using a potter's wheel as a form (Figure 8-1). They realized that by changing the size of the wheel at different levels they could make interesting shapes. This circular shape is often found in early architecture. Circular archways have been used often, both for aesthetic reasons and because they form an extremely strong support. You can see many uses of the circular shape in the structures pictured in Figure 8-2.

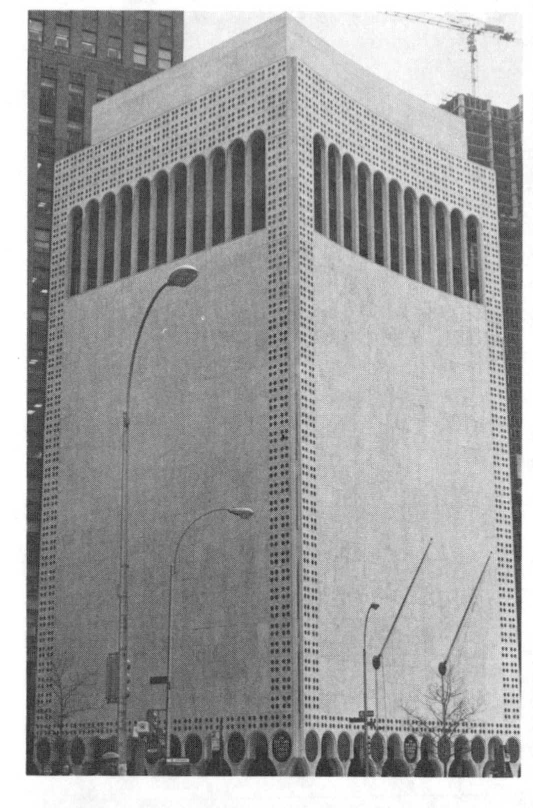

Figure 8-2

Forming a Circle

The circle is different from most of the figures we have been dealing with. Take a look at Figure 8-3. What makes the circle different from the other figures? As you know from Chapter 1, each of these

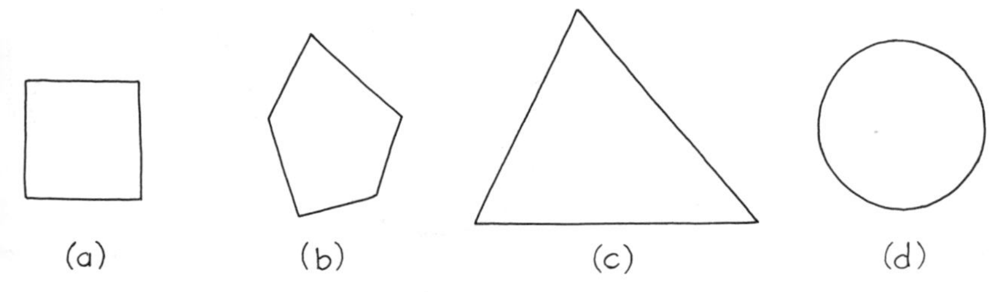

Figure 8-3

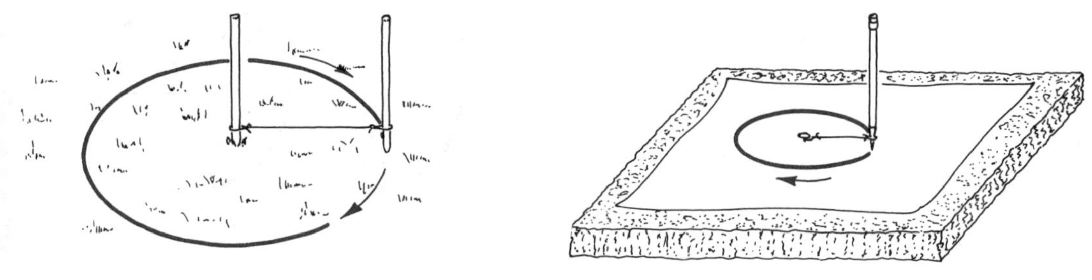

Figure 8-4

Figure 8-5

figures is a *curve,* is *simple* (does not intersect itself), and is *closed* (end points are joined). Each can be called a simple closed curve. But shapes (a), (b), and (c) are made up entirely of line segments. They are *polygons.* The circle (d) is *not* made up of line segments: it is a simple closed curve that is *not* a polygon.

The circle has some special properties we can discover by making and examining circles. How can we make a circle? If you wanted to make a circular garden you might take two sticks and a piece of cord and proceed as in Figure 8-4. One stick is driven into the ground. This will be the center. Two loops are formed, one at each end of the cord. One loop is put around the stick at the center while the other is put around the other stick. Keeping the sticks vertical to the ground and keeping the cord taut at all times, you can trace the path of a circular shape as indicated.

You might try the same thing with a piece of paper, a pencil, a thumbtack, and a piece of string. To avoid marring the surface under the paper, use a backing of wood or corkboard. Again, holding the pencil so that it forms a right angle with the paper and keeping the string taut, you can draw the shape of a circle as in Figure 8-5.

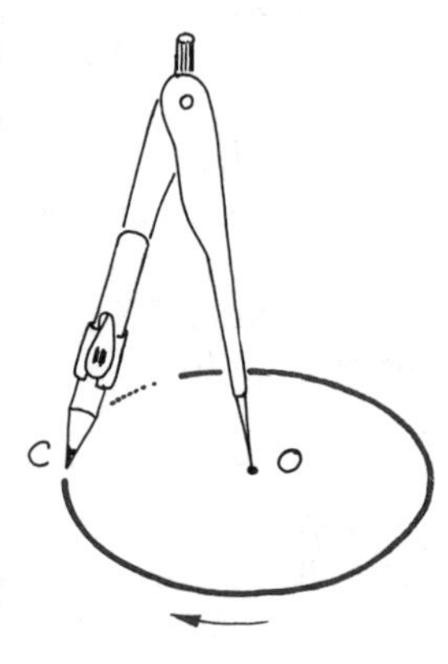

Figure 8-6

A pair of compasses is very useful for making circles. As you can see in Figure 8-6, this instrument employs the same principles we used with the sticks and cord or thumbtacks and string: there is one stationary point (O), which is the center, and one moving point (C). These points are always kept the same distance apart. The pencil is attached to the compasses so that its tip is at C. The path of point C is traced as we rotate the compasses—keeping point O stationary and the distance between points O and C constant.

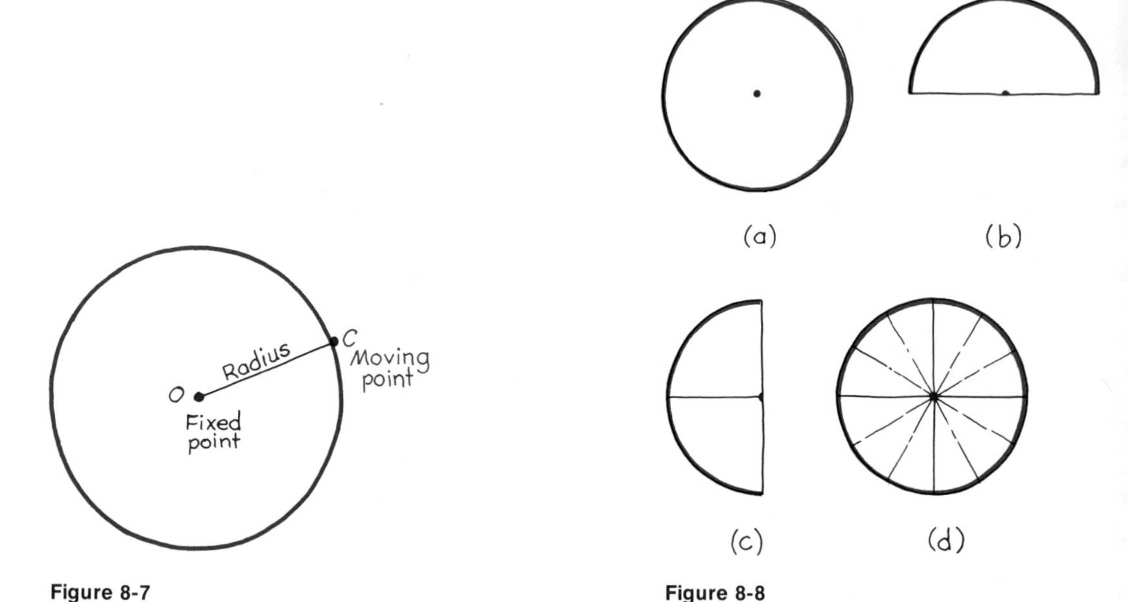

Figure 8-7

Figure 8-8

We are now ready to describe a circle more precisely. What is a circle? We know it is a simple closed curve. In making a circle, we formed that curve as a **set of points on a flat surface that are a certain distance from a fixed point.** In Figure 8-7, the fixed point O is called the **center** of the circle. It is not a point in the set of points making up the circle. Only the points that form the circular boundary (brown curve) belong to the set of points called the circle. Each point of the circle is the same distance from the center. This distance can be represented by a line segment from the center (O) to any point on the circle. That line segment is called the **radius** of the circle. Only one point on any radius belongs to the set of points that form the circle—the end point of the radius which touches the circle.

Suppose we make a circular shape out of paper, like the one in Figure 8-8(a). Mark the center with a dot on both sides of the paper shape. The *edge* of this circular shape (outlined in brown) represents a circle. Now let us fold this circular shape in half so that the crease passes through the center point [Figure 8-8(b)]. Make several creases in the same way [Figure 8-8(c) and (d)]. You will be able to make as many of these creases as you want. Each crease can be thought of as a line segment extending from a point on the circle, through the center, to another point on the circle [Figure 8-9(a)]. Such a line segment is called a **diameter** of the circle. A part of the diameter extending from the center to a point of the circle is the same as a radius. How much longer is the diameter than the radius? As you can see from Figure 8-9(b) and can verify from your own model, the measure of the length of the diameter, d, is twice the measure of the length of the radius, r. We can say that $d = 2r$.* Another way of describing this relationship is to say that

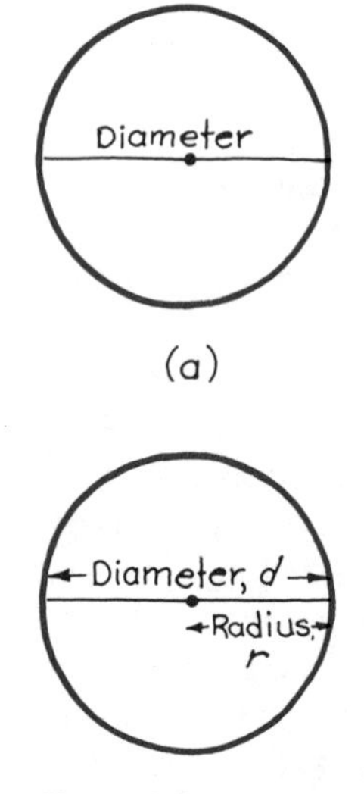

Figure 8-9

*$d = 2r$ is another way of writing $d = 2 \cdot r$ algebraically.

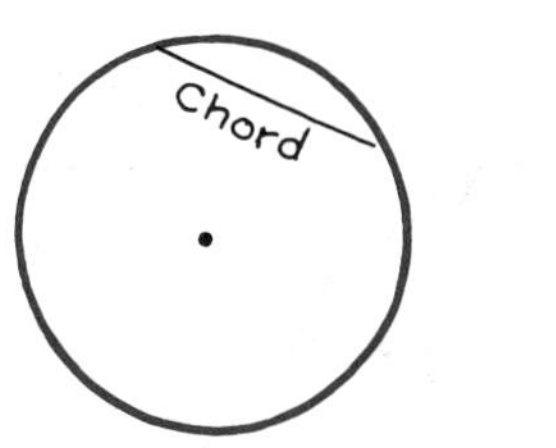

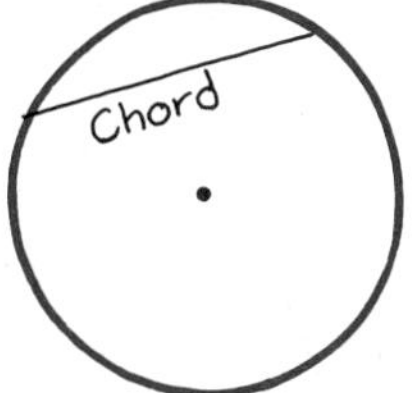

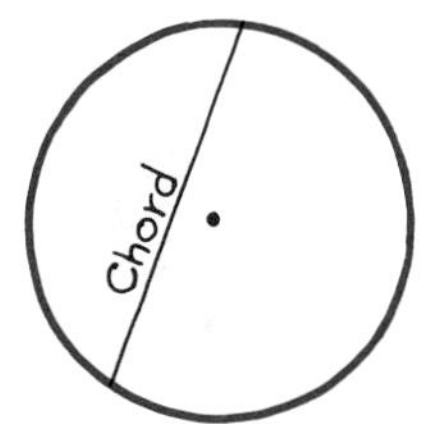

Figure 8-10

the measure of the length of the radius is one-half the measure of the length of the diameter: $r = \frac{1}{2}d$.

Look at the way the circular shapes were folded in Figure 8-10. The crease formed in each case can be considered a **chord.** A chord is a line segment that extends from one point on a circle to another point on that circle. It partitions the circle into two parts. A diameter is a special chord: it is a chord that passes through the center of the circle. It is also the longest chord in a circle. Why?

Points, Circles, and Inscribed Polygons

Take a piece of paper and make a dot on it with a pencil. Label that dot as point *A*. Now take a pair of compasses and draw a circle that includes (passes through) point *A* [Figure 8-11(a)]. Keeping the opening of the compasses unchanged, draw other circles that include point *A*. Make many circles in this manner and indicate the center of each circle with a dot [Figure 8-11(b)]. What do you

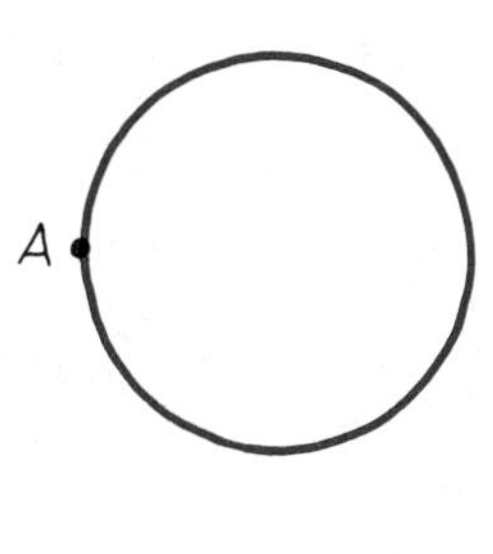

Figure 8-11

Figure 8-12

notice? The centers of these circles seem to belong to another circle whose center is *A* and whose radius is the same measure as the radii (plural of *radius*) of all the other circles!

In a similar way, if we change the size of the opening of our compasses, we can draw another set of circles each of which contains *A* (Figure 8-12). In fact, we can draw any number of sets of circles containing point *A*. We can conclude that given any point, it is possible to find an unlimited number of circles which include that point.

A B

Figure 8-13

Now suppose we make two dots on the paper and call them points *A* and *B* (Figure 8-13). Can you draw a circle that includes both these points? How?

You might draw that circle which has line segment $\overline{AB}$ as its diameter and whose center is the midpoint of $\overline{AB}$ [Figure 8-14(a)]. Are there any other circles that include both *A* and *B?* Try drawing some. If you move the fixed point of the compasses to another point *C* that is equidistant from *A* and *B* [Figure 8-14(b)], you can draw another circle (with a larger radius) that passes through *A* and *B*. By moving further upward you can find other points like point *D*, which are equidistant from *A* and *B* and can be used as the centers for even larger circles that include *A* and *B*. Similarly, we can draw many other circles that include *A* and *B*. The circles with centers *C′* and *D′* in Figure 8-14(b) have their centers the same distance from $\overline{AB}$ as *C* and *D*, respectively.

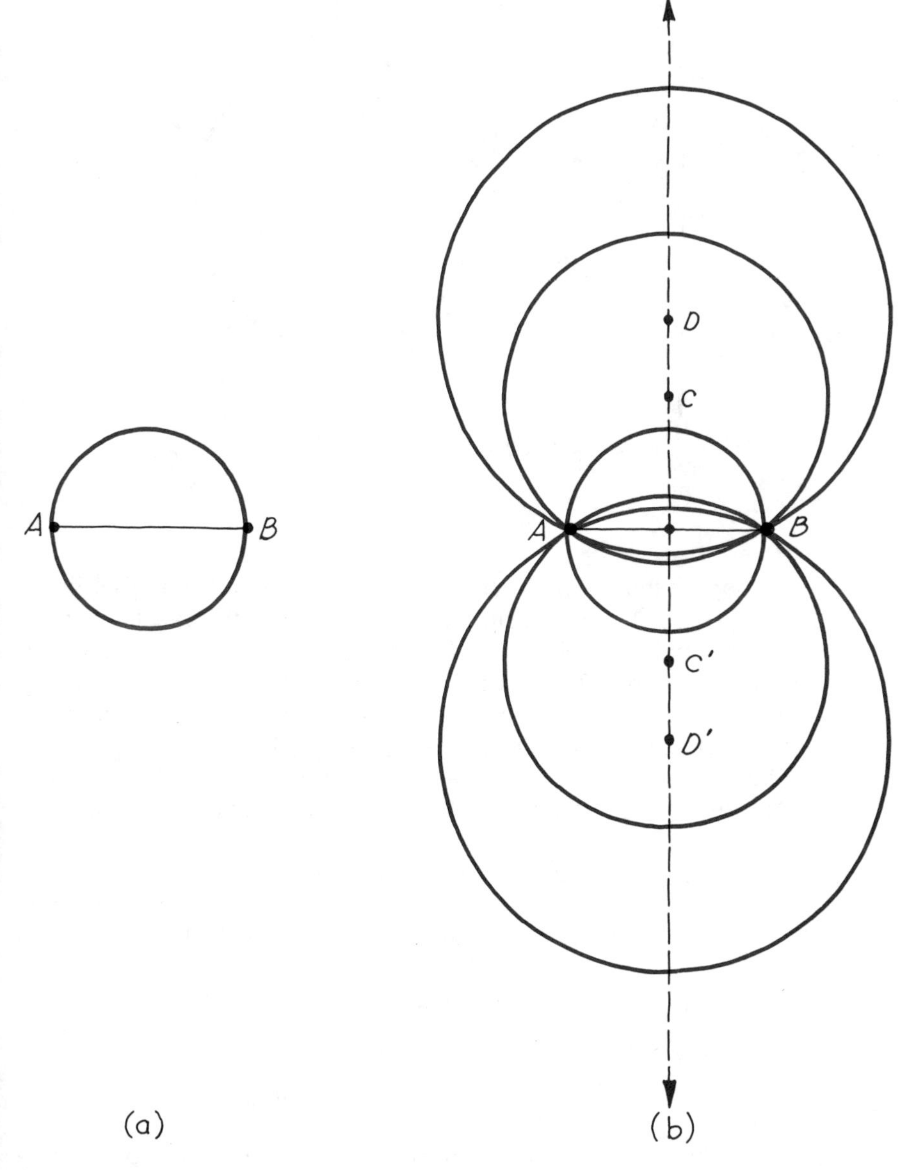

Figure 8-14

In this way we can draw an unlimited number of circles that include *A* and *B*. What about the centers of this set of circles? Note that these centers are contained in the line that is perpendicular to $\overline{AB}$ at its midpoint. Every circle we can draw that passes through *A* and *B* has a center point on that line.

Now let us draw *three* dots on the paper (not along the same line) and label them points *A*, *B*, and *C*. Are there an unlimited number of circles that include all three points?

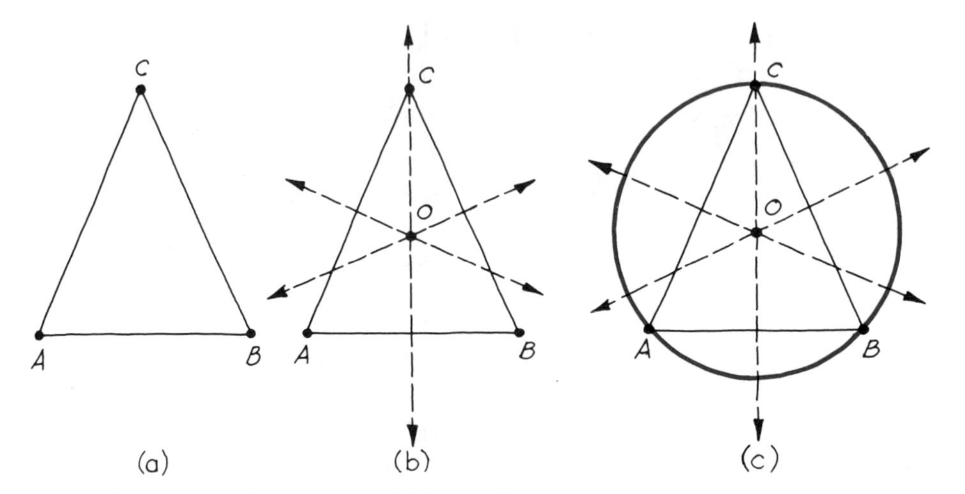

Figure 8-15

Imagine that points *A*, *B*, and *C* are vertices of a triangle [Figure 8-15(a)]. In the preceding pages we found that all the circles which include points *A* and *B* have centers on the line that is perpendicular to $\overline{AB}$ at its midpoint. Similarly, all the circles that include points *A* and *C* have their centers on the line perpendicular to $\overline{AC}$ at its midpoint and all the circles that include points *C* and *B* have their centers on the line perpendicular to $\overline{CB}$ at its midpoint. As you can see from Figure 8-15(b), these lines meet at one point, labeled *O*.

Point *O* is the intersection of the three lines determined by the center points. That is, it is the only point which belongs to all three lines. Point *O* is the *only* point that is simultaneously:

1. The center of a circle that includes *A* and *B*
2. The center of a circle that includes *A* and *C*
3. The center of a circle that includes *B* and *C*

Briefly, point *O* is the only point that is the center of a circle which includes *A, B, and C.* If we open our compasses so that the opening is the same length as the distance between *O* and *A* (or *O* and *C*, or *O* and *B*), we can draw that circle [Figure 8-15(c)]. It is the only circle that includes *all* three points *A*, *B*, and *C*.

Since the vertices of triangle *ABC* belong to the circle, we can call the triangle an **inscribed triangle.** Also, the circle can be called the circle **circumscribed** about the triangle, and point *O* can be named the **circumcenter** of the triangle.

This brings us to a consideration of whether, given *four* points, we can find a circle that includes all four points. What do you think? Draw four points that might be vertices of a quadrilateral. See if you can find a circle that includes all four points. As you can see in Figure 8-16, it is only possible *sometimes*.

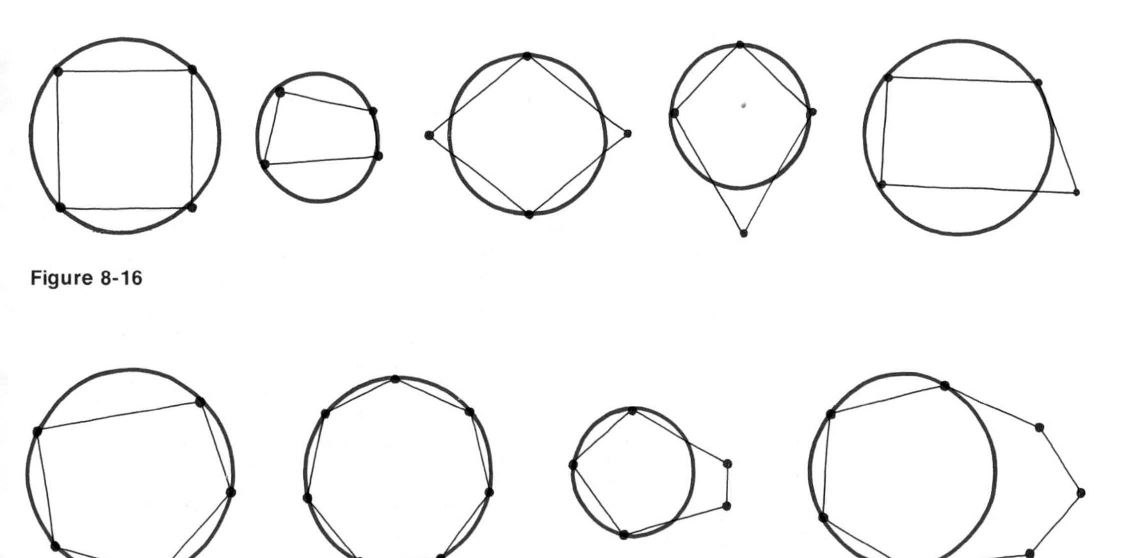

Figure 8-16

Figure 8-17

We would discover the same results if we began with *more* than four points. Considering these points as vertices of a polygon, it would only sometimes be possible to draw a circle which includes all the vertices of that polygon. If such a circle exists, it is called the circle *circumscribed* about the polygon. The corresponding polygon is then called an *inscribed* polygon of the circle [Figure 8-17(a) and (b)].

Let us summarize our findings about points and circles in a plane:

1. *Given one point,* we can find an unlimited number of sets of circles which include that point.
2. *Given two points A* and *B,* we can find an unlimited number of circles which include *A* and *B.* The centers of these circles are included in the line that is perpendicular to $\overline{AB}$ at its midpoint.
3. *Given three points A, B,* and *C* that can form the vertices of a triangle, we can find exactly *one* circle which includes *A, B, and C.* The center point of the circle is the intersection point of the three lines that are perpendicular to the sides of the triangle at their midpoints.
4. *Given more than three points,* it is only *sometimes* possible to find a circle which includes all the given points.

Lines, Circles, and Circumscribed Polygons

We just discovered some relationships between points and circles. Now let us take a look at relationships between lines and circles. As indicated in Figure 8-18, if we draw a line and a circle, the line can be an **outside** line to the circle (a), a **tangent** line that touches or intersects with the circle at one point (b), or a **secant** line that intersects with the circle at two points (c). When a circle and a line have only one point in common as in (b), they are called tangent to each other. In this section we will explore this tangent relationship between lines and circles.

Imagine a line like the one represented in Figure 8-19. How many circles can we draw that touch the line at only one point (are tangent)?

Take a pair of compasses and draw a circle tangent to the line. Keeping the opening of the compasses unchanged, you will find you can draw many other circles on each side of the line as in Figure 8-20. What do you notice about the centers of these congruent circles tangent to the line? The center points of these circles seem to belong to one of two lines that are parallel to the original line. Make as many other congruent circles tangent to the original line as you wish, and note that their centers always belong to one of these two parallel lines. These congruent circles can be considered a set with an unlimited number of circles tangent to the line.

What happens if we change the opening of the compasses? As shown in Figure 8-21, any number of different circles can be drawn

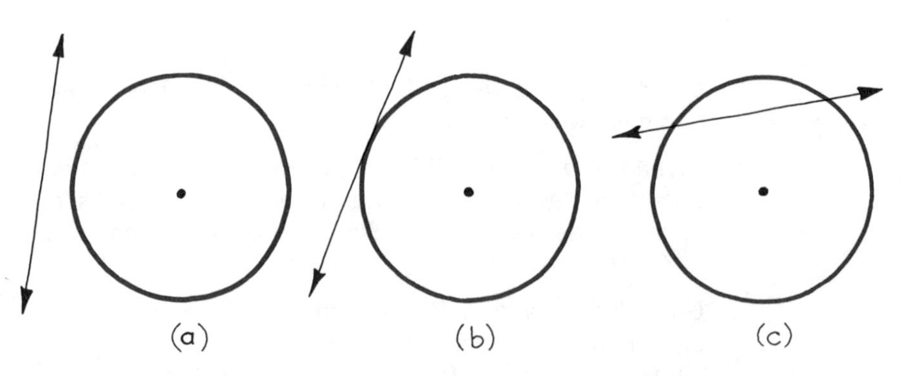

Figure 8-18

Figure 8-19

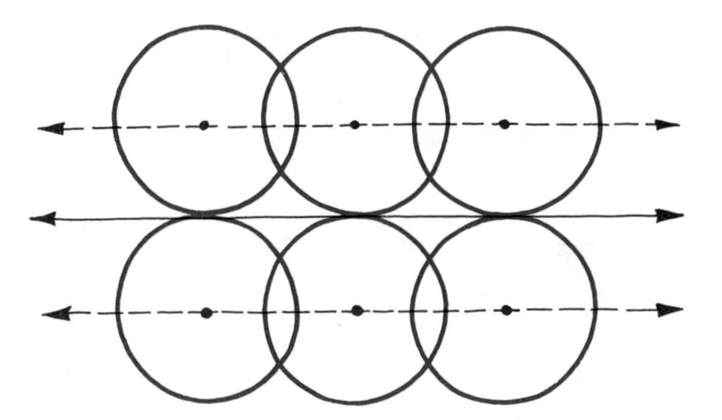

Figure 8-20

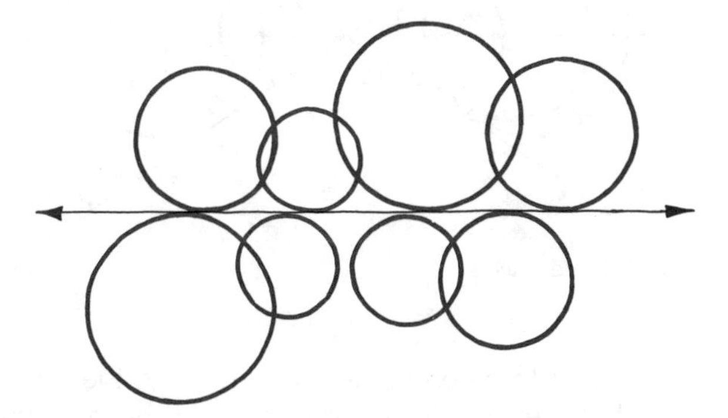

Figure 8-21

tangent to the original line. Each different opening allows us to draw another set containing an unlimited number of circles tangent to the line.

Now suppose we begin with *two* lines in a plane. The lines might be parallel or intersecting (Figure 8-22). How many circles can we draw tangent to two lines at the same time?

If the lines are parallel, we can find a set with an unlimited number of circles tangent to the lines. The centers of these circles

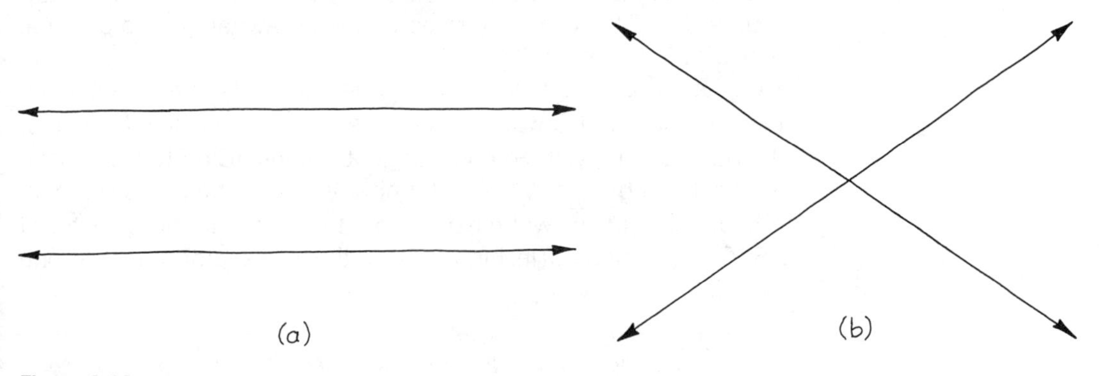

Figure 8-22

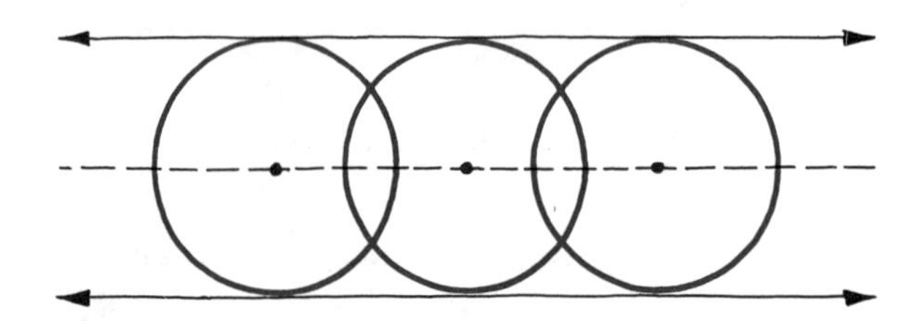

Figure 8-23

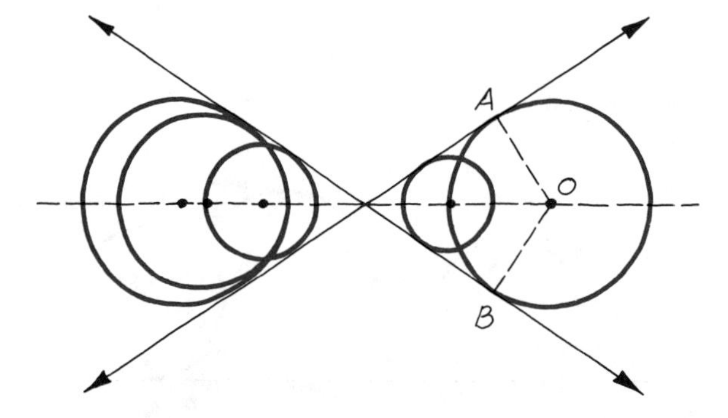

Figure 8-24

seem to belong to a line parallel to the original lines and midway between them (equidistant from the original lines—Figure 8-23). Note that the diameter of each circle is the same length as the distance between the original lines.

If we start with intersecting lines, we can also find a set that contains an unlimited number of circles tangent to the original lines, as suggested by Figure 8-24. What do you notice about the centers of these circles? The center point of each circle is the same distance from the two original lines. For example, $\overline{OA} = \overline{OB}$ because these are radii of the same circle. We can see that the centers of these tangent circles belong to a line that **bisects** the angles formed by the original intersecting lines (the **bisector** of that angle).

Let us consider three lines. Suppose the three lines intersect at three points that can be considered the vertices of a triangle *ABC* (Figure 8-25). How many circles can we draw tangent to all three lines at the same time?

We just discovered that all the circles tangent to two intersecting lines like $\overleftrightarrow{AB}$ and $\overleftrightarrow{BC}$ will have centers which belong to the bisector of $\angle ABC$. Similarly, the circles tangent to lines $\overleftrightarrow{AC}$ and $\overleftrightarrow{BC}$ will have centers belonging to the bisector of $\angle ACB$, and the circles tangent to lines $\overleftrightarrow{AB}$ and $\overleftrightarrow{AC}$ will have centers belonging to the bisector of $\angle BAC$. As indicated in Figure 8-26, these bisectors meet at one point *O*.*

**You may recall from Chapter 2 that the point where the three bisectors meet is called the incenter of the triangle. Do you now see why it has that name?*

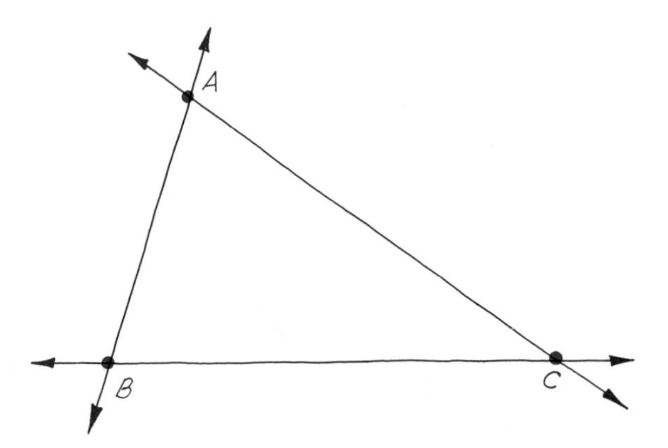

Figure 8-25

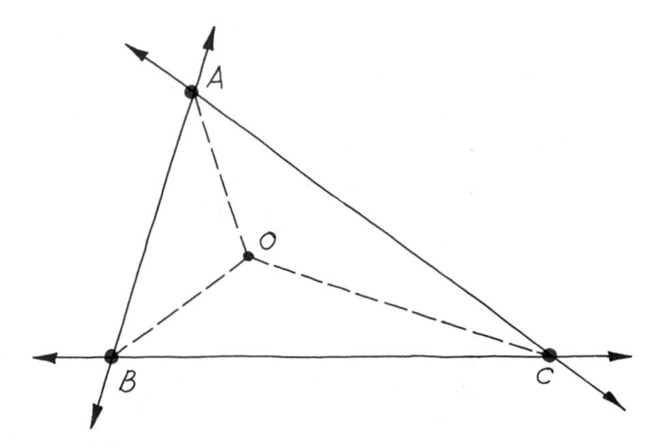

Figure 8-26

Point *O* is the intersection of the three bisectors. It is the only point that belongs to all three bisectors at the same time. Point *O* is the only point that is simultaneously:

1. The center of a circle tangent to $\overleftrightarrow{AB}$ and $\overleftrightarrow{BC}$
2. The center of a circle tangent to $\overleftrightarrow{AC}$ and $\overleftrightarrow{BC}$
3. The center of a circle tangent to $\overleftrightarrow{AB}$ and $\overleftrightarrow{AC}$

Therefore, point *O* is the only point that is the center of a circle tangent to $\overleftrightarrow{AB}$, $\overleftrightarrow{BC}$, and $\overleftrightarrow{AC}$. With *O* as the center, open your compasses so that the amount of opening equals the distance from *O* to one of these lines. You can now form the one (and only) circle tangent to all three original lines (Figure 8-27).

Looking at triangle *ABC* of Figure 8-27, you will note that the point of intersection of the bisectors of the angles (the *incenter*) is the center of the circle which can be drawn inside the triangle so that it is tangent to each side. The circle is said to be *inscribed*

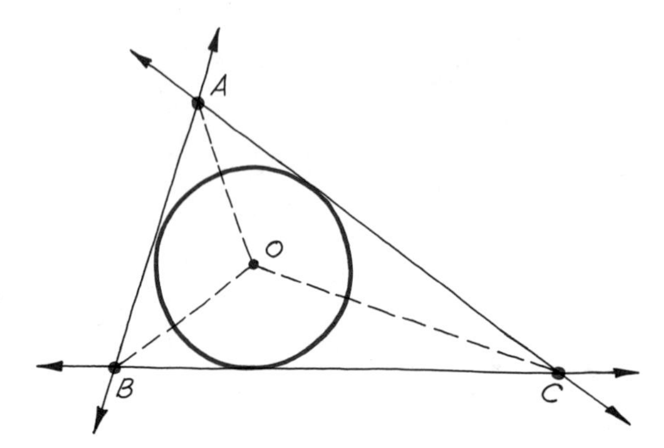

Figure 8-27

in triangle *ABC,* while the triangle is referred to as *circumscribed* about the circle.

Suppose we begin with more than three different lines. Can we find a circle that is tangent to all the lines? Draw several lines and try to draw a circle tangent to all the lines. As Figure 8-28 indicates, it is *not* always possible to find a circle tangent to every line in a given set.

Let us summarize our findings about lines and circles in a plane:

1. *Given one line,* we can find an unlimited number of circles tangent to that line.
2. *Given two parallel lines,* we can find an unlimited number of circles tangent to both lines. The centers of these circles belong to a line that is parallel to the given lines and midway between them.
3. *Given two intersecting lines,* we can find an unlimited number of circles tangent to both lines. The centers of these circles belong to the *bisectors* of the angles formed by the original lines.

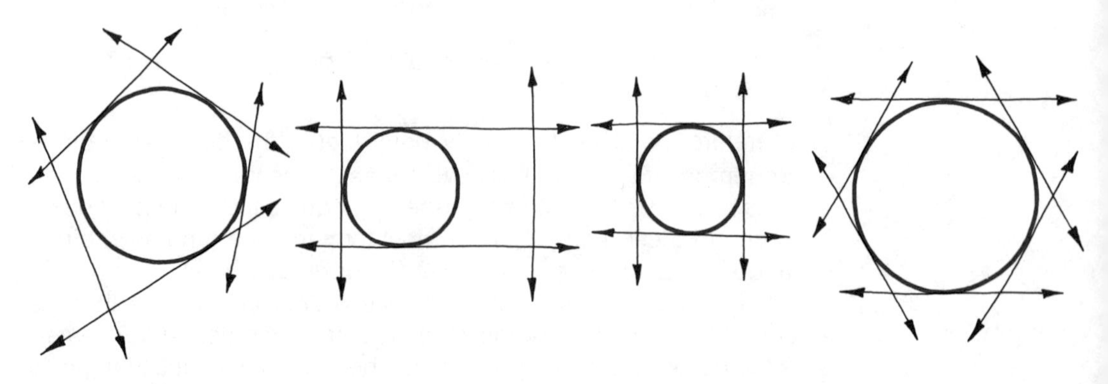

Figure 8-28

4. *Given three intersecting lines* that determine a triangle, we can find only one circle tangent to all three lines. The center of that circle is the point of intersection of the three lines which bisect the angles of the triangle (bisectors).
5. *Given more than three lines,* it is only *sometimes* possible to find a circle tangent to all the lines.

Compare this summary to the one about points and circles in the previous section. How are they alike?

Finding the Circumference of a Circle

When we examined polygons we saw that the perimeter of a polygon can be found by calculating the sum of the measures of its sides, which are line segments and can easily be measured. How do we find the *perimeter* of a circle? It is not a polygon and does not have line segments as its sides.

Let us take a circular object (like the top of a jar) and some string. If we carefully wrap the string around the jar, we can form a circular shape with the string [Figure 8-29(a)]. Now mark the point where the string begins to overlap. The distance from the end of the string on the object to the marked-off point represents the length of the circle [Figure 8-29(b)]. The length or perimeter of a circle is usually called the **circumference** of the circle.*

Since a circular shape is curved, it is not as easy to measure the circumference of a circle as it is to measure something with a polygonal shape. Fortunately, we can use an easier method to find the circumference of a circle.

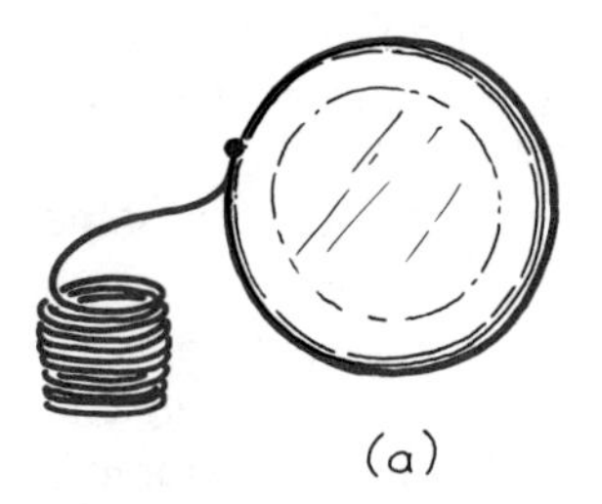

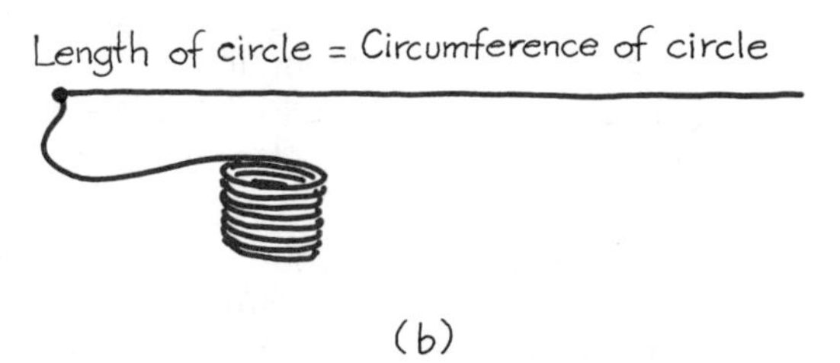

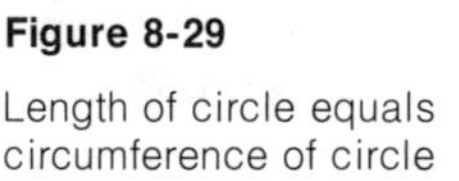

Figure 8-29

Length of circle equals circumference of circle

**Sometimes the term circumference is also used to mean circle. For example, "draw a circumference" might be used to mean "draw a circle."*

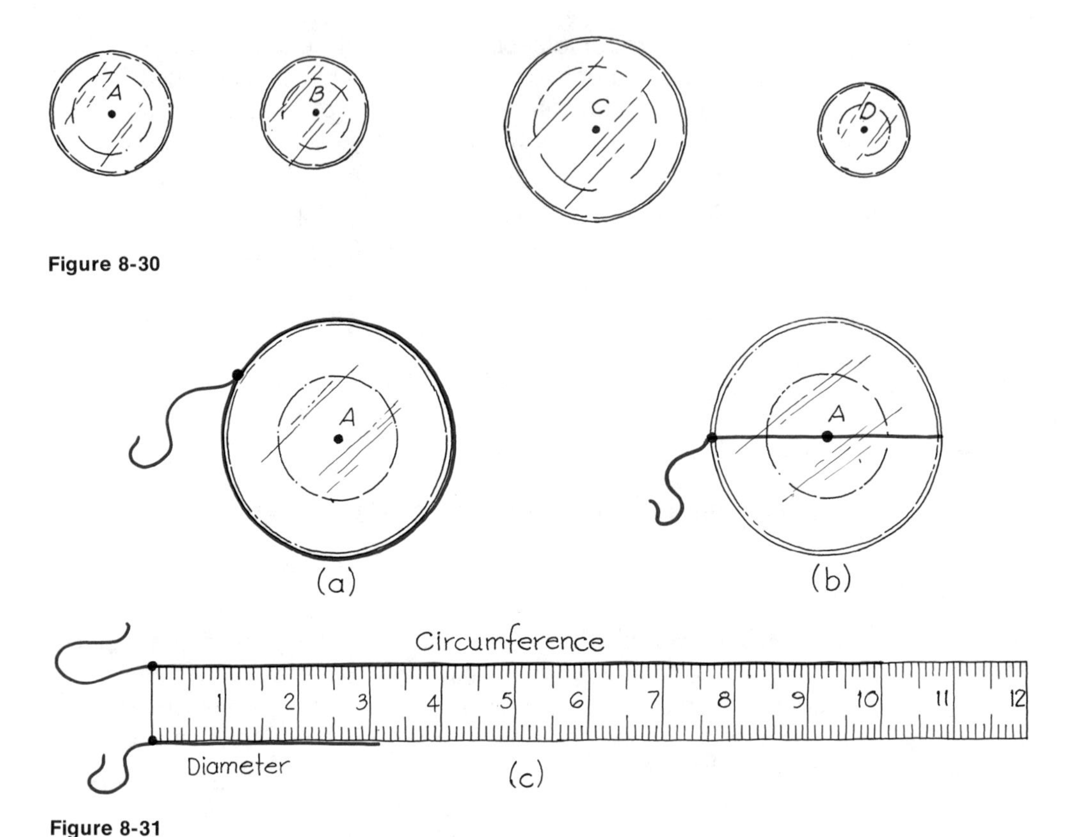

Figure 8-30

Figure 8-31

For this activity you will need some circular objects of different sizes (jar tops will do), string, a marking pen, and a ruler marked off in millimeter units. Label each top with the letters *A, B, C,* and so forth as in Figure 8-30, and mark the center of each top with a dot. Starting with the top marked *A,* wrap a piece of string around the top as in Figure 8-31(a), mark it, and use the ruler to measure the portion of the string that corresponds to the circumference of *A.* Now take another piece of string and stretch it across the top to form a diameter as in Figure 8-31(b), mark the end point, and measure the piece of string that corresponds to the diameter of *A.* Find these measurements to the nearest millimeter (mm).

Measure the circumference and diameter of each top in the same way. Keep a record of your measures as in Table 8-1.

Take a look at your record of the pairs of measurements of circumference and diameter for each circular top. What do you notice? In Table 8-1 the smallest diameter (15) corresponds to the smallest circumference (48), and the largest diameter (40) corresponds to the largest circumference (127). As we might suspect, the larger the diameter of a circle, the larger its circumference.

Let us discover something less apparent by forming the ratio of the measure of the circumference to the measure of the diameter in each case. Table 8-2 includes a column for that ratio—$c:d$ or c/d.

Table 8-1

Top	Measure of circumference (mm)	Measure of diameter (mm)
A	93	30
B	75	24
C	127	40
D	48	15
.	.	.
.	.	.
.	.	.

Table 8-2

Top	Measure of circumference, *c* (mm)	Measure of diameter, *d* (mm)	$\frac{c}{d}$
A	93	30	$\frac{93}{30} = 3\frac{1}{10} = 3.1$
B	75	24	$\frac{75}{24} = 3\frac{1}{8} = 3.125$
C	127	40	$\frac{127}{40} = 3\frac{7}{40} = 3.175$
D	48	15	$\frac{48}{15} = 3\frac{1}{5} = 3.2$
.	.	.	.
.	.	.	.
.	.	.	.

Look what happens! Although the tops are of different sizes, when we form the ratio of the measure of the circumference to the measure of the diameter of each top, we always get a result close to 3. In fact, if we measure the circumference and the diameter of *any* circular object—regardless of how big or how small—and form this ratio c/d, we will *always* end up with a number close to 3 (provided we measure carefully and perform the calculations accurately!). We might have expected this result. Since all circles have the same shape, they are similar, and, as we found in Chapter 7, a ratio formed by the measure of corresponding parts of similar figures remains the same. This special number that is close to 3 and represents the ratio of the measure of the circumference to the measure of the diameter of a circle is known by the symbol π (the Greek letter pi, pronounced "pie").

Now that we know this relationship is true for any circle, we have an easier way to find the measure of the circumference of a circle without measuring it directly:

$$\text{If } \frac{c}{d} = \pi, \qquad \text{then } c = \pi \cdot d \qquad \text{or} \qquad \pi d$$

(Just as, if $6/2 = 3$, then $6 = 3 \cdot 2$.) So if we want to know the measure of the circumference of a circle, we simply measure its diameter and multiply that number by π.

Of course, if we know the radius of a circle, we can easily find its circumference. Since a diameter is twice the length of a radius ($d = 2r$), the formula $c = \pi d$ becomes $c = \pi(2r)$ or:

$$\underset{\text{circumference of circle}}{c} = 2\pi r$$

We can use this formula to find the measure of the circumference of a circle when we know the measure of its radius.

But what is the value of π? We said it is a number "close to 3." Actually π is a special number that cannot be represented exactly in fraction or decimal form. In this activity we measured to the nearest millimeter, and we got an approximate value for the length of the circumference, diameter, and consequently the ratio c/d. Since this type of measurement (along a line) is always approximate, we were able to get only an approximation for those measures. If we need a more accurate approximation, we can use a unit smaller than a millimeter and get a closer approximation. With smaller and smaller units, our approximation becomes more and more accurate. For most practical situations we can use the approximations 3.14 or $22/7$ as π. It should be stressed that 3.14 and $22/7$ are *convenient approximations* for the value of π.

Now whenever we know the measure of the diameter of a circle, we can find the approximate measure of the circumference. If the measure of the diameter of a circle is 8, then since $c = \pi d$,

$$c \approx 3.14 \cdot 8$$
$$c \approx 25.12$$

We now know a way to find the approximate value of the measure of the circumference of a circle when we know the measure of

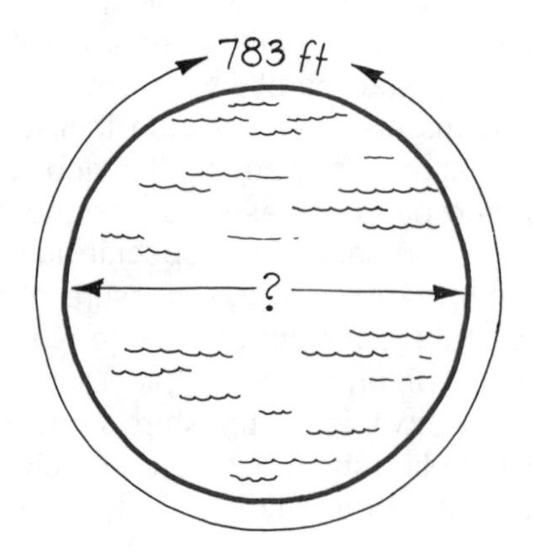

Figure 8-32

the diameter. Suppose, however, we happen to know the circumference of a circle and we need to know its diameter. How can we do it?

Imagine we have measured the circumference of a circular lake as 783 ft (Figure 8-32) and want to know the approximate length of its diameter. We don't need to swim across the lake with a tape measure!

$$\text{If } c = \pi d, \qquad \text{then } \frac{c}{\pi} = d$$

(Just as, if $6 = 3 \cdot 2$, then $6/3 = 2$.) Substituting values for c and π we have

$$\frac{c}{\pi} = d$$
$$\frac{783}{3.14} \approx d$$
$$249.36 \approx d$$

The approximate measure of the diameter of the lake (to the nearest hundredth) is 249.36 ft.

Area within a Circle

Now that we know how to find the circumference of a circle, we are ready to discover how to find the area of a region bounded by a circle (area within a circle) just as we were able to determine ways to find the area within polygons in Chapter 5. To find area, we calculated the number of square units it would take to cover the region bounded by the polygon.

Using a pair of compasses, draw several circles with different radii on a piece of squared paper as in Figure 8-33. Which circle has the most squares within it? Note that there is a direct relationship between the length of the radius and the amount of area

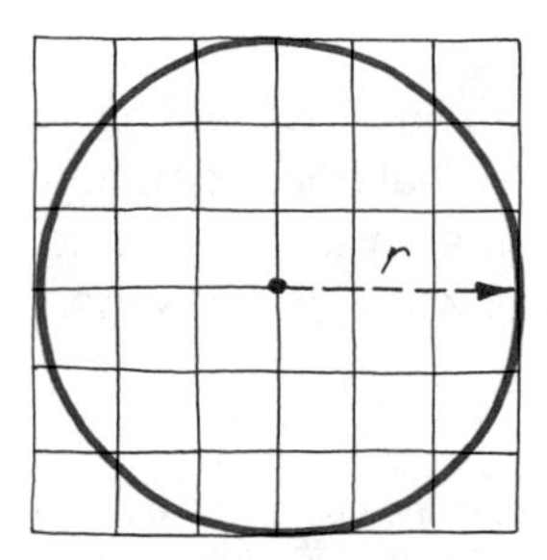

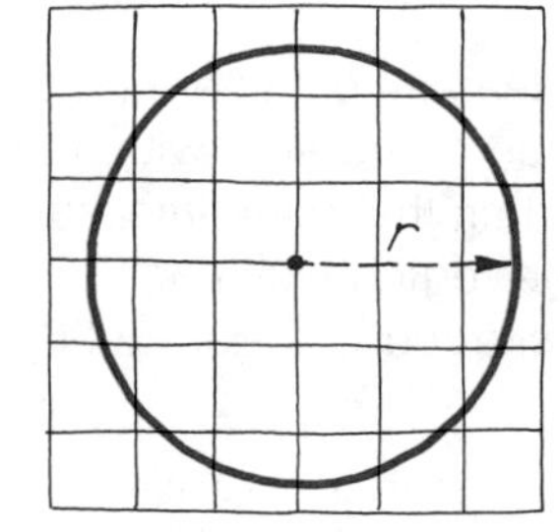

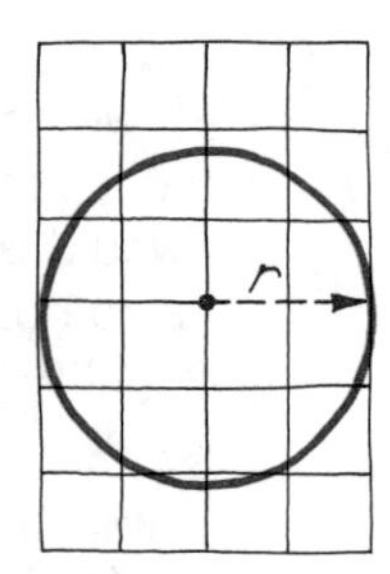

Figure 8-33

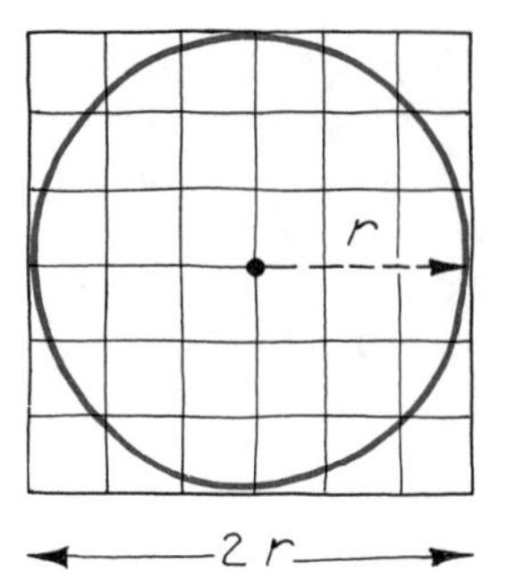

Figure 8-34

within a circle: the longer the radius, the greater the area within the circle.

How can we find the area within each circle? We can find an approximation of these areas by counting the number of square units and parts of square units within each circle. This will give us an estimate of the areas, but it is rather tedious. We can find better ways to determine the area.

On the squared paper, draw a square and an inscribed circle. The diameter of that circle will be the same size as the length of a side of the square. Let us call the radius of the circle r. What can we call the length of the side of the square? Since it is equivalent to the diameter of the circle, we can call it $2 \cdot r$ or $2r$ (Figure 8-34).

We know how to find the area A within the square. Remember:

$$A_{\text{square}} = s^2$$

where s represents the measure of the length of a side. In this case, the area within the square equals $(2r)^2$ or $(2r \cdot 2r)$ or $4r^2$. So we find that the area within the square equals four times the radius squared.

As you can see, the area within the inscribed circle is *less than* the area within the square since some square units included within the square are *not* included within the circle. Now we can estimate:

$$A_{\text{circle}} < 4r^2$$

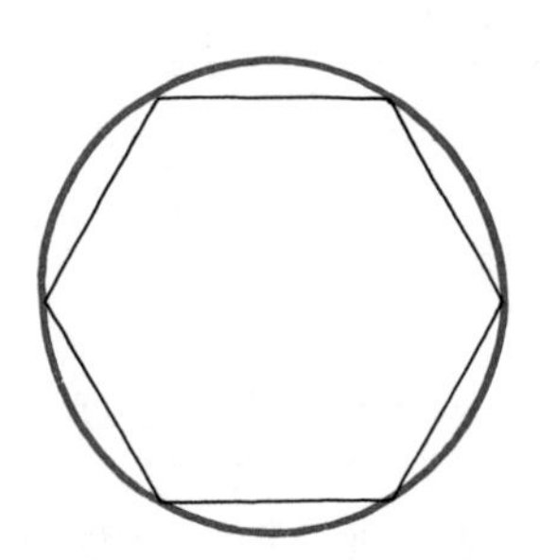

Figure 8-35

If we have a circle with radius 2 units, we know the area within it is less than $4 \cdot 2^2 = 4 \cdot 4 = 16$ square units. If a circle has a radius of 3 units, the area within it is less than 36 square units.

But we can find a better way for estimating the area within a circle. Suppose we start with a regular hexagon inscribed in a circle (Figure 8-35). We developed a way for finding the area within the regular hexagon in Chapter 5. By partitioning the hexagonal region into congruent triangles (Figure 8-36), we see that the area, A, within the hexagon is equal to one-half the product of measures of the altitude, h, of one of the triangles* and the perimeter, p:

$$A_{\text{hexagon}} = \frac{h \cdot p}{2}$$

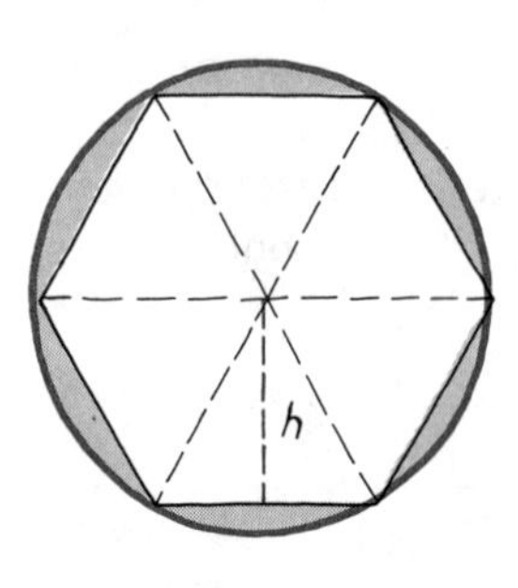

Figure 8-36

Circle with inscribed regular hexagon (6 sides)

If we find the measure of the area within the hexagon, we will have an approximation of the area within the circle. That approximation would be less than the actual measure of the area within the circle since part of the region within the circle is not within the hexagon (that region is shaded in Figure 8-36).

When a polygon is inscribed in a circle, the distance from the center of the circle to one of the sides of the polygon (which is the same as that of the altitude of a triangle formed) is known as the* ***apothem.*

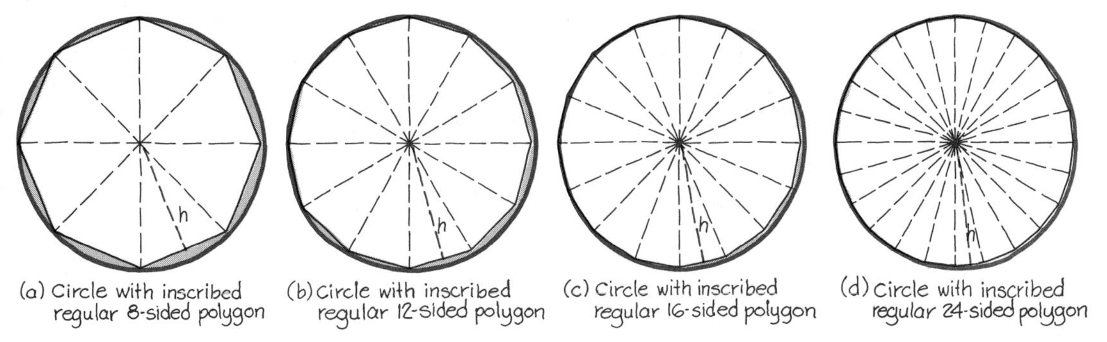

(a) Circle with inscribed regular 8-sided polygon (b) Circle with inscribed regular 12-sided polygon (c) Circle with inscribed regular 16-sided polygon (d) Circle with inscribed regular 24-sided polygon

Figure 8-37

But suppose we inscribe a regular polygon with a greater number of sides as in Figure 8-37. What happens? As we increase the number of sides, the inscribed polygon covers more and more of the area within the circle. As we increase the number of sides, the area within the regular polygon becomes a better and better approximation of the area within the circle.

Imagine that we continue to increase the number of sides indefinitely. How does the measure of the altitude of the triangles (the apothem) change? How does the measure of the perimeter change? You can see from Figure 8-37 that the measure of the apothem increases and gets *closer and closer* to the measure of the *radius* of the circle. The measure of the perimeter of the inscribed polygons also increases and gets *closer and closer* to the measure of the *circumference* of the circle. Using the symbol $\longrightarrow$ to mean "approaches or gets closer and closer to," we can say that:

$$h \longrightarrow r$$
$$p \longrightarrow 2\pi r$$

As we increase the number of sides of the inscribed polygon indefinitely, the measure of the area within that polygon also changes:

$$\frac{h \cdot p}{2} \longrightarrow \frac{r \cdot 2\pi r}{2} \qquad \text{or} \qquad \pi \cdot r^2$$

Since the measure of the area within the polygons is getting closer and closer to the measure of the area within the circle, we can see that the formula πr^2 can be used to find the measure of the area within the circle. In general, we can find the area within any circle when we know the measure of its radius:

$$\underset{\text{circle}}{A} = \pi r^2$$

As the formula indicates, the measure of the area within the circle is directly related to the measure of the radius of that circle (we saw that previously). If the measure of the radius of a circle is 2 units, then the area within that circle will be $\pi \cdot 4$ or (using $\pi \approx 3.14$) about 12.56 square units. In more concrete terms, if the measure of the radius of a circular table top is 3 ft, the measure of the area of that surface is $\pi \cdot 9$ or about 28.26 sq ft.

Measuring Arcs

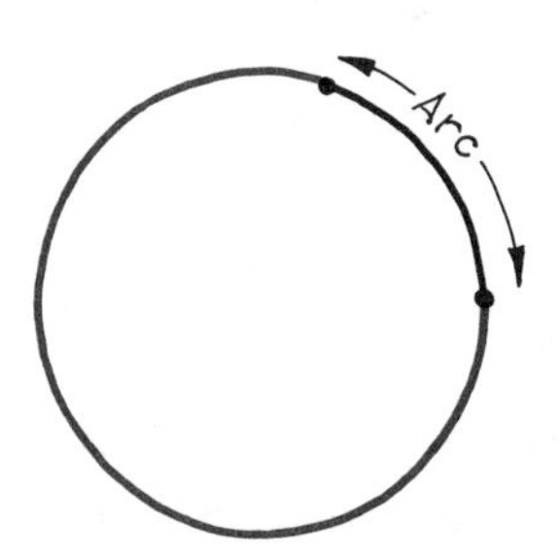

Figure 8-38

Whenever we know the measure of the radius r of a circle, we are able to find the measure of its circumference ($2\pi r$). But suppose we want to know the measure of the length of a portion of a circle. A portion of a circle, as illustrated in Figure 8-38, is called an **arc.** How can we find the measure of the length of an arc?

Examine the arcs in Figure 8-39. Radii have been drawn from the center to the end points of each arc. If you imagine these radii to be rays, you can think of them as sides of an angle. Such an angle formed by the radii of a circle is called a **central angle.**

Each arc, then, corresponds to a central angle formed by the radii drawn to its end points. Notice that the longer the arc, the larger the corresponding central angle. In Figure 8-39, arc (a) is the longest arc and corresponds to the largest central angle.

Imagine that the interior of a circle is partitioned into 360° by its radii (like the clock face partitioned in Chapter 1). The circle would then be partitioned into 360 arcs. Each arc corresponds to a central angle of 1° (Figure 8-40). We can say that each arc contains 1°. Look at the arc illustrated in Figure 8-41. Since the central angle determined by the arc contains 90°, the arc contains 90°.

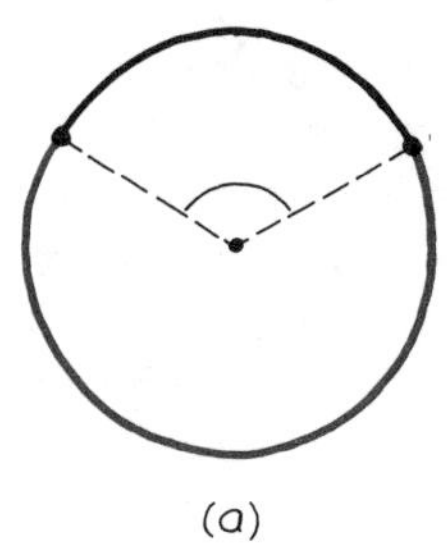

(a)

(b)

(c)

Figure 8-39

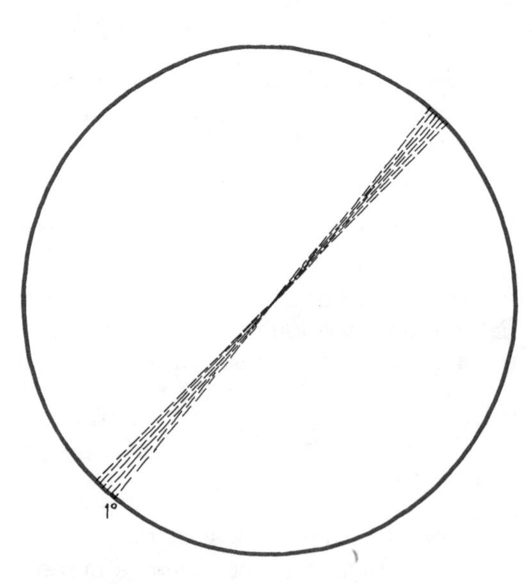

Figure 8-40

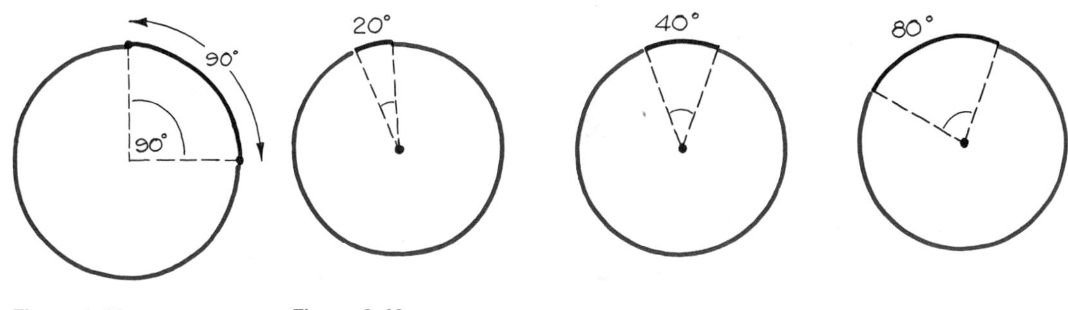

Figure 8-41 **Figure 8-42**

To find the measure of the length of an arc, we need to know how many degrees it contains. As you can see from these congruent circles in Figure 8-42, the greater the number of degrees contained by an arc, the larger the arc. In other words, the size of an arc is directly related to the number of degrees it contains.

But we also need other information. If we simply know that an arc contains 90°, we cannot determine its length. It is *not* 90° long. That just tells us the size of its corresponding central angle. The degree is a unit of *angular* measurement. For length we need a unit of *linear* measurement. Figure 8-43 illustrates several arcs of different length. Each arc contains 90°. What else must we know to find the length of the arcs? The size of the circle as determined by its radius. The length of an arc, then, depends upon the size of the circle *and* the number of degrees the arc contains. We will now find a way to calculate the measure of that length.

Make three congruent circular shapes out of paper and fold them as indicated in Figure 8-44. The creases can be thought of as radii of the circles. We will call the measure of a radius in these circles *r*. The crease formed in (a) partitions the circle into two congruent arcs. Similarly, (b) is partitioned into four congruent arcs and (c) into eight congruent arcs. What is the measure of the lengths of the arcs?

In (a) each arc contains 180°. That is one-half of all the degrees contained in the circle ($^{180}/_{360} = {}^{1}/_{2}$). We can see that the length of the arc is one-half the circumference of the circle. Similarly, in

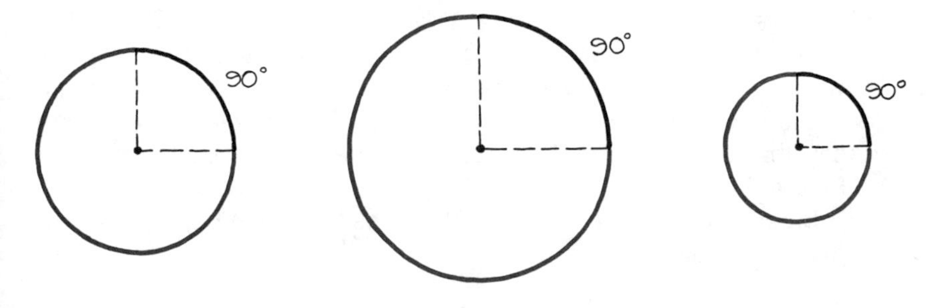

Figure 8-43

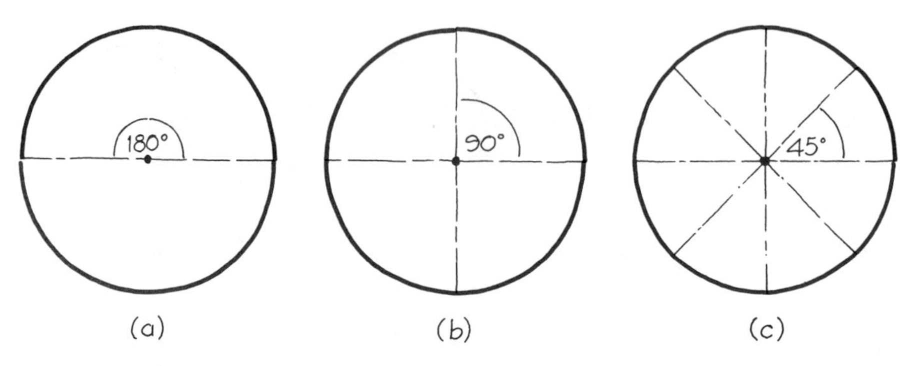

Figure 8-44

(b) each arc contains 90°, which is one-fourth of all the degrees contained in the circle ($^{90}/_{360} = ^{1}/_{4}$), and thus each arc is one-fourth the circumference of the circle. In (c) each arc contains 45°, or one-eighth of all the degrees contained in the circle ($^{45}/_{360} = ^{1}/_{8}$), and is thus one-eighth the circumference of the circle.

In other words, the ratio of the number of degrees in an arc to the total number of degrees in the circle (360) tells us what fractional part of the circumference is the measure of the length of the arc. If c represents the length of the circumference in Figure 8-44, then we can summarize as in Table 8-3.

Table 8-3

Figure	Degrees in arc	Ratio: $\frac{\text{degrees in arc}}{360}$	Length of arc
(a)	180	$\frac{180}{360} = \frac{1}{2}$	$\frac{1}{2} \cdot c = \frac{c}{2}$
(b)	90	$\frac{90}{360} = \frac{1}{4}$	$\frac{1}{4} \cdot c = \frac{c}{4}$
(c)	45	$\frac{45}{360} = \frac{1}{8}$	$\frac{1}{8} \cdot c = \frac{c}{8}$

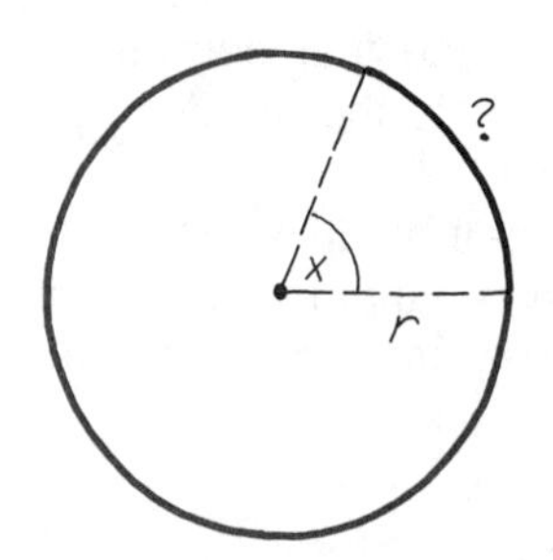

Figure 8-45

If arc contains x degrees and r is radius of circle,

Length of arc

$= \frac{x}{360} \cdot (2 \cdot \pi \cdot r)$

$= \frac{x \cdot \pi \cdot r}{180}$

In general we can say that if an arc contains x degrees, the measure of the length of the arc will be represented by $^{x}/_{360} \cdot c$, where c is the measure of the length of the circumference for that circle. If r is the measure of the radius of the circle, then we can say that the measure of the length of the arc can be found as in Figure 8-45.

In terms of r, the measures of the arcs formed in Figure 8-44 are given in Table 8-4.

Table 8-4

Figure	Degrees in arc	Ratio: $\frac{\text{degrees in arc}}{360}$	Length of arc
(a)	180	$\frac{1}{2}$	$\frac{1}{2} \cdot (2\pi r) = \pi r$
(b)	90	$\frac{1}{4}$	$\frac{1}{4} \cdot (2\pi r) = \frac{\pi r}{2}$
(c)	45	$\frac{1}{8}$	$\frac{1}{8} \cdot (2\pi r) = \frac{\pi r}{4}$

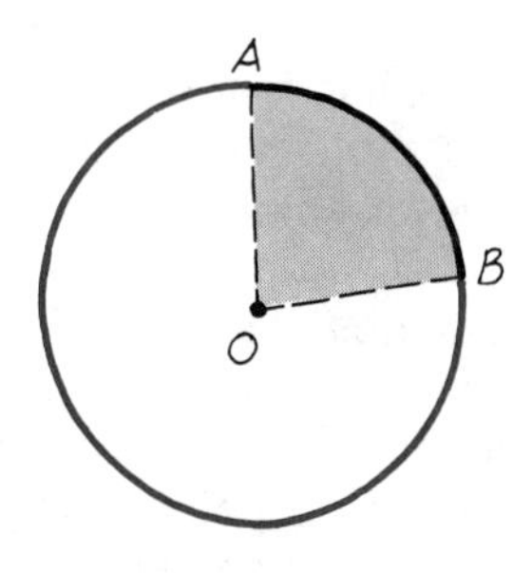

Figure 8-46

Measuring Sectors

Let us turn our attention to Figure 8-46. Look at the shaded region within the circle. It is bounded by the arc AB and the radii $\overline{OA}$ and $\overline{OB}$. This portion of the region within the circle is called a circular **sector.** How can we find its area?

As in the previous discussion about lengths of arcs, the size of the sector depends upon the size of the circle *and* the size of the arc that forms part of the boundary of the sector. As you can see from Figure 8-47, the larger the arc, the larger the corresponding sector. The arc of the sector in (a) contains one-fourth of all the degrees contained in the circle ($^{90}/_{360} = {}^{1}/_{4}$) and the area of the sector is one-fourth of the area within the circle. In (b) the arc of the sector contains one-eighth of all the degrees contained in the circle ($^{45}/_{360} = {}^{1}/_{8}$) and the area of the sector is one-eighth of the area within the circle.

In general, we can say that if the arc of a sector contains x degrees, then the area, A, of the sector is $^{x}/_{360}$ of the area within the circle:

$$\text{Area of sector} = \frac{x}{360} \cdot A$$

We know how to find the area within a circle in terms of its radius:

$$A = \pi \cdot r^2$$

The formula for the area of a sector becomes

$$\frac{x}{360} \cdot A = \frac{x}{360} \cdot (\pi r^2)$$

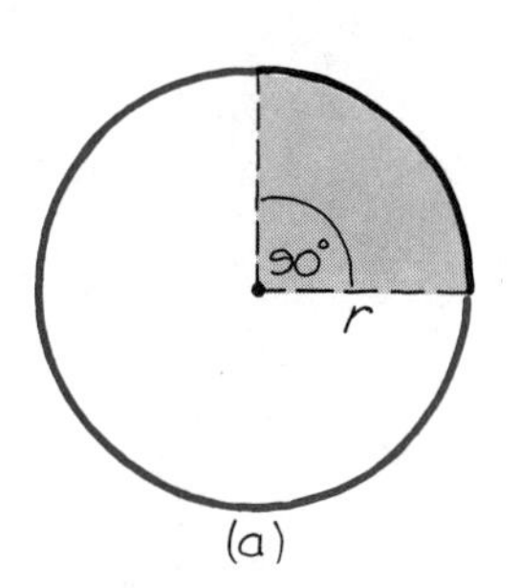

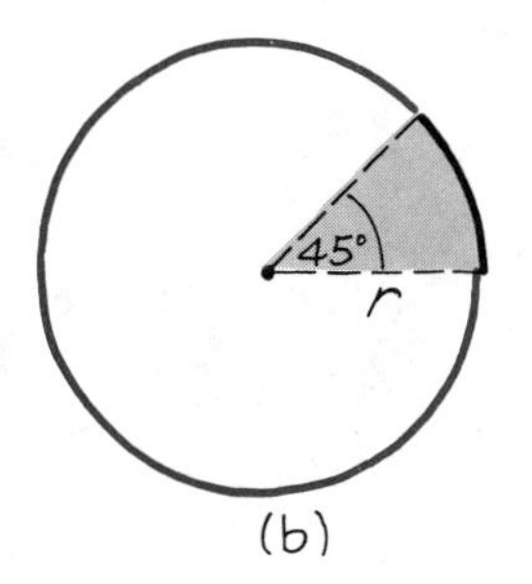

Figure 8-47

If the measure of the radius of the circle in Figure 8-47 is represented as r, we can say that the area of the sector in (a) is

$$\frac{x}{360} \cdot (\pi r^2) = \frac{90}{360} \cdot (\pi r^2) = \frac{1}{4} \cdot (\pi r^2)$$

For the figure in (b),

$$\frac{x}{360} \cdot (\pi r^2) = \frac{45}{360} \cdot (\pi r^2) = \frac{1}{8} \cdot (\pi r^2)$$

Using Circles to Find Angle Bisectors

In this chapter and in Chapter 2, we referred to a line that bisects an angle—an angle bisector. If we are given an angle like the one drawn in Figure 8-48, how can we find its angle bisector? Circles can help us.

Draw an angle like the one in Figure 8-48 and label its vertex O. Take a pair of compasses. Draw any circle by using the vertex of the angle as the stationary point and intersecting the side of the angle at points A and C [Figure 8-49(a)]. Next, using point A as the stationary point, draw an arc of another circle. That arc should be drawn in the interior of the angle. Then, *keeping the opening of*

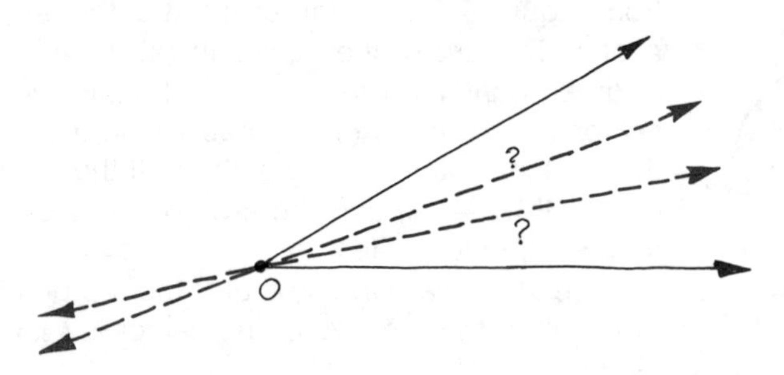

Figure 8-48

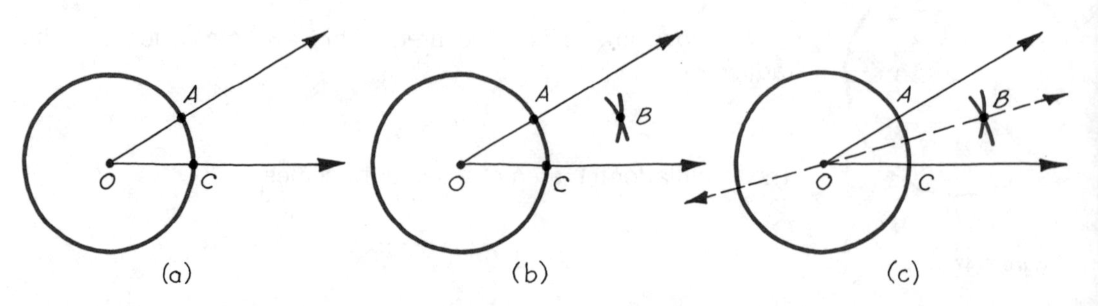

Figure 8-49

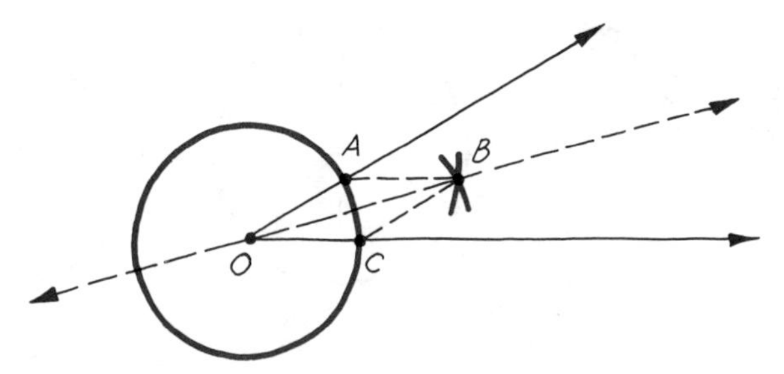

Figure 8-50

the compasses unchanged, use C as a stationary point and draw another arc that intersects the first arc at point B [Figure 8-49(b)].

If you draw the line that passes through the vertex of $\angle O$ and point B [Figure 8-49(c)], you will have located the angle bisector of the original angle!

How do we know line $\overleftrightarrow{OB}$ is the angle bisector? Look at Figure 8-50. Line segments $\overline{AB}$ and $\overline{BC}$ have been drawn and you can see two triangles, ABO and CBO. We can be certain $\overleftrightarrow{OB}$ is the angle bisector if $\angle AOB$ is congruent to $\angle BOC$ (since that shows $\angle AOC$, our original angle, has been bisected).

But how can we show that $\angle AOB$ is congruent to $\angle BOC$? As we saw in Chapter 7, if we can show that triangle ABO is congruent to triangle CBO, then $\angle AOB$ can be seen to be congruent to $\angle BOC$ because they are corresponding parts of congruent triangles. So all we have to do is show that triangle ABO is congruent to triangle CBO. Can you do it? (*Hint:* Can you show that the corresponding sides must have the same measure?)

Using Circles to Make Regular Polygons

The circle can be a very useful shape. Beginning with a circle, we can easily form many different regular polygons.* Cut several circular shapes out of paper. Mark the center of each with a dot. Take one of these circular shapes and fold it in half so that the crease goes through the center. Fold in half again to form right angles and then reopen the paper shape (Figure 8-51). You have now partitioned the circle into four equivalent parts. As indicated, if you join adjacent end points of the creases with line segments you can form a square, which is a regular polygon (regular quadrilateral).

**As you will remember from Chapter 1, a regular polygon is a polygon with all sides and all angles congruent.*

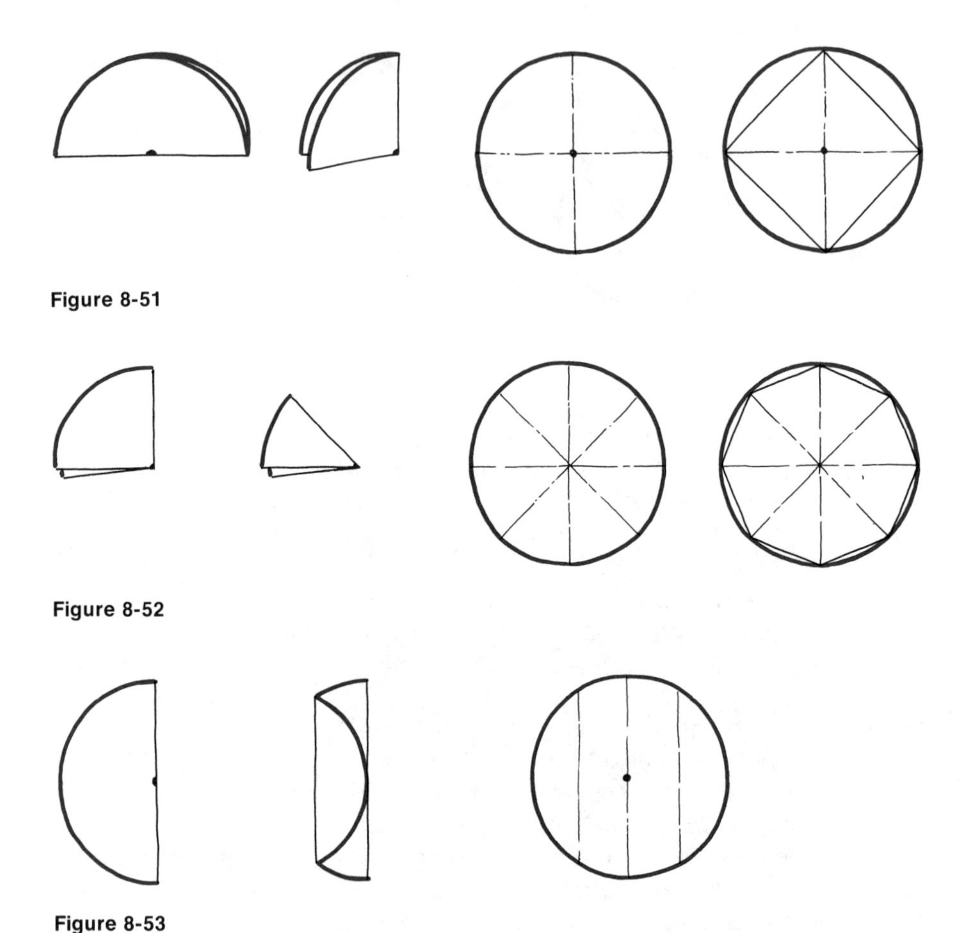

Figure 8-51

Figure 8-52

Figure 8-53

Now take another circular shape and fold as before. Then fold in half once more to partition the circle into eight equivalent parts. This time, if you join adjacent end points of the creases, you can form a regular octagon (Figure 8-52). Similarly, by continually folding in half you can develop a regular polygon with 16, 32, 64, . . . , sides.

Let us try folding a different way. Take another circular shape and fold it in half as you did before. Then fold it again so that an edge of the shape just touches the crease, forming another crease parallel to the first one (Figure 8-53). When you reopen the paper shape, you will find it has been partitioned into four parts. Now fold the same shape in half again, making a crease that is perpendicular to the other creases (Figure 8-54). Then fold it again as before so that the edge of the shape just touches the new crease. When you reopen the paper shape, you will find it has been partitioned into 16 parts.

There are now 12 end points of the creases formed. By joining the end points with line segments in different ways as in Figure 8-55, you can form many different regular polygons. Each polygon

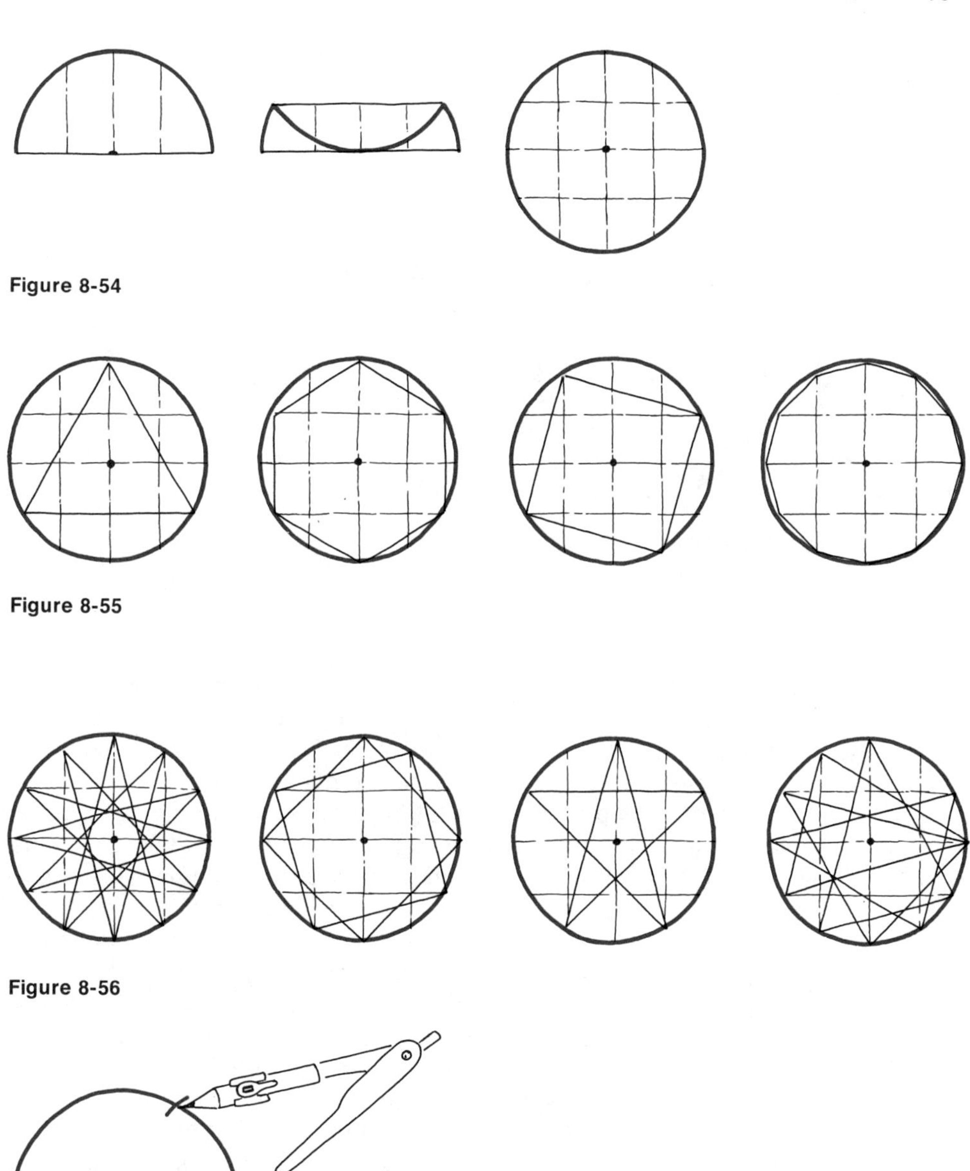

Figure 8-54

Figure 8-55

Figure 8-56

Figure 8-57

can be thought of as being inscribed in the circle. You can also form many interesting designs as shown in Figure 8-56.

A very easy way to make an equilateral triangle or a regular hexagon is to use a pair of compasses. Start by drawing a circle. Keeping the opening of the compass unchanged, place the stationary point of the compasses on the circle and mark off an arc that intersects the circle (Figure 8-57). Now, still keeping the opening of the compasses unchanged, place the stationary point of the compasses on the point where the arc intersects the circle and draw another arc that intersects the circle. In this way, you will

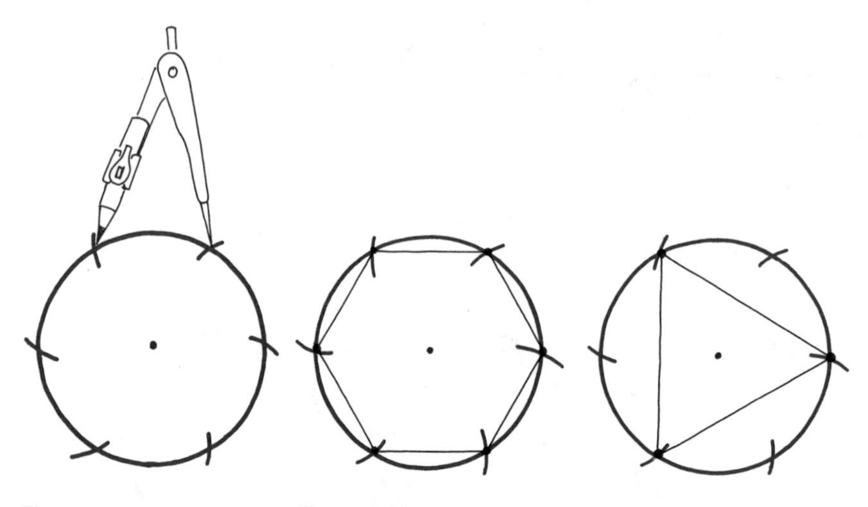

Figure 8-58

Figure 8-59

be able to draw exactly six arcs that intersect the circle and partition it into six (approximately congruent) parts (Figure 8-58).

Why does this work? Think of the relationship between the radius and the circumference of a circle. If the length of the circumference of a circle is about three times the length of its diameter, it must be about six times the length of its radius. Since the opening of the compasses corresponded to the radius of the circle, we were able to mark off the circle into six parts.

By using the points of intersection as end points for line segments (chords), we can form a regular hexagon or an equilateral triangle (Figure 8-59). In fact, many different designs are possible once we have partitioned the circle into six parts. The designs in Figure 8-60 offer a few examples. Can you make others?

All sorts of other interesting patterns can be made with circles. Can you discover how the patterns in Figure 8-61 were made? Try to draw these patterns with a pair of compasses. Invent some patterns of your own! Isn't the circle a remarkable shape?

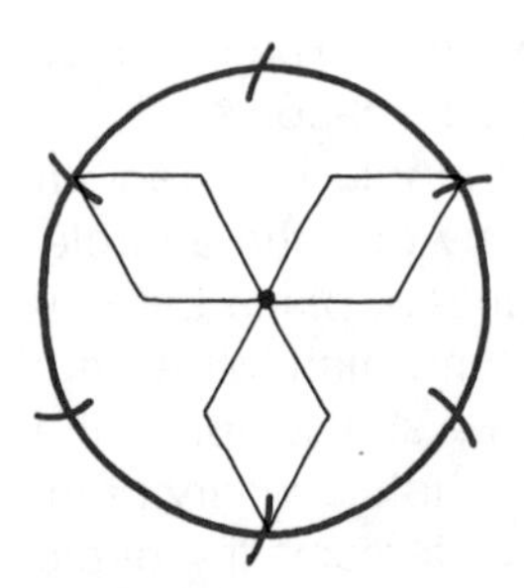

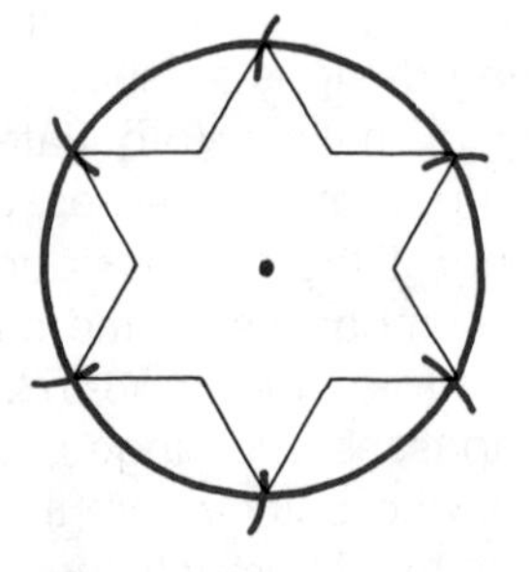

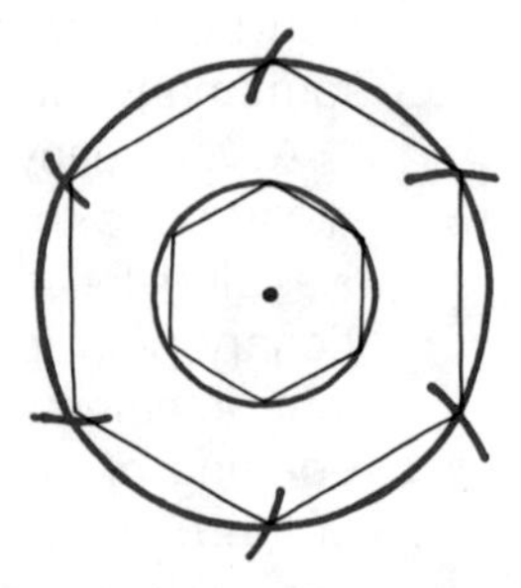

Figure 8-60

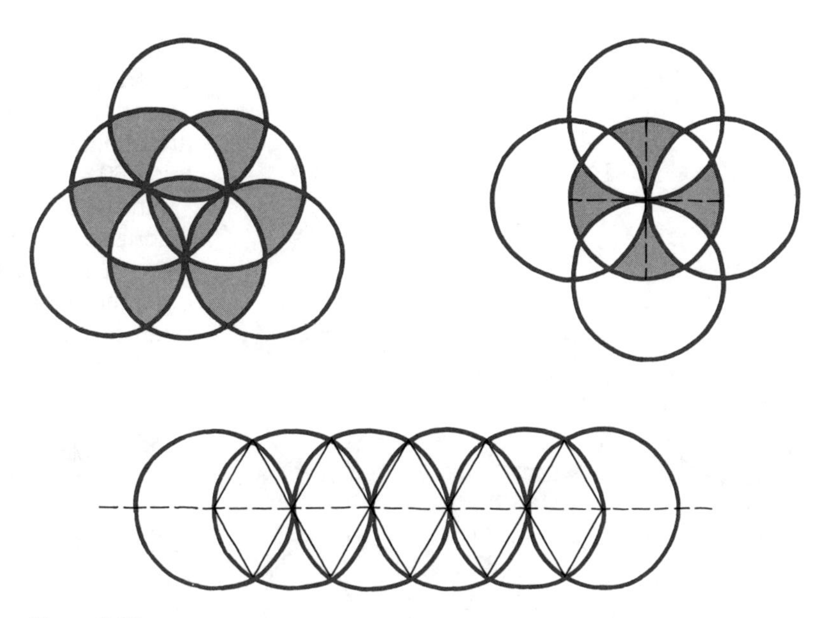

Figure 8-61

More Experiences

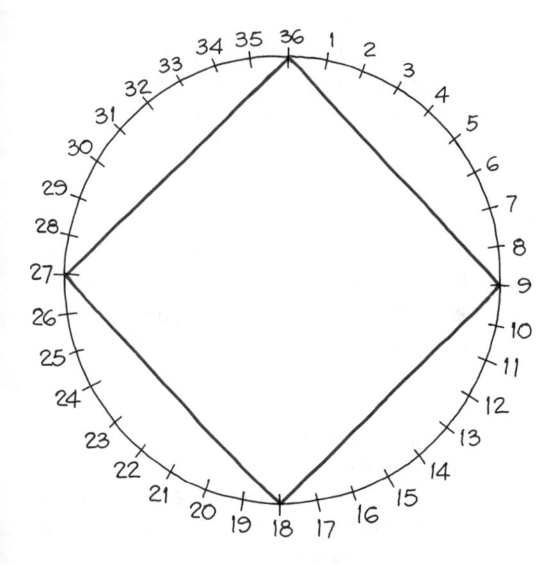

8-1. The circumference of the circle at the left has been partitioned into 36 congruent arcs. Points 9, 18, 27, and 36 can be the vertices of a square. Find a set of four other points that can be the vertices for a square. How many different sets of points can you find?

What sets of points on the circumference might be the vertices of an equilateral triangle? How many such sets are there?

What points might form the vertices of the following:

(*a*) Regular hexagon
(*b*) Regular octagon
(*c*) Regular dodecagon

Can you find a set of points on the circumference that are vertices for a regular pentagon? Explain.

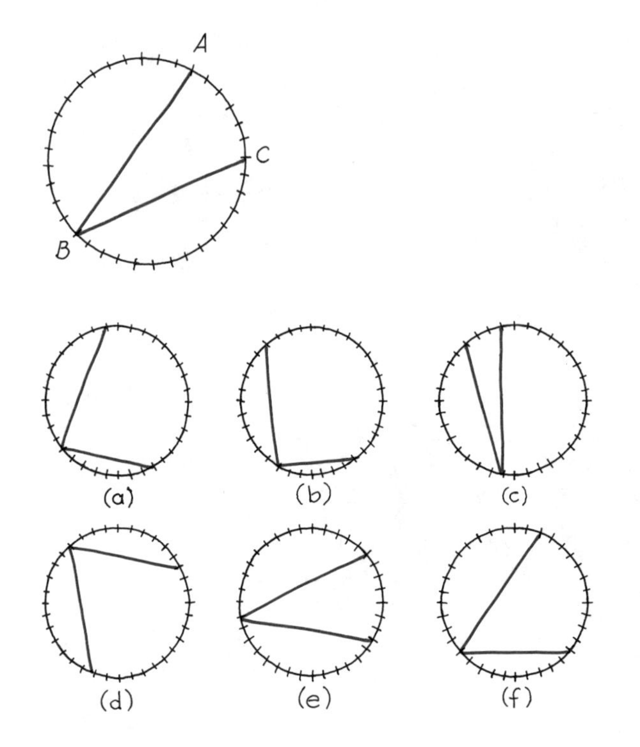

8-2. The angle shown at the left is called an inscribed angle because its vertex is on the circle and its sides are chords of the circle. Arc *AC* contains 60°. Find the measure of $\angle B$.

Find the measures of each of the inscribed angles shown at the left. Keep a record of your results. (The circumferences of the circles have been partitioned into congruent arcs of 10° in each case.)

Circle	Measure in degrees of inscribed angle	Measure in degrees of corresponding arc
(a)		
(b)		
(c)		
(d)		
(e)		
(f)		

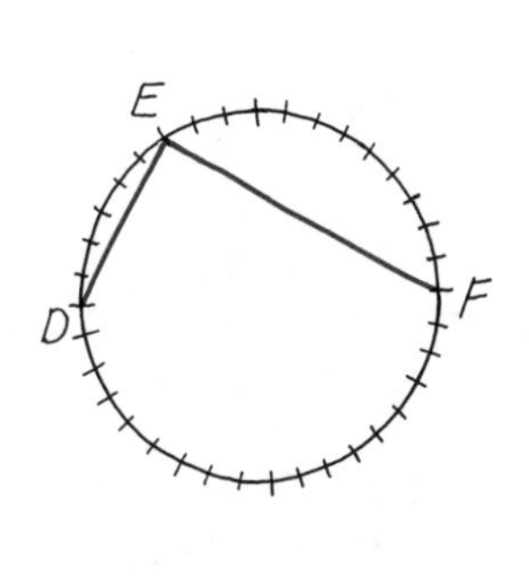

What seems to be the relationship between the measure of an inscribed angle and the measure of its corresponding arc?

Angle *DEF* at the left is an inscribed angle; $\overline{DF}$ is a diameter of the circle. What must be the measure of $\angle E$?

8-3. Suppose you want to construct a line perpendicular to the line on the left at point *P*. Using a compass and straightedge (unmarked ruler), you might proceed as follows:

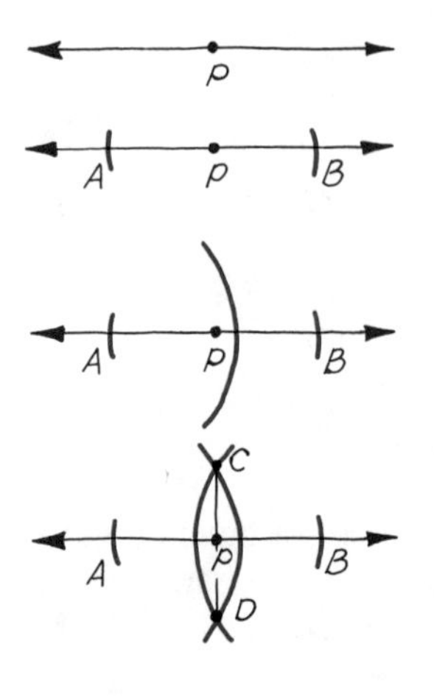

1. Use the compass to mark off two points, *A* and *B*, on the line equidistant from point *P*.
2. Using *A* as a center and opening the compass wide enough so that the distance between its end points is greater than the distance from *A* to *P*, draw an arc.
3. Using *B* as a center and keeping the opening as in step 2, draw another arc intersecting the first arc at points *C* and *D*.

Line segment $\overline{CD}$ is perpendicular to line $\overleftrightarrow{AB}$. How do we know this? If points *ACBD* are the vertices of a quadrilateral, that quadrilateral has congruent sides. (Why?) Then *ACBD* is a rhombus. What is true about rhombuses?

In general, how can you construct the perpendicular bisector of a segment $\overline{MN}$ with compass and straightedge?

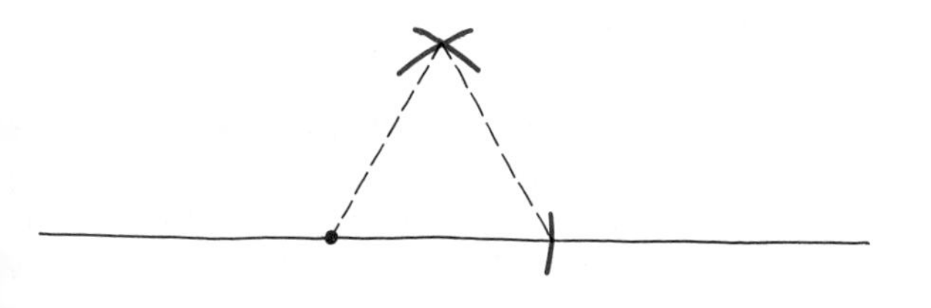

8-4. The diagram at the left shows an equilateral triangle constructed with compass and straightedge. Can you explain what was done? Why *must* the resulting triangle be an equilateral triangle?

8-5. Suppose you want to construct a perpendicular line segment from point *A* to the line segment below it using a compass and straightedge. One way is illustrated below. How can you show that $\overline{AD}$ must be perpendicular to $\overline{BC}$?

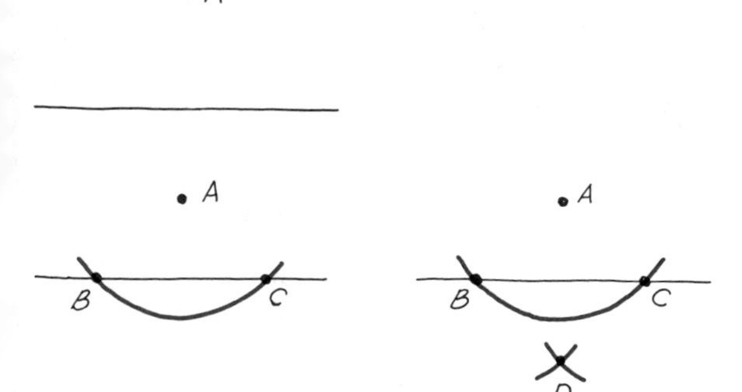

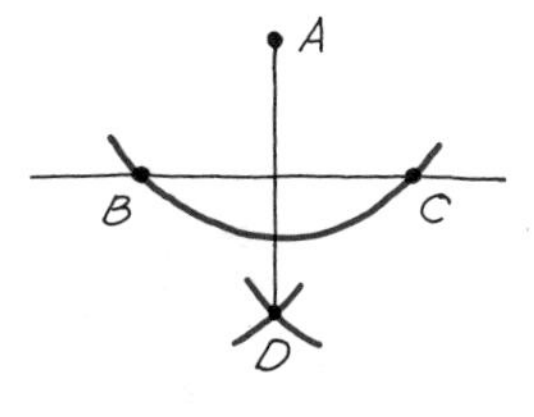

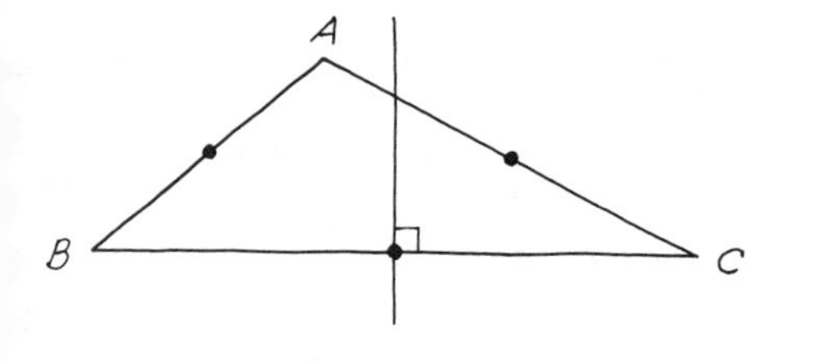

8-6. Draw a triangle like *ABC* at the left and locate the midpoint of each side. Construct a perpendicular to each side at these midpoints. The three perpendicular lines you draw will meet at a point *O*. Use a compass to draw a circle that includes point *A* and has point *O* as its center. If you have made these constructions carefully, points *A*, *B*, and *C* will be on the circle. Why?

8-7. Draw a triangle. Using a compass and straightedge, *inscribe* a circle in that triangle. How will you locate the center for the inscribed triangle?

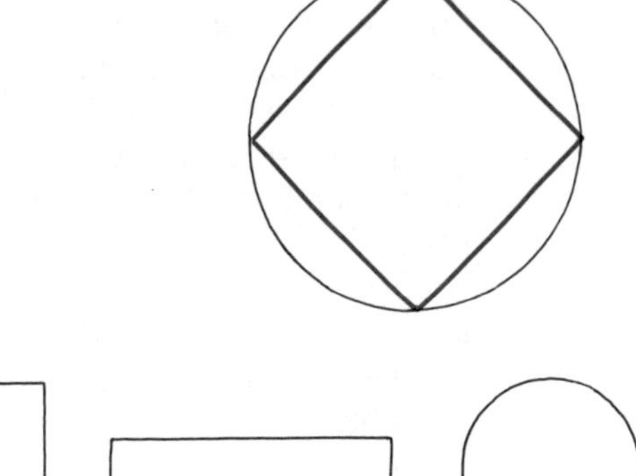

8-8. In the diagram at the left, the square is inscribed in the circle. The measure of the radius of the circle is 6 cm. How long is a side of the square? What is the area of the square?

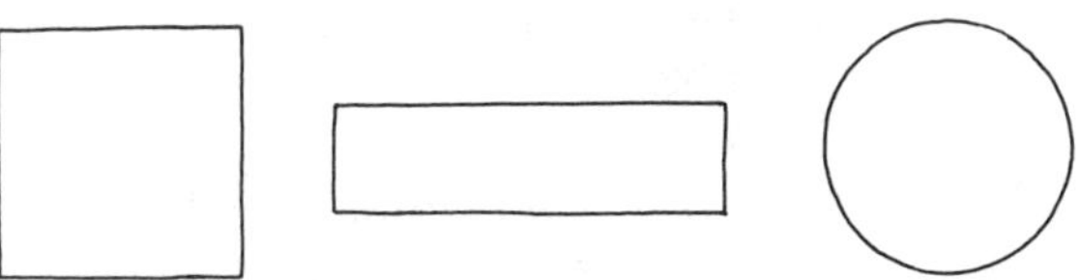

8-9. Cut three pieces of string, each 24 cm in length. Use one piece to make a square, another to make a rectangle, and the third to make a circle, as in the diagram at the left. What is the area of each shape?

Each of these shapes has the same perimeter. Do they all have the same area? Explain.

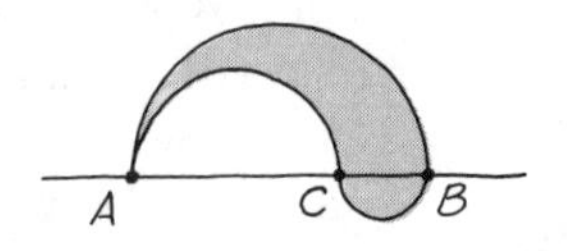

8-10. In the diagram at the left, the length of $\overline{AB}$ is 24 cm and the length of $\overline{CB}$ is one-third the length of $\overline{AB}$. Find the area of the shaded region.

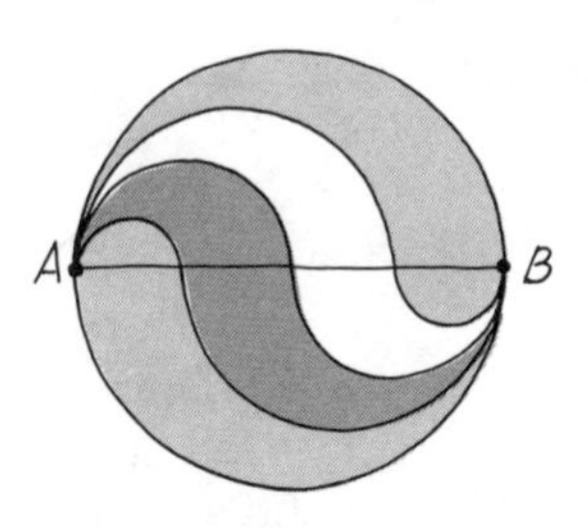

8-11. Show that the four regions indicated in the diagram at the left have the same area and equal perimeters. Let $m(\overline{AB}) = 8$ and assume it has been partitioned into four congruent parts.

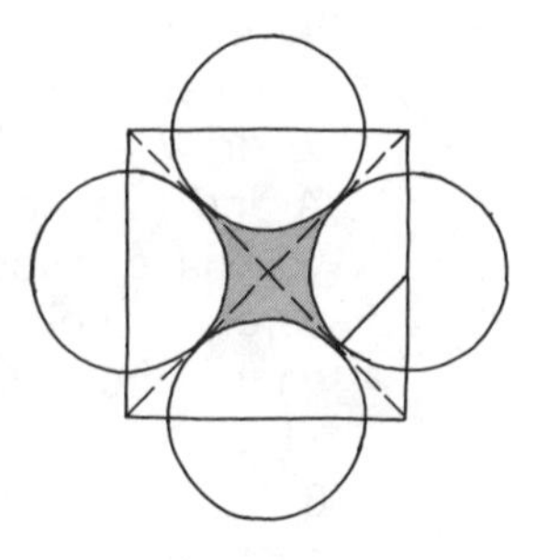

8-12. In the figure shown at the left, the length of a diagonal of the square is 8 cm and each circle has a radius that is one-fourth the length of a diagonal. Can you find the area of the shaded portion of the diagram? Explain how you did it.

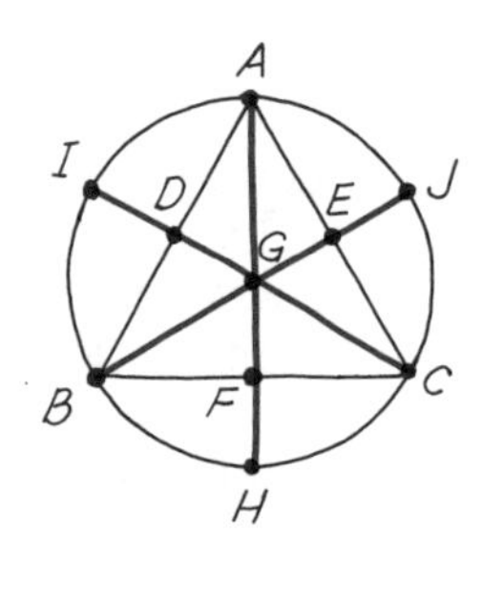

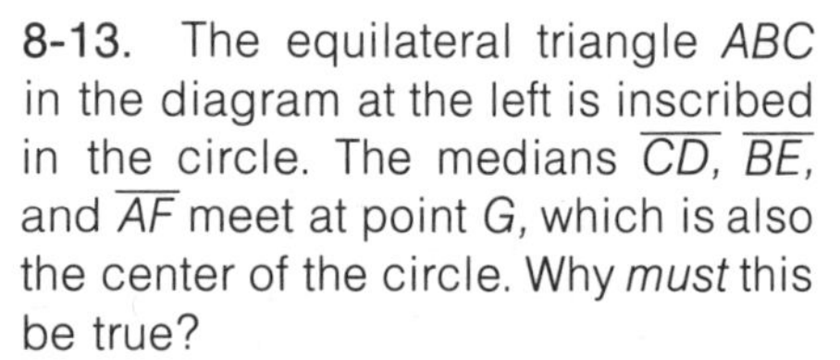

8-13. The equilateral triangle *ABC* in the diagram at the left is inscribed in the circle. The medians $\overline{CD}$, $\overline{BE}$, and $\overline{AF}$ meet at point *G*, which is also the center of the circle. Why *must* this be true?

If the length of the radius of the circle is 8 cm, what is the length of $\overline{GF}$? How long is $\overline{FH}$? How long is $\overline{GE}$? $\overline{EJ}$? $\overline{GD}$? $\overline{DI}$?

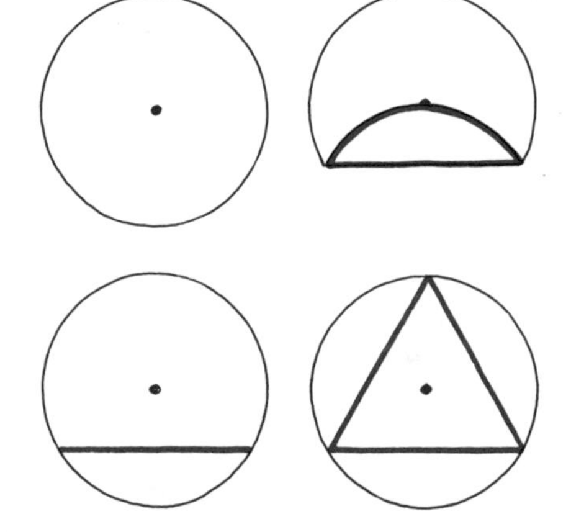

8-14. One way to make an equilateral triangle inscribed in a circle is to fold a circular piece of paper as indicated in the diagram at the left. Try to make an equilateral triangle in this way. Can you explain why this method works?

8-15. The arc of a circle corresponds to a central angle of 40° and has a length of 4π. Can you find the length of the radius of this circle?

8-16. The area of a circle is 9π. Find the length of its radius and the length of its circumference.

8-17. The length of the circumference of a circle is 8π. What is the length of a radius of this circle? Find the perimeter of a square inscribed in this circle.

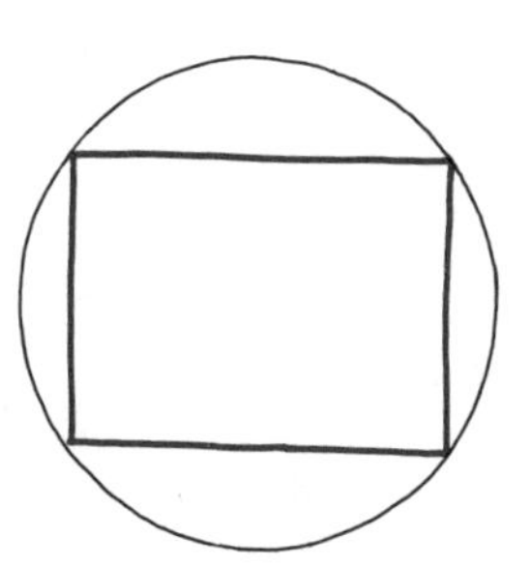

8-18. A rectangle with dimensions 9×12 cm is inscribed in a circle. Find the circumference and area of the circle in terms of π.

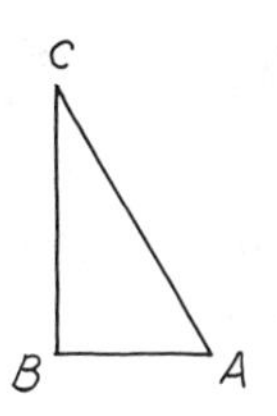

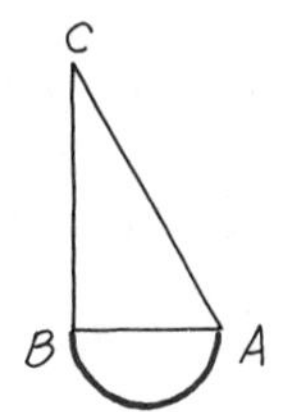

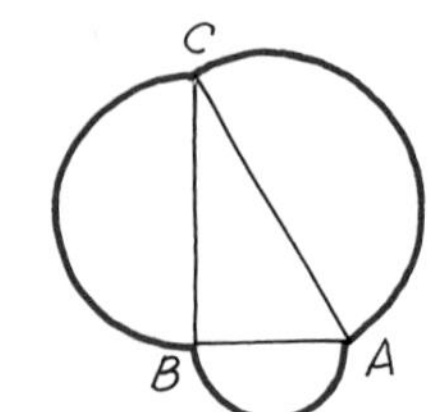

8-19. Draw a right triangle like triangle *ABC* at the left. Using a compass with the midpoint of side $\overline{AB}$ as center and radius one-half the measure of $\overline{AB}$, draw a semicircle (half circle) as shown. Repeat the procedure on sides $\overline{BC}$ and $\overline{CA}$.

If $m(\overline{BA}) = 6$ and $m(\overline{BC}) = 8$, show that the sum of the measures of the areas of the two smaller semicircles is the same as the measure of the area of the largest semicircle. This can be considered an extension of the Pythagorean relationship.

Geometric Transformations

Introduction

In the preceding chapters we became familiar with many different shapes, properties of shapes, and relationships among shapes. Often we used previously developed ideas to discover new ideas. We are now going to draw upon many of these concepts to look at geometry in a different way. We will examine some special geometric relationships we can see in the world around us even though we may not realize they are there.

Have you ever seen a row of trees like the ones pictured in Figure 9-1? Look at the shadows of the trees on the road. These shadows look like parallel lines, just as the trees seem to be arranged parallel to each other. Let us try a little experiment. Take two pencils and stick the points into a piece of styrofoam so that the pencils can stand vertically. Take this arrangement outdoors on a sunny day and look at the shadows formed. You will find that the shadows of the pencils vary in length according to the time of day, but those shadows look like two lines parallel to each other just

Figure 9-1

Figure 9-2

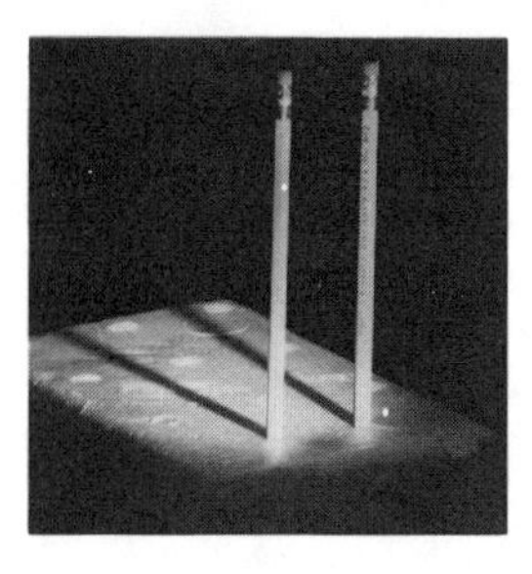

Figure 9-3

as the pencils are parallel (Figure 9-2). Now place the arrangement indoors several feet away from an artificial light source (like a table lamp or high-intensity lamp). What do the shadows look like now? Again the shadows are like two lines, but you will notice that they are *not* parallel (Figure 9-3). Why not?

Sunlight produces a different effect than artificial light. The transformation of the shape of the pencils by sunlight produces parallel shadows, while the transformation by artificial light results in shadows that are *not* parallel. The rays from the sun come from a very great distance—so great we can consider it nearly infinite. Rays from the sun, then, can be considered parallel. The rays from the artificial light source come from a limited distance and are not parallel. This explains why the resulting shadows differ.

In this chapter we will consider some transformations of shapes that occur in our environment. We will investigate space illuminated by the sun and then space illuminated by an artificial light source. Finally, we will study movements in space that do not change the form or size of an object but only its position.

Using Sunlight for Affine Transformations

Figure 9-4

Try the following experiment outdoors on a sunny day. Make models of a square, a rectangle, and a parallelogram like those shown in Figure 9-4, using cardboard and brass fasteners. You can use a paper punch to make the holes. Now take the shapes out into the sunlight. Hold each shape upright against the ground and look at the shadows formed. As you change the angle your model makes with the ground, you will be able to form different shadows. You can think of the shadows as transformations of the shapes onto the ground.

Let us take a closer look at these shadows or transformations. Figure 9-5 illustrates some different shadows that result when the models are held at different angles to the ground. As you can see, the sunlight has transformed the parallelogram into different shapes (the shadows) in (c) and (d). How are the shadows different from the original shapes? In general, the angles of the original shapes have been transformed into *angles of different sizes* and the *shapes are different from their shadows.*

But how are the shapes and their shadows *alike?* What do they have in common? A rather obvious, but important, observation is that each line segment in an original shape corresponds to a line segment in its shadow. That is why this type of transformation is called a **linear transformation.** Not all transformations are linear. Have you ever seen the image of an object in a concave mirror?

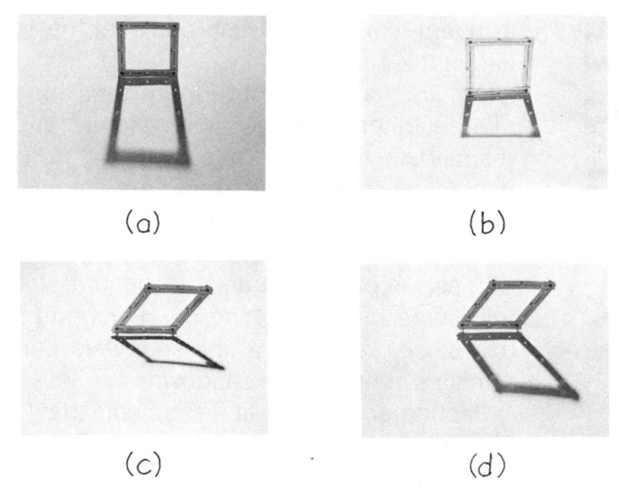

Figure 9-5

If the object has straight sides, its image will not have straight sides. The transformation produced by the concave mirror is non-linear.

Let us get back to the shapes in Figure 9-5. What else remains the same when the original shapes are transformed into their shadows? Notice that when two sides are parallel in the original shape, they remain parallel in the transformed shape. A transformation that maintains parallelism is an **affinity** or an **affine transformation**—terms we encountered in Chapter 3 when we used an affine transformation to develop a set of parallelograms from a rectangle (Figure 9-6).

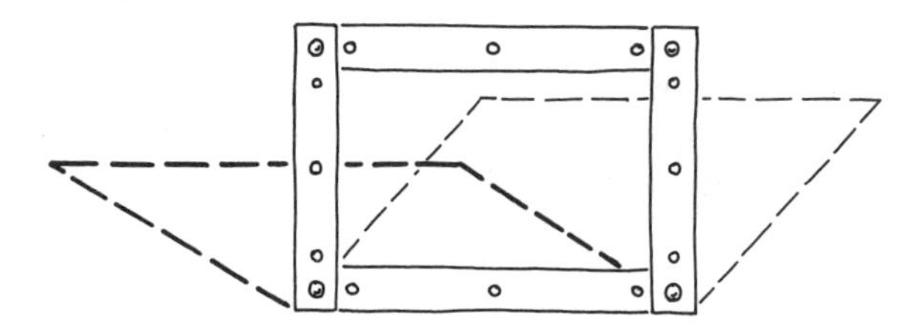

Figure 9-6

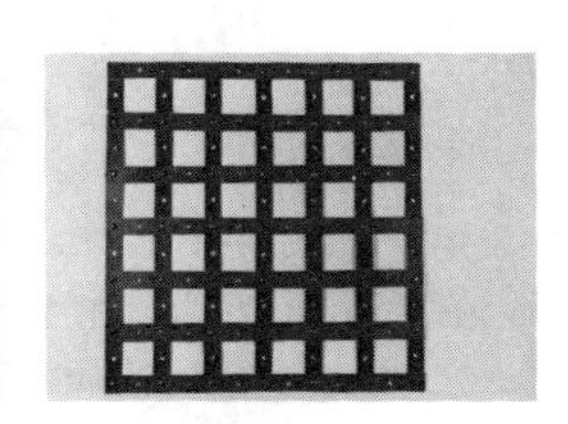

Figure 9-7

Another Affine Transformation: The Shadow of a Square Grid

If you join 14 strips together as in Figure 9-7, you can form a square grid. We can imagine the strips to be line segments. The grid has six rows of six squares for a total of 36 small squares. Make

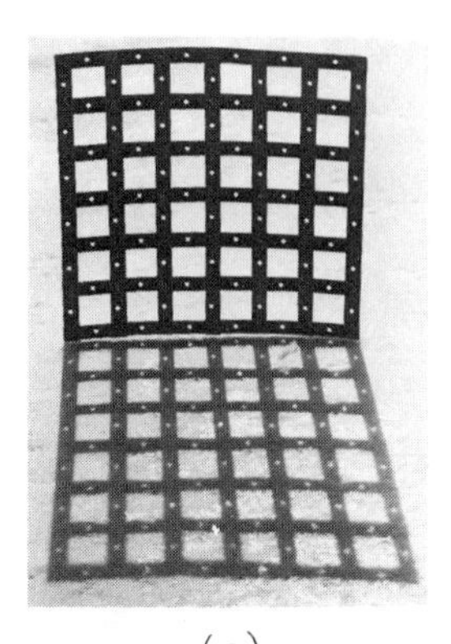

(a)

(b)

Figure 9-8

a model like this and find out what the shadows of this grid look like in the sunlight.

As you can see from Figure 9-8, this shape and its shadows offer another example of an affinity. Again line segments correspond to line segments, and all the line segments parallel to each other in the original grid are also parallel in its shadows. We can also make three other observations.

First, the squares in the original grid become rhombuses, rectangles, or parallelograms. In general, the square is transformed into some member of the parallelogram family. We might expect this since, as we saw, line segments parallel in the original grid remain parallel in the shadows.

Second, look at the line segments that form the sides of the grid. The length of these sides varies as the grid is transformed into its shadows. But something remains the same. What is it? Note that a side of the original grid is partitioned into six equivalent line segments, each one-sixth the length of the side. In the shadows of the grid we see the same relationship! The sides of the grid (though a different length) are partitioned into six line segments, each one-sixth the length of the side of that shadow. In other words, the ratio of the length of a part of a line segment to the length of the entire line segment (1:6) remains the same with this affinity.

Third, the interior of the original grid is partitioned into 36 congruent squares. Each square can be considered $1/36$ the area within the grid. What happens when the grid is transformed? As you can see from Figure 9-8, the interiors of the resulting shadows are partitioned into 36 congruent parts (rhombuses, rectangles, or parallelograms) and each part is $1/36$ the total area within the shadow. In other words, the ratio of the area of a part of the grid to the area of the entire grid (1:36) remains the same with the affine transformation.

Let us summarize the properties that remain the same between a shape and its transformation when that shape undergoes an *affine* transformation:

1. Line segments correspond to line segments.
2. Parallel line segments correspond to parallel line segments.
3. The ratio of the length of a part of a line segment to the entire segment remains the same.
4. The ratio of the area of a part of the shape to the area of the entire shape remains the same.

Suppose that instead of examining the shadows formed by the grid we transform the grid by pushing two opposite corners toward each other (Figure 9-9). Each little square becomes a rhombus. Parallel segments remain parallel. We can see this is another type of affine transformation. What other properties remain the same?

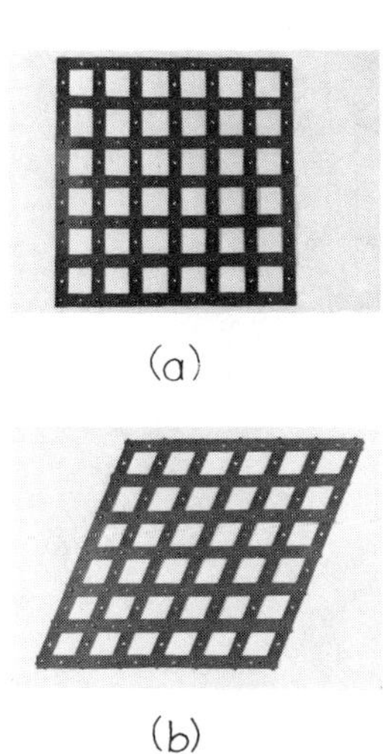
(a)

(b)

Figure 9-9

Let us go back into the sunlight with our grid. Could the grid and its shadow be the same shape and size (congruent)? If we wait until the sun is directly overhead and hold the grid parallel to the plane of the ground, the resulting shadow is identical in size and shape to the original grid. In this special case, the affine transformation produces a congruence between the grid and its shadow. Congruence can be considered a special kind of affinity.

Now let us take a look at the transformation of a shape that is not a polygon. Make a circular strip out of cardboard. Use string to make two perpendicular diagonals. What do the shadows of this circular shape look like? The shadows shown in Figure 9-10 are shaped like ellipses. The circle has been transformed into an ellipse!

Since there were no parallel line segments in the original shape—the circle—we cannot check to see if parallelism is maintained. But notice that the diameters of the original circle bisect each other (each segment is one-half the length of the diameter) and in the *shadow* of the circle—the ellipse—the corresponding line segments (the axes) bisect each other (each segment of an axis is one-half the length of that axis). Also, the diameters of the circle partition its interior into four sectors. The area within each sector is one-fourth the total area within the circle. Correspondingly, the axes of the ellipse partition its interior into four parts. The area within each part is one-fourth the total area within the ellipse. This is indeed an affine transformation.

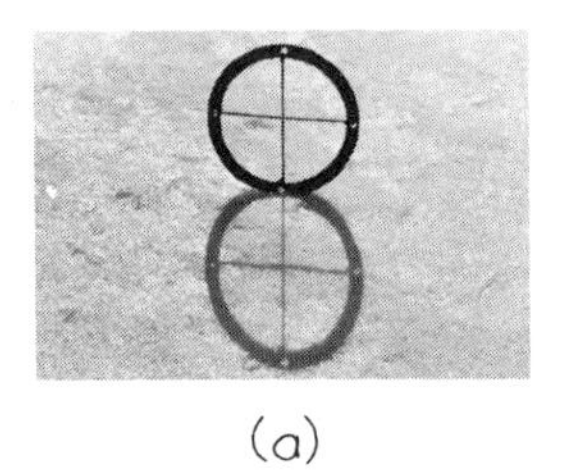
(a)

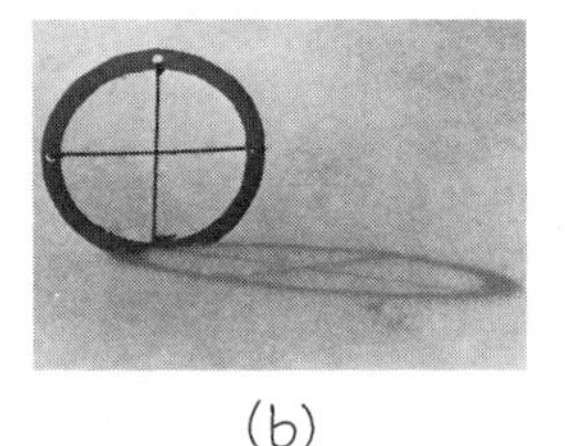
(b)

Figure 9-10

More Affine Transformations

With the help of some flat pieces of elastic, we can discover other affine transformations. Get some flat pieces of elastic, draw a square grid on each piece, and then draw a different design on each grid. Now stretch the elastic so that the line segments of the grid remain parallel. As you can see from Figure 9-11, the original design can be transformed into all kinds of interesting variations. Since we have maintained parallelism, these are affine transformations.

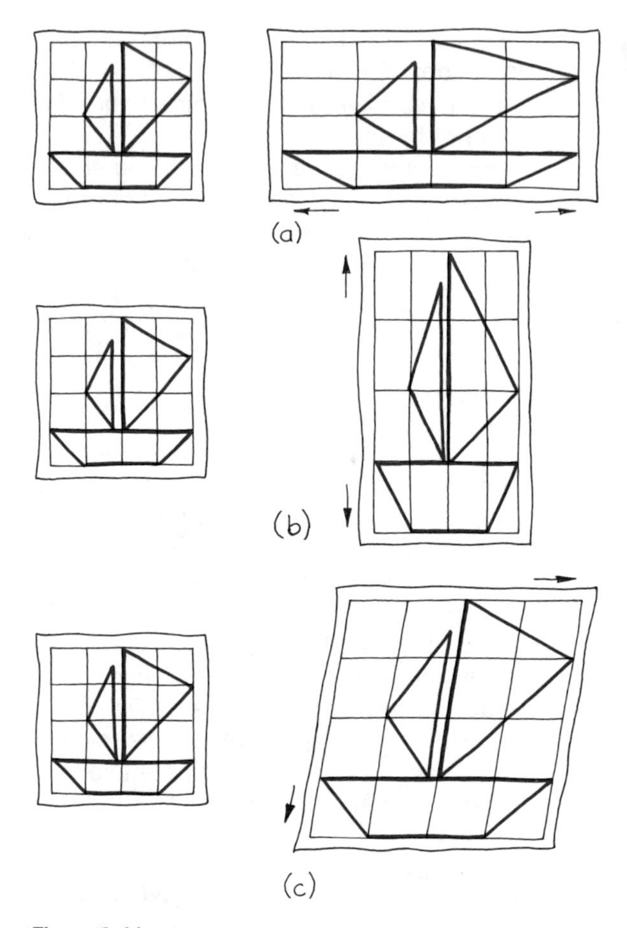

Figure 9-11

Using Artificial Light for Other Transformations

Let us make use of our square grid once more. This time we will examine some shadows formed indoors with a table lamp.* Hold the square grid on a flat surface (the floor or a large table top) a few feet away from the lamp. Look at the shadows that are formed as you alter the position of the grid. What do you notice? How does the shape of the square grid change? What remains the same? Is this an affine transformation?

**A high-intensity lamp works well. Otherwise, cover the lamp bulb with aluminum foil, leaving one hole in the foil for the rays of light to escape. This arrangement will create clear shadows.*

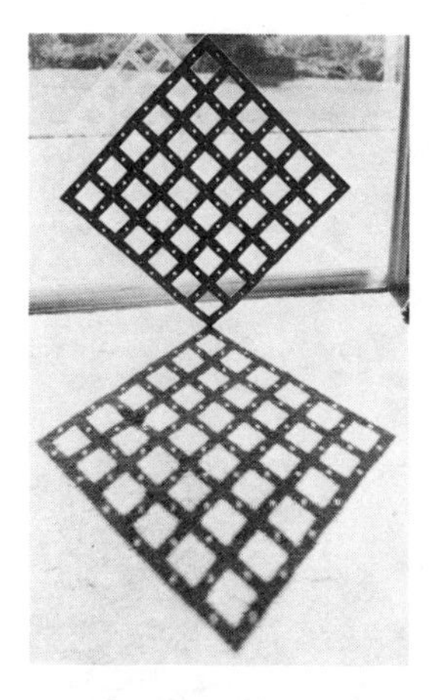

Figure 9-12

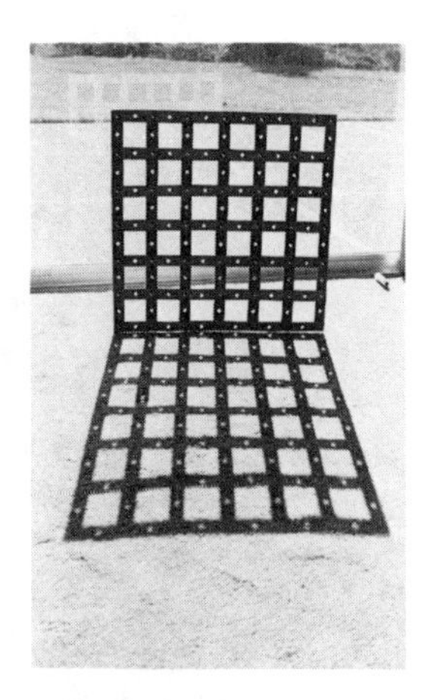

Figure 9-13

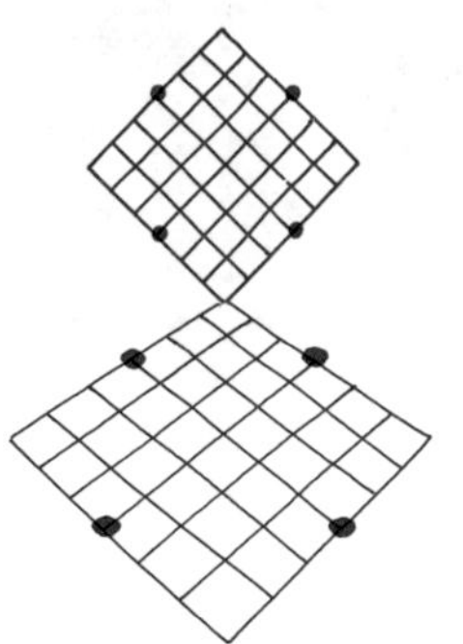

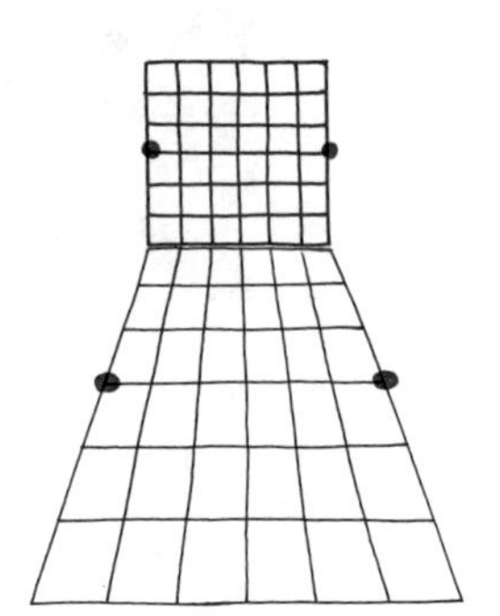

Figure 9-14

As you can see in Figure 9-12, the squares of the original grid are transformed into a different shape. That shape is not a parallelogram, but it is a quadrilateral. This transformation has changed line segments into line segments—it *is* a linear transformation—but line segments parallel in the original grid are *not* necessarily parallel in its shadow. This is *not* an affine transformation. This transformation caused by the rays of artificial light is called a **projective** transformation.

Let us lean an entire side of the square grid on the flat surface and look at the shadow formed. As you can see in Figure 9-13, this is another projective transformation. This time another type of quadrilateral—a trapezoid—corresponds to the square. Notice that this is a linear transformation: lines correspond to lines. It is not an affinity: parallelism is not maintained. Note also that a point in the shadow which corresponds to the midpoint of a side of the square grid is not the midpoint of the shadow of that side (Figure 9-14). Further, while each square in the original grid represents $^1/_{36}$ the area within that grid, the area within each quadrilateral in the shadow of the grid varies and is not necessarily $^1/_{36}$ the area of the interior of the shadow.

Figure 9-15

The Wedding of the Madonna by Raphael.

This type of projection is often used in art. You can see in the famous masterpiece of Raphael in Figure 9-15 how rectangular sections of the pavement were designed as trapezoidal shapes. This gives perspective to the painting: a feeling for the depth of the scene results.

Projections, Affinities, and Similitudes

Suppose we could move our lamp very far away—infinitely far away—from the grid. The shadow of the grid would indicate that we have an affine transformation (that is, parallelism would be maintained). The lamp would be so far away that the rays from the lamp would be like the rays from the sun. We can think of an affinity as a special kind of projection.

Try this. Leave the lamp near the grid but hold the grid so that it is parallel to the wall. What does the shadow look like? How does it compare to the original grid? This time the shadow has the *same shape* as the original grid but is larger. The shadow is *similar* to

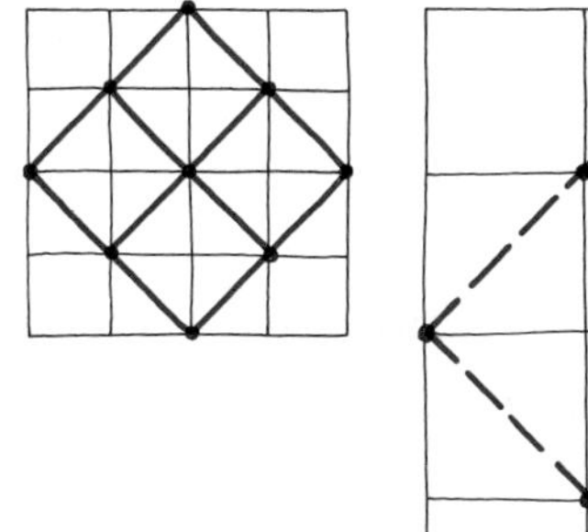

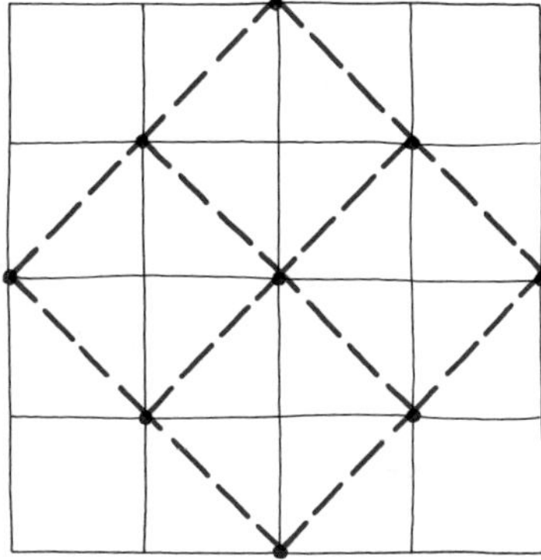

Figure 9-16

Figure 9-17

the original grid. This transformation, called a **similitude,** is a special kind of projection. We make use of similitudes whenever we use a slide projector. When the screen is parallel to the projector, the image on the screen is the same shape (only larger) as the figure on the slide.

Figure 9-16 shows the projection of the grid so that a similitude results. The shape of the image is the same as the original shape (squares). Notice that the size of the angles does *not* change (otherwise the shape would change). The original shape and its image are similar figures and have all the properties of similar figures (for example, corresponding parts have the same ratio).

It is easy to draw a figure that is similar to another figure. Take two pieces of squared paper, one with 16 small squares and the other with 16 larger squares. Draw a figure on one sheet. Now mark off the vertices of that figure at the corresponding points of the second sheet. These are the vertices for the similar figure (Figure 9-17).

Congruence

Let us try one more experiment with the grid. Hold the grid parallel to a wall and gradually move the lamp further and further away from the grid. What happens? The shadow of the grid is similar to the original grid, since we are holding the grid parallel to the wall. You will see that the shadow gets smaller and smaller although it is always larger than the grid.

If we could move our lamp infinitely far away, the shadow would become the same size as the grid. Then the rays from the lamp might be considered parallel, like the rays from the sun. This shows us that **congruence is a special kind of similitude.**

Congruence and Movements

Until now we have talked about movements that change the size or shape of a figure in some way. In this section we will explore movements that do *not* alter the form or size of a figure but only change its position. Such movements or transformations are called **isometries.** The study of isometries is closely connected to the study of congruence. When we wanted to show that two figures were congruent, we showed that we could place one on top of the other and they would coincide exactly. Actually we were moving the figure (without changing its size or shape) from one position to another. These transformations were isometries. We will explore isometries of plane figures on a flat surface or plane.

Translations, Rotations, and Reflections*

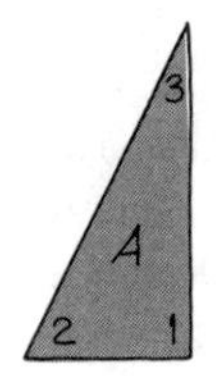

Figure 9-18

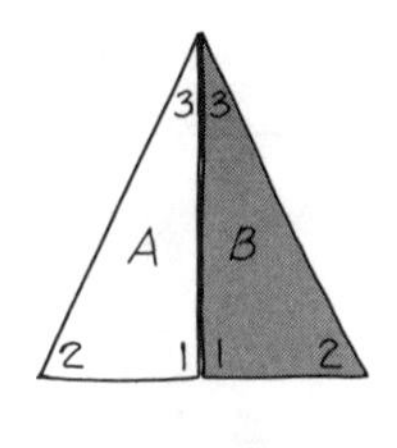

Figure 9-19

Make a right triangular shape out of cardboard like the one in Figure 9-18. We can consider the boundary of this shape to be a right triangle. Call the triangle *A*. The triangle has three angles. Identify the angles as follows: right angle 1, larger acute angle 2, smaller acute angle 3. Now *flip the cardboard shape over from left to right* and trace its outline on a piece of cardboard. Cut out this second cardboard triangular shape and call the right triangle that forms the boundary triangle *B*. Again, we will identify the angles in the same way: right angle 1, larger acute angle 2, smaller acute angle 3 (Figure 9-19). Mark the names of the angles on both sides of the cardboard. As you can see, triangles *A* and *B* are congruent (same shape and size), but something is different. How are they different?

By flipping triangle *A* over we did not change its size or its shape. We did change the *orientation* of the angles. Naming the angles in *A* in a clockwise fashion, we have angles 1, 2, and 3. Naming the angles in *B* in a clockwise fashion, we have angles 1, 3, and 2. We can say that *A* and *B* are congruent but *B* has an "opposite sense" from *A*. The movement or transformation that took place in flipping the triangle over is called a **reflection.** Since only the position of the triangle changed and not its shape or size, a reflection is an isometry. We became acquainted with reflections when we investigated line symmetry in Chapter 4. In this case the line or axis of symmetry is a side of the triangle.

**You might find it useful to review Chapter 4 on symmetry before reading this section.*

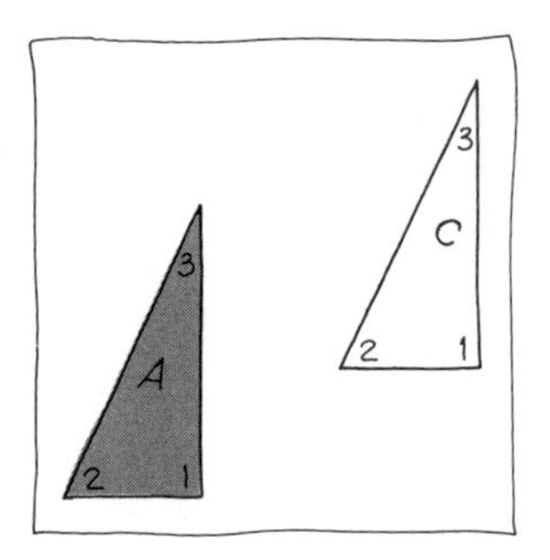

Figure 9-20

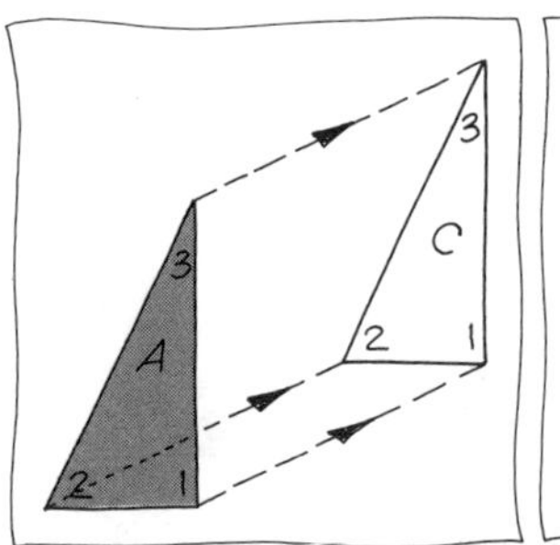

Figure 9-21

Using the cardboard shape as a template, draw the outline of triangle *A* on a piece of paper. Label that outline triangle *C*. Since you made an exact copy of *A* in size and shape, triangles *A* and *C* are congruent. Place triangle *A* on the paper as in Figure 9-20. Suppose we want to make the triangles coincide to show they are congruent. How can we make *A* coincide with *C*?

As you can see from Figure 9-21, by *sliding* *A* toward *C* so that the corresponding sides of *A* and *C* remain parallel we can make triangle *A* coincide with triangle *C*. We have changed the position of *A* on the paper but not its shape or size. This is another type of isometry. This sliding movement is called a **translation.**

Now place triangle *A* on the paper as in Figure 9-22. How can we move *A* so that it will coincide with *C*? We cannot use a translation. For a translation, we need to keep the corresponding sides of *A* parallel as in Figure 9-23. But to make *A* coincide with *C* we cannot keep the sides parallel. We need to *rotate* *A* by using one of its vertices as a center of rotation (Figure 9-24). We came across this rotation movement in Chapter 4 when we investigated rotational symmetry. **Rotation,** then, is a third type of isometry. By rotating the triangle, we do not change its size or shape. Again we are merely changing its position on the paper.

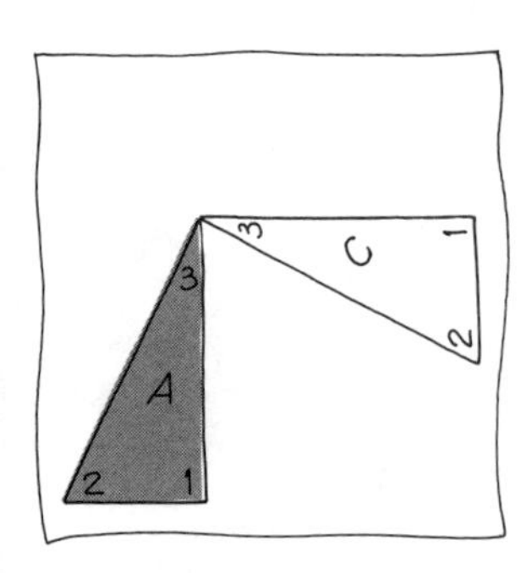

Figure 9-22

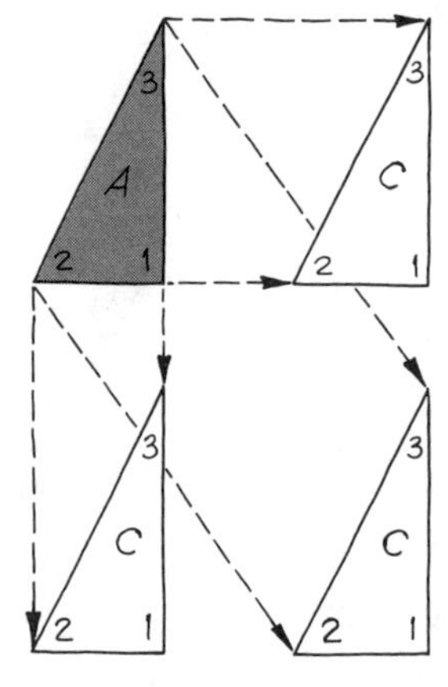

Figure 9-23

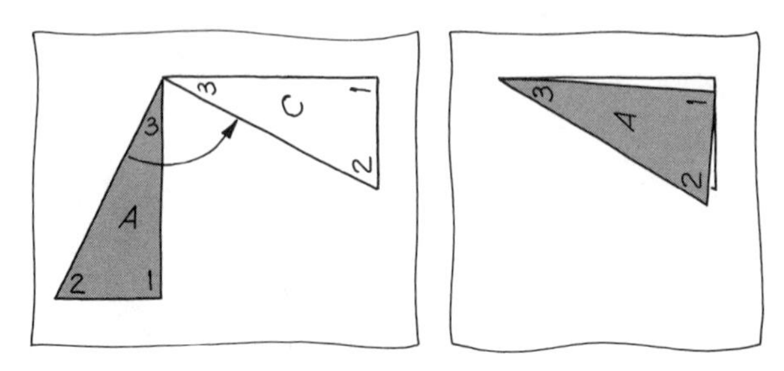

Figure 9-24

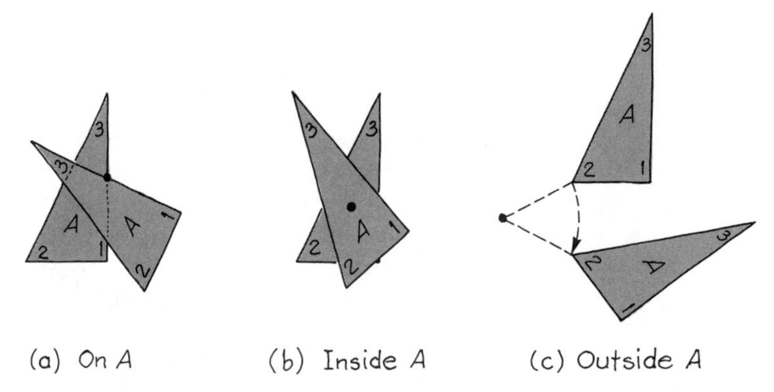

Figure 9-25
Centers of rotation

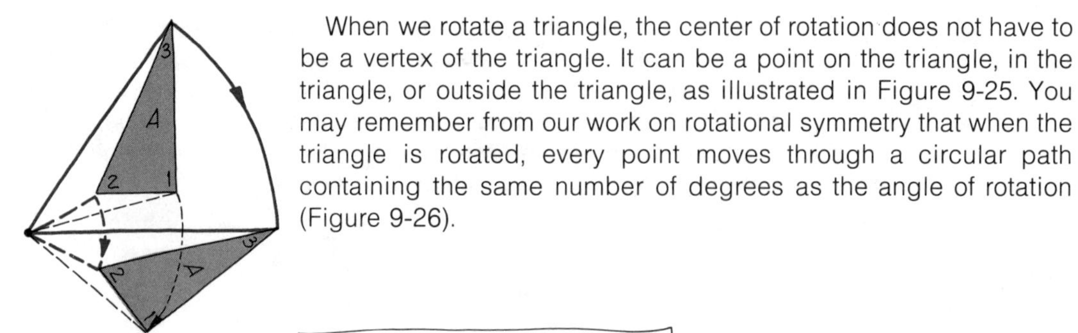

When we rotate a triangle, the center of rotation does not have to be a vertex of the triangle. It can be a point on the triangle, in the triangle, or outside the triangle, as illustrated in Figure 9-25. You may remember from our work on rotational symmetry that when the triangle is rotated, every point moves through a circular path containing the same number of degrees as the angle of rotation (Figure 9-26).

Figure 9-26

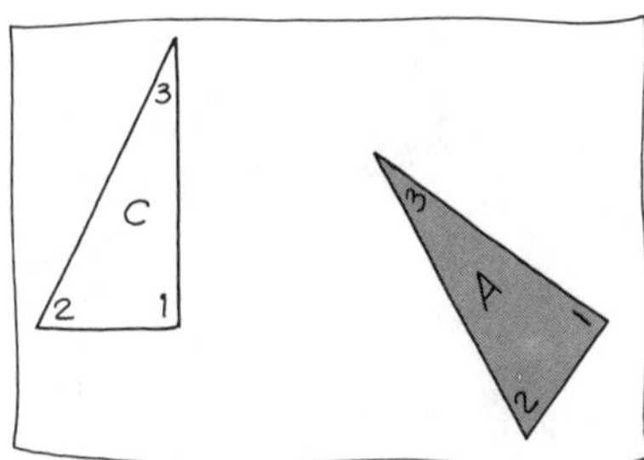

Figure 9-27

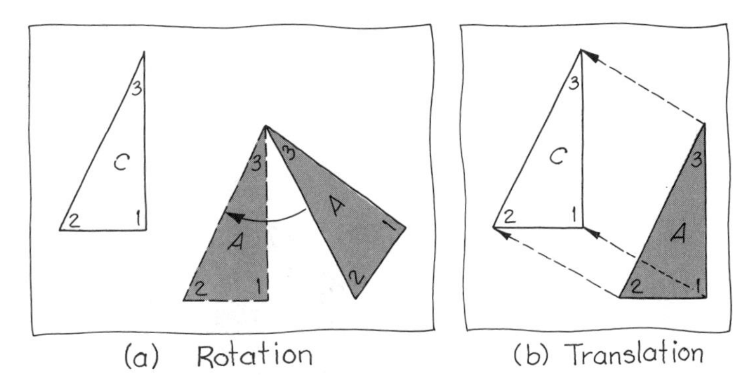

Figure 9-28

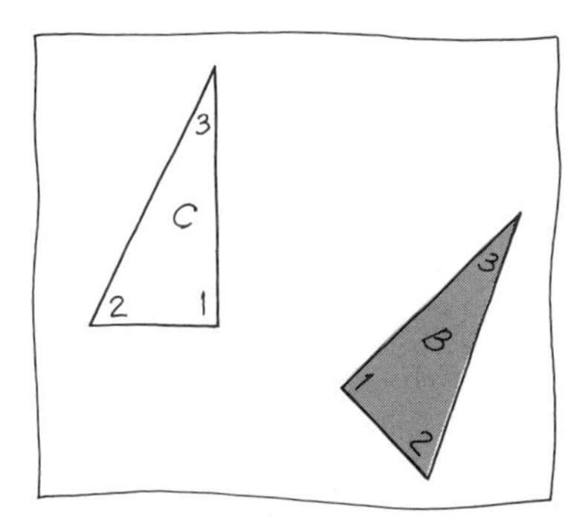

Figure 9-29

Suppose triangles *A* and *C* are arranged in a plane as shown in Figure 9-27. Using a rotation and a translation, we can transform triangle *A* so that it coincides with triangle *C* (Figure 9-28).

Sometimes rotations and translations are not enough. For example, place triangle *B* on the paper near triangle *C* as in Figure 9-29. How can we make triangle *B* coincide with triangle *C?*

We might start by *rotating* *B* about a vertex so that two of its sides are parallel to the corresponding two sides of *C*. Next, we can *translate* *B* so that one side of *B* coincides with the corresponding side of *C*. Finally, by flipping *B* over, or *reflecting* it, we can make triangle *B* coincide with triangle *C* (Figure 9-30).

A reflection is the only movement which requires that the triangle leave the plane. It is also the only movement that changes the orientation of the angles (for example, the clockwise order of angles 1, 2, 3 becomes 1, 3, 2). This is why a reflection is called an **indirect movement** while a translation or a rotation is called a **direct movement** (the orientation of the angles stays the same).

The three movements—translation, rotation, and reflection—are all we need to make any triangle in a plane coincide with another congruent triangle somewhere else in that plane.

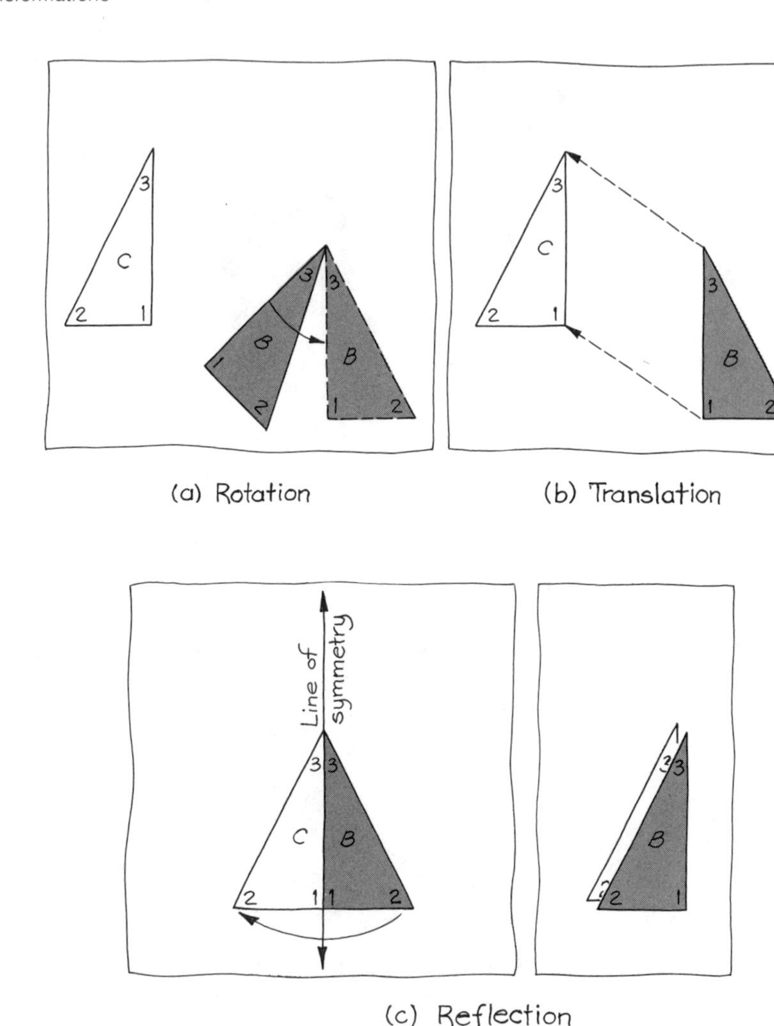

Figure 9-30

An Experience with Rotations and Reflections

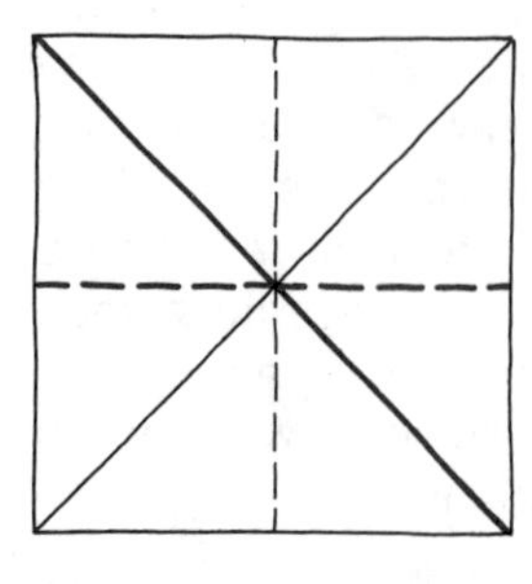

Figure 9-31

Take a large paper square and fold it to find all its axes of symmetry. Draw all these axes as illustrated in Figure 9-31. Next, refold the paper square along the axes of symmetry so that the square of paper becomes a small right triangle (Figure 9-32). Now draw triangle *A* on the paper, as in Figure 9-33, and pierce through all the folds of the paper at the vertices of the triangle (you can use a pin or the point of a compass). Reopen the paper square and use the perforations to draw seven more triangles congruent to the one in position *A* but in different positions within the square. Label

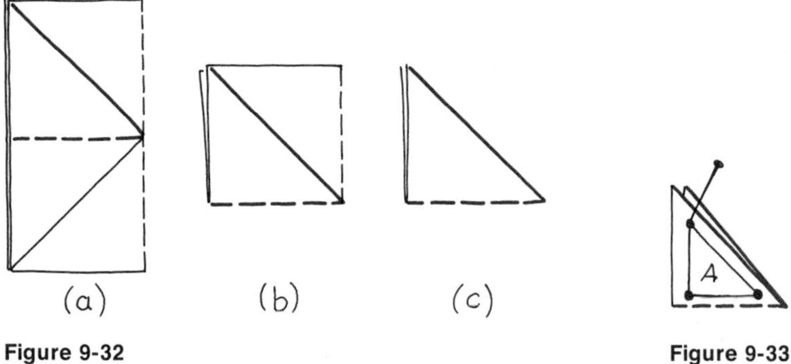

Figure 9-32

Figure 9-33

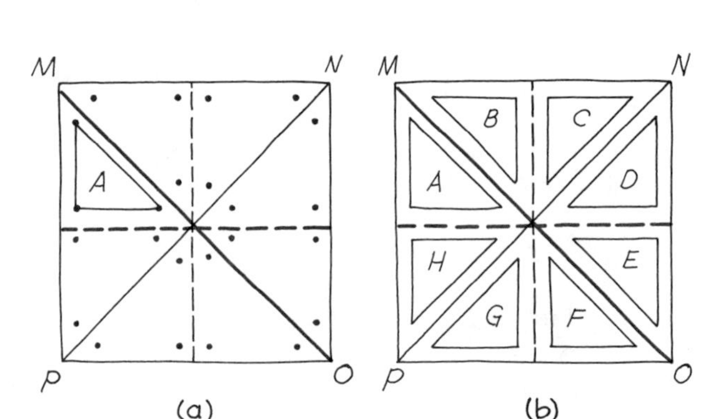

Figure 9-34

the positions of these triangles *B, C, D, E, F, G,* and *H,* as in Figure 9-34(b).

Let us examine this configuration of congruent triangles. What kind of movement would transform a triangle in position *A* to position *B?* A reflection! If we reflect the triangle in position *A* along the brown diagonal axis of symmetry *MO,* it is transformed to position *B.* We can verify this by folding the square along the brown axis *MO.*

If the triangle in *A* is reflected along the brown dotted horizontal axis, it is transformed to position *H.* Similarly, if the triangle in position *B* is reflected along the brown dotted axis, it is transformed to position *G.* This paper square helps us verify the many reflections possible. Can you complete Table 9-1?

Suppose we perform *two* reflections. Starting with the triangle in position *A,* reflect it along the dotted vertical axis, and then reflect it again along the brown dotted horizontal axis. The resulting position is *E.* When reflected along the dotted vertical axis, the triangle in *A* is transformed to position *D.* Then the reflection along the dotted horizontal axis transforms it once more to position *E.* Can you find the resulting positions in Table 9-2?

Table 9-1

Original position	Reflected along axis	Resulting position
B	/	*E*
B	¦	*C*
H	\	?
F	– – – –	?
E	/	?
?	¦	*H*
?	\	*F*
?	– – – –	*G*

Table 9-2

Original position	Reflected along axis	Then reflected along axis	Resulting position
A	\	– – – –	?
C	\	/	?
E	¦	– – – –	?
F	\	/	?

This model shows different reflections of the triangle. Now let us examine some rotations. Make a model exactly like the one you have been using [Figure 9-34(b)], but make this one on clear plastic (you need only trace it onto the plastic). Place the plastic model over the original so that they coincide, and stick a pin at the center (where the diagonals intersect) as in Figure 9-35. As you rotate the plastic model in a clockwise fashion, the positions of the triangles are transformed. The triangle in position *A* can be rotated to position *C, E,* or *G*. Where can the triangle in position *B* be rotated? It can be rotated to *D, F,* or *H.*

Let us examine these transformations more closely. How can the triangle in position *A* be transformed to position *C?* As we saw, it can be *rotated* to *C*. But there is also another way. It can get to

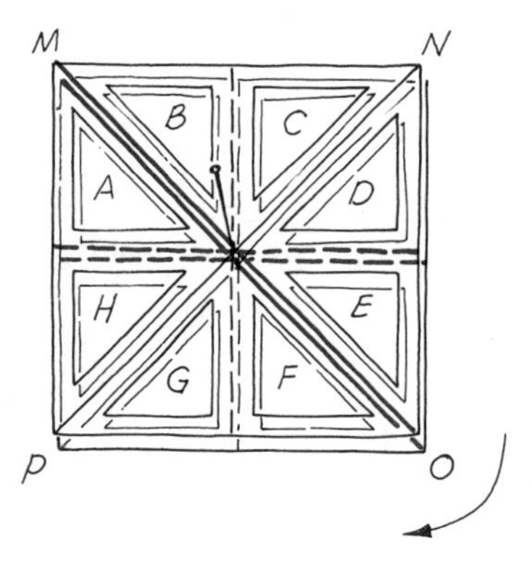

Figure 9-35

position C by two successive reflections: by reflecting A along the brown diagonal axis MO and then along the dotted vertical axis. We can say that a rotation is the same as the *product* of two reflections—it accomplishes the same thing. Using F for reflection, R for rotation, and the symbol $\circ$ to mean "followed by," we can write $F \circ F = R$.

What one movement can accomplish the same thing as two rotations? Rotating the triangle in A to position C and then rotating it to position E is the same as *one* rotation from position A to position E. So the product of a rotation and a rotation is another rotation:

$$R \circ R = R$$

Suppose we have a rotation followed by a reflection. Is there one movement that does the same thing? Rotating the triangle in position A to C and following that movement with a reflection along the diagonal axis NP brings the triangle to position D. But these two movements are the same as one reflection of A along the dotted vertical axis, transforming it directly to position D. Therefore,

$$R \circ F = F$$

Similarly, if we start with a reflection and follow with a rotation, we will move the triangle to a position that could have been reached by a single reflection. Try it and you will see. Thus,

$$F \circ R = F$$

We can summarize our findings in the table of compositions below, which reads like a multiplication table:

$R \circ R = R$
$R \circ F = F$
$F \circ R = F$
$F \circ F = R$

$\circ$	R	F
R	R	F
F	F	R

This table has the same structure as the addition table for odd and even numbers! That is, if O is an odd number and E is an even number, then the following is true:

$$E+E=E$$
$$E+O=O$$
$$O+E=O$$
$$O+O=E$$

+	E	O
E	E	O
O	O	E

Although this composition of isometries appears to have no relationship to addition of odd and even numbers, we can see a similarity of structures. Recognizing similar structures is an important part of thinking mathematically.

Isometries in Everyday Life

We have investigated the three isometries—translation, rotation, and reflection—as they apply to plane figures on a flat surface. Actually, examples of these movements appear quite often in our everyday lives. Isn't a rotation involved when we turn a door handle or open a door? Certainly a reflection is involved every time we look in a mirror or use a rubber stamp. Doesn't rearranging books on a shelf involve translation?

Often, more than one of these isometries is involved or one type of movement causes another kind to occur. For example, when we use a screwdriver we are using a rotation to cause a translation (point of screw into the wood). When you roll up a car window, you are using a rotation to create a translation. What other examples of translations, rotations, or reflections can you think of? How were these movements used to create the patterns in Figure 9-36?

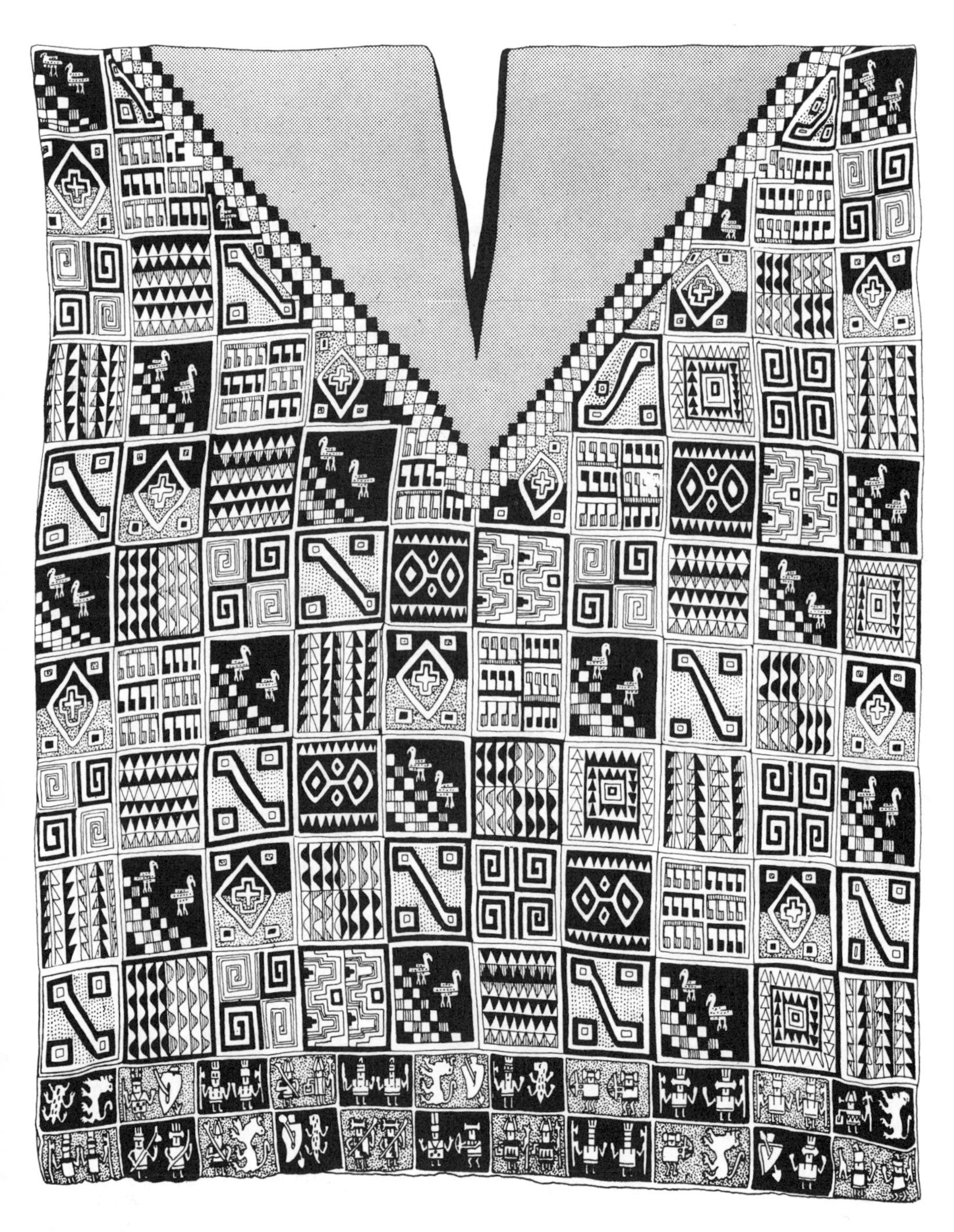

Figure 9-36

Le Roy H. Appleton, *American Indian Design and Decoration,* Dover Publications, Inc., New York, 1971. Reprinted through the permission of the publisher.

More Experiences

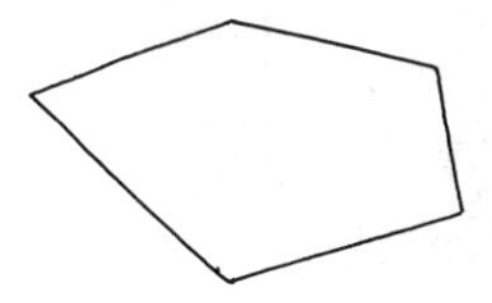
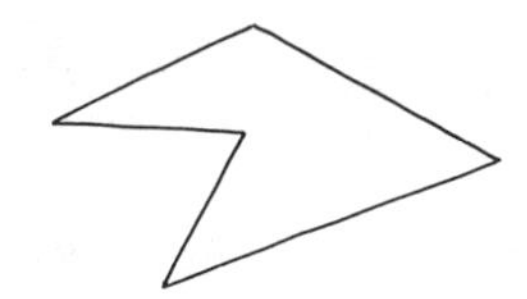

9-1. Cut some different shapes out of cardboard (see diagrams at the left for examples). Make some with convex boundaries and others with nonconvex boundaries. Then examine the shadows that can be cast by these shapes in sunlight. Can a convex shape cast a nonconvex shadow? Can a nonconvex shape cast a convex shadow? Draw some of your results.

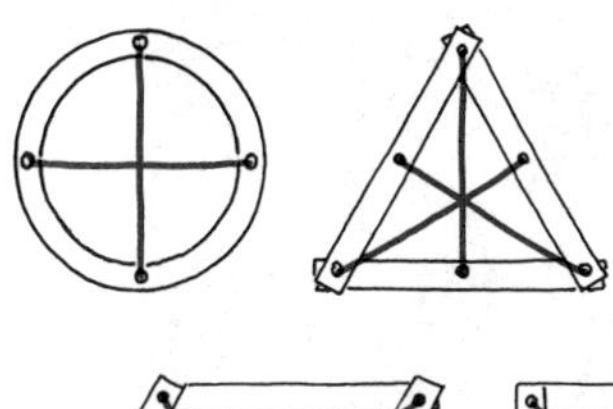
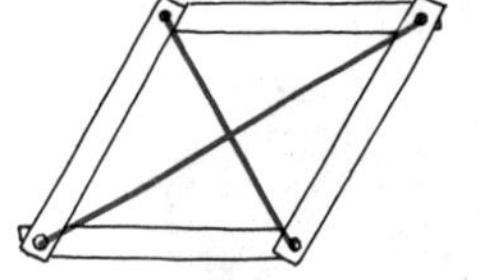
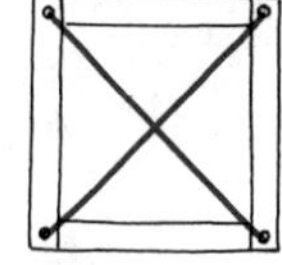
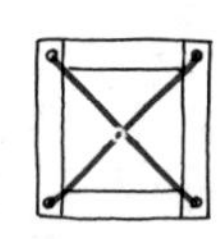

9-2. Make shapes like the ones at the left out of cardboard or strips, and use elastic thread or string to form diagonals, medians, or diameters as shown.

(*a*) Examine the shadows of these shapes in sunlight. What characteristics remain unchanged? What changes? How do you know the shadows represent an affine transformation of the original shape?

(*b*) You can examine the shadows of these shapes formed by artificial light as follows:

1. Remove the shade from a table lamp and cover the bulb with aluminum foil except for a small opening.
2. Hold each shape in the path of the light rays and examine the shadows formed on the floor or the wall.

How is the transformation of the shape by artificial light different from that caused by sunlight?

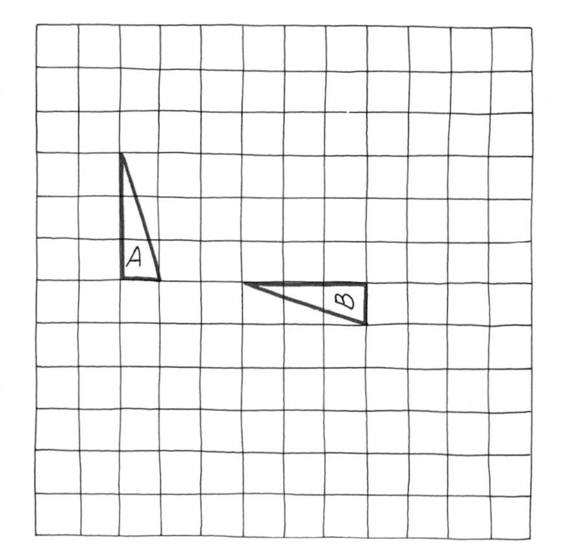

9-3. Triangles *A* and *B* on the grid at the left are congruent. One way to show they can be made to coincide is as follows:

1. Slide *A* six spaces to the right horizontally.
2. Rotate 90° clockwise about the vertex *A*.
3. Flip along the shortest side.

Can you find other sets of isometries that will make the triangles coincide? How many different ones can you find?

Can you make them coincide without using a sliding movement? Without rotating? Without flipping or reflecting?

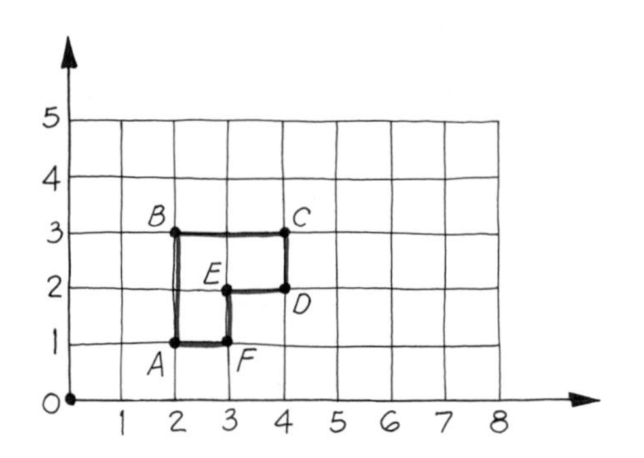

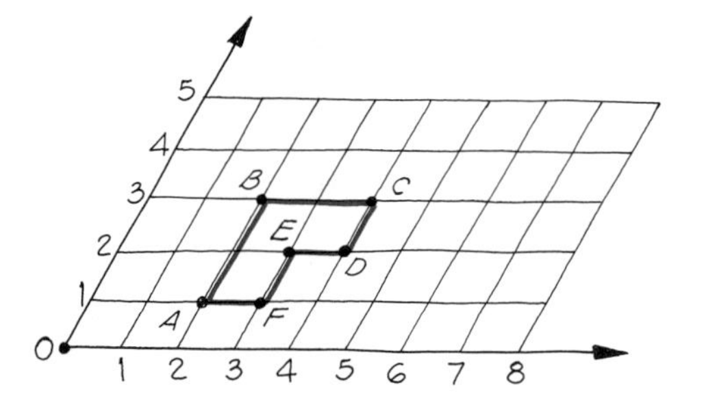

9-4. The polygon in the diagram at the left is drawn on a square grid with horizontal and vertical axes partitioned into congruent segments and numbered as shown. Each point of the polygon has a set of coordinates: *A* (2, 1), *B* (2, 3), and so on. Name the coordinates for all the vertices of this polygon.

Suppose the vertices of this polygon are located on a different grid like the one at the left. How will the polygon change? Locate the vertices on the grid. What remains the same in this transformation? What changes? What kind of transformation is this?

9-5. Draw rectangle *ABCD* on a 1-cm-square grid as in the diagram at the top of the next page and locate the coordinates of the vertices of the rectangle.

(*a*) Coordinates of *A*: (   )
(*b*) Coordinates of *B*: (   )
(*c*) Coordinates of *C*: (   )
(*d*) Coordinates of *D*: (   )

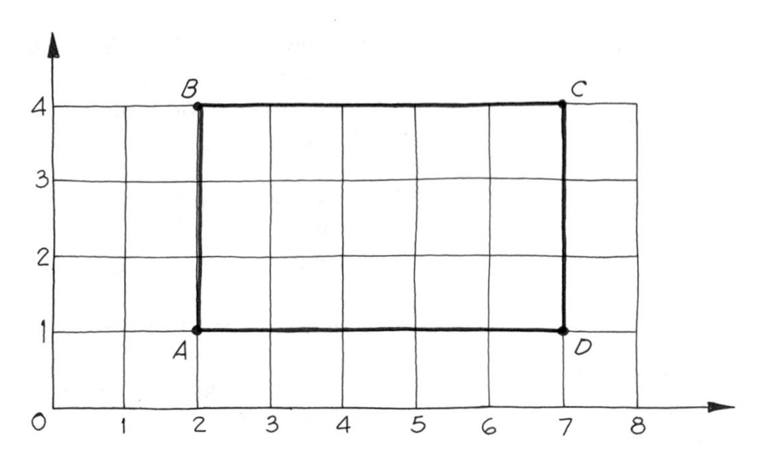

What is the length of $\overline{AB}$ in centimeters? $\overline{AD}$? What is the area of *ABCD* in square centimeters?

Now draw *ABCD* on a 2-cm-square grid (each square 2 × 2 cm). Locate each coordinate of *A*, *B*, *C*, and *D*. What is the length of $\overline{AB}$ in centimeters? $\overline{AD}$? What is the area of *ABCD* in square centimeters?

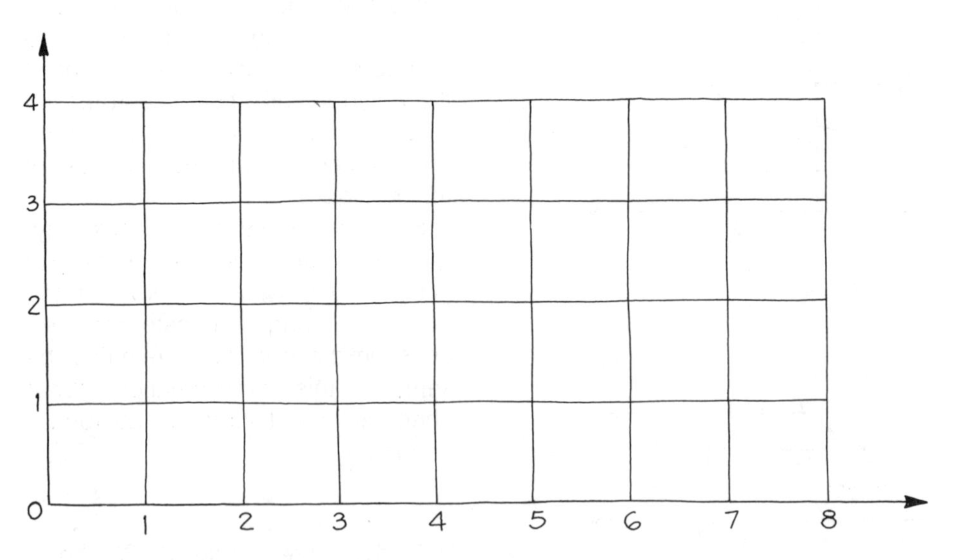

Grid	Length $\overline{AB}$ (in cm)	Length $\overline{AD}$ (in cm)	Area *ABCD* (in sq cm)
½ × ½			
1 × 1			
2 × 2			
3 × 3			
4 × 4			

What would be the result if *ABCD* were drawn on a 3-cm-square grid? A 4-cm-square grid? A ½-cm-square grid? Summarize your results in a table such as the one at the left.

What can you conclude about the way the size of *ABCD* changes in this transformation?

Suppose you draw rectangle *ABCD* on a rectangular grid. For example, draw *ABCD* on a 1×2 cm grid (each box 1×2 cm). This is like stretching the original rectangle. What is the length of $\overline{AB}$ in centimeters? $\overline{AD}$? What is the area of *ABCD* in square centimeters?

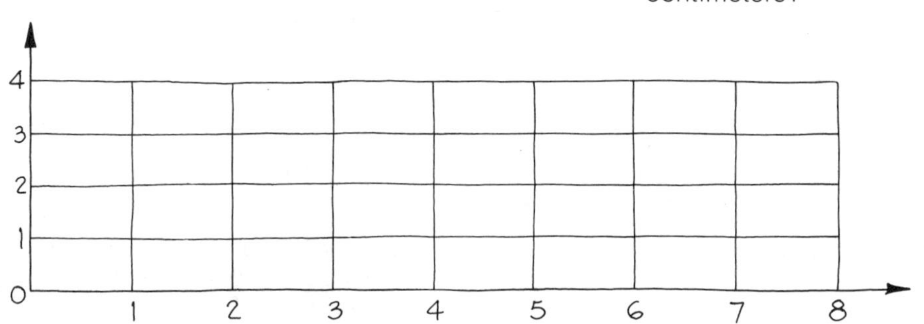

Draw *ABCD* on the following grids: $1 \times 1\frac{1}{2}$, $1 \times 2\frac{1}{2}$, 1×3 cm. How does the length of $\overline{AB}$ vary? How about the length of $\overline{AD}$? How does the area change?

Summarize your results in a table such as the one at the left.

Grid	Length $\overline{AB}$ (in cm)	Length $\overline{AD}$ (in cm)	Area *ABCD* (in sq cm)
1×1			
$1 \times 1\frac{1}{2}$			
1×2			
$1 \times 2\frac{1}{2}$			
1×3			

What can you conclude about the way the size of *ABCD* changes in this stretching transformation?

9-6. The two triangles at the left are similar and their corresponding sides are parallel ($\overline{AC}$ parallel to $\overline{DF}$ and so on).

If you draw a line through *A* and *D*, another line through *B* and *E*, and a third line through *C* and *F*, these three lines meet at a point *L* as shown in the diagram.

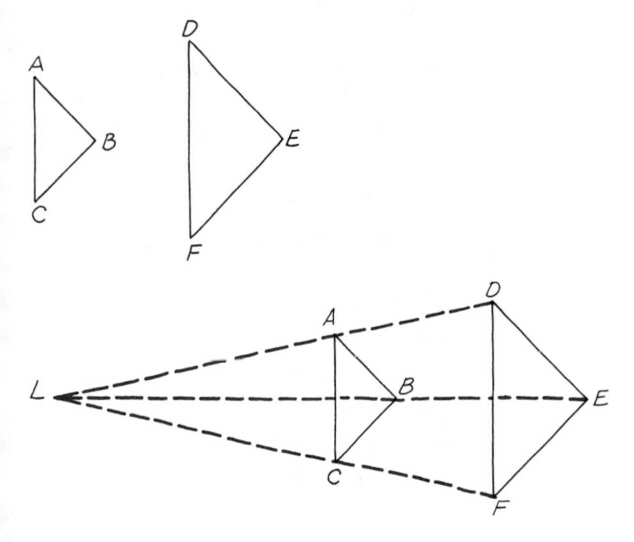

Point *L* is like an artificial light source, and *DEF* can be thought of as the shadow of a triangular region *ABC* on a wall as *ABC* is held parallel to the wall.

What is the ratio of the measures of the corresponding sides of the two similar triangles?

What is the measure of $\overline{LA}$? $\overline{LD}$? What is the ratio of m ($\overline{LA}$) to m ($\overline{LD}$)?

Measure $\overline{LB}$, $\overline{LE}$, $\overline{LC}$, and $\overline{LF}$, and form the same kind of ratios. What do you discover?

9-7. Exercise 9-6 suggests a way to transform a shape into similar shapes of a different size:

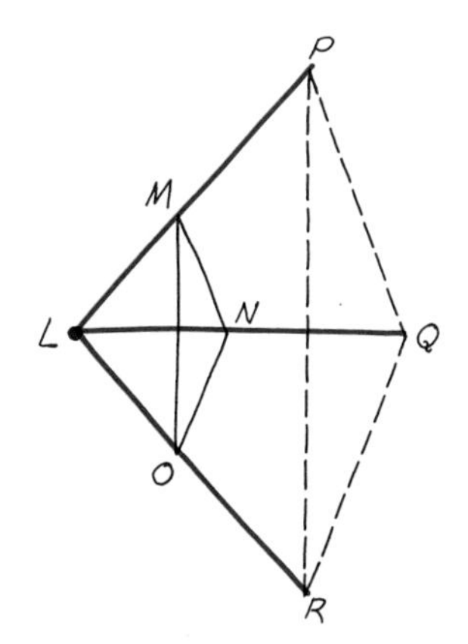

1. Draw triangle *MNO.*
2. Choose some point *L* in the plane.
3. Draw line segment $\overline{LM}$. Measure $\overline{LM}$.
4. Extend $\overline{LM}$ to *P* so that m ($\overline{LM}$) = m ($\overline{MP}$).
5. Draw $\overline{LN}$ and $\overline{LO}$ and proceed as in steps 3 and 4, extending $\overline{LN}$ to *Q* and $\overline{LO}$ to *R.*
6. Draw line segments $\overline{PQ}$, $\overline{QR}$, and $\overline{PR}$.
7. Triangle *PQR* is similar to triangle *MNO.*

What is the ratio of the measures of corresponding sides?

Suppose m ($\overline{LP}$) is three times m ($\overline{LM}$). What would be the ratio of the measures of the corresponding sides?

Draw some shapes and try to transform them into similar shapes by using this method.

9-8. Which movements (rotation, reflection, translation) are demonstrated in the use of the following:

(*a*) Radio dial
(*b*) Car window
(*c*) Pulley
(*d*) Key
(*e*) Water faucet
(*f*) Phonograph
(*g*) Book
(*h*) Light switch
(*i*) Wheels of car
(*j*) Piston

Make a list of other examples of these movements in everyday life.

9-9. If you cut out a design on cardboard, such as the one at the left, you can use it as a template to make all kinds of interesting patterns. One example is shown here.

What movements were used to make this pattern? In what order? Make a template with a different design and create your own patterns.

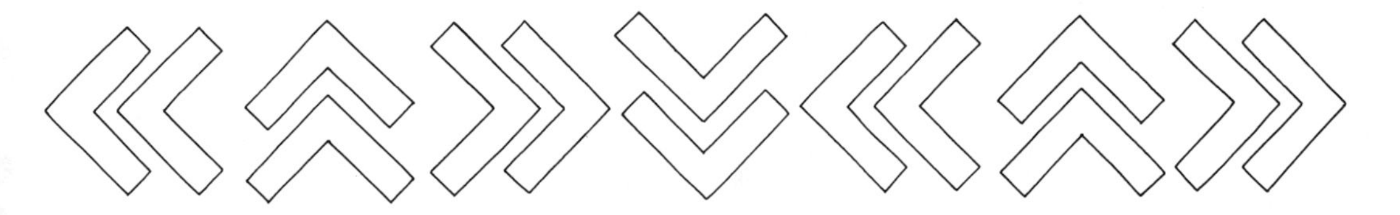

10 Solid Figures

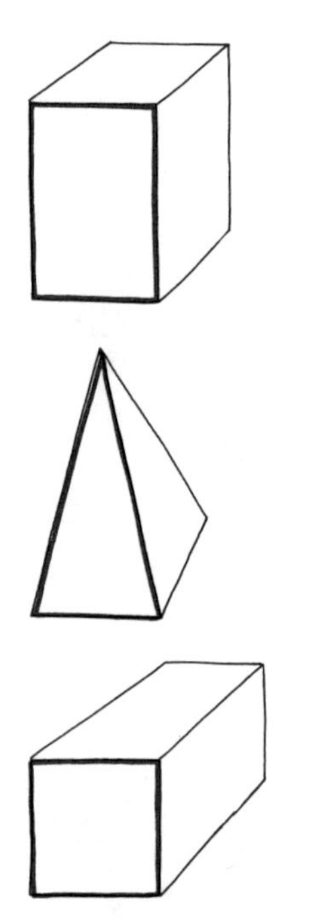

Figure 10-1

Plane Figures and Solid Figures

In previous chapters we became acquainted with many shapes—such as triangles, quadrilaterals, circles—and their properties. These shapes can be considered sets of points on a flat surface or plane; they are called **plane figures.**

We now turn our attention to shapes whose points are on more than one plane. In the next three chapters we will explore properties of three-dimensional figures, which can be called **space figures** or **solid figures.** In the real world, triangles, quadrilaterals, circles, and other plane figures are found as parts of solid figures (Figure 10-1). As we study solid figures, we will be drawing upon much of the knowledge we have accumulated about plane figures.

Squares and Cubes

You will recall from Chapter 1 that a square, like the one pictured in Figure 10-2, is an example of a special plane figure—a simple

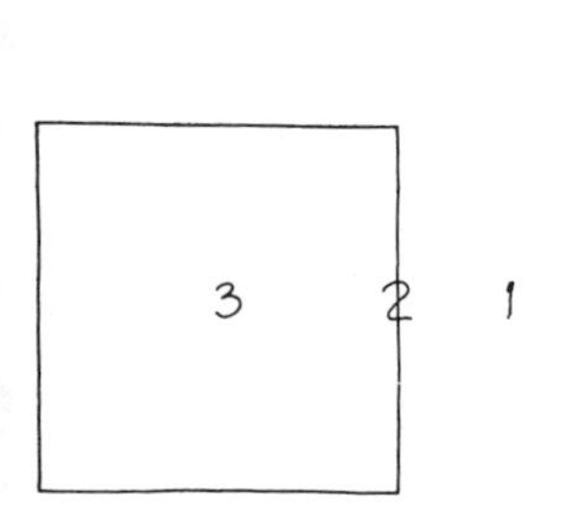

Figure 10-2

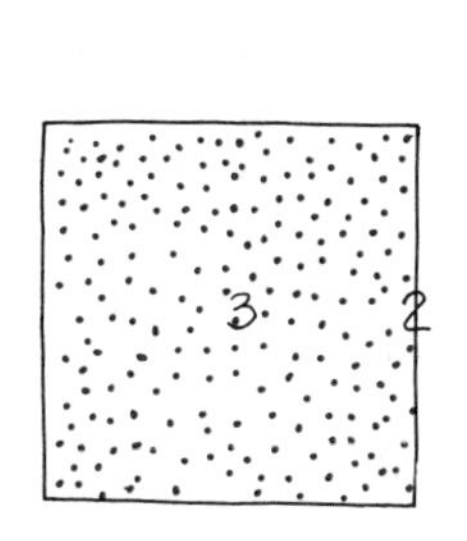

Figure 10-3

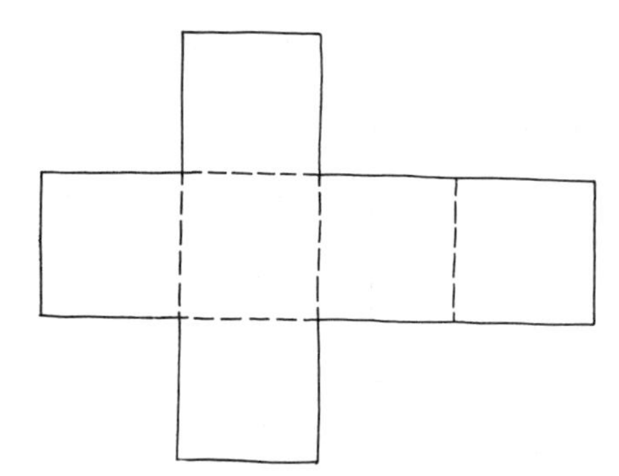

Figure 10-4

closed curve. It partitions the plane into three sets of points: (1) points outside the figure, (2) points that belong to the figure, and (3) points inside the figure. The square as a plane figure is composed of only that set of points forming the figure (set 2). When we discussed the area within the plane figure, we were referring to the portion of the plane bounded by the plane figure, or the **plane region.** In Figure 10-3 the plane region (in this case a square region) is composed of all the points that belong to the square (set 2) in addition to all the points inside the square (set 3).

Now let us look at the situation in three dimensions. Copy the pattern of six squares shown in Figure 10-4 onto a piece of oaktag (make the squares larger than those in this diagram). Cut along the solid lines and fold along the dotted lines to form a model of a cube.* You can use transparent tape to hold it together (Figure 10-5).

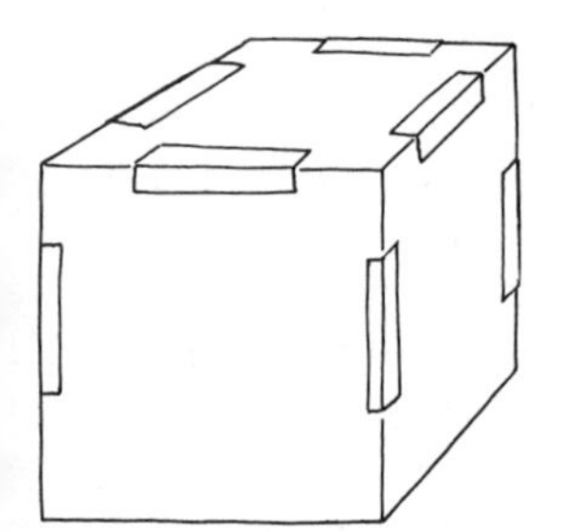

Figure 10-5

The cube is a model of a solid figure. This figure can be thought of as a set of points in space forming a *simple closed surface.* The simple closed surface *partitions all the points in space into three sets:* (1) points outside the figure, (2) points that belong to the figure, and (3) points inside the figure. Although the cube is called a solid figure, when we refer to the solid we will mean its *surface only* (set 2). The portion of space within the cube, along with the cube that forms its boundary (sets 2 and 3), will be considered a **solid region.**

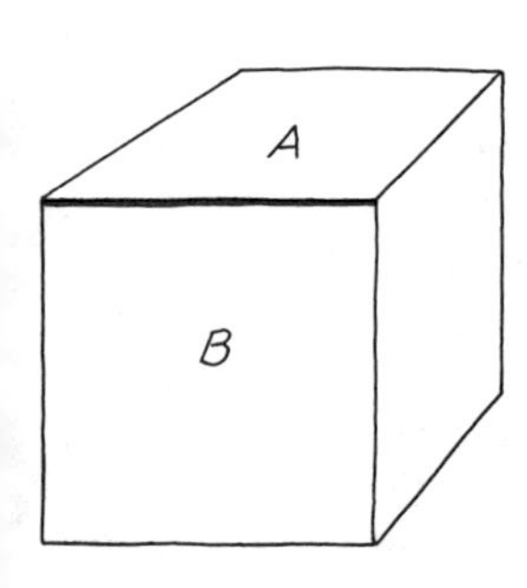

Figure 10-6

Parts of solid figures offer us good models of plane figures. The six squares we used form the six **faces** of the cube. These faces are actually square regions. In Figure 10-6, we can think of face *A* and face *B* as two sets of points. What points do the two sets have in common? The common points would be the intersection of the two sets. As indicated, the intersection can be represented by the

---

**You will find you can fold the cardboard more easily if you score it by running a sharp knife or scissor blade along the dotted lines (using a metal straightedge as a guide) and bending the oaktag away from the cut. The oaktag is difficult to fold without scoring.*

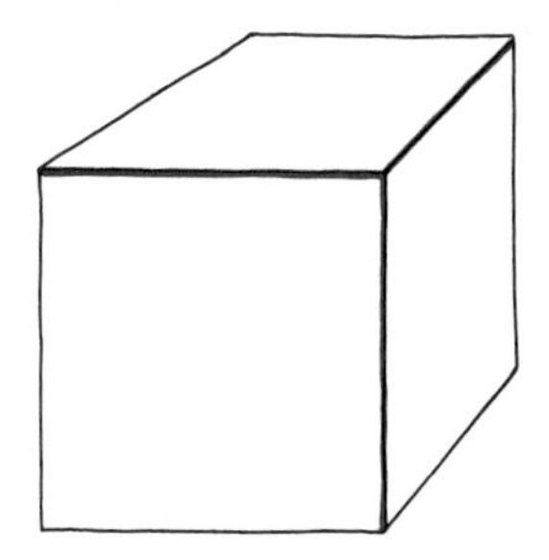

Figure 10-7

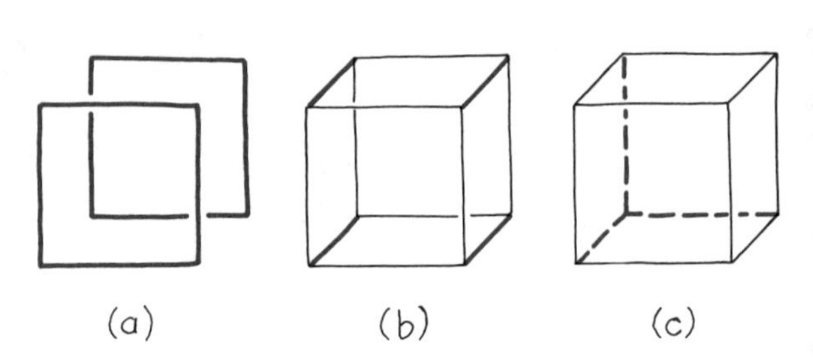

Figure 10-8

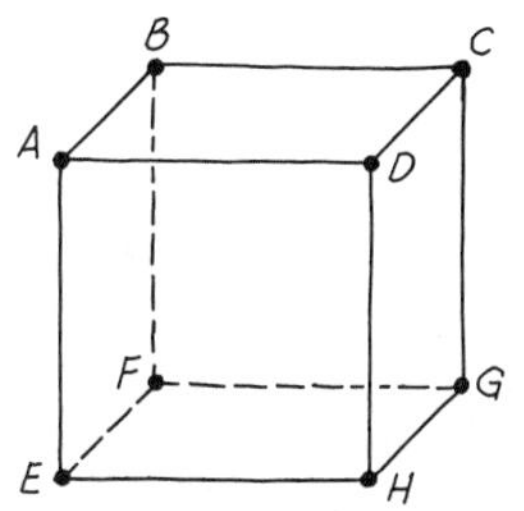

Figure 10-9

brown line segment. The intersection of two faces of a solid is called an **edge.** How many edges does a cube have?

Suppose we consider each edge of the cube as a set of points. The three edges represented by the brown line segments in Figure 10-7 intersect at one point. We often think of that point as a corner of the cube. The intersection of three or more edges of a solid is called a **vertex** (plural: vertices). How many vertices does a cube have?

Because solid figures are three-dimensional, in representing them on the two-dimensional surface of paper we must give the illusion of a third dimension. Notice how the cube is drawn in Figure 10-8. First, two squares are drawn that intersect and have corresponding sides parallel (a). Then corresponding vertices are joined by line segments (b). Finally, the edges that would not be visible when you look at the cardboard model from one direction are drawn as dotted line segments (c). The result looks like the three-dimensional cube. Note that in the drawing each square face of the cube is represented either as a square or as a parallelogram, which (as we found in Chapter 9) is an affine transformation of the square.

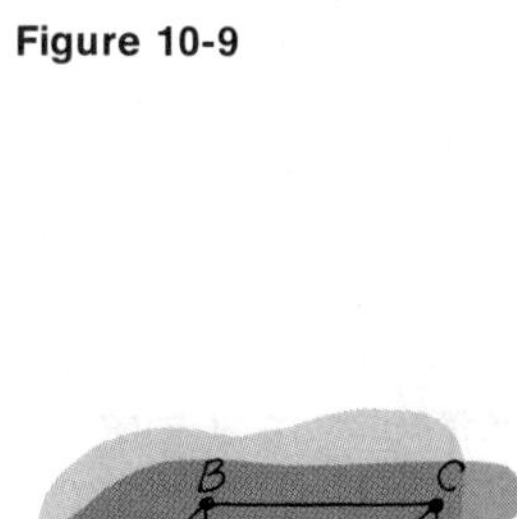

Figure 10-10

Imagine the cube drawn in Figure 10-9 with vertices labeled *A, B, C, D, E, F, G,* and *H.* Think of these vertices as points in space. Now imagine that the face containing points *A, B, C,* and *D* extends indefinitely to form the flat surface we call a plane. In the same way, we can think of each face of the cube as being part of a plane.

We know that the intersection of face *ABCD* and face *ADHE* is the edge represented by $\overline{AD}$. But if we imagine faces *ABCD* and *ADHE* extending indefinitely (forming planes as in Figure 10-10), then their intersection will also extend indefinitely. Line segment $\overline{AD}$ will become line $\overleftrightarrow{AD}$. Planes *ABCD* and *ADHE* are called **intersecting planes.** If two planes intersect, they intersect in a line (in this case $\overleftrightarrow{AD}$) that includes all the points the two planes have in common.

Sometimes two planes, like planes *ABCD* and *FGHE* in Figure 10-11, do not intersect (have no points in common). These are called **parallel planes.** Can you identify other parallel or intersecting planes in the models of cubes in Figure 10-12?

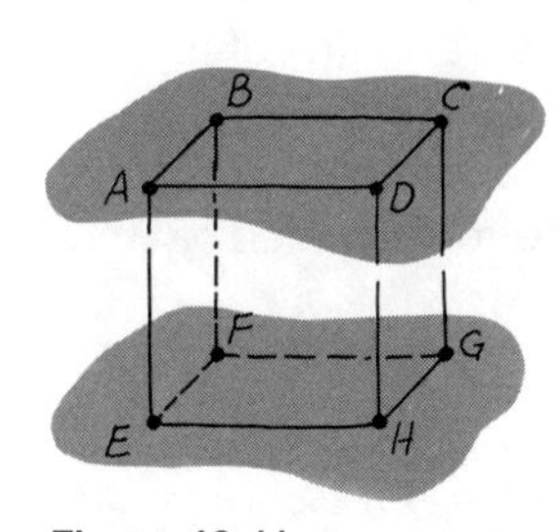

Figure 10-11

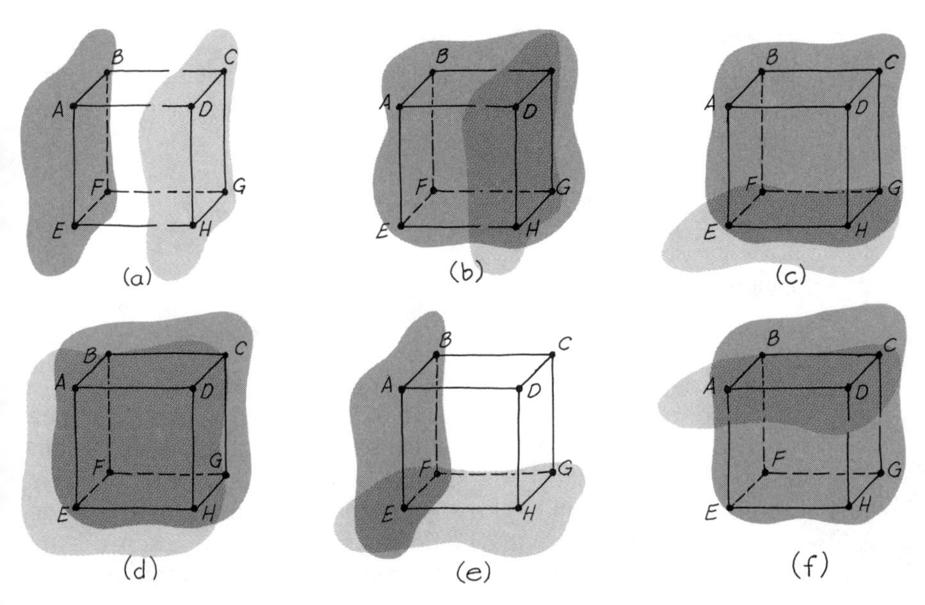

Figure 10-12

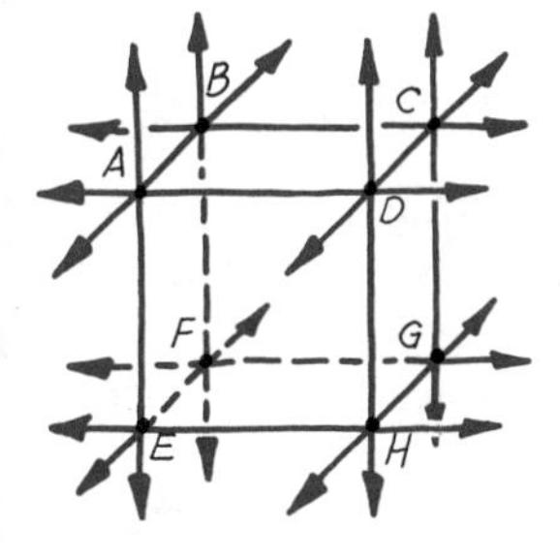

Figure 10-13

Let us take another look at the edges of the cube. Suppose each edge is extended indefinitely. We can think of $\overline{AD}$ as being part of the line that goes through A and D ($\overleftrightarrow{AD}$). Similarly, each edge of the cube can be considered a line segment that is part of a particular line (Figure 10-13).

The lines suggested by the edges of the cube show us three possible relationships between any two lines in space (Figure 10-14). As in cube (a), the lines might be in the same plane and intersect. Lines $\overleftrightarrow{AD}$ and $\overleftrightarrow{AB}$ intersect at point A in plane $ABCD$. They are called *intersecting lines.* As in cube (b), the lines might be in the same plane and *not* intersect. Lines $\overleftrightarrow{AB}$ and $\overleftrightarrow{CD}$ in plane $ABCD$ will never meet, no matter how far they are extended. They are called *parallel lines.* As in cube (c), the lines might be in different planes. There is no plane that contains both $\overleftrightarrow{AB}$ and $\overleftrightarrow{DH}$. That

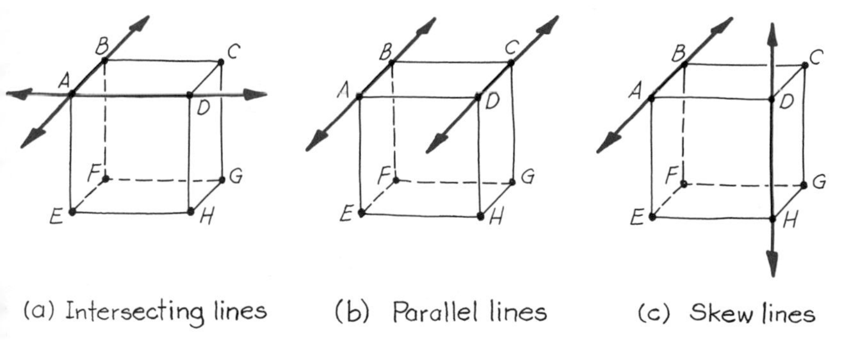

Figure 10-14

is, there is no flat surface which can include all the points of $\overleftrightarrow{AB}$ and $\overleftrightarrow{DH}$. These lines never meet and are in different planes.* They are called *skew lines.* Can you find other examples of intersecting lines, parallel lines, or skew lines in Figure 10-13?

Cylinders

Figure 10-15

A cylinder is a very familiar solid shape. We often see cylindrical shapes in our environment, as illustrated in Figure 10-15. Usually, when we talk about cylinders we are referring to circular cylinders. The pattern shown in Figure 10-16(a) can be used to make a model of a cardboard circular cylinder as shown in (b). Each circular region is called a **base** of the cylinder. The rectangular region forms the curved surface along the side of the cylinder, which is called the **lateral surface.** Notice that the upper and lower bases of this cylinder are just alike; that is, they have the same size and shape and are therefore congruent. If you imagine that these bases extend indefinitely, you can think of them as parts of planes. The planes determined by the upper and lower bases of this cylinder are parallel planes (Figure 10-17).

Figure 10-18 shows another model for a circular cylinder you can make as follows. First cut two pieces of heavy cardboard into congruent circular regions. Then punch equally spaced holes along the borders of the circular regions. Finally, thread elastic string from one hole on a border to a corresponding hole on the other border to form parallel line segments as shown (knotting the ends of the elastic string). If you imagine many parallel line segments joining corresponding points of the upper and lower bases, these line segments will form the lateral surface of the cylinder. *Keeping*

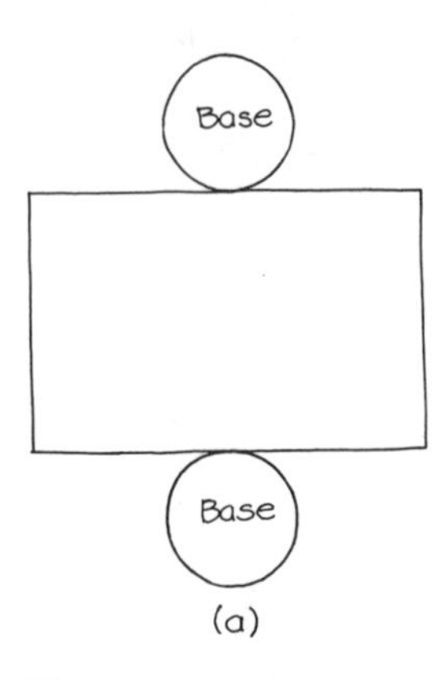

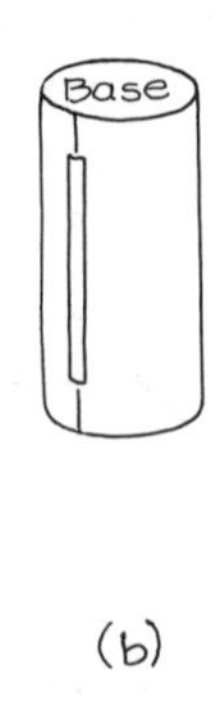

Figure 10-16

In Figure 10-14(c) it may appear that lines $\overleftrightarrow{AB}$ and $\overleftrightarrow{DH}$ will meet. This is only because we are using a two-dimensional diagram to represent a three-dimensional figure. If you locate the corresponding edges on a model of a cube and imagine that those edges extend indefinitely, you will realize they can never meet.

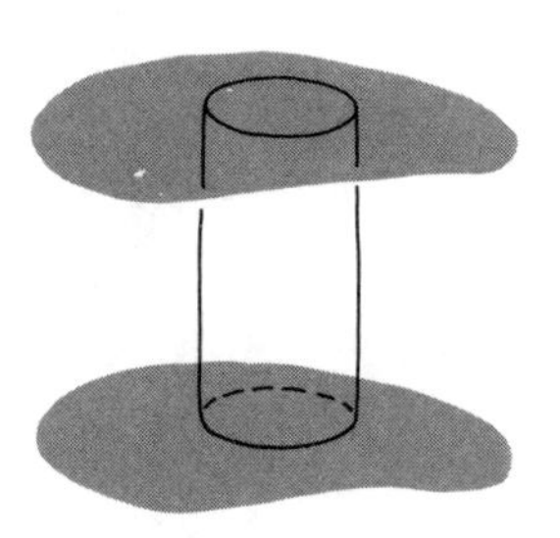

Figure 10-17

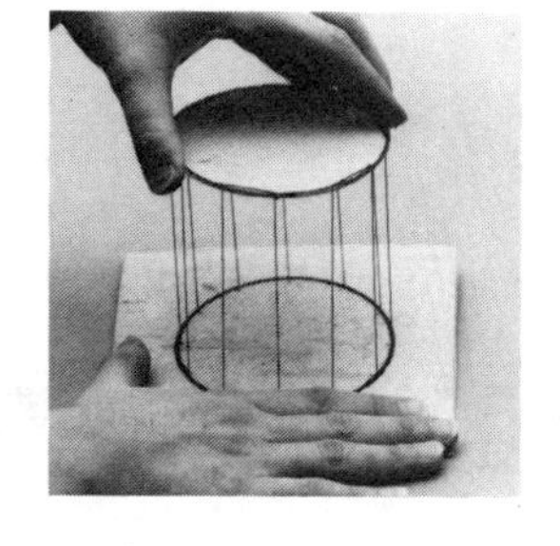

Figure 10-18

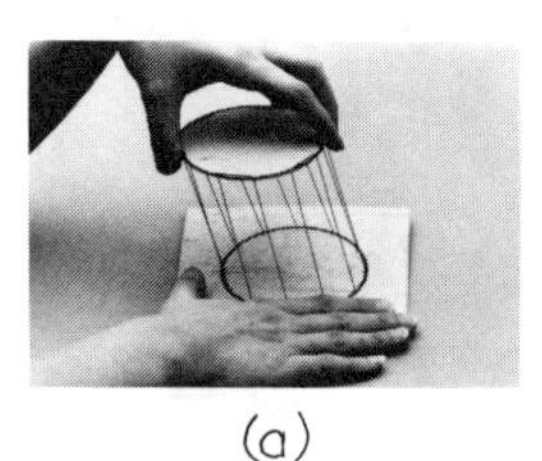

(a)

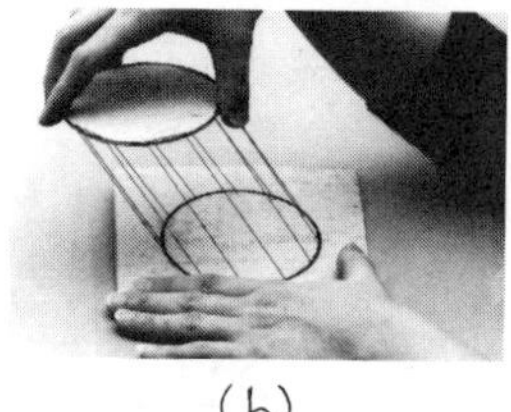

(b)

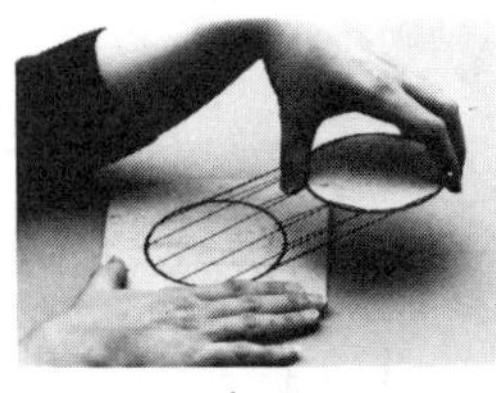

(c)

Figure 10-19

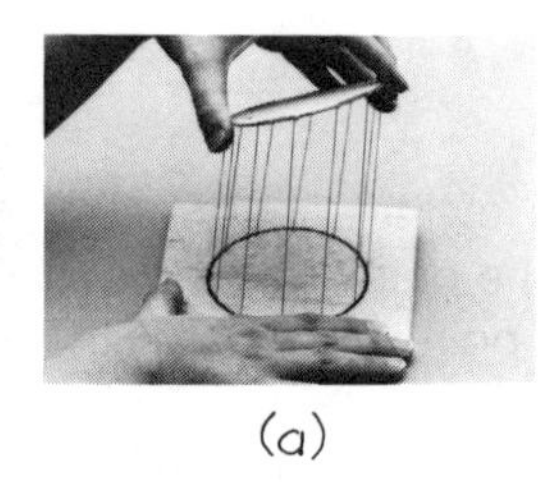

(a)

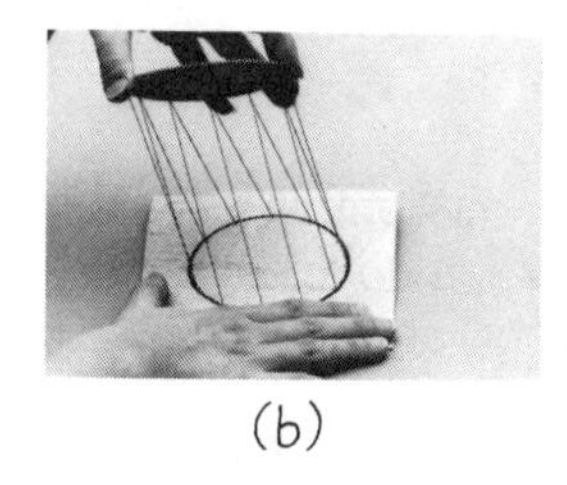

(b)

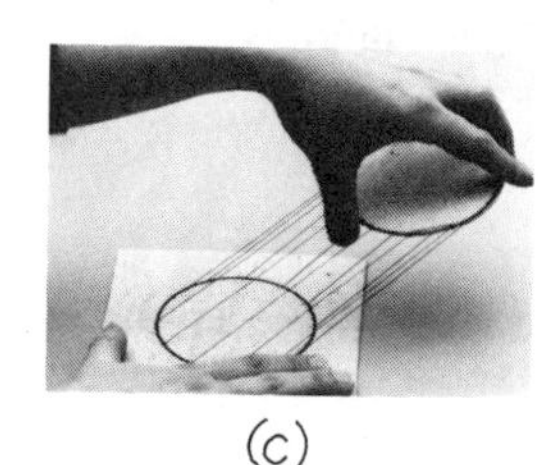

(c)

Figure 10-20

the bases in parallel planes, you can move the upper base to form different circular cylinders (Figure 10-19). Notice that the models formed in Figure 10-20 *cannot* represent cylinders since the bases are *not* in parallel planes.

Actually, a circular cylinder is just one kind of cylinder. In mathematics, the term *cylinder* usually has a broad meaning and includes certain solids that do not have circular bases. In fact, it is possible to imagine many different kinds of cylinders.

Think of two parallel planes. As in Figure 10-21, suppose a simple closed curve is drawn on one plane (a). Imagine that parallel line segments extend from the points of the curve in one plane to the other plane (b). As we increase the number of parallel line segments, we begin to see that the end points of these segments form a simple closed curve in the other plane (c). Notice that the two regions bounded by the simple closed curves are congruent. Now imagine a solid shape (d) that has congruent upper and lower bases in parallel planes and a lateral surface formed by parallel line segments whose end points are corresponding points on the boundaries of the bases. This solid is a cylinder. Since this

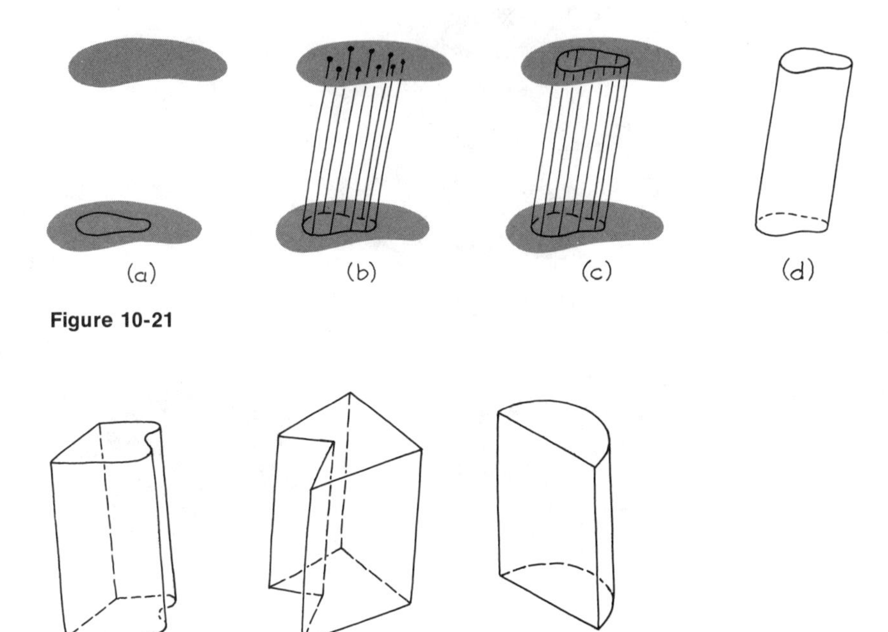

Figure 10-21

Figure 10-22

solid figure is "empty," we can think of this cylinder as a set of points in space consisting of the union of all the points that belong to the upper and lower bases and all the points that form the lateral surface.

In a similar way, we can develop models of other cylinders like the ones in Figure 10-22. Notice that the following is true for each cylinder:

1. Upper and lower bases are bounded by simple closed curves.
2. Upper and lower bases are congruent.
3. Upper and lower bases are in parallel planes.
4. The lateral surface can be thought of as being made up of *parallel line segments* whose end points are corresponding points on the boundaries of the bases. Each such line segment is *parallel* to *every* other line segment formed in the same way.

The solid shapes in Figure 10-23 cannot qualify as cylinders. Why not?

We imagined the lateral surface of a cylinder to be made up of an infinite number of line segments. The end points for such line segments were corresponding points on the boundaries of the upper and lower bases of the cylinder. Each of these line segments is called an **element** of the cylinder. In some special cylinders the

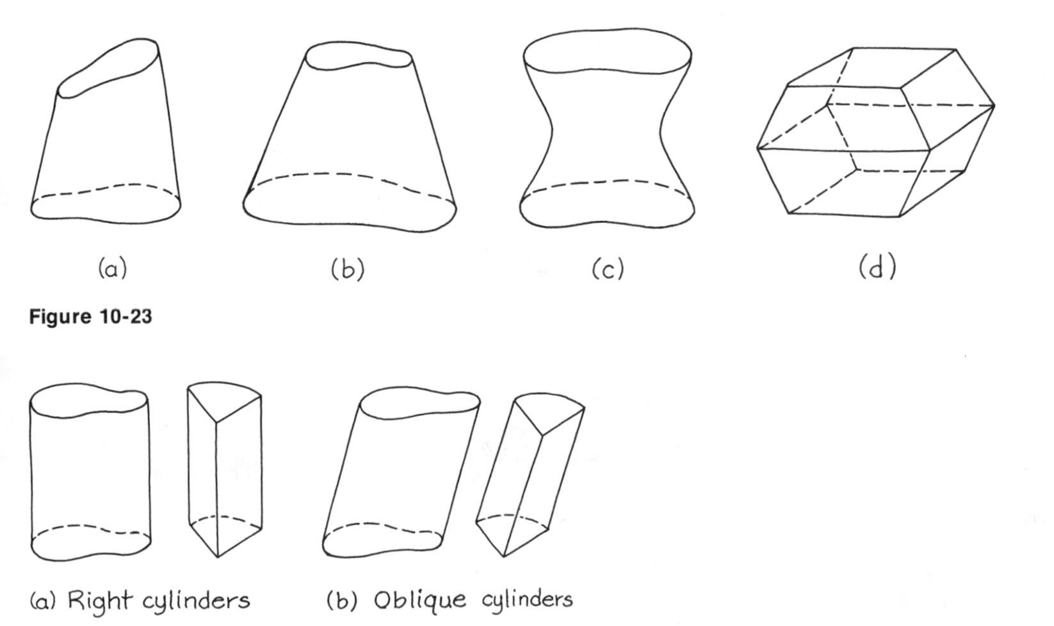

Figure 10-23

Figure 10-24

Figure 10-25

elements are perpendicular to the bases. These cylinders are called **right cylinders,** while cylinders whose elements do *not* meet the bases at right angles are called **oblique cylinders** (Figure 10-24).

A tin can like the one pictured in Figure 10-25 can be considered the model of a cylinder. Yet it is a special cylinder. The elements are perpendicular to the bases, so it is a *right* cylinder. The bases are bounded by special simple closed curves—circles. It can be called a *circular* cylinder. Its complete name might be a right circular cylinder. Why do you think cans have a circular cylindrical shape? What are some advantages of that shape?

Special Cylinders: Prisms

Some well-known solid shapes belong to the cylinder family. Each qualifies as a cylinder but has other special characteristics. We can think of these shapes as subsets of the set of all cylinders.

If a cylinder has bases bounded by polygons, the cylinder is called a **prism.** Several prisms are illustrated in Figure 10-26. Notice that we can identify a prism according to the polygonal shape of its bases: a prism whose bases are bounded by triangles is called a triangular prism, a prism whose bases are bounded by pentagons is called a pentagonal prism, etc.

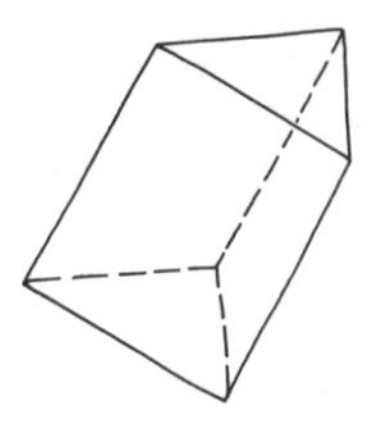

(a) Triangular prism

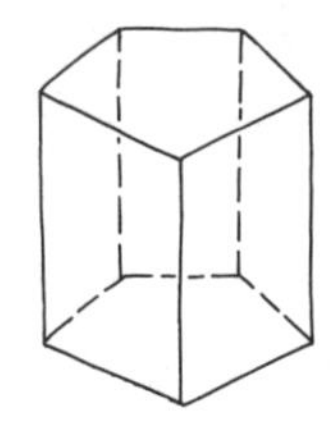

(b) Pentagonal prism

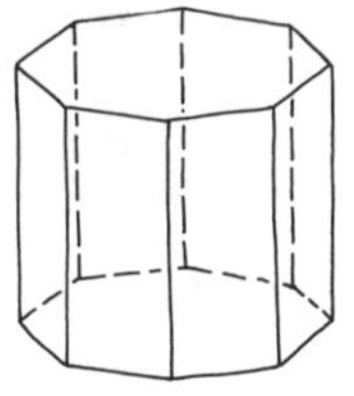

(c) Octagonal prism

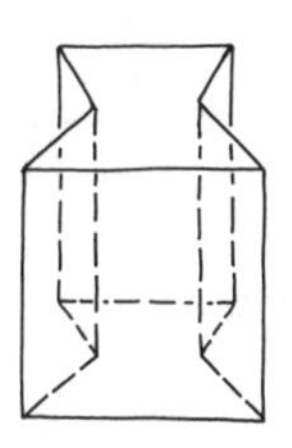

(d) Hexagonal prism

Figure 10-26

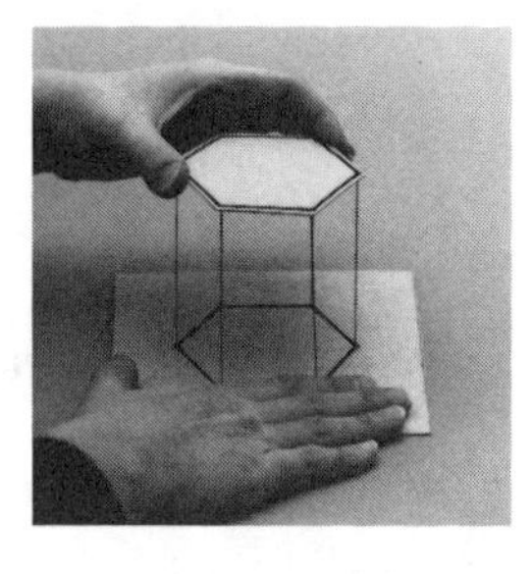

Figure 10-27

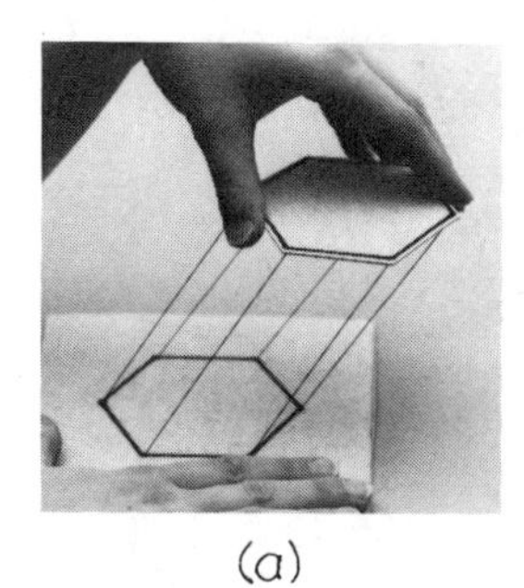

(a)

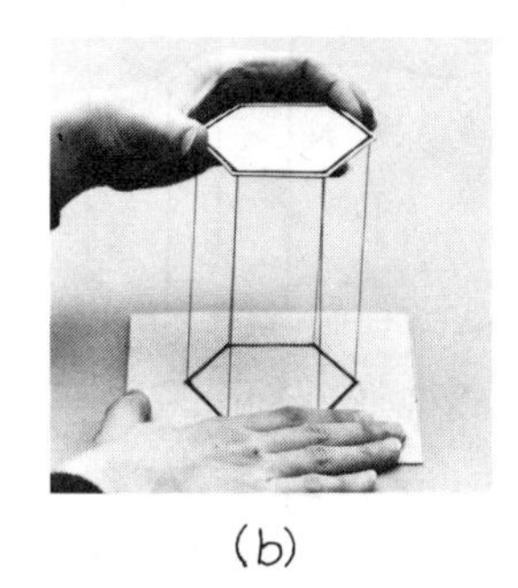

(b)

Figure 10-28

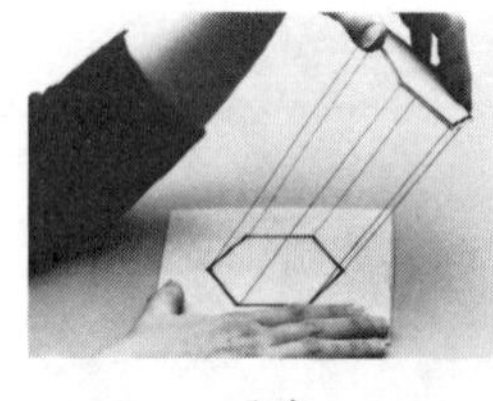

(a)

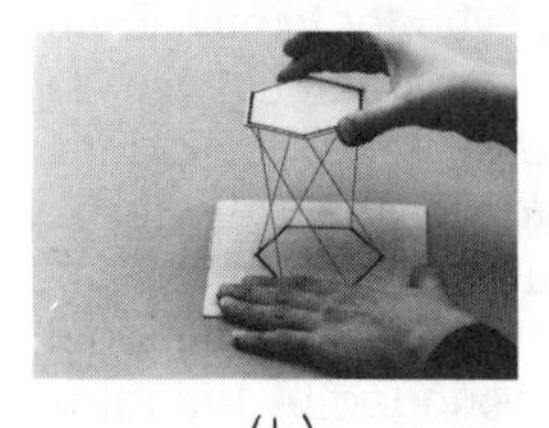

(b)

Figure 10-29

You can see that the lateral surfaces of the prisms consist of regions bounded by parallelograms. Since the bases of the prism are bounded by polygons, the lateral surface is no longer curved but consists of many faces bounded by parallelograms.

The model in Figure 10-27 was made like the model of the circular cyclinder by using heavy cardboard and elastic string. In this case the upper and lower bases are bounded by congruent hexagons and the string forms parallel line segments whose end points are corresponding vertices of the hexagons as shown. If you move the upper base, *keeping it in a plane parallel to the plane of the base* so that the lateral faces of this model are bounded by parallelograms, a model of different hexagonal prisms can be formed (Figure 10-28).

When you turn the upper base or do not keep it in a plane parallel to the plane of the lower base, the resulting models are *not* prisms (Figure 10-29), since either the bases are *not* in parallel planes (a) or the lateral faces are not bounded by parallelograms (b).

The solids in Figure 10-30 seem as if they might qualify as prisms, but they are not prisms (nor are they cylinders). Why not? The lateral surfaces are *not* composed of parallelogram regions. Instead, those surfaces contain triangular regions. These solids *cannot* qualify as cylinders or prisms. However, they do have a special name—**antiprisms.**

Let us return to solids that do qualify as prisms. Sometimes the lateral faces of prisms are bounded by special parallelograms—

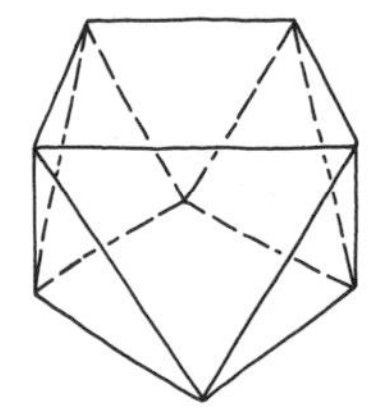

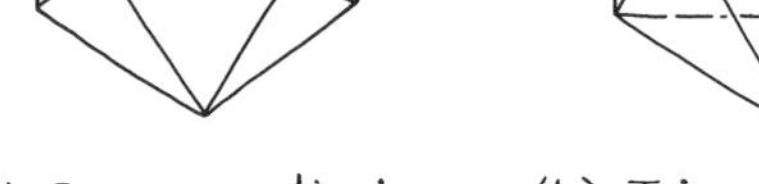

Figure 10-30

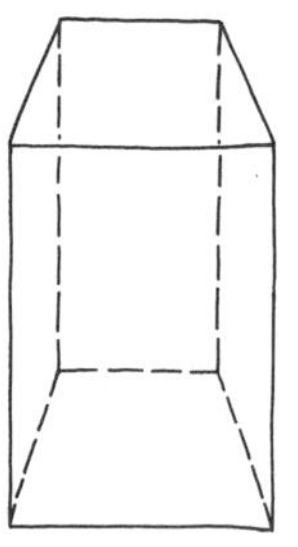

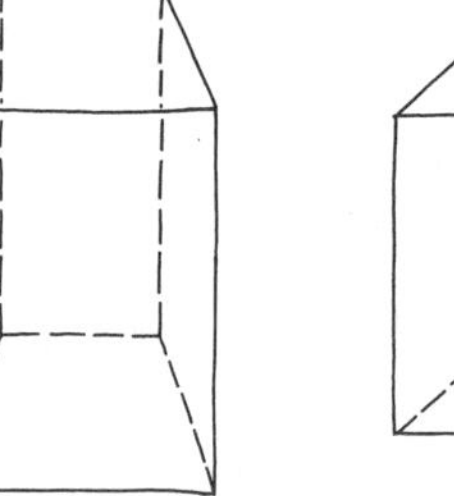

Figure 10-31

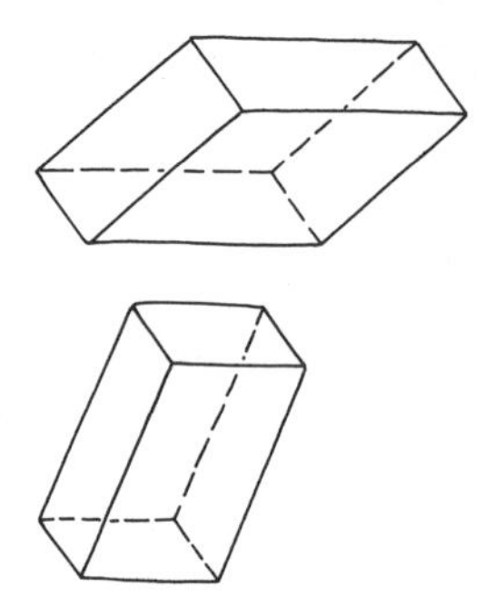

Figure 10-32

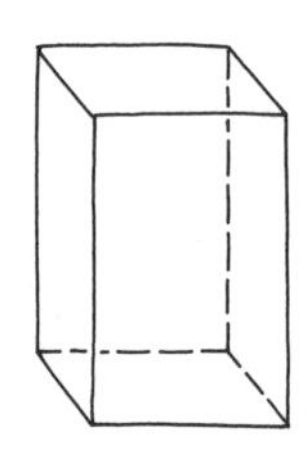

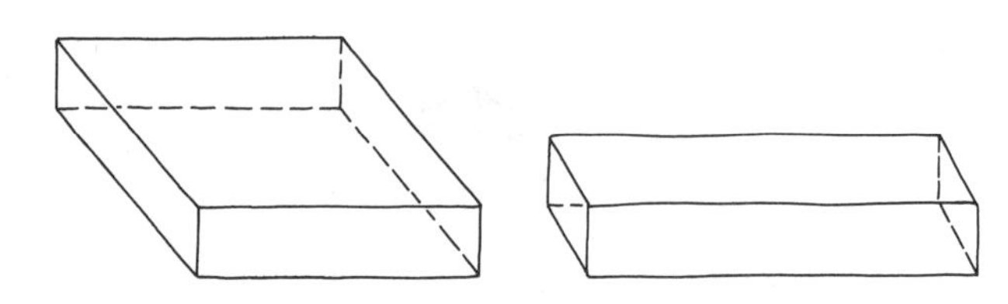

Figure 10-33

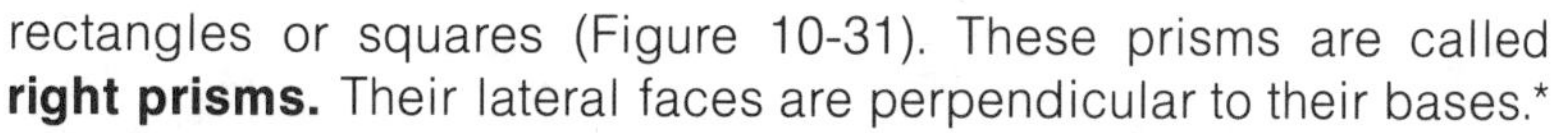

rectangles or squares (Figure 10-31). These prisms are called **right prisms.** Their lateral faces are perpendicular to their bases.*

Sometimes the upper and lower bases are bounded by parallelograms. Then all the faces of the prism are bounded by parallelograms and the prism can be called a **parallelepiped.** The solids in Figure 10-32 are examples of parallelepipeds since they are prisms with all faces bounded by parallelograms.

The solids in Figure 10-33 are special parallelepipeds. Their faces are not only bounded by parallelograms, but the parallelograms are rectangles. These solids are often called **cuboids.** They are also known as **rectangular parallelepipeds** or **rectangular solids.** Note that a rectangular parallelepiped also qualifies as a *right rectangular prism.* (Why?) Finally, we come to a very special prism. All the faces of this prism are squares. This is a familiar solid, the *cube.*

Figure 10-34

If you make another model with heavy cardboard and elastic string like the one in Figure 10-34 with square bases, you can transform it so that different types of prisms are formed. Can you transform the model into different types of prisms? Name the kinds of prisms formed.

**When lateral faces are perpendicular to the bases, they meet the bases at right angles (like the intersecting faces of a cube).*

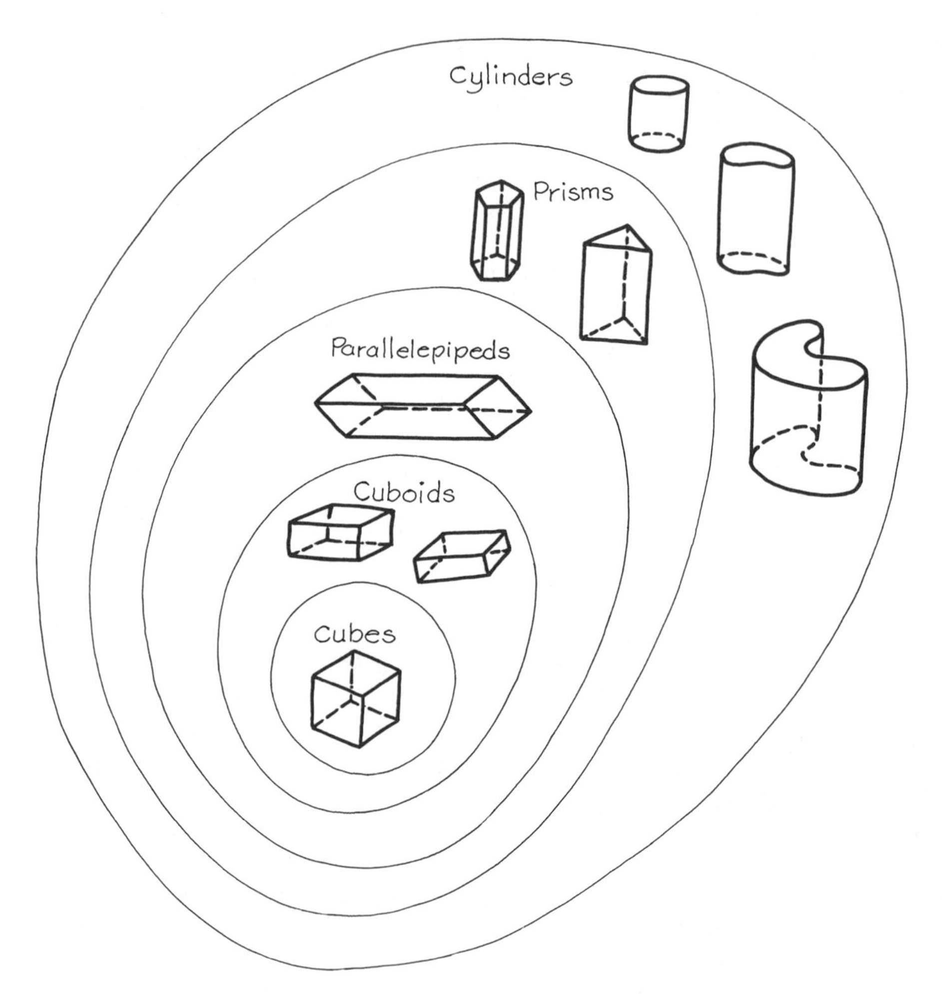

Figure 10-35

The cylinder family

We can use a Venn diagram to summarize our classification of cylinders (Figure 10-35). From the Venn diagram we see that prisms are a subset of the set of all possible cylinders. They are cylinders whose bases are bounded by simple closed curves that are polygons. Parallelepipeds form a subset of the set of prisms since they are prisms with a special feature: *all* their faces are bounded by parallelograms. Cuboids are a subset of parallelepipeds since they are parallelepipeds with an extra feature: all their faces are bounded by parallelograms that are rectangles. Finally, cubes form a subset of the set of cuboids because they are cuboids with the extra characteristic that all their faces are bounded by rectangles that are squares. The cube, then, is a very special solid. It qualifies as a cuboid, as a parallelepiped, as a prism, and as a cylinder!

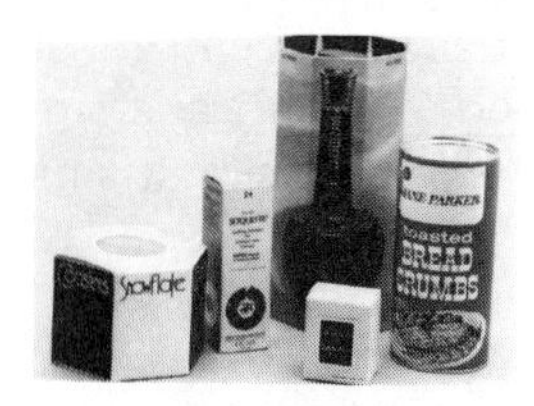

Figure 10-36

Many boxes we use in everyday life are shaped like cylinders of different types. Can you name the kinds of cylinders suggested by the boxes in Figure 10-36? Try to find boxes that are models of other types of cylinders.

(a)

(b)

(c)

Figure 10-37

Cones

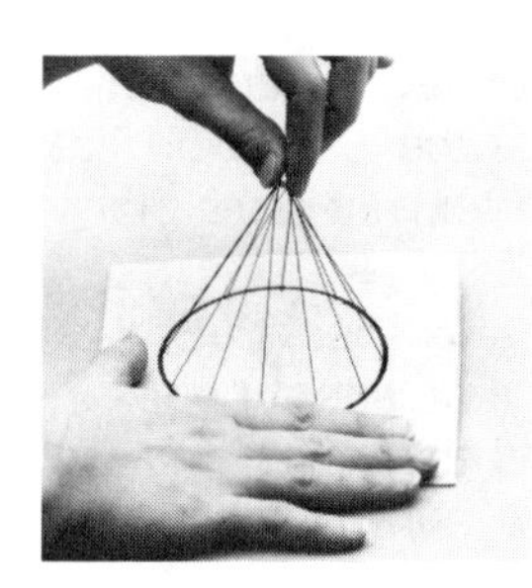
Figure 10-38

Another shape we often see (especially those of us who love ice cream) is the **cone.** Some cones are suggested in Figure 10-37. You can make a model of a circular cone with heavy cardboard and elastic string as suggested in Figure 10-38. First, a circle is drawn on the base. Then holes are punched at equal distances around the circumference of the circle. Next, congruent pieces of elastic string are extended from the holes to a common point. Knots are made on the other side of the holes so that the ends of the strings do not go through the holes. The strings can be joined at a common point with a cup hook or metal ring as indicated. By moving the common end point from side to side or up and down, you can transform the model into many different circular cones (Figure 10-39). In each case the line segments represented by the elastic string are called *elements* and are part of the *lateral surface* of the cone. The common end point is the *vertex*, and the region bounded by the circle is called the *base.*

When the vertex lies directly above the center of the base as in Figures 10-38 and 10-39(a) and (b), the cone is a *right* cone. Otherwise the cone is an *oblique* cone. Notice that in a right cone a line

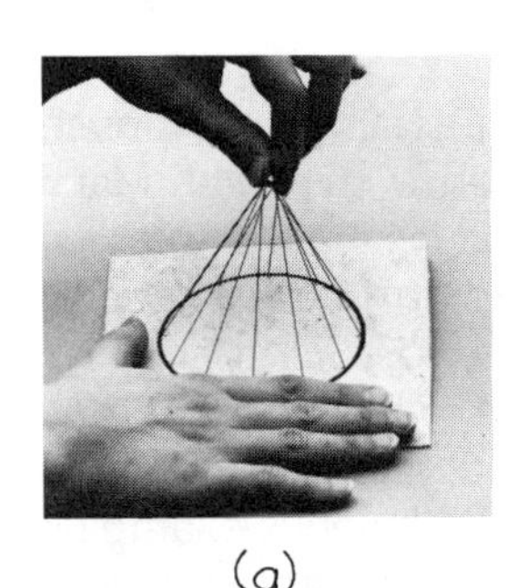
(a)

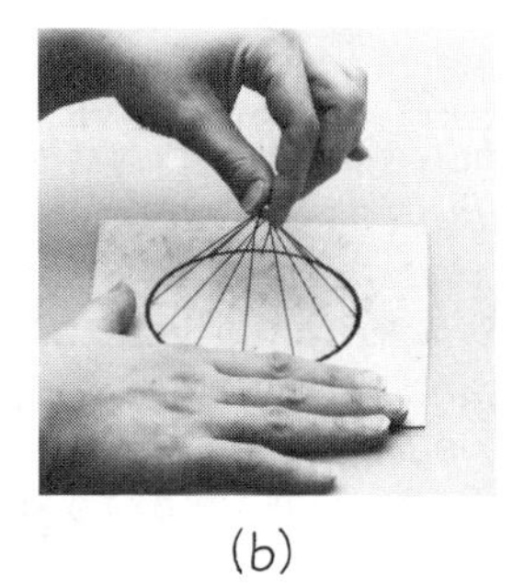
(b)

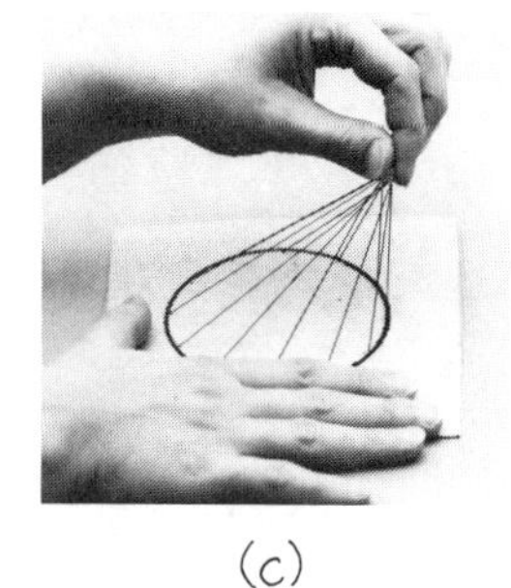
(c)

Figure 10-39

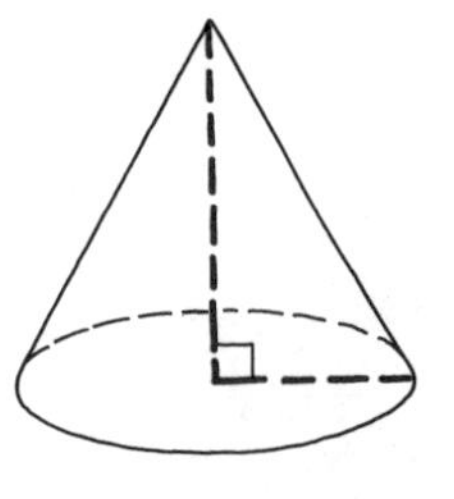

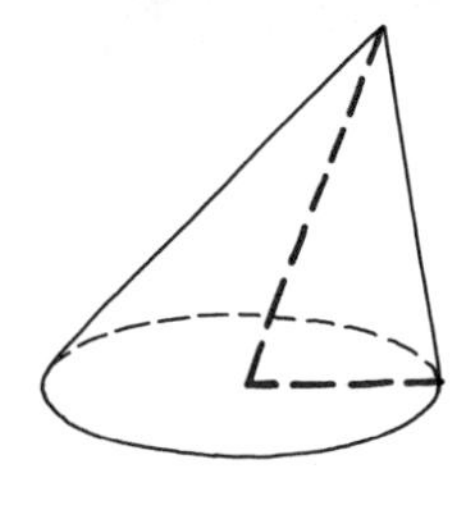

Figure 10-40

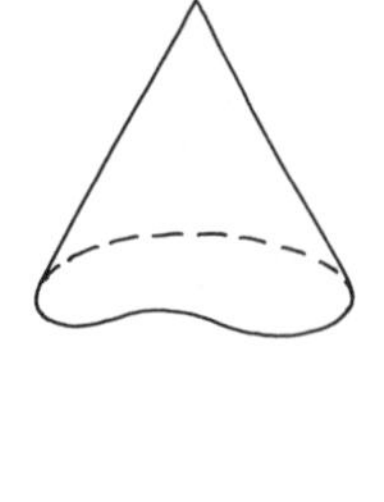
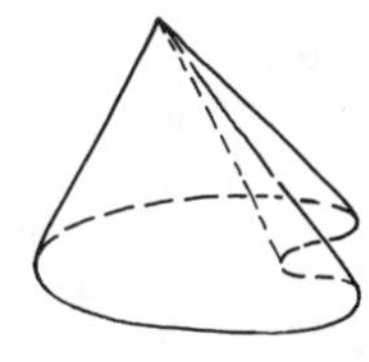
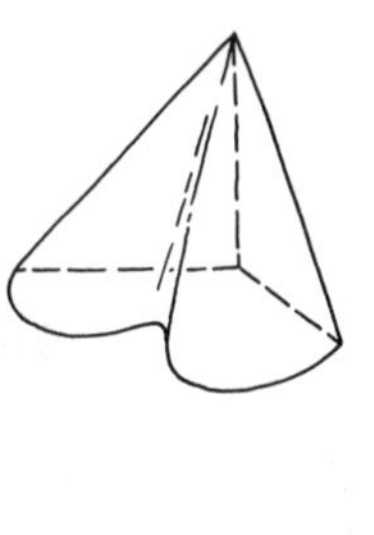

Figure 10-41

segment from the vertex to the center of the base makes a right angle with a radius of the circular base while a right angle cannot be formed in that way with an oblique cone (Figure 10-40). You can also see from Figure 10-39 that the elements of a right cone are all the same size (congruent) while the elements of an oblique cone vary in length.

Although we usually think of cones as having a circular base (that is, circular cones), in mathematics we can make the term *cone* more general. We can include solids with shapes like the ones suggested by the diagrams in Figure 10-41. The bases of these cones are plane regions bounded by simple closed curves. The lateral surface can be thought of as being formed by an infinite number of line segments which extend from the points of that simple closed curve to a common point (the vertex) not in the plane of the base of the cone.

A cone, then, has a lateral surface and a base bounded by a simple closed curve. We can think of a cone as a set of points in space consisting of the union of all the points that belong to the base and all the points that belong to the lateral surface. How does a cone differ from a cylinder?

Special Cones: Pyramids

When the base of a cone is bounded by a special simple closed curve—a polygon—the cone is usually called a **pyramid.** Many kinds of pyramids are shown in Figure 10-42. Notice that precise names can be assigned to pyramids according to the shape of their bases. A pyramid whose base is bounded by a triangle is called a triangular pyramid, a pyramid whose base is bounded by a hexagon is called a hexagonal pyramid, etc. You often see the pyramid shape as the form of a spire of a church. Famous examples of pyramids in real life are the ancient Great Pyramids in Egypt, with their huge square bases (which makes them square pyramids),

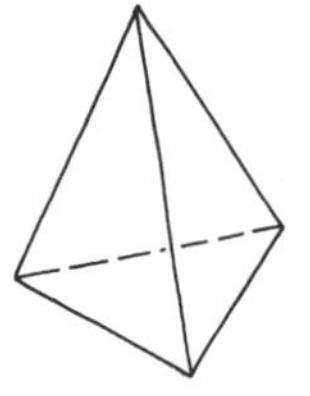
(a) Triangular pyramid

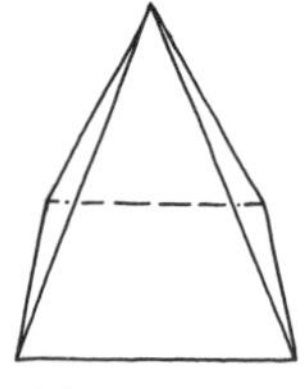
(b) Square pyramid

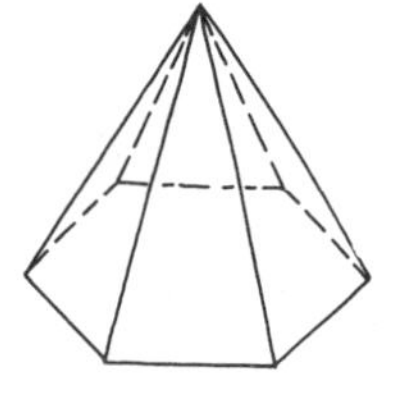
(c) Hexagonal pyramid

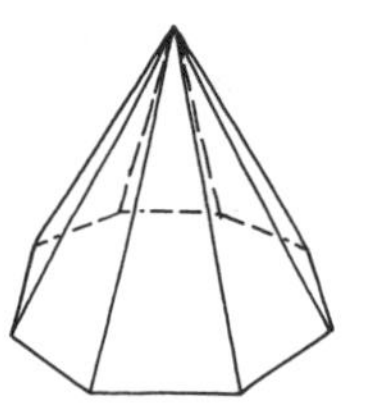
(d) Octagonal pyramid

Figure 10-42

Figure 10-43

and the new Transamerica building in San Francisco (Figure 10-43). We mentioned these pyramids in Chapter 8 when we discovered how the Egyptians made the base of the pyramid a square region.

Figure 10-44 indicates how a model for a pyramid can be constructed. Again, heavy cardboard and elastic string are used. The segments of string extend from the vertices of the base to the common vertex, which is a point outside the plane of the base. In this case the base is a square region, and we can imagine the model being transformed into many square pyramids as the vertex is moved to different positions. The lateral surface of these pyramids always consists of triangular regions. When the common vertex is directly above the center of the base of the pyramid as in Figure 10-44(a), these lateral surfaces are bounded by congruent *isosceles* triangles.

The models shown in Figure 10-45 suggest other kinds of pyramids. In each case the pyramid can be thought of as consist-

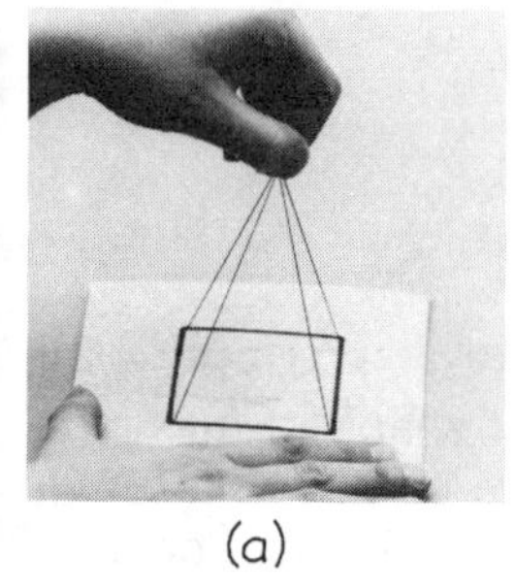
(a)

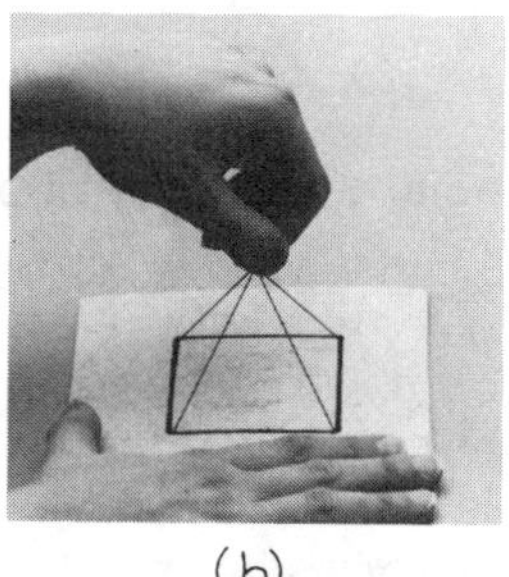
(b)

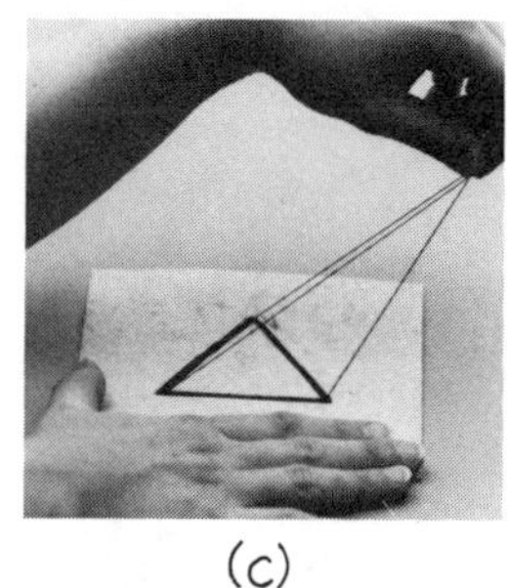
(c)

Figure 10-44

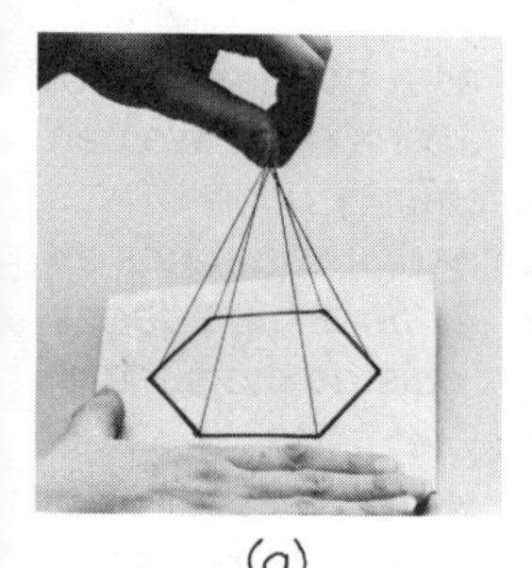
(a)

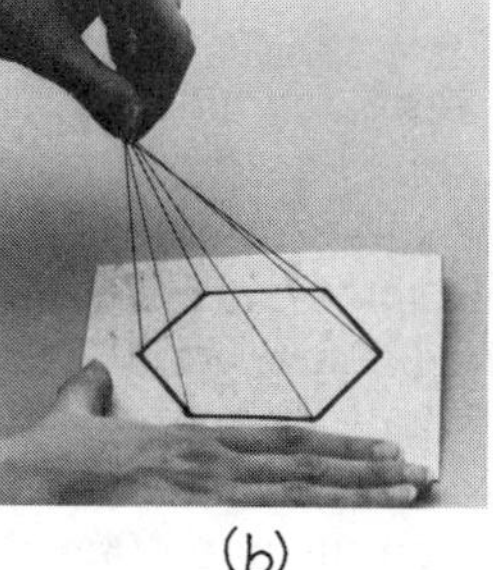
(b)

(c)

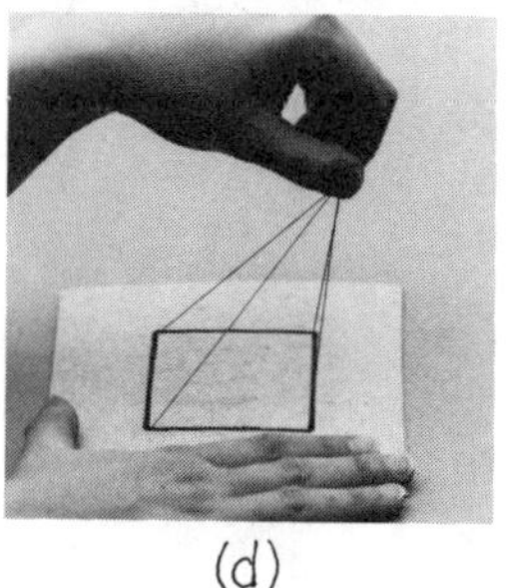
(d)

Figure 10-45

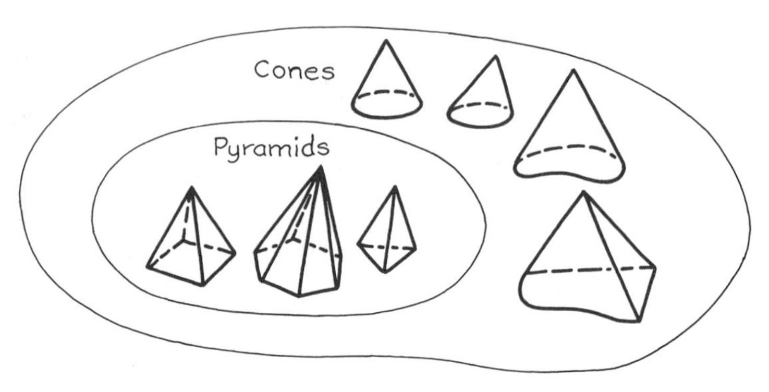

Figure 10-46

ing of the union of all the points of the region bounded by the polygon (base) and all the points of the triangular regions (lateral faces) forming the lateral surface.

Pyramids are a subset of the set of cones as expressed in the Venn diagram in Figure 10-46. Every pyramid qualifies as a cone since its base is bounded by a simple closed curve, and you can think of its lateral surface as being made up of an infinite number of line segments which extend from the points of that simple closed curve to a common point (the vertex) not in the plane of the base. But the base of the pyramid is bounded by a special simple closed curve—a polygon.

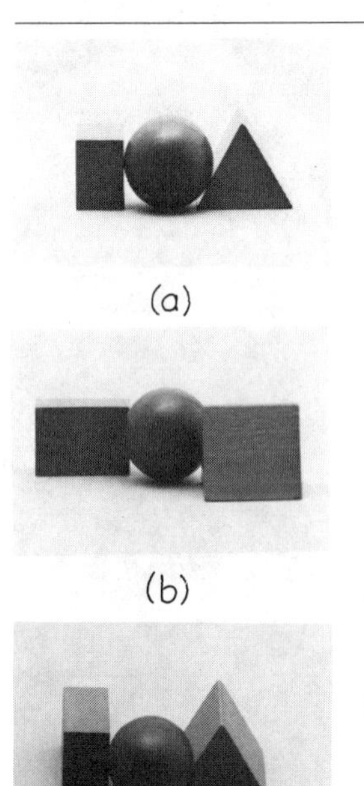

(a)

(b)

(c)

Figure 10-47

A Very Special Shape: The Sphere

Probably the most familiar three-dimensional shape of all is the **sphere.** It is different from all the shapes we have discussed. It does *not* qualify as a cylinder, as a prism, as a cone, or as a pyramid. It has no bases, no edges, and no vertices. Unlike other shapes, the sphere looks the same from every direction (Figure 10-47).

The surface of a hollow rubber ball serves as a good model of a sphere. Why is a sphere such an ideal shape for a ball? We can think of the surface of a ball as a set of points forming a continuous curved surface. Like the other solids we examined, it separates all the points in space into three sets of points: (1) points inside the sphere, (2) points on the sphere, and (3) points outside the sphere.

In a way, you can think of a sphere as being related to a circle. When we talked about circles, we described them as a set of points *in a plane* that are a given distance (the radius) from a fixed point (the center). Similarly, a sphere can be described as a set of points

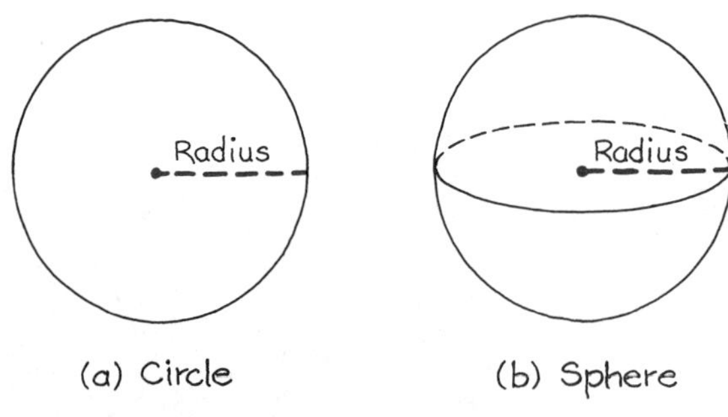

Figure 10-48

in space that are a given distance (the radius of the sphere) from a fixed point (the center of the sphere). Every point on a sphere is the same distance from the center of the sphere. A line segment that represents the distance from a point on the sphere to the center is called a radius of the sphere (Figure 10-48).

Intersections of Planes and Solids

We saw that the faces of many solids have the shapes of plane figures such as squares, rectangles, triangles, and circles. There are other ways we can see plane shapes as a part of solid shapes.

Look at the glass jar containing sand pictured in Figure 10-49. The jar is shaped like a circular cylinder. But look at the upper surface of the sand in the jar. The boundary of that surface forms a circle. If you imagine that jar to be a cylinder and the upper surface of the sand to be a plane, you can think of that boundary as the intersection of the cylinder and the plane (Figure 10-50). Similarly, if you put a lid on the jar of sand and tilt it, the boundary of the upper surface of the sand forms other shapes like the ellipses in Figure

Figure 10-49

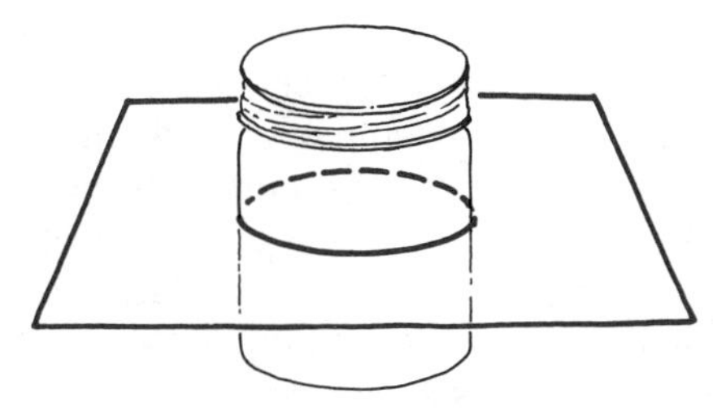

Figure 10-50

Figure 10-51

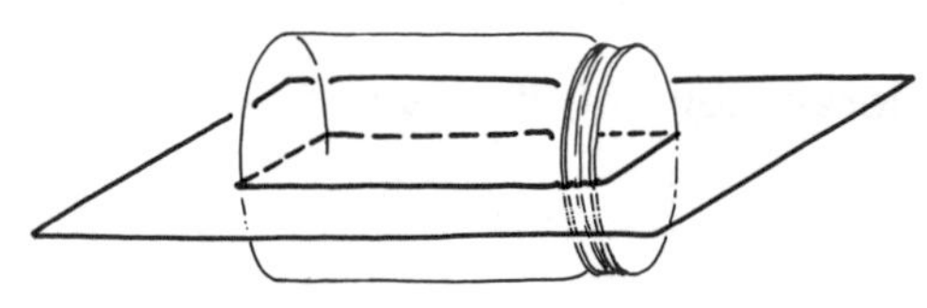

Figure 10-52

10-51. If you tilt the jar horizontally, the surface of sand is bounded by a rectangle (Figure 10-52).

A sand timer offers another example of the intersection of planes and solids. The glass sand timer pictured in Figure 10-53 looks like two cones joined at their vertices. Study the shape of the boundary of the surface of the sand as the sand timer is placed in different positions. In (a) that surface is bounded by a *circle,* in (b) it is bounded by an *ellipse,* in (c) the surface takes the form of a shape we met in Chapter 5 called a *parabola,* and in (d) there are two parts to the shape (that is, two symmetrical shapes are formed) called a *hyperbola,* another shape we met in Chapter 5. Shapes formed as the intersection of a plane and two conical surfaces joined at their vertices are called **conics.**

Another way to visualize the intersection of planes and solid shapes is to use models of solids (made from elastic string and cardboard) and a slide projector as follows: First, make a model of a cylinder like the one shown in Figure 10-54 out of two cardboard circular regions (bases) with holes spaced equally along the

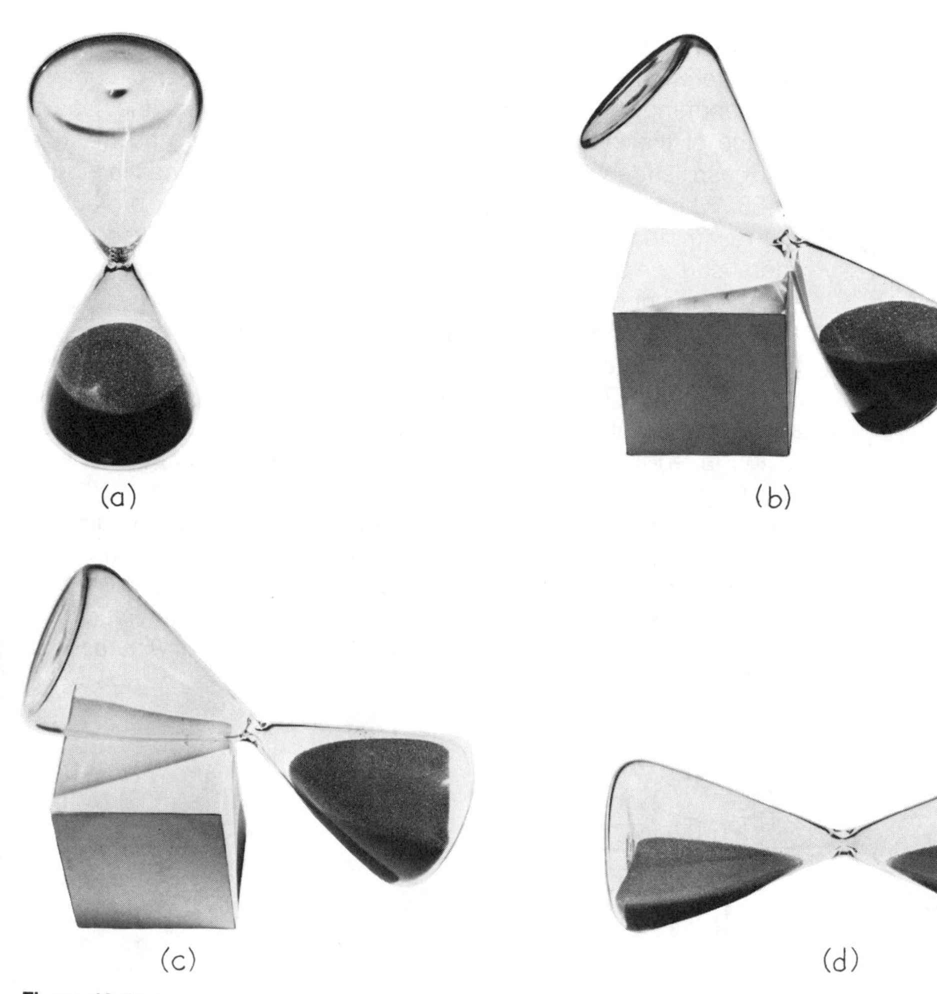

Figure 10-53

Figure 10-54

Figure 10-55

circumferences. The elastic string is threaded from a hole on one circumference to the corresponding hole on the other circumference forming parallel line segments that represent the lateral surface of the cylinder. Next, take a slide projector and make a cardboard slide with a horizontal slit across the center as in Figure 10-55. When you place this slide in the projector, your projection will resemble a plane of light. As the plane of light intersects the lateral surface of the cylinder (parallel segments of string), you will see a series of dots of light that seem to form a plane shape (Figure 10-56). That plane shape represents the intersection of this plane of light and the lateral surface of the cylinder.

In Figure 10-57 we see different intersections that can be formed as the cylinder is held in different positions with respect to the projector. In (a) the intersection is a *circle,* in (b) it is an *ellipse,* and in (c) two parallel line segments appear.

If you hold the lower base of the cylinder and carefully turn the upper base (keeping the bases in parallel planes), you can transform this model of a cylinder into a model of two cones joined at their vertices (Figure 10-58). Again, many different intersections can be seen as we change the position of these cones with respect

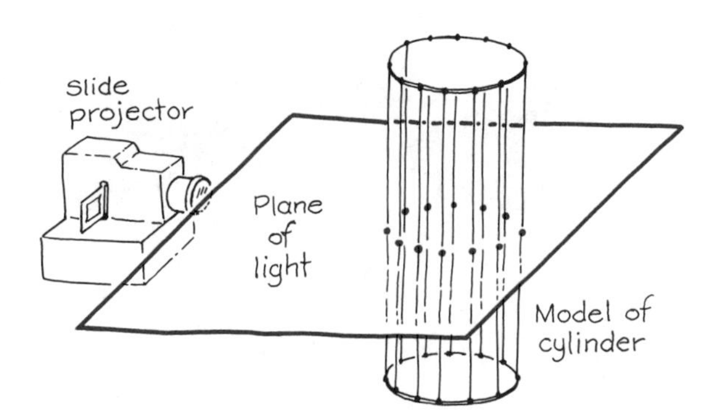

Figure 10-56

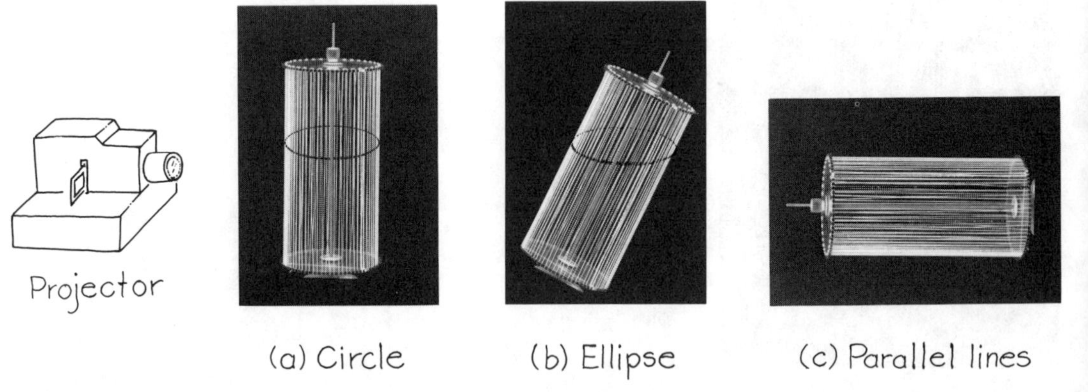

Figure 10-57

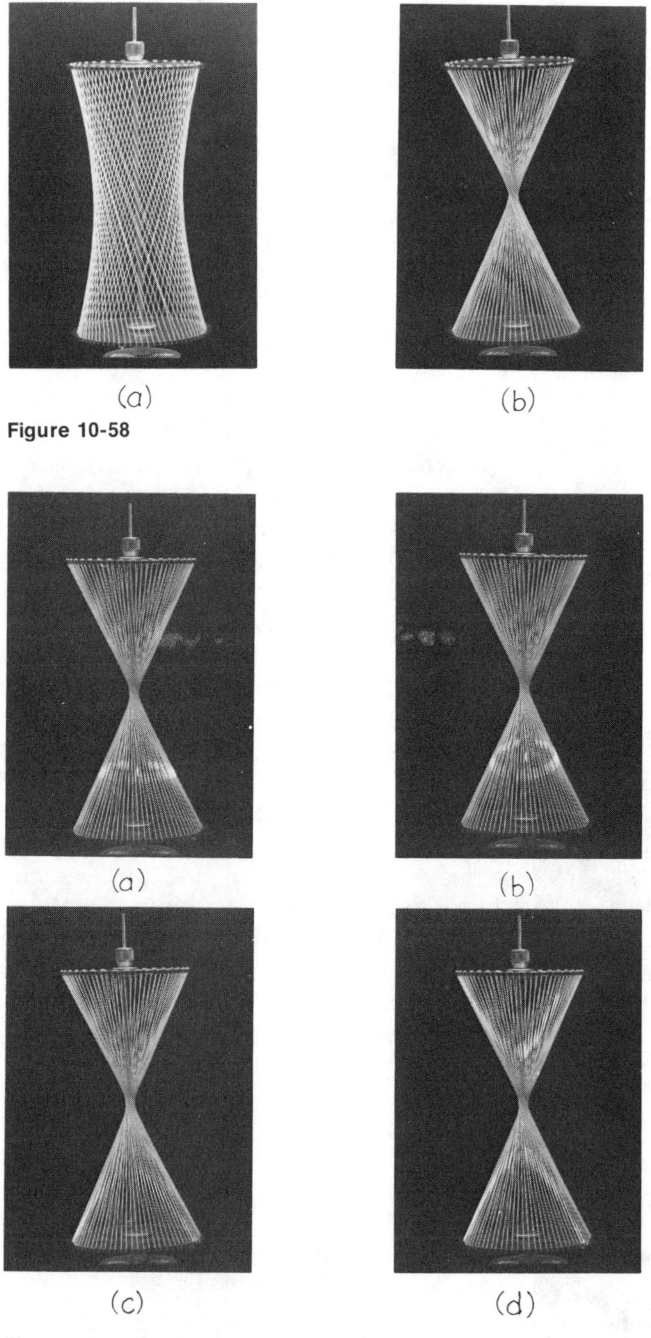

(a) (b)

Figure 10-58

(a) (b)

(c) (d)

Figure 10-59

to the projector (Figure 10-59). In each case the intersection of the plane of light and the lateral surfaces of the cones is as follows: (a) a circle, (b) an ellipse, (c) a parabola, and (d) a hyperbola. These curves all belong to that family of curves known as the conics, which represent the intersection of a plane and two conical surfaces joined at their vertices.

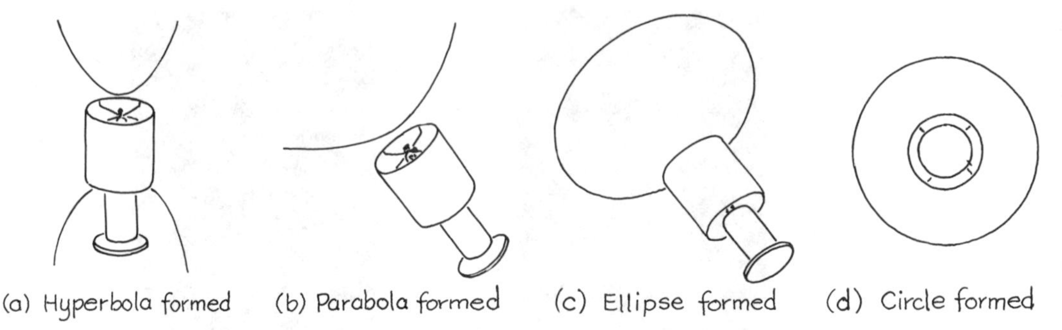

Figure 10-60

Another simple way to see the conics is to look at the outlines formed on a wall by a table lamp held in different positions as in Figure 10-60. If you hold the lamp close to the wall and parallel to it, the outline looks like the two branches of a hyperbola (a). If you hold the lamp at an angle to the wall, a parabola (b) or an ellipse (c) can be formed. When the lamp is perpendicular to the wall, the outline of a circle is formed (d).

Figure 10-61

Let us get back to the slide projector for another experience. The six faces of the model of the cube shown in Figure 10-61 are made up of parallel line segments formed by elastic string. The 12 edges are congruent strips of wood. When the cube is held in front of the projector so that one face is parallel to the front of the projector, the intersection of the plane of light and the surface of the cube forms a square [Figure 10-62(a)]. As the cube is moved into different positions, it is possible to obtain different shapes at the intersections. Rectangles and triangles seem to be formed as the intersection of the plane of light and the model of the cube in Figure 10-62(b) and (c). What is the largest rectangle that can be formed? What is the largest triangle? It is possible to position the cube so that the plane of light intersects six edges at their midpoints as in Figure 10-63. In that case the outline of a regular hexagon is formed!

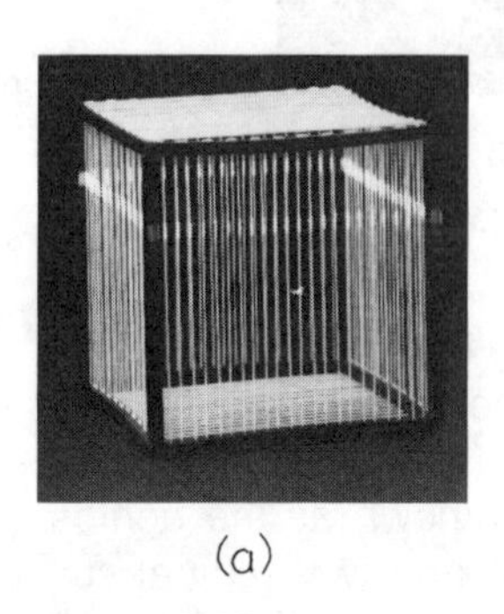

(a)

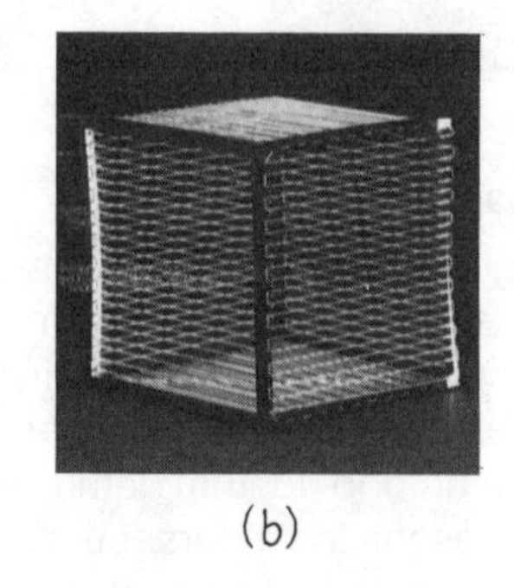

(b)

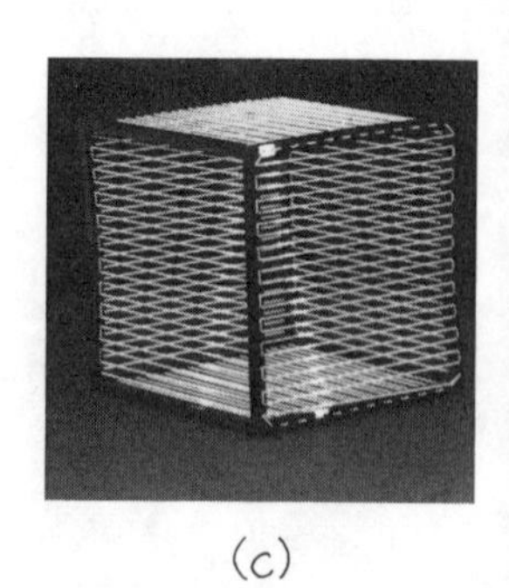

(c)

Figure 10-62

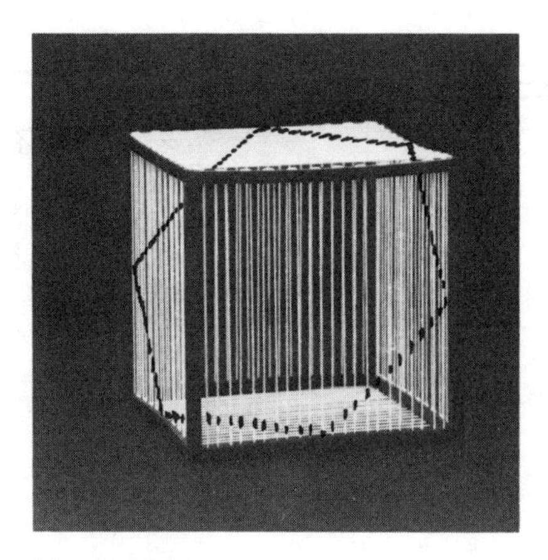

Figure 10-63

Symmetry Revisited

Another way to study the intersection of planes and solids is to make solid shapes out of plasticene. Make several cylinders, cones, cubes, cuboids, pyramids, and spheres. Then cut them with a sharp knife in different ways and examine the shapes of the cuts. Figure 10-64 suggests some possibilities. Can you find other ways planes can intersect the solids?

Cut a cube into two halves as shown in Figure 10-65, and place one half against the surface of a mirror. You will notice that the half cube plus its reflection look like your original cube (Figure 10-66). We can say that the cut lies in a **plane of symmetry** of the cube (Figure 10-67).

When we discussed *lines of symmetry* of a plane figure in Chapter 4, we found that when a figure has a line of symmetry,

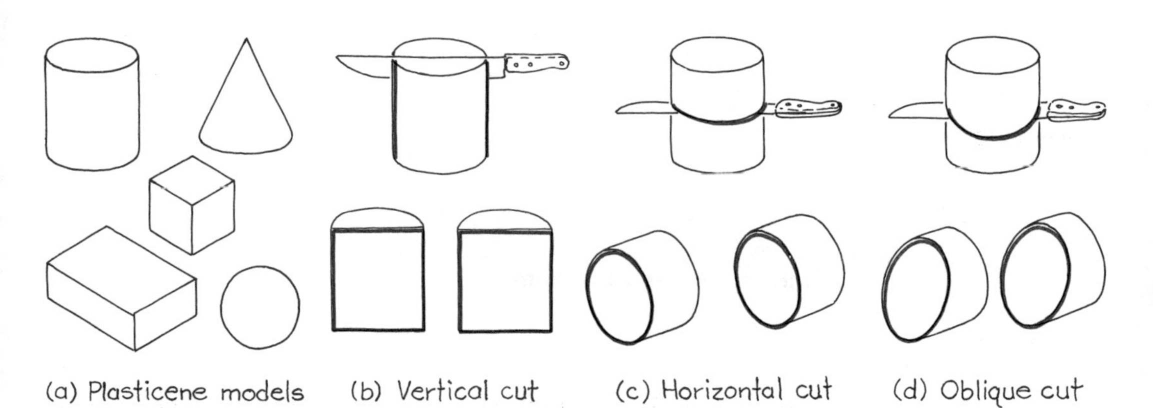

Figure 10-64

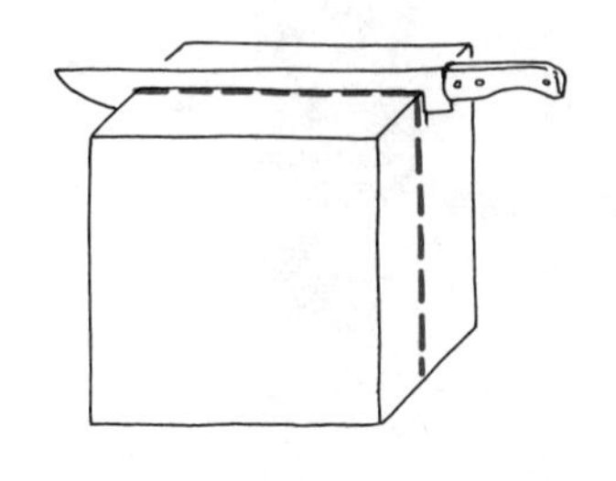

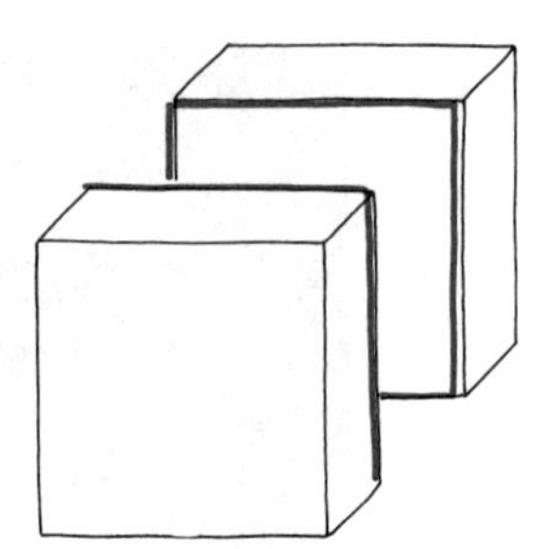

Figure 10-65

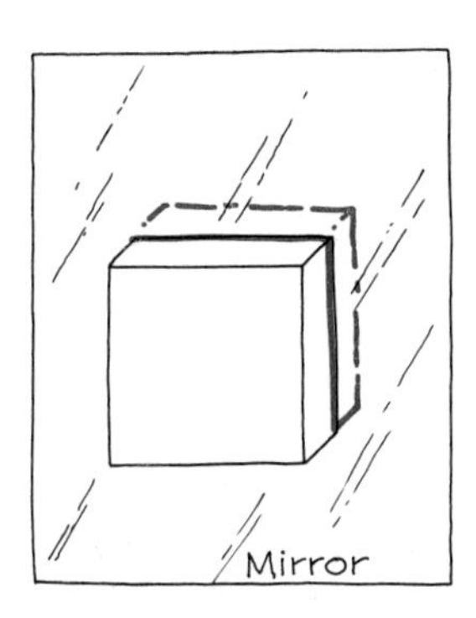

Figure 10-66

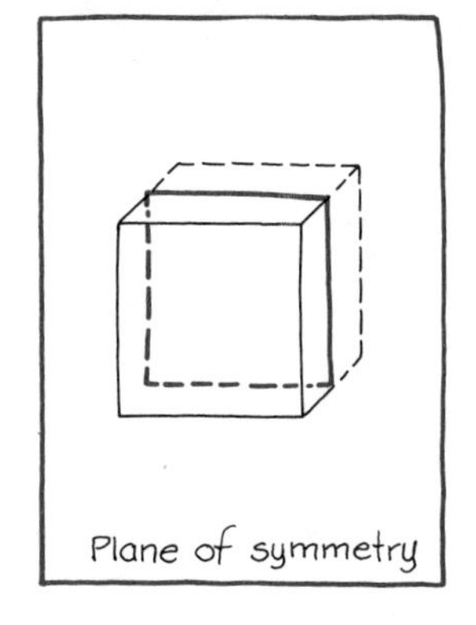

Figure 10-67

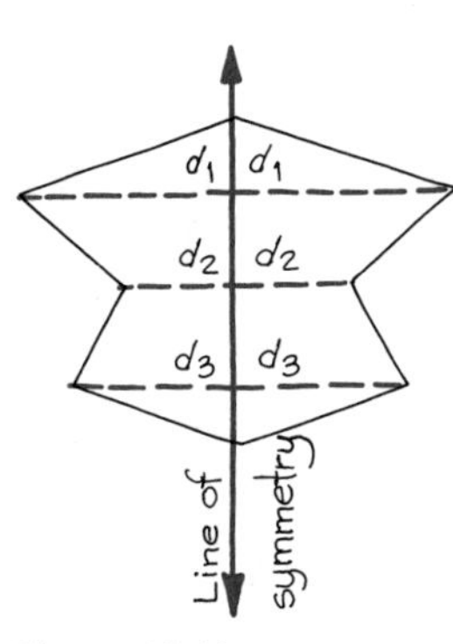

Figure 10-68

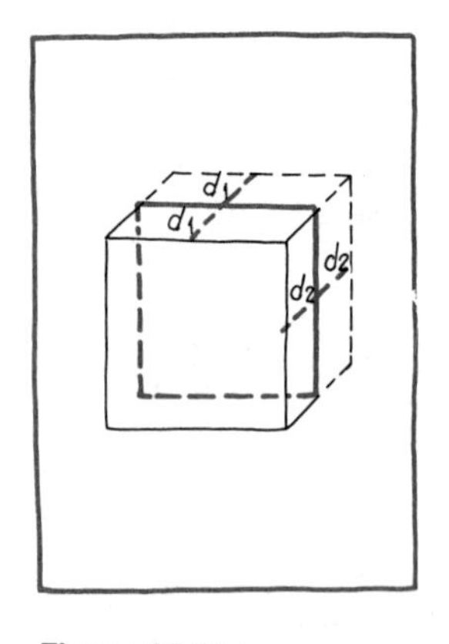

Figure 10-69

every point of the figure on one side of the line of symmetry has a corresponding point on the other side. Corresponding points are the same distance from the line of symmetry (Figure 10-68).

Similarly, when a solid in three dimensions has a *plane of symmetry,* every point of the solid on one side of the plane has a corresponding point on the other side. These corresponding points are the same distance from the plane of symmetry (Figure 10-69).

Finding Planes of Symmetry

The cube in Figure 10-70(a) was *not* cut along a plane of symmetry, although it was cut in half. If you place half of the cube

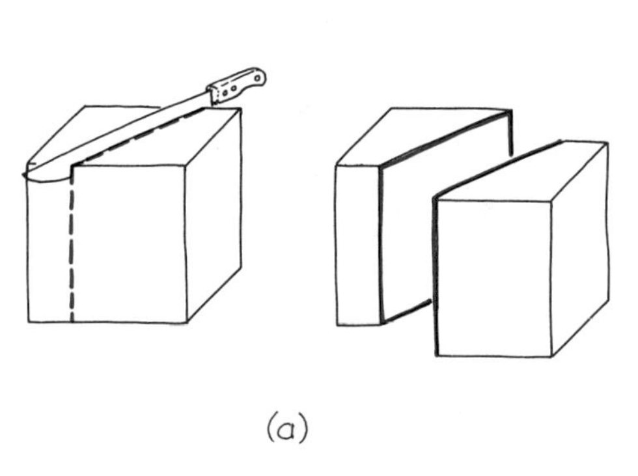

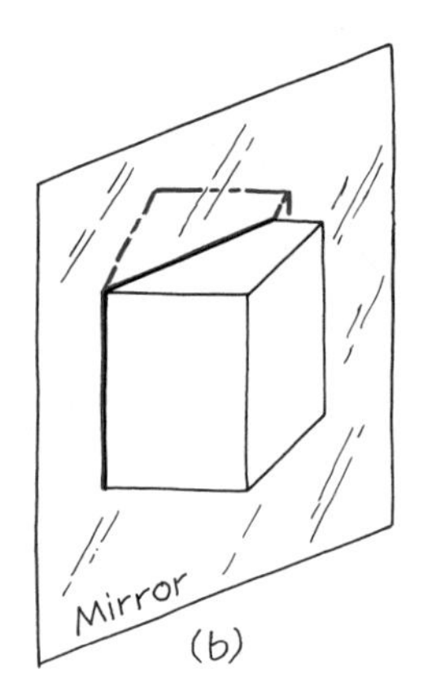

Figure 10-70

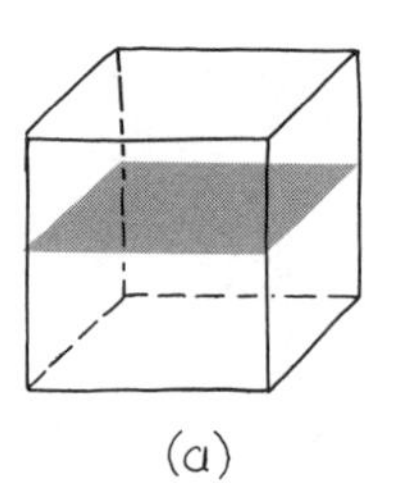

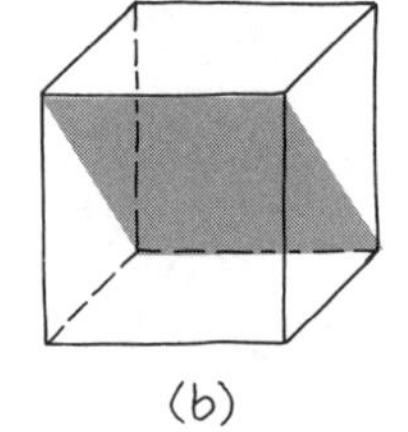

Figure 10-71

against a mirror, you will *not* see a shape that looks like the original cube [Figure 10-70(b)].

Figure 10-71 shows other possible planes of symmetry of the cube. How many other planes of symmetry can you find? How many planes of symmetry does a cube have altogether? It is possible to find *nine* planes of symmetry. Three of them are like the one in Figure 10-71(a), which intersects faces of the cube. Six of them are like the plane of symmetry in Figure 10-71(b), which intersects edges of the cube. Can you locate all *nine* planes of symmetry?

Now try to cut the plasticene cuboids, cylinders, pyramids, and cones that you made earlier along their planes of symmetry. Test each one by holding one piece against a mirror to see if the original shape can be seen. How many planes of symmetry can you find for the cuboid? For the circular cylinder? For the square pyramid? For the triangular pyramid (with base bounded by an equilateral triangle)? For the circular cone? Which intersections shown in Figure 10-72 represent planes of symmetry?

Let us examine the plasticene model of a sphere now. How can we cut it along a plane of symmetry? Which plane in Figure 10-73 is a plane of symmetry? Notice that as a sphere is cut by a plane, the intersection forms a circle. The largest circle is formed when the plane passes through the center of the sphere [Figure 10-73(c)]. Any plane that passes through the center of the sphere can be considered a plane of symmetry (Figure 10-74).

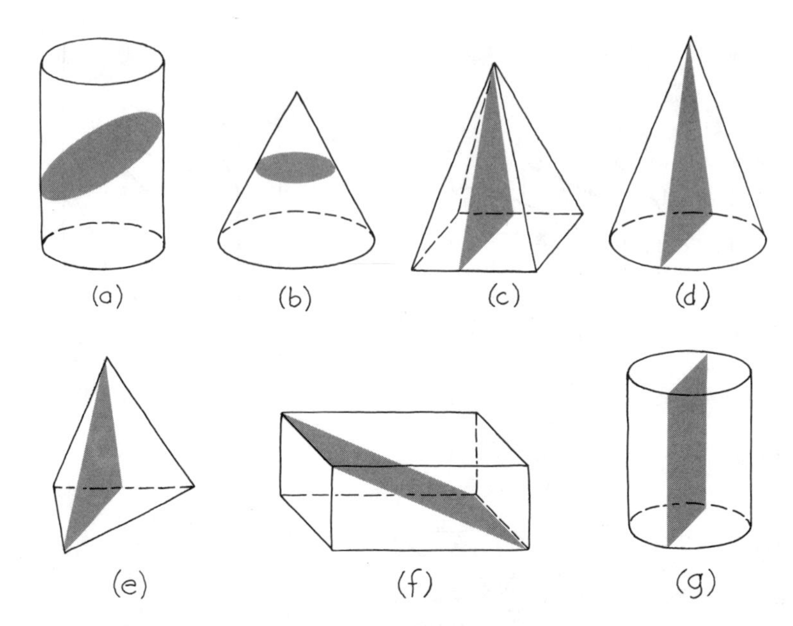

Figure 10-72

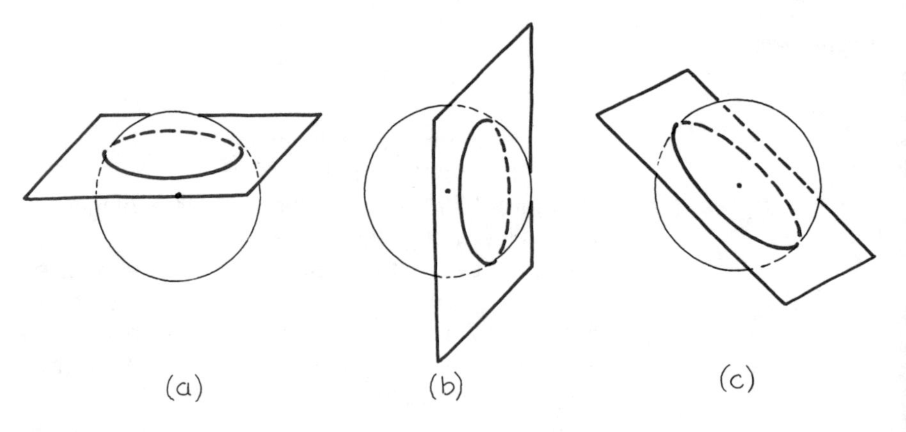

Figure 10-73

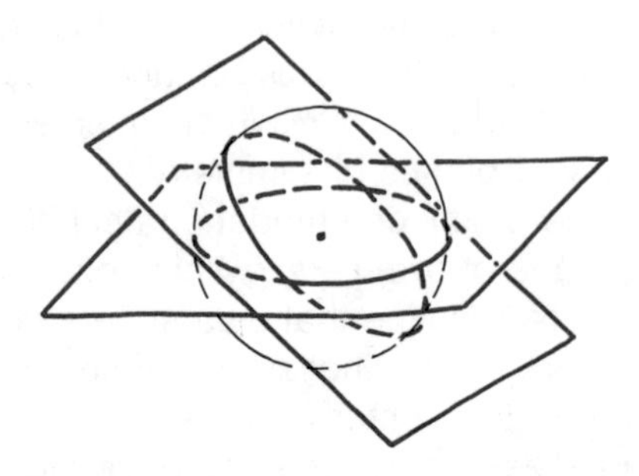

Figure 10-74

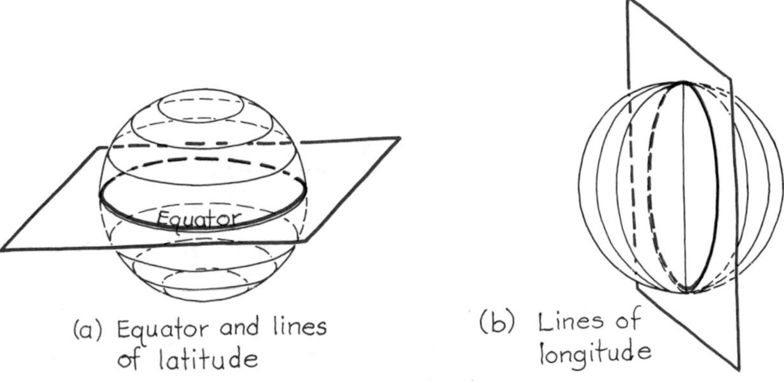

Figure 10-75

If you consider a world globe as a sphere, the Equator can be represented as the intersection of a plane of symmetry and the sphere (Figure 10-75). Similarly, all the lines of longitude can be thought of as the intersections of different planes of symmetry and the sphere. The lines of latitude are smaller circles formed by the intersection of the sphere and planes that are parallel to the plane of the Equator.

Rotational Symmetry in Solids

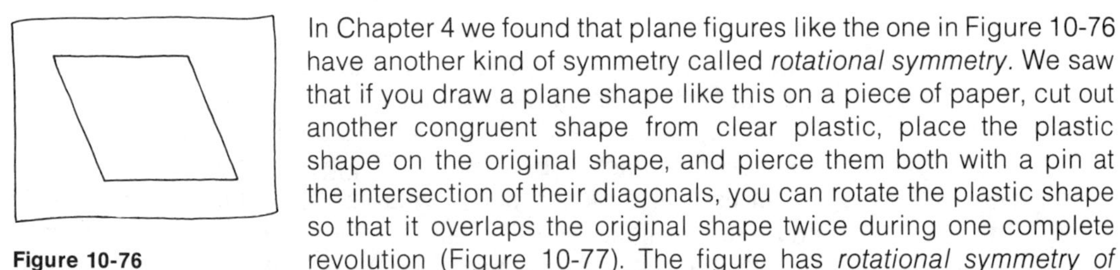

Figure 10-76

In Chapter 4 we found that plane figures like the one in Figure 10-76 have another kind of symmetry called *rotational symmetry*. We saw that if you draw a plane shape like this on a piece of paper, cut out another congruent shape from clear plastic, place the plastic shape on the original shape, and pierce them both with a pin at the intersection of their diagonals, you can rotate the plastic shape so that it overlaps the original shape twice during one complete revolution (Figure 10-77). The figure has *rotational symmetry of order 2*.

Can solid figures have rotational symmetry? How? Let us explore this possibility. First, make a cube with oaktag and tape as in Figure 10-78. Label the faces of the cube 1, 2, 3, 4, 5, and 6 as indicated on the pattern. Draw the diagonals on each face. Then take a long knitting needle and pierce the oaktag cube with the needle, intersecting the cube at the midpoints (intersections of diagonals) of faces 5 and 6 as indicated in Figure 10-79. Holding the needle at either end, you can rotate the cube with the needle as an axis.

Hold the needle so that face 1 of the cube is toward you and the needle is in a vertical position. Now gradually turn the cube by rotating the axis in a clockwise direction as in Figure 10-80. Notice that when faces 2, 3, and 4 are in the same position as face 1

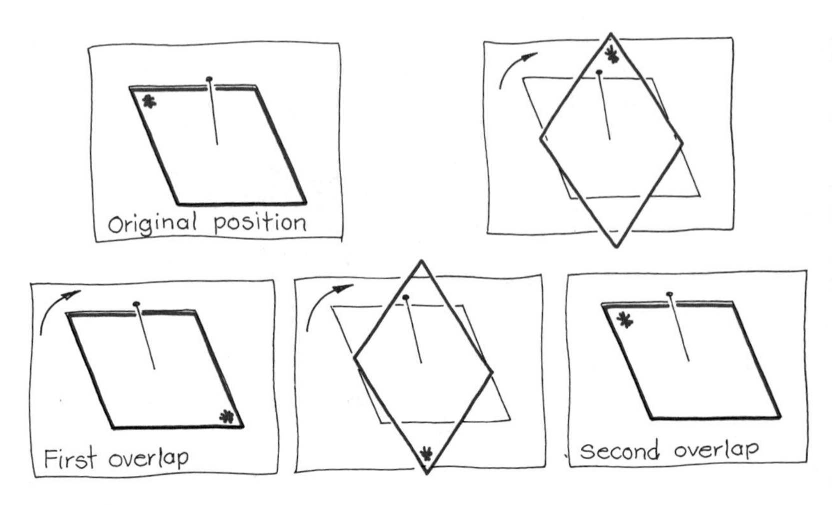

Figure 10-77

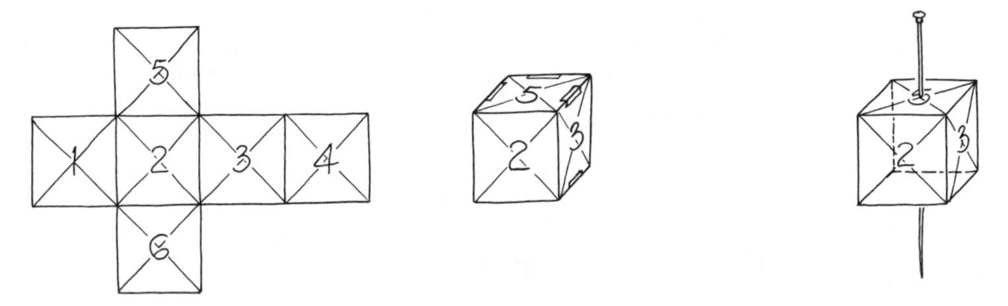

Figure 10-78

Figure 10-79

was originally, the cube *occupies the same position in space.* Except for the numerals on the faces, the cube looks just the same in (a), (c), (e), (g), and (i) of Figure 10-80. The cube is displaying *rotational symmetry* about the axis suggested by the needle.

Let us take a closer look at what is happening. In Figure 10-81 the line segment from the axis to a vertex of face 5 is drawn in brown. The diagram describes the movement of that line segment as the cube is rotated about this axis. As you can see, from position (a) to position (b) the segment has made a 90° turn. Similarly, it turns by 90° from (b) to (c), from (c) to (d), and from (d) to (e), where (e) is the same as (a). This shows that as we make one complete revolution of the axis (360°), the cube occupies the same position in space four times, at 90° intervals. We can say that the cube has *rotational symmetry of order 4* about this axis and that the angle of rotation is 90°.

Does the cube have other axes of symmetry? It certainly does! It has two more, each of which intersects the cube at the midpoints of pairs of opposite faces. Can you locate them on your oaktag model? They are shown in Figure 10-82. The cube has *six* faces and *three* axes of symmetry, each of which intersects the cube at the midpoints of pairs of opposite faces.

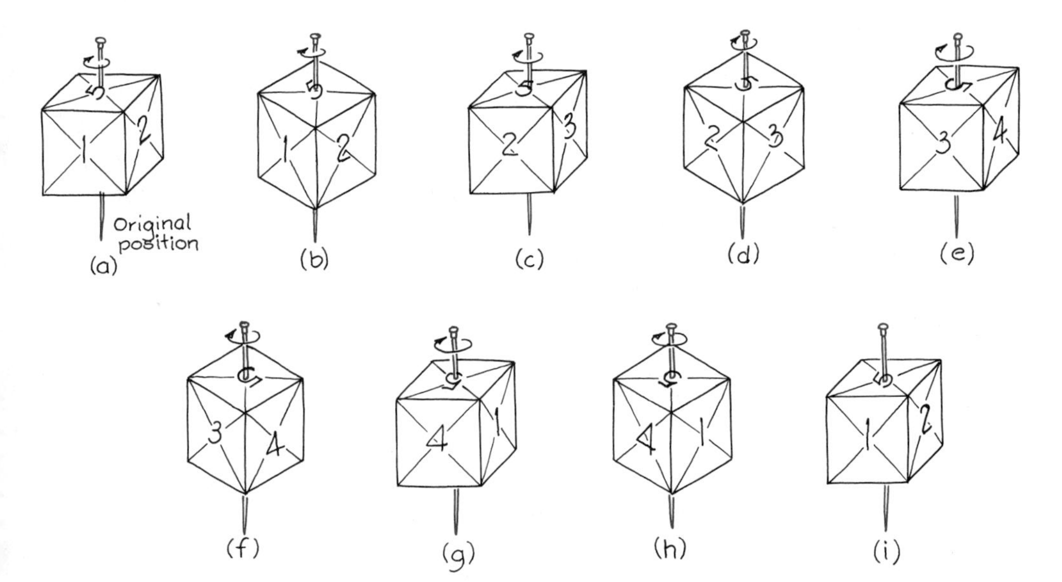

Figure 10-80

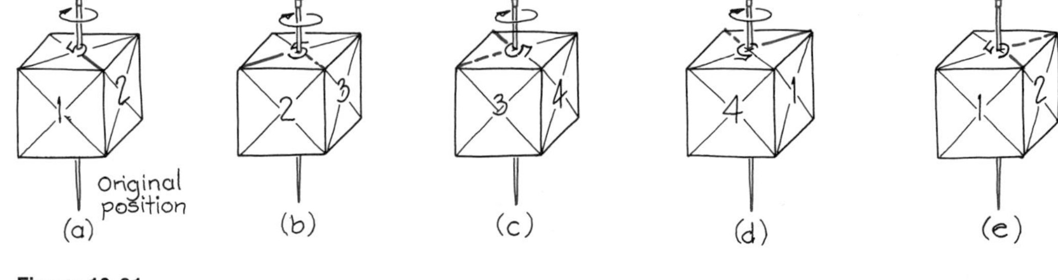

Figure 10-81

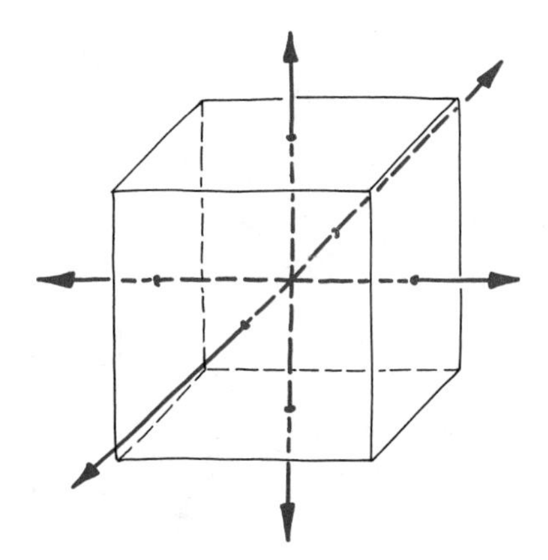

Figure 10-82

But the cube has other axes of symmetry. If you pierce the cube at a vertex of the upper base and extend the needle through the diagonally opposite vertex of the lower base [Figure 10-83(a)], you will locate another axis of symmetry. As the cube is rotated about this axis, it occupies the same position in space in three different instances; it looks the same as the original position in (c), (e), and (g). We can say that the cube has *rotational symmetry of order 3* about this axis.

There are several other axes of symmetry like this one, piercing the cube at diagonally opposite vertices. Can you locate them on the oaktag model? How many are there of this type? You should be able to locate three more. Figure 10-84 illustrates all four axes of symmetry of this type. The cube has *eight* vertices and *four* axes of symmetry intersecting the vertices.

Are there any more axes of symmetry? We have pierced the cube through its vertices and through its faces. Let us try having the needle intersect the cube at its edges. Pierce the oaktag cube with the needle so that it intersects at the midpoint of an edge of the upper base and the midpoint of an opposite edge of the lower base as in Figure 10-85. Rotate the cube about that axis. How many times does the cube occupy the same position in space during one complete revolution? What is the order of rotational symmetry about this axis?

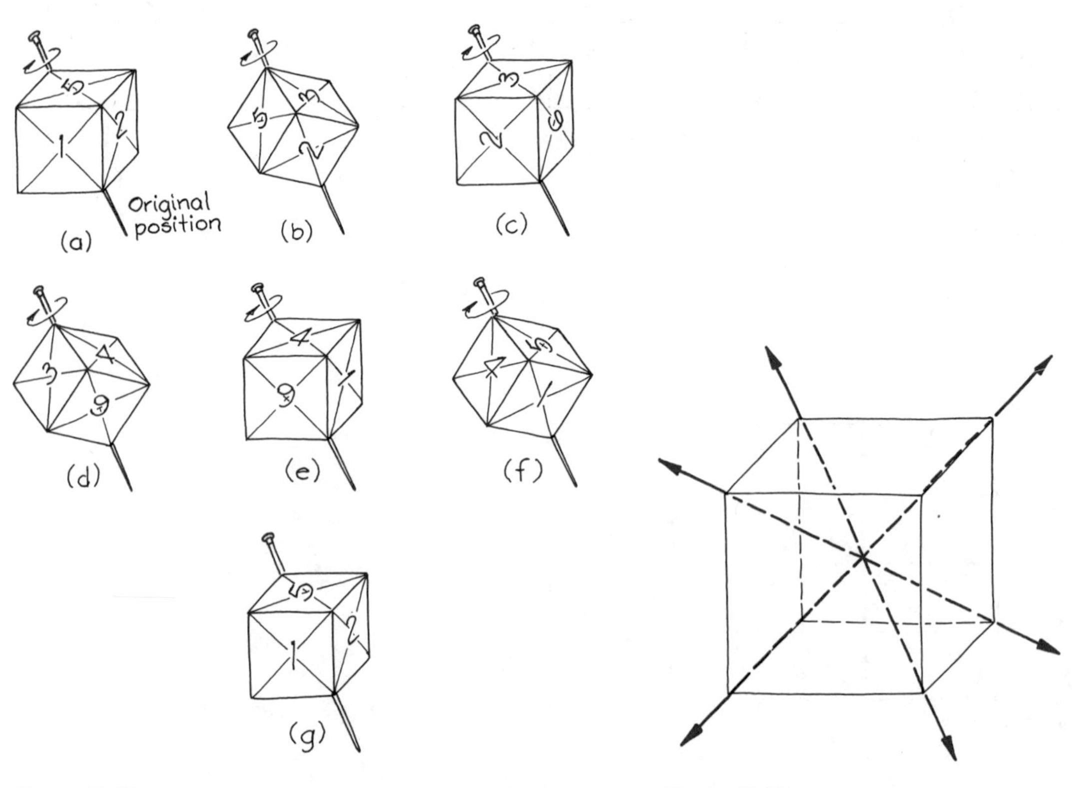

Figure 10-83

Figure 10-84

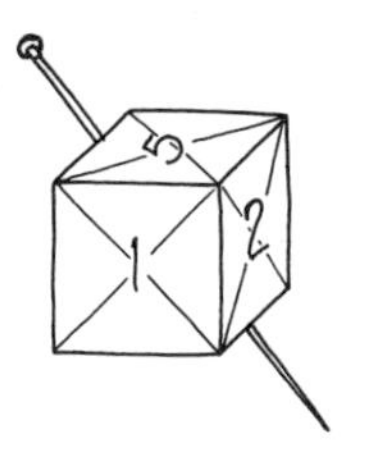

Figure 10-85

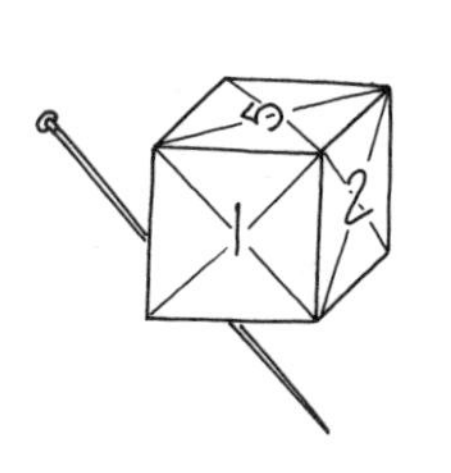

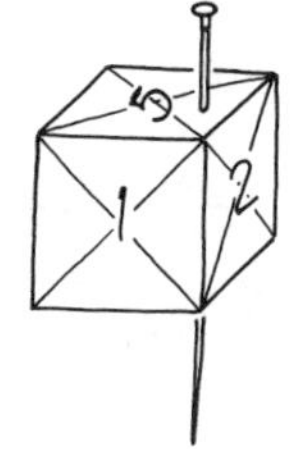

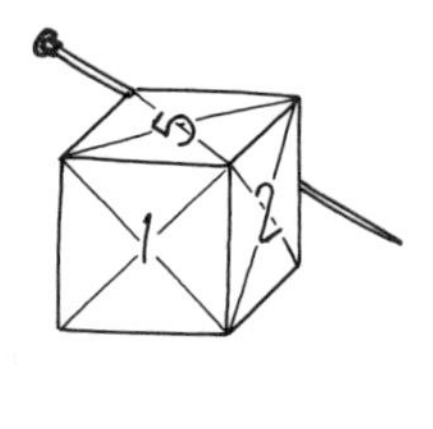

Figure 10-86

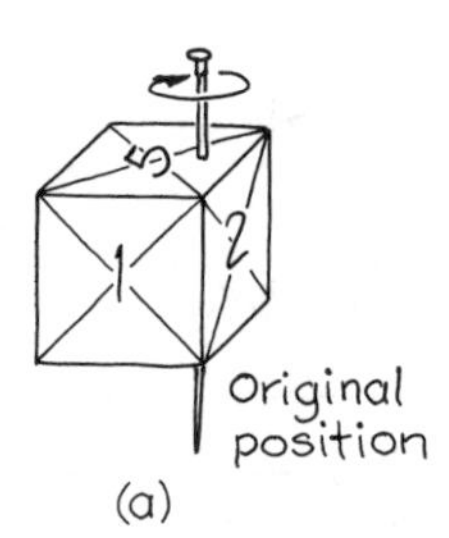

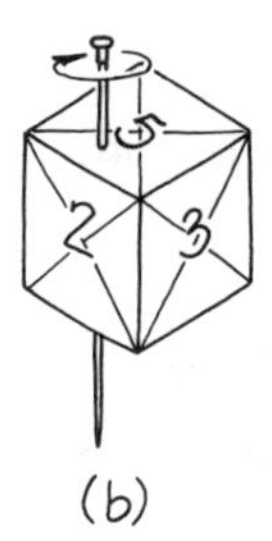

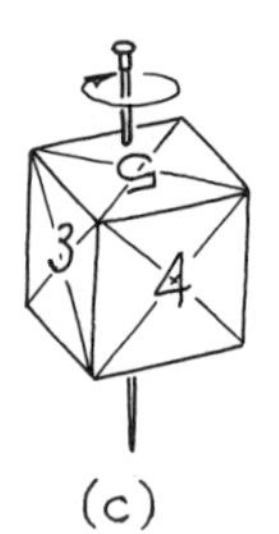

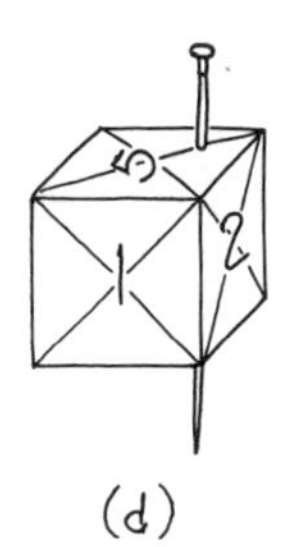

Figure 10-87

As you might have guessed, there are other axes of symmetry that intersect the cube through its edges. In fact there are five more. Can you locate them on your oaktag model?

Can you locate any other axes of symmetry? Try piercing the cube with the needle in various ways as in Figure 10-86. Rotate it about each axis and test for rotational symmetry. No matter where the axis is, when you rotate the cube it will return to its original position after one complete revolution (Figure 10-87). But does it occupy the same position in space at any other time *during* that revolution?

You will find there are no other axes of symmetry. To summarize, the cube has 13 axes of symmetry:

1. Six axes of symmetry of order 2 (through edges)
2. Four axes of symmetry of order 3 (through vertices)
3. Three axes of symmetry of order 4 (through faces)

Do other solids have this property of rotational symmetry? You can make oaktag models of other solids as suggested by the patterns in Figure 10-88 and use a knitting needle to investigate rotational symmetry as you did with the cube. The total number of axes of symmetry for each solid shown in Figure 10-88 is listed in Table 10-1.

Can you locate *all* the axes of symmetry for each solid? Some of these axes are suggested in Figure 10-89.

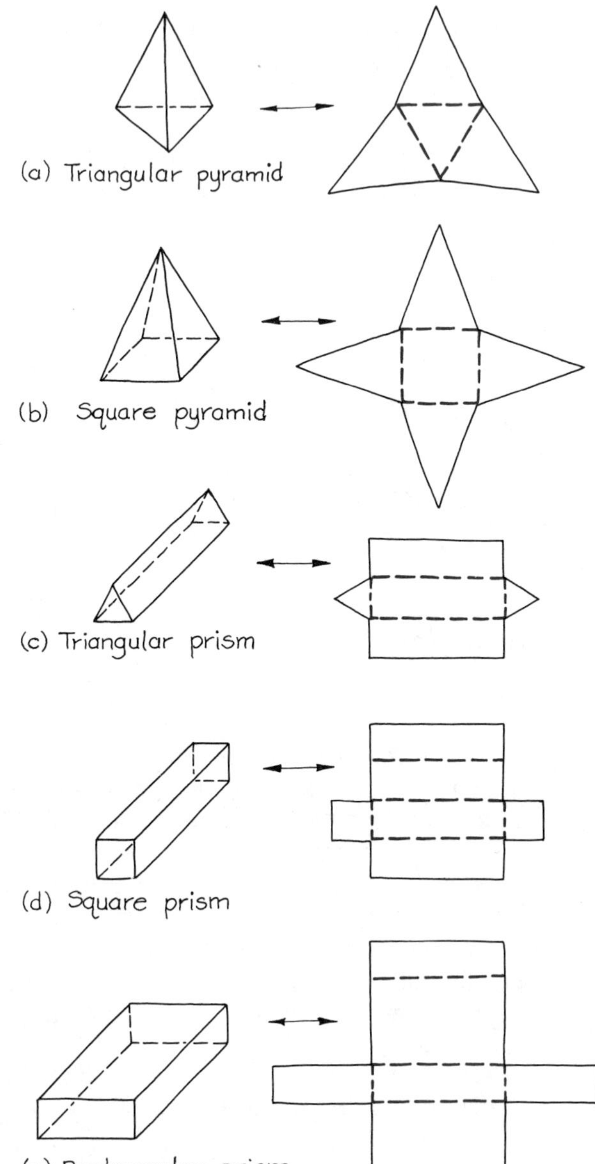

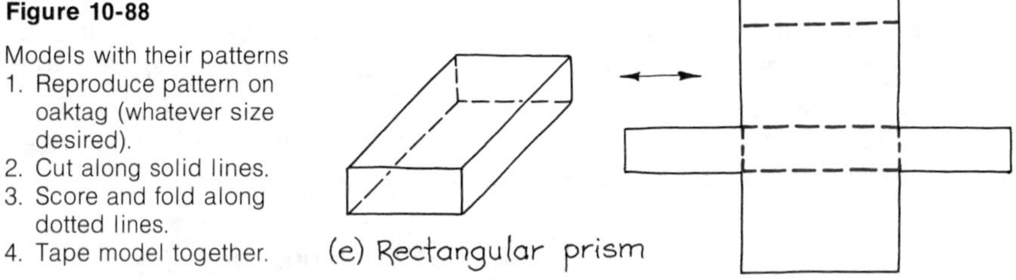

Figure 10-88

Models with their patterns
1. Reproduce pattern on oaktag (whatever size desired).
2. Cut along solid lines.
3. Score and fold along dotted lines.
4. Tape model together.

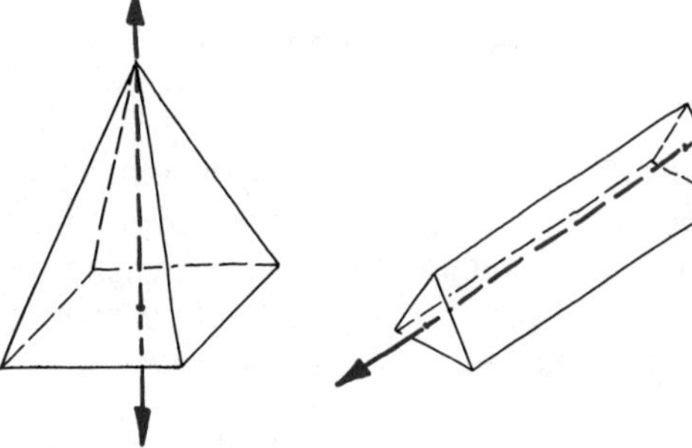

Figure 10-89

Table 10-1

Solid	Axes of symmetry
(a) Triangular pyramid	1
(b) Square pyramid	1
(c) Triangular prism	4
(d) Square prism	5
(e) Rectangular prism	5

More Experiences

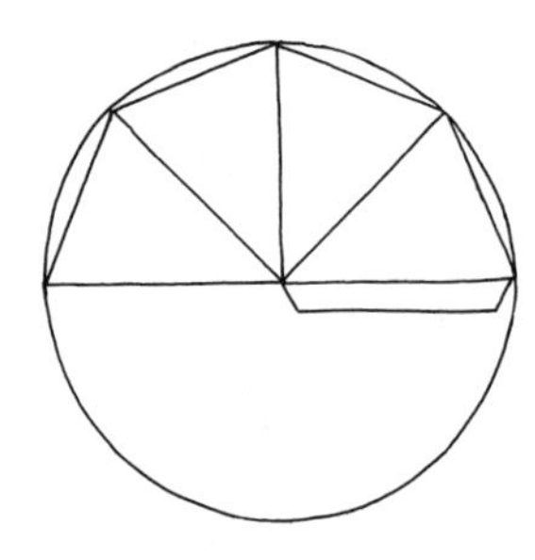

10-1. The diagram at the left illustrates a pattern, or **net,** for a model of a square pyramid (with open base). Can you explain how the net was made?

(*a*) Make a net for a square pyramid with all edges 5 cm in length.
(*b*) In a similar way, can you make the net of a pentagonal pyramid? A hexagonal pyramid?

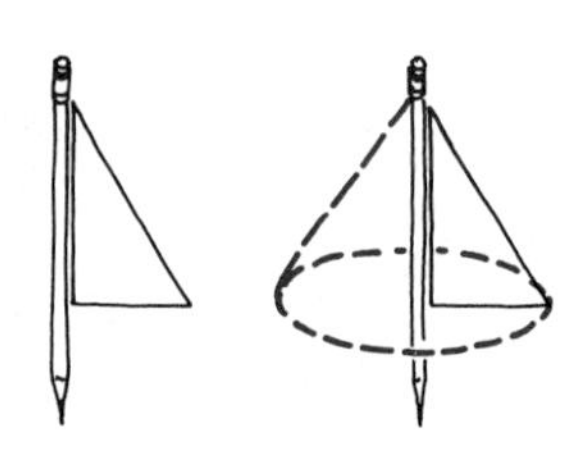

10-2. Cut out a triangular shape from oaktag or cardboard and attach one edge to a pencil as shown at the left. With the pencil as axis, rotate the shape through 360°. What kind of solid shape does the path of the unattached edges of the triangle suggest?

Try this procedure with other plane shapes like the ones at the left. What kinds of solid shapes result? Can you form a rectangular solid in this way? Explain.

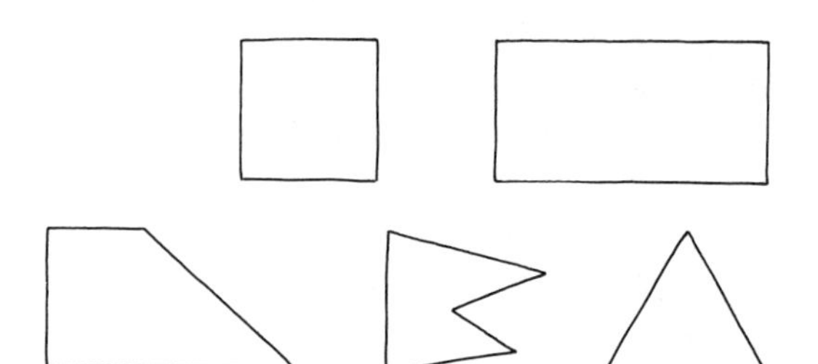

10-3. One way to classify solids into three sets is as follows:

1. Solids with no edges and no corners
2. Solids with edges but no corners
3. Solids with edges and corners

Draw diagrams of the different solid shapes that might belong to each set. Can a solid belong to more than one of these sets?

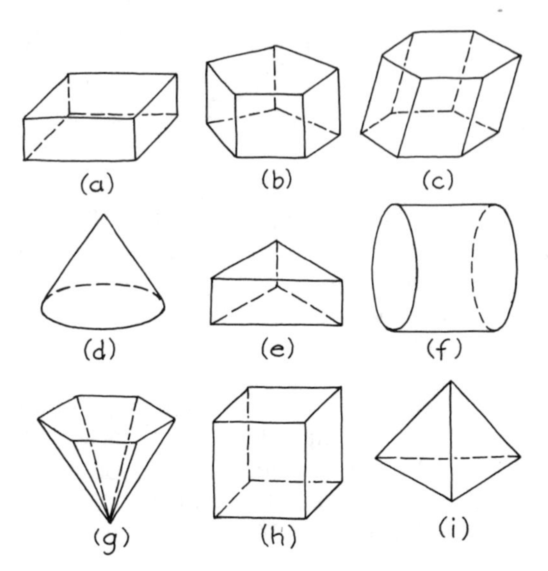

10-4. Select as many of the following words as you can to describe each solid suggested in the diagrams at the left:

1. Parallelepiped
2. Pyramid
3. Cone
4. Cylinder
5. Prism
6. Cuboid
7. Cube

10-5. The relationship between pyramids, cones, prisms, and cylinders might be expressed as follows: *Pyramids are to cones as prisms are to cylinders.* Can you explain this statement?

10-6. Find the number of axes of symmetry and the order of rotational symmetry for each axis of the figures listed in the table at the left.

Solid	Axes of symmetry	Order of rotational symmetry
Square pyramid	1	4
Pentagonal pyramid		
Square prism		
Rectangular prism		
Circular cylinder		
Circular cone		
Sphere		

10-7. Collect some common objects like the following: pencils, books, paper clips, empty milk containers or other containers, ashtrays, coins, empty bottles with assorted shapes. For each object, locate its axes of symmetry and the order of rotational symmetry for each axis. Keep a record. Next, locate all the planes of symmetry for each object and record this information. Is there a relationship between the number of axes of symmetry and the number of planes of symmetry an object has?

10-8. Two lines in space can intersect, be parallel, or be skew. Two planes in space can intersect or they can be parallel. Might two planes be skew? Explain. Can a line be skew to a plane? Explain.

11 Measuring Solids

Which Is Larger?

Look at the boxes pictured in Figure 11-1. Which is larger, box *A* or box *B?* Box *A* is wider, but box *B* is taller. Which box is *larger?* This problem is similar to the one we discussed in Chapter 7 when we attempted to compare the sizes of rectangles. We found there were different ways to measure a rectangle. Look at the two rectangles in Figure 11-2. Which is larger? If you measure the *perimeter* of each rectangle, you will find that rectangle *A* has a greater perimeter than rectangle *B*. Yet the area within *B* is greater than the area

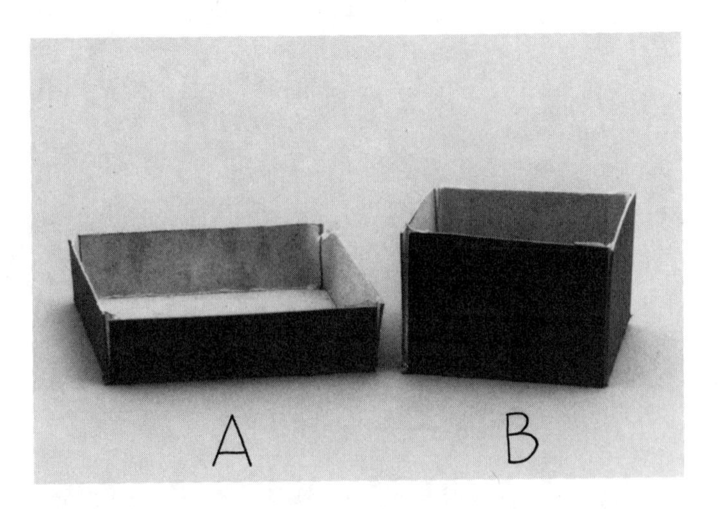

Figure 11-1

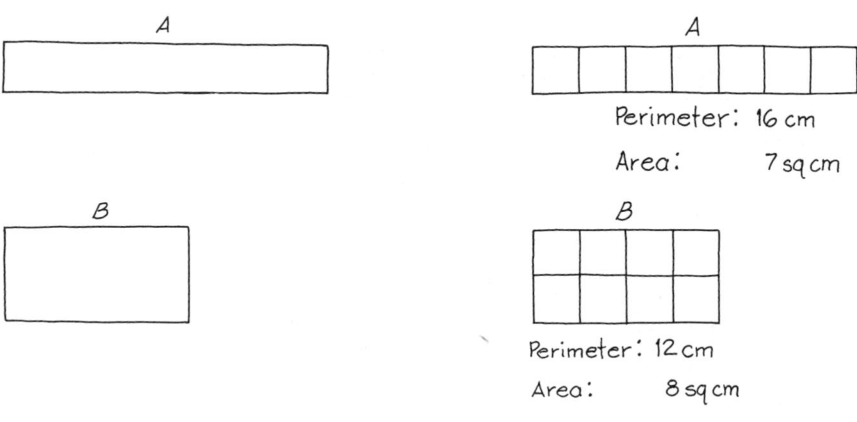

Figure 11-2

Figure 11-3

within *A* (Figure 11-3). We can conclude that rectangle *A* is larger *if we are comparing perimeters* and rectangle *B* is larger *if we are comparing areas.*

Now let us get back to the boxes. We can solve our dilemma by making models of the boxes with oaktag ruled into squares as shown in Figure 11-4. Cellophane tape can be used to hold the models together. The squared face of the oaktag is kept on the

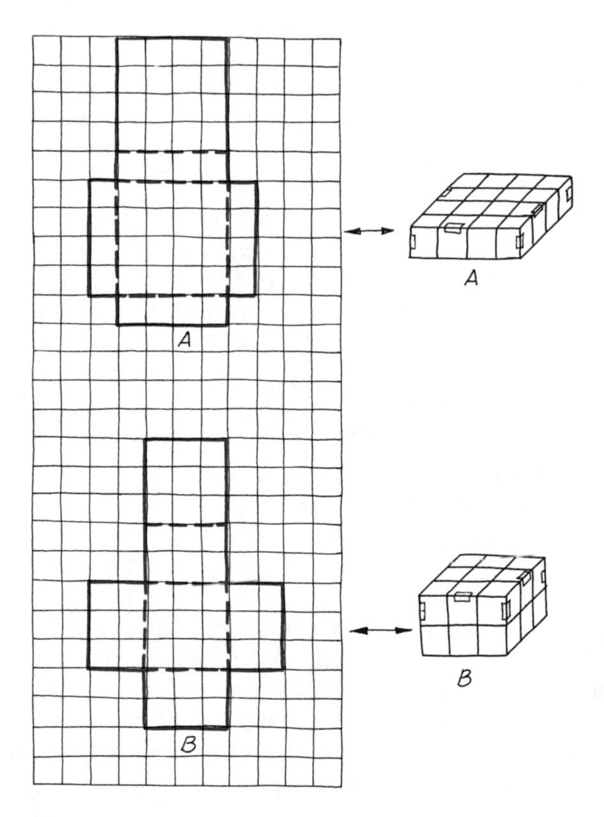

Figure 11-4

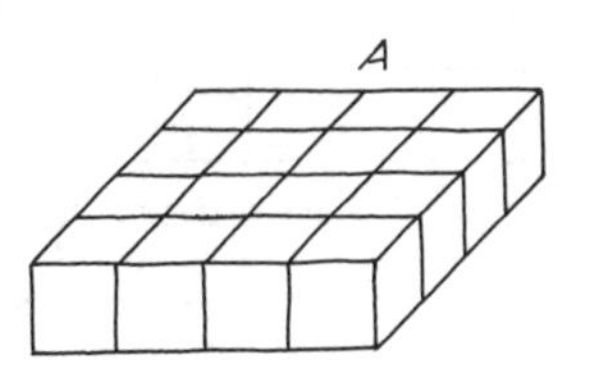

Top + bottom:
2 × 16 = 32 squares
Sides:
4 × 4 = 16 squares
Total surface area:
48 squares

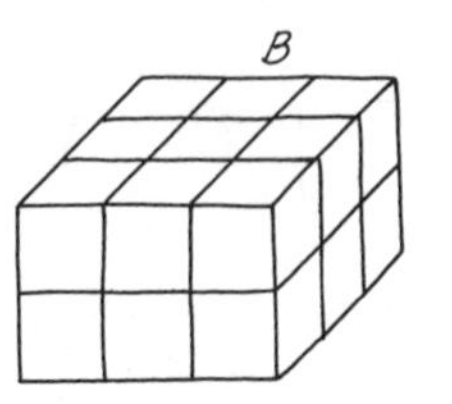

Top + bottom:
2 × 9 = 18 squares
Sides:
4 × 6 = 24 squares
Total surface area:
42 squares

Figure 11-5

(a)

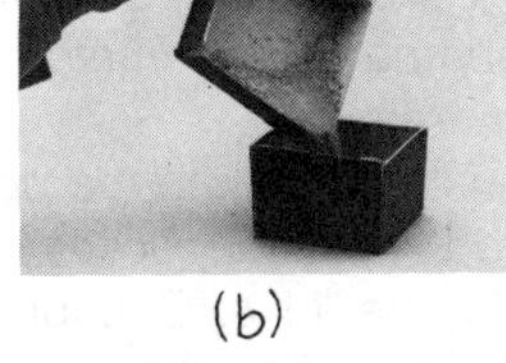

(b)

(c)

Figure 11-6

outside of the boxes. Count the total number of squares that appear on the base, sides, and lid of each box. Which box has the most squares?

The base, sides, and lid of each box have the shape of rectangles. Counting the squares is like finding the area within each rectangle. The total number of squares represents the area of the base, sides, and lid or the area of the entire surface of the box. We can call this the **surface area** of the box. As Figure 11-5 indicates, box *A* has a greater surface area than box *B*.

Next, fill box *A* completely with sand, salt, or a similar substance. Then pour the contents of box *A* into box *B*. What do you notice? Box *B* is *not* filled completely (Figure 11-6). Box *B* can hold more sand than box *A*. We can say that box *B* has a greater **capacity** or **volume** than box *A*.

We can now answer the original question. Which box is larger? It depends upon what we are measuring. We have used two ways to measure the boxes. Box *A* is larger *if we are comparing total surface areas*. Box *B* is larger *if we are comparing volumes*.

Figure 11-7

In a similar way, we can compare the sizes of other kinds of boxes or containers (or structures) like those illustrated in Figure 11-7. These can have many different shapes. In this chapter we will explore ways of finding the surface area and the volume of some well-known solid shapes.

Surface Area

Suppose we want to make oaktag models of the solids in Figure 11-8. The diagrams at the right of the shapes show patterns of geometric shapes that become the faces of the corresponding

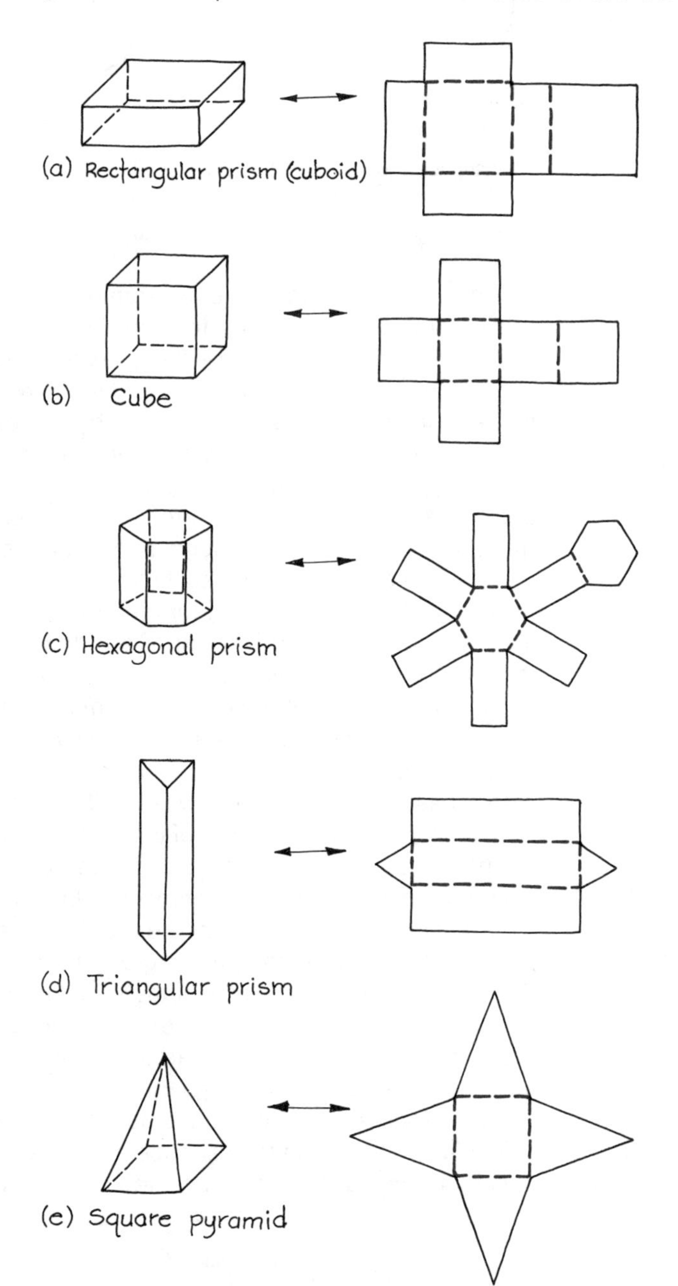

Figure 11-8

Solid models with their patterns

1. Reproduce pattern on oaktag (whatever size desired).
2. Cut along solid lines.
3. Score and fold along dotted lines.
4. Tape model together.

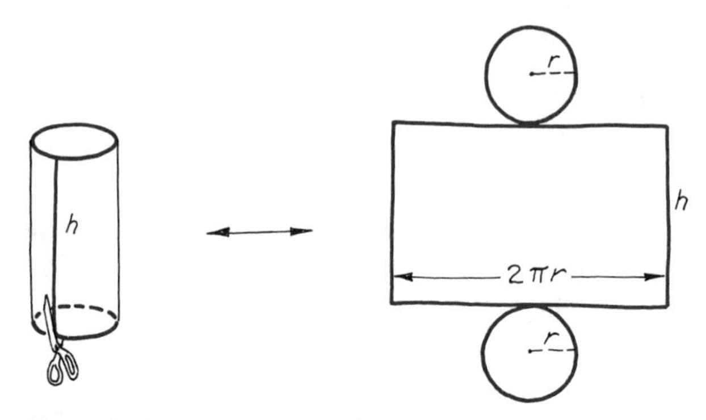

Figure 11-9

Circular cylinder

Total surface area $= 2 \cdot (\pi r^2) + (2\pi r) \cdot h$
$= 2\pi r^2 + 2\pi rh$

models. In each case the two-dimensional pattern should be folded and taped together to form the three-dimensional model.

Which model will require the greatest amount of oaktag? To answer this question we need to find the total *surface area* of each model. Since the faces are bounded by polygons, finding the total surface area is easy. We simply calculate the sum of the areas of each polygon that forms the boundary of a face. In Chapter 5 we discovered ways to find the areas within these familiar polygons (rectangles, squares, hexagons, and triangles).* The total surface area for the *cube* becomes six times the area of one of the square faces. The total surface area for the *hexagonal prism* is the sum of the areas of the hexagonal upper and lower bases and six times the area of a rectangular lateral face, etc.

The surface area of a *circular cylinder* poses a slightly different problem since the faces are not all bounded by polygons. But, as shown in Figure 11-9, the model of a circular cylinder can be considered as being made up of two circular regions and a rectangular region. Notice that the length of the rectangular region is the same as the circumference of the circular regions. If the radius of the cylinder is r, then the circumference† of the circle (and length of the rectangular region) is $2 \cdot \pi \cdot r$. The problem, then, is to find the sum of the area of the rectangular region and the areas of the bases. We need to find the radius r of the circular base and the height h of the cylinder (which is the height of the rectangular region) in order to find the total surface area. The area of each circular base is πr^2 and the area of the rectangular lateral surface is $(2\pi r) \cdot h$. Thus the total surface area becomes $2 \cdot (\pi r^2) + (2\pi r \cdot h)$.

Similarly, to find the total surface area of a *circular cone,* we can consider the base of the cone to be a circular region and the

See pages 79 to 103 for a review of finding areas within polygons.

†*See pages 163 to 170 for a review of finding the circumference and area of a circle.*

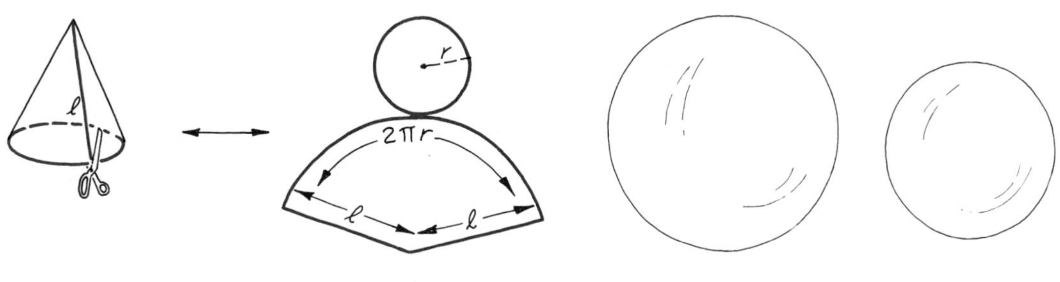

Figure 11-10

Circular cone

Total surface area $= \pi r^2 + \dfrac{2\pi r \cdot l}{2}$

$= \pi r^2 + \pi r l$

Figure 11-11

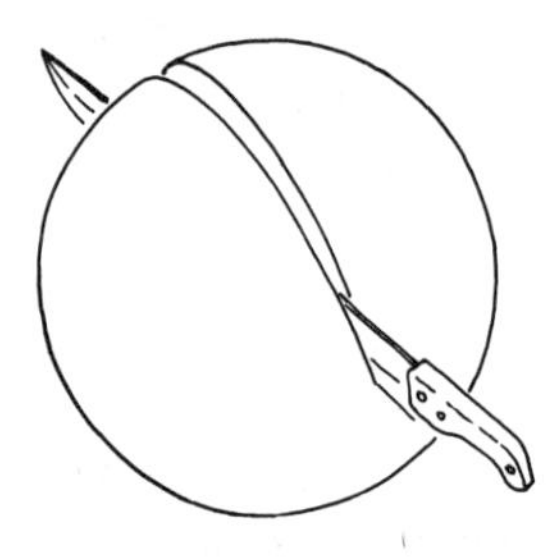

Figure 11-12

lateral surface to be a sector of a circular region (Figure 11-10). The arc of this sector has the same length as the circumference of the circular base of the cone ($2\pi r$). The radius of the sector is the same length as a line segment whose end points are the apex of the cone and a point on the circle that forms the boundary of the base (l). The total surface area, then, is the sum of the area of the base (circular region: πr^2) and the area of the lateral surface [circular sector: $(2\pi r \cdot l)/2 = \pi \cdot r \cdot l$].*

Finding the surface area of a *sphere* seems to offer a real challenge. A model of a sphere cannot be made out of paper or oaktag like the other models. Unlike the other solids discussed, the sphere has no faces. How can we find the surface area of a sphere?

Two balls (which are models of spheres) are pictured in Figure 11-11. Suppose we want to compare the surface areas of these balls. It is easy to see that the ball on the left has the greater surface area. But how much greater? One way to find out might be to glue dried beans onto the surfaces of each ball, count the beans on each surface, and find the difference. This would give us an idea of the difference in surface area between the two balls. But it is a rather tedious way to solve the problem. And beans are not a standard unit of measurement! Can we find the surface areas in square units?

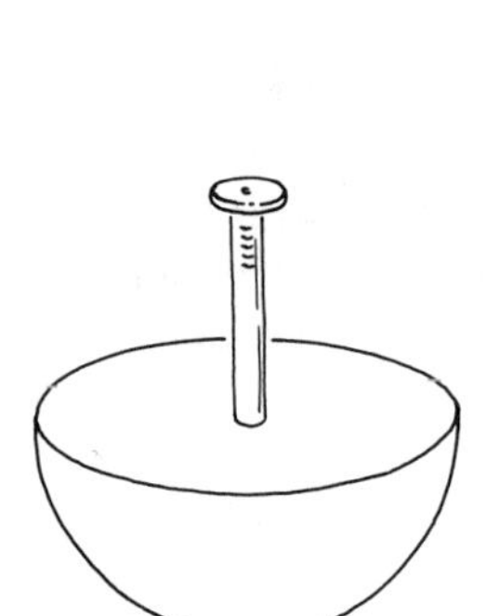

Figure 11-13

Let us try an experiment. You will need a solid rubber ball, a sharp knife, a large nail, some pliable rope (like lightweight clothesline), and a pencil or pen. Carefully cut the rubber ball in half so that you have two congruent half spheres or hemispheres. Your knife should pass through the center of the sphere (Figure 11-12). A face of a hemisphere will look like a circular region. Next, push the nail into a hemisphere so that it is perpendicular to the circular region at its center (Figure 11-13). Then take the rope and, starting at one end, carefully wind it around the nail in a

*See pages 173 to 174 for a discussion of finding the area of a sector of a circle.

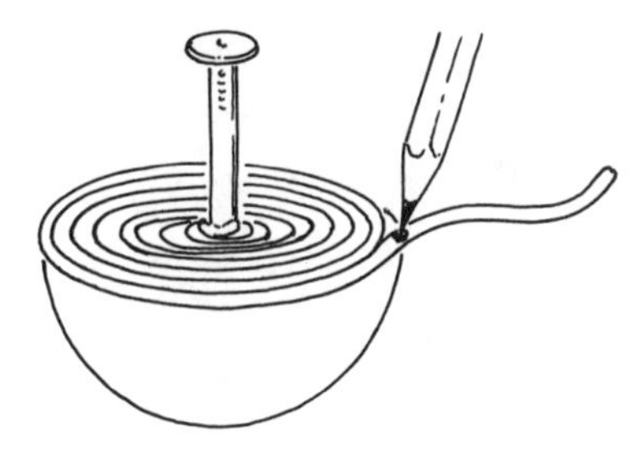

Figure 11-14

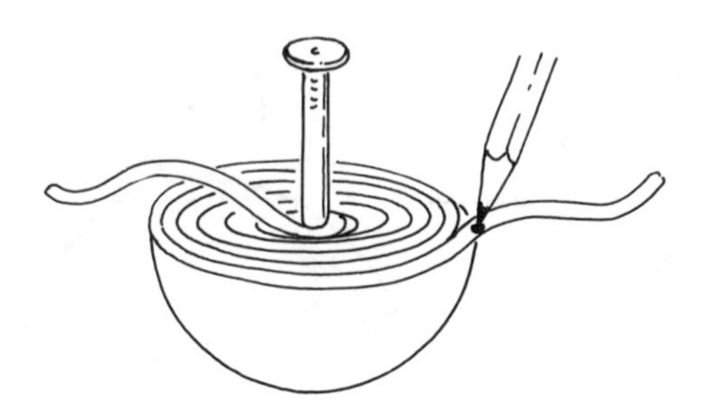

Figure 11-15

spiral fashion until it covers the circular region* Using the pencil, mark the rope at the point where it completely covers the circular region (Figure 11-14). The length of the rope in this spiral represents the amount of rope needed to cover the circular region.

Now unwind the rope and do the same thing all over again. But this time start the spiral where you stopped on the rope (see the pencil mark), and again cover the circular region with the rope (Figure 11-15). Make another pencil mark on the rope at the point where it covers the circular region the second time.

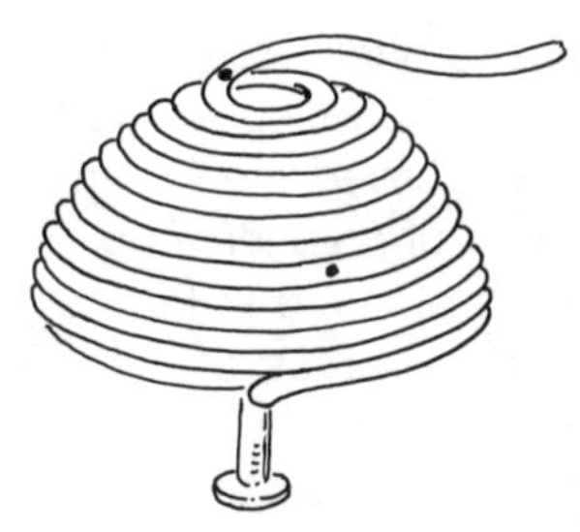

Figure 11-16

Unwind the rope. Turn the hemisphere over and, using that same piece of rope, cover the surface of the hemisphere as indicated in Figure 11-16. Look what happens! It takes *just as much* rope to cover the hemisphere as it did to cover the circular region *twice!*

How much rope would we need to cover the entire surface of the sphere? It takes *four* times as much rope as is needed to cover the circular region. This tells us that the surface area of the sphere is the same as four times the area of the circular region. We *know* how to find the area of a circular region. If r is the radius, the area is πr^2. So the surface area of the sphere must be four times πr^2, or $4\pi r^2$.

With this discovery, we can find the surface area of our original models of spheres (Figure 11-11). We will need to know the measures of the radii of these spheres. Figure 11-17 shows one way to find the diameters: using two books and a ruler to form a homemade **caliper.** † The length of the diameter of the larger ball is 6 cm (radius 3 cm) and the length of the diameter of the smaller ball is about 5 cm (radius $2\frac{1}{2}$ or $\frac{5}{2}$ cm). Then the area of the spherical surfaces would be about 113 and 79 sq cm respectively, using the approximation of $\frac{22}{7}$ as the value of π (Figure 11-18). Conclusion: The surface area of the larger ball is about 34 sq cm greater than the surface area of the smaller ball.

**If it is difficult to hold the rope in place, use cellophane tape to keep it down.*

†*A caliper is an instrument that can be used to measure the diameter of spherical objects.*

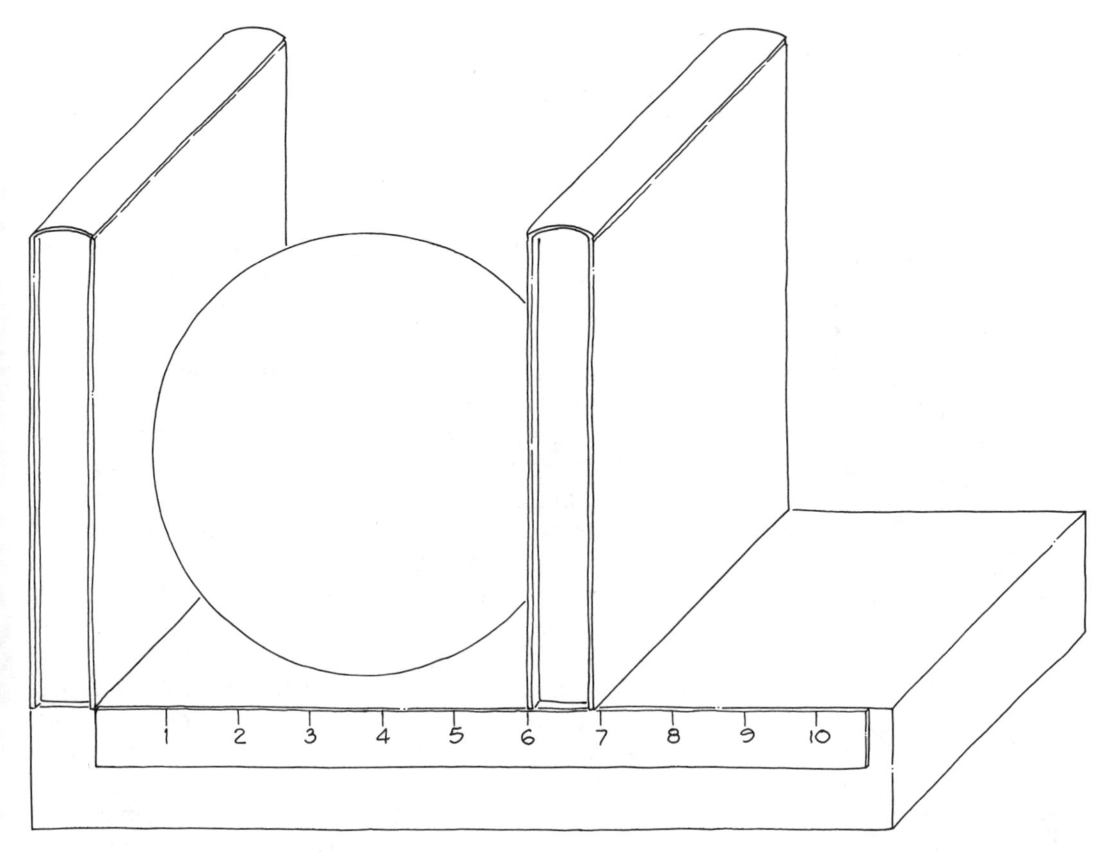

Figure 11-17

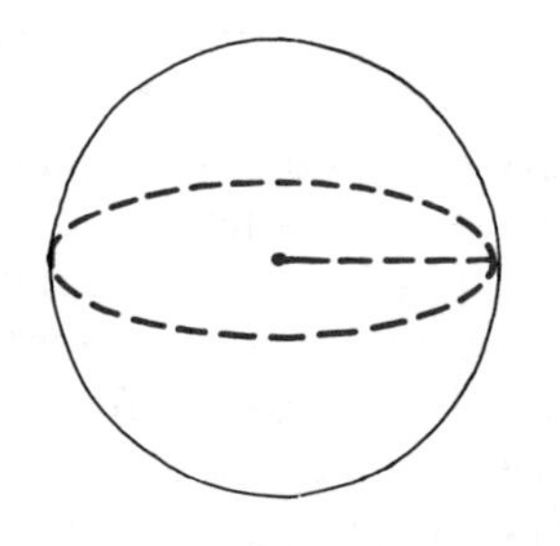

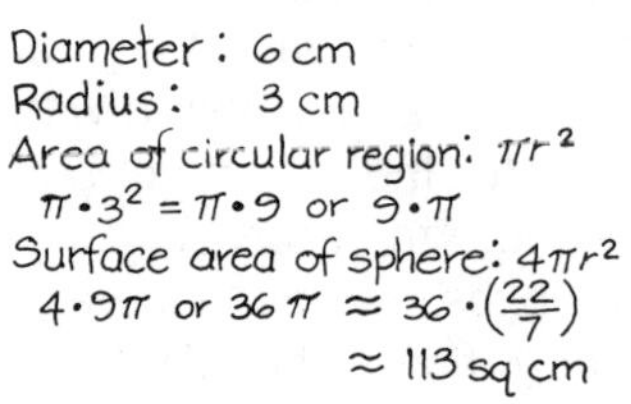

Diameter: 6 cm
Radius: 3 cm
Area of circular region: πr^2
$\pi \cdot 3^2 = \pi \cdot 9$ or $9 \cdot \pi$
Surface area of sphere: $4\pi r^2$
$4 \cdot 9\pi$ or $36\pi \approx 36 \cdot \left(\frac{22}{7}\right)$
≈ 113 sq cm

Diameter: 5 cm
Radius: 5/2 or 2½ cm
Area of circular region: πr^2
$\pi \cdot \left(\frac{5}{2}\right)^2 = \pi \cdot \frac{25}{4}$ or $\frac{25}{4} \cdot \pi$
Surface area of sphere: $4\pi r^2$
$4 \cdot \pi \cdot \frac{25}{4}$ or $25\pi \approx 25 \cdot \left(\frac{22}{7}\right)$
≈ 79 sq cm

Figure 11-18

Volumes of Cubes and Other Rectangular Solids

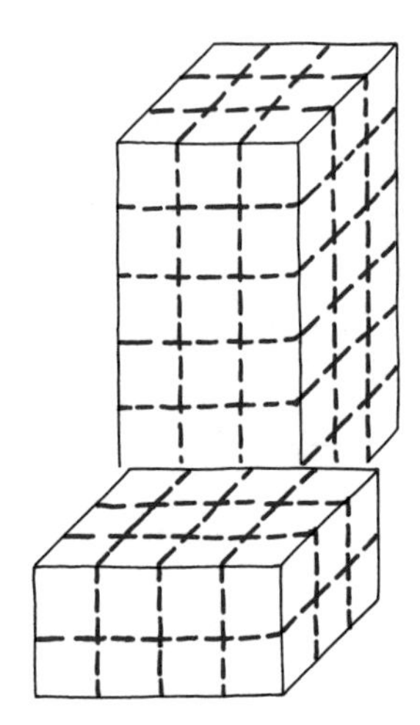

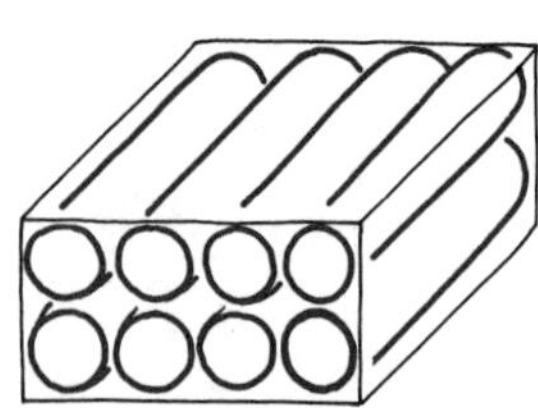

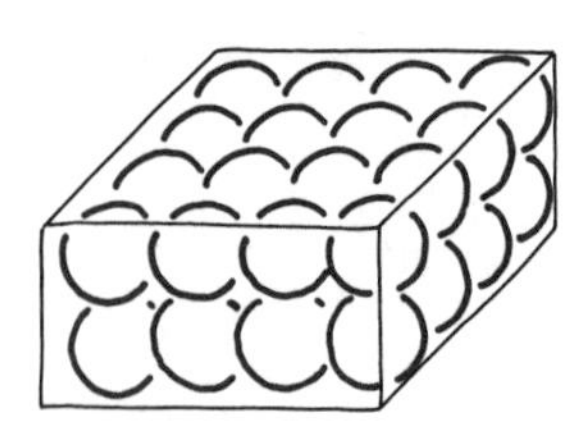

Figure 11-19

We now turn our attention to the *capacity* or *volume* of solids. Earlier in this chapter we compared the volumes of two boxes by filling one with sand and then pouring the contents into the other box. That was one way to find out which box had a greater capacity.

We might have used other materials such as rice, beans, or marbles to fill the boxes. We would have been able to discover which box has a greater capacity by using any of these materials. But when we are asked to give the capacity of box *A*, it is rather ambiguous to reply that box *A* holds 1283 beans, since beans are a nonstandard unit. How large are the beans? Do they vary in size? How much space is there between them?

As we saw when we used other types of measurement (linear, area, angle), it is useful to use a *standard* unit that does not create ambiguity. The cube is often used as a standard unit for volume measurement. The capacity or volume of boxes, tanks, houses, etc., might be measured in cubic inches, cubic feet, or cubic yards, or in metric units such as cubic centimeters, cubic decimeters, or cubic meters.

The cube is a convenient shape for volume measurement because cubes fit compactly into rectangular solids (cuboids) of all types—unlike spherical or cylindrical units, which do not fit compactly but leave holes (Figure 11-19). You will recall it was convenient to use squares for area measurement for a similar reason: squares fit compactly into rectangular regions.

Let us make use of box *A* and box *B* again. This time imagine we have many small cubes all of whose faces are congruent to the squares drawn on the surfaces of the boxes (Figure 11-20).* Now suppose we fill the boxes with the small cubes. Box *A* holds 4 rows of 4 cubes, or 16 cubes altogether. You can put 3 rows of 3 cubes, or 9 cubes, on the floor of box *B*, and there is still room for another level (Figure 11-21). You can put 3 rows of 3 cubes, or 9 more cubes, on top of the first 9. Box *B* holds 18 cubes altogether.

We can say that the volume of box *A* is 16 cubic units and the volume of box *B* is 18 cubic units. This is another way of saying that box *B* has a greater volume than box *A*, as we discovered with the sand. But now we know *how* much greater the volume of box *B* is. Box *B* holds 2 cubic units of volume more than box *A*.

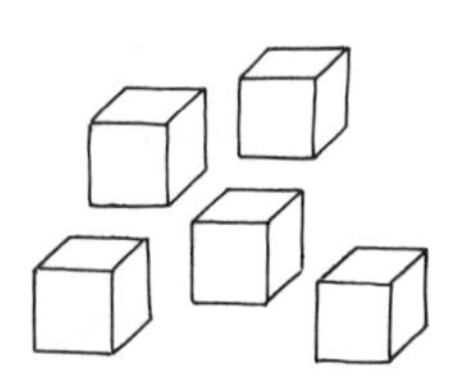

Figure 11-20

If the edges of the cubes used to fill the boxes have a measure of 1 cm, each cubic unit is a cubic centimeter (cu cm). Then if we say that box *B* has a volume of 18 cu cm, we mean that it takes 18 cubes with 1-cm edges to fill box *B* completely. Similarly, if the

**You may want to make the cubes and fill the boxes with cubes for a more concrete picture of volume.*

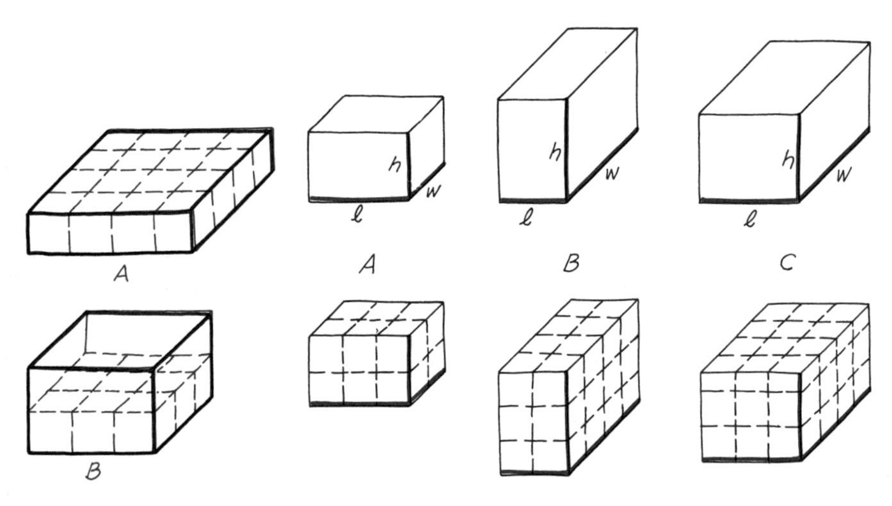

Figure 11-21 **Figure 11-22**

edges of the cubes are 1 in., then the cubic unit is a cubic inch. An edge of 1 ft forms a cubic foot, an edge of 1 m forms a cubic meter, and so on.

These boxes have the shape of a rectangular solid or cuboid (all faces are bounded by rectangles). Filling the boxes with cubes suggests a general way for finding the volume of a cuboid. The solids in Figure 11-22 have edges labeled l for length, w for width, and h for height. Filling the solids with unit cubes provides the information given in Table 11-1.

Table 11-1

	Number of cubes that fit along l	Number of cubes that fit along w	Number of cubes that fit along h	Total number of cubes that fit in solid
Solid *A*	3	2	2	12
Solid *B*	2	4	3	24
Solid *C*	3	4	$2\frac{1}{2}$	30

Notice that for solid *C* there are two layers of 12 cubes, or 24 cubes, plus one layer of 12 half cubes (which make 6 more cubes) for a total of 30 cubes.

In each case it is possible to find the volume in cubic units *without* filling the solids with unit cubes: simply find the product of l, w, and h, which represent the number of linear units* for each dimension of the solid. We can say that the volume of a cuboid

**Our implicit agreement is that if the edges are measured in inches, the volume is expressed in cubic inches; if the edges are measured in centimeters, the volume is expressed in cubic centimeters, etc.*

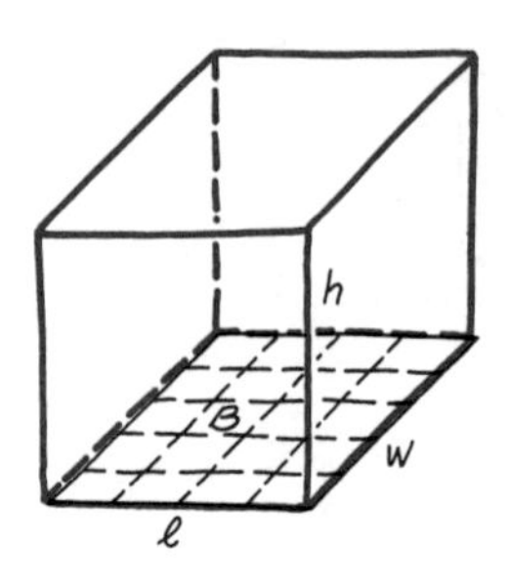

$V_{cuboid} = B \cdot h$

Figure 11-23

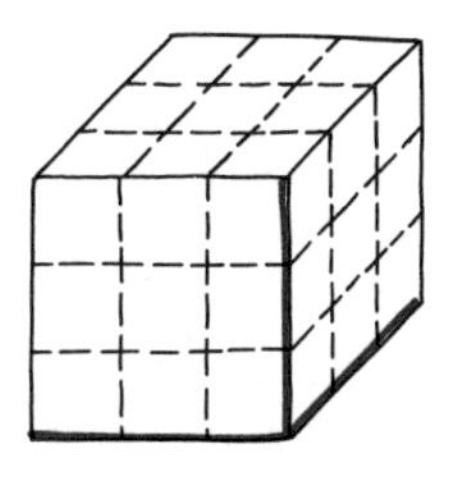

Figure 11-24

with dimensions l, w, and h can be found by using the following formula:

$$V_{\text{cuboid}} = l \cdot w \cdot h$$

In Figure 11-23 the base of the cuboid is bounded by a rectangle and the area within that rectangular base B is $l \cdot w$. If B is the area of the base and h is the height of a solid, then the volume can be expressed as

$$V_{\text{cuboid}} = B \cdot h$$

where B = measure of area of base
h = measure of height

If that rectangular solid happens to be a cube, we have a special case where $l = w = h$. If we label the measure of the edge of the cube e, the volume can be expressed as follows:

$$V_{\text{cube}} = e \cdot e \cdot e$$

$$V_{\text{cube}} = e^3$$

For example, if the edge has a linear measure of 3, then the cube has a volume measure of 9 (Figure 11-24).

Cavalieri's Principle

Suppose we take 12 wooden cubes and arrange them in different ways as in Figure 11-25. Imagine the solid figures suggested by the blocks. Notice that the figures do not have the same surface area. The surface area of (a) is 40 square units (area of faces: 12 + 12 + 2 + 2 + 6 + 6) and the surface area of (b) is 32 square

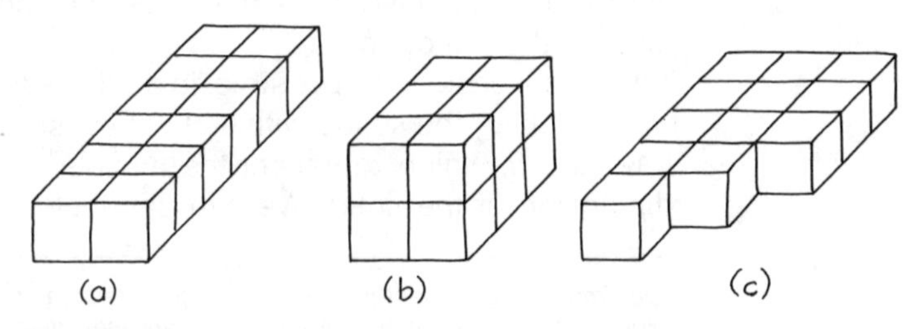

Figure 11-25

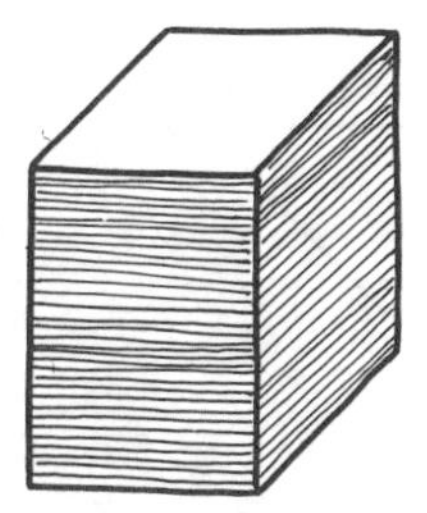

Figure 11-26

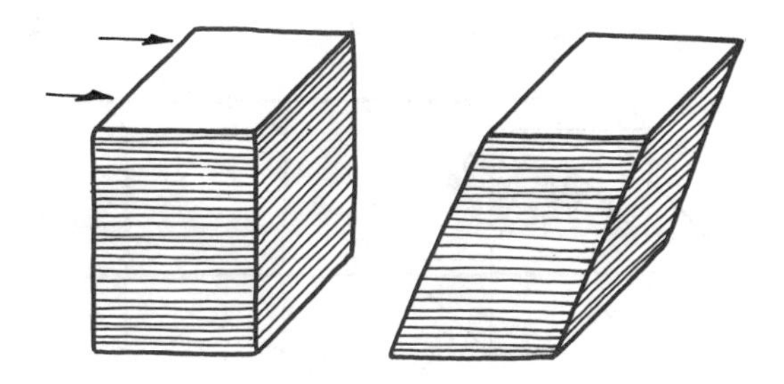

Figure 11-27

units (area of faces: $6 + 6 + 6 + 6 + 4 + 4$). What is the surface area of (c)? The figures do have the same *volume*. Each is composed of 12 cubic units. When solid figures have the same volume, they are called *equivalent* solids. We used the world *equivalent* before when we described equivalent polygons as polygons that have the same area (and when we described equivalent line segments as segments having the same length).

Suppose you have a stack of sheets of paper (all the same size) like the one shown in Figure 11-26. The shape of that stack resembles a cuboid. You can push the stack over in such a way that it is transformed into a different shape (Figure 11-27). What has changed in this transformation? What remains the same? Clearly, the volume of this new shape is the same as that of the original shape since both shapes are made up of the same amount of paper. The areas of the upper and lower bases are still the same (size of one sheet of paper). The height of the stack remains the same. The figure is still a prism, but it is not a cuboid since all the faces are not bounded by rectangles. The lateral faces are bounded by parallelograms that are not rectangles.

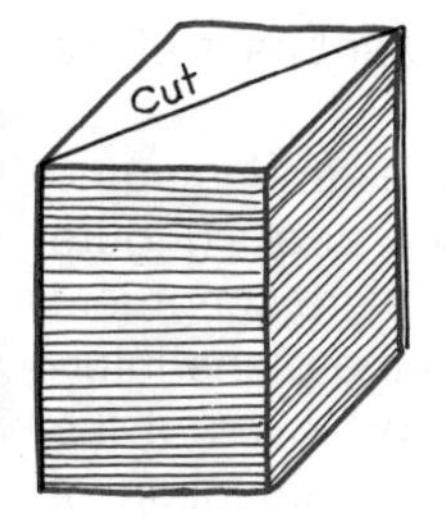

Figure 11-28

Let us try something else with this stack of paper. Suppose we cut the stack in such a way that each sheet is cut in half diagonally (Figure 11-28). We can arrange the two stacks of paper formed so that the shape of the stack is transformed from a cuboid to a triangular prism (Figure 11-29). Again the volume remains the same with this transformation, since each stack has just as much paper. The upper and lower bases of the triangular prism are

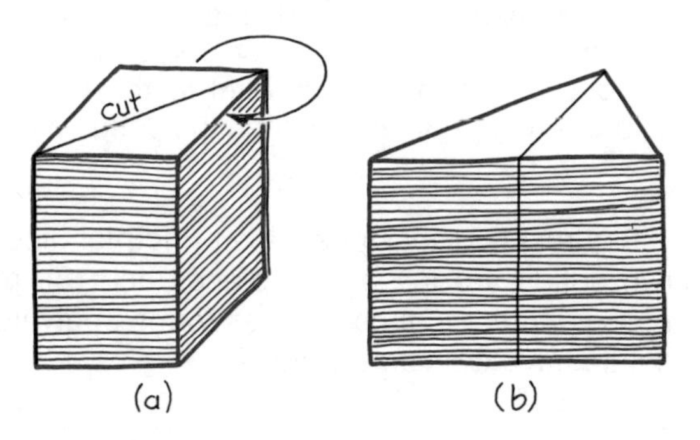

Figure 11-29

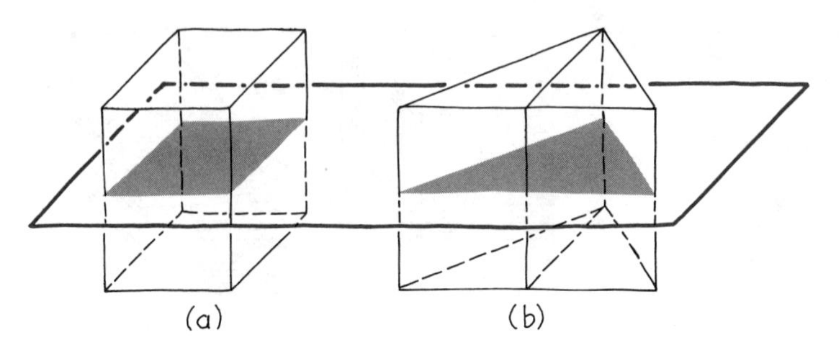

Figure 11-30

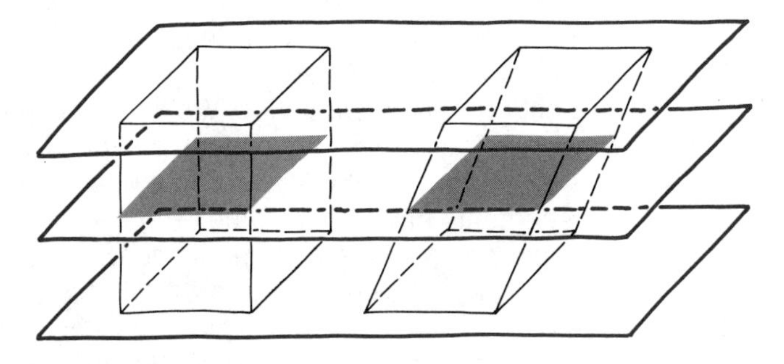

Figure 11-31

congruent, and they are equivalent to the rectangular bases of the original stack of paper.

In fact, if you think of these stacks as solid shapes and imagine a plane parallel to the bases and intersecting the two stacks (Figure 11-30), the intersections of the plane with the solids are congruent plane regions. The triangular region formed when the plane intersects the triangular prism is composed of the two halves of the corresponding rectangular region formed by the intersection of the plane and the cuboid.

It was observations such as these that led the Italian mathematician Bonaventura Cavalieri* to develop the following principle:

Cavalieri's Principle

If the bases of two solids are in the same parallel planes and if the corresponding plane regions formed by the intersections of any planes parallel to those planes and the solids are always equivalent, then the solids have the same volume.

Figure 11-31 illustrates this principle. Since (1) the bases of the two solids are in parallel planes and (2) the intersection of any plane parallel to those planes forms equivalent plane regions, the solids have the same volume.

**Cavalieri was a disciple of Galileo. He lived during the seventeenth century.*

Volume of Prisms

Let us examine the shapes of the stacks of paper in Figure 11-29 again. The original stack (a) has the shape of a cuboid and the transformed stack (b) has the shape of a triangular prism. We know that the corresponding bases of (a) and (b) are equivalent (have the same area) and that the heights of (a) and (b) are the same. We also know that the figures have the *same volume:*

$$V_{\text{cuboid}} = V_{\substack{\text{triangular}\\ \text{prism}}}$$

If

$$V_{\text{cuboid}} = B \cdot h$$

where B = area of base and h = height, then

$$V_{\substack{\text{triangular}\\ \text{prism}}} = B \cdot h$$

We can make this result more general by applying Cavalieri's principle. Suppose we have *any* prism whose base has an area B and whose height is h. Imagine a corresponding cuboid with an equivalent base B and the same height h, as in Figure 11-32(a). The corresponding plane regions formed by the intersection of these figures with any plane parallel to their bases are equivalent [Figure 11-32(b)]. The conditions of Cavalieri's principle are satisfied,* so we know that the volumes of the two figures are the same:

$$V_{\text{cuboid}} = V_{\text{prism}}$$

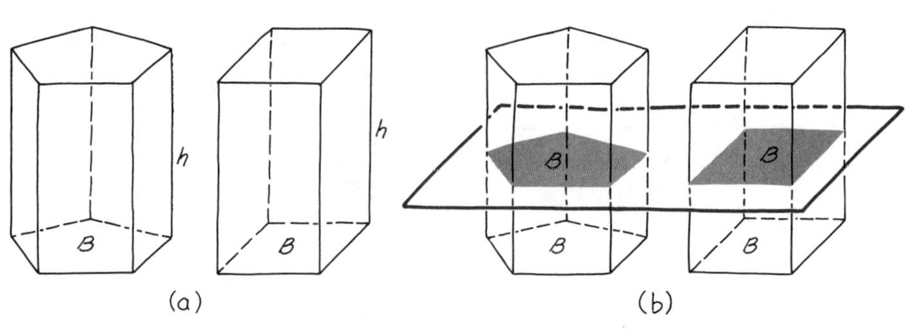

Figure 11-32

Since their height is the same, we can consider the upper and lower bases to be in parallel planes.

This means that if

$$V_{\text{cuboid}} = B \cdot h$$

then

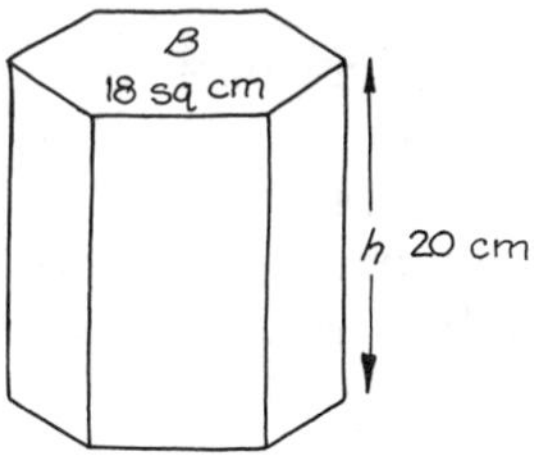

$$V_{\text{prism}} = B \cdot h$$

where B = measure of area of base
h = measure of height

Figure 11-33

This formula tells us that to find the volume of any prism, we multiply the area of the base of the prism by its height and express the result in cubic units. For example, the box in Figure 11-33 has the shape of a hexagonal prism. It has a base B of 18 sq cm and a height h of 20 cm. The volume is found as follows:

$$\begin{aligned} V_{\text{prism}} &= B \cdot h \\ &= 18 \cdot 20 \\ &= 360 \text{ cu cm} \end{aligned}$$

Volume of Cylinders

As you might expect, the preceding formula for finding the volume of a prism is also valid for finding the volume of a cylinder, since a prism is a type of cylinder (with polygonal bases). If we begin with any cylinder and imagine a corresponding prism with the same height and an equivalent base, by Cavalieri's principle the figures would have the same volume (Figure 11-34). The volume of the cylinder can then be calculated by using this formula:

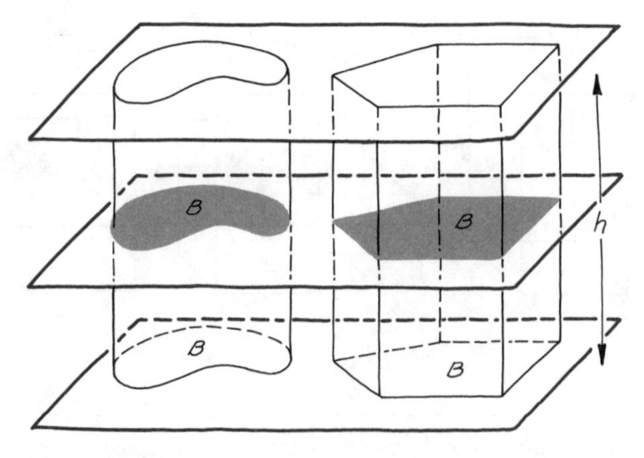

Figure 11-34

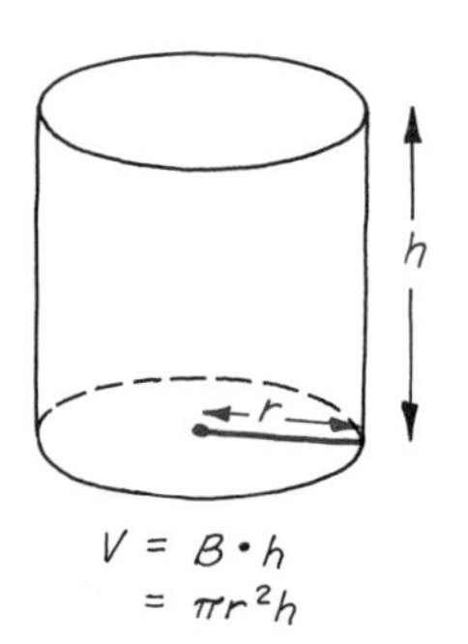

Figure 11-35

$$\underset{\text{cylinder}}{V} = B \cdot h$$

where B = measure of area of base
h = measure of height

When the cylinder is a circular cylinder, the formula can be written in terms of the radius of the base. Look at the cylinder in Figure 11-35. The base is bounded by a circle. The length of the radius is represented as r. The area B of that circular region is πr^2. The formula becomes

$$\underset{\text{circular cylinder}}{V} = (\pi r^2) \cdot h$$

Volume of a Pyramid

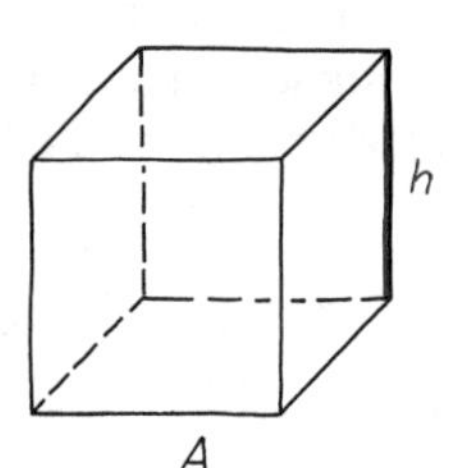

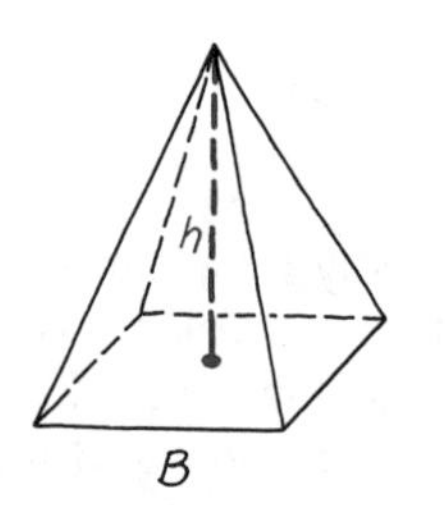

Figure 11-36

The solids in Figure 11-36 have the same height and their bases are congruent. Solid A is a cube while solid B is a square pyramid. It is apparent that the cube has a greater volume than the pyramid. But how much greater is the volume of the cube?

Suppose you could fill the pyramid with colored water and pour the contents into the cube (Figure 11-37). How many times would you have to repeat this to fill the cube completely? Twice? Two and one-half times? Three times? Four times?

Many people estimate two times. Actually the pyramid is filled three times. The volume of the cube is *exactly three times* the volume of the pyramid or, expressed another way, the volume of the pyramid is one-third the volume of the cube.

Can we find a formula for calculating the volume of the pyramid? We know that the volume of the cube V can be expressed as $V = B \times h$, where B is the measure of the area of the base and h is the measure of the height.

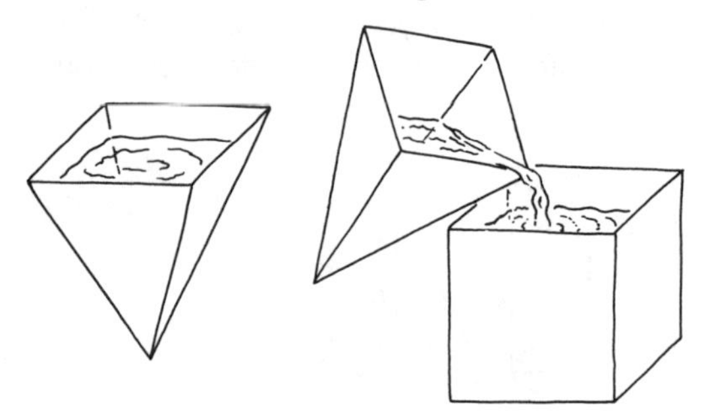
Figure 11-37

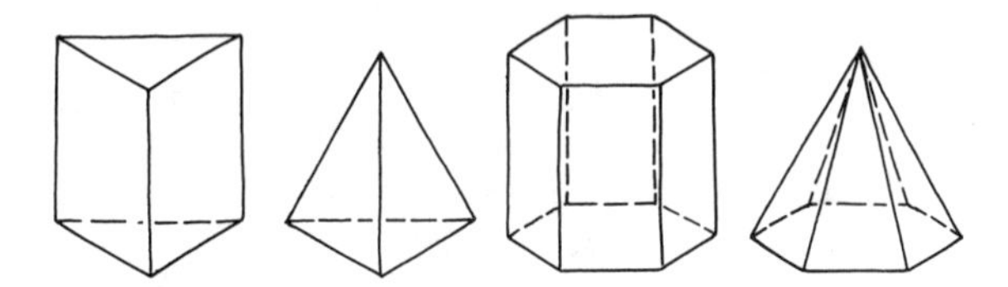

Figure 11-38

Since the pyramid has the same height as the cube (h), the bases are congruent, and the volume of the pyramid is one-third the volume of the cube, we can express the volume of the pyramid as follows:

$$V_{\text{pyramid}} = \frac{1}{3}(B \cdot h)$$

where B = measure of area of base
h = measure of height

Actually, we only considered the square pyramid in the preceding experience. However, if you made pairs of solids like those in Figure 11-38, where the prisms and corresponding pyramids have the same height and congruent bases, in each case you could show that the volume of the pyramid is one-third the volume of the corresponding prism.

Volume of a Circular Cone

Since the pyramid is a special kind of cone (with polygonal base), it seems reasonable that the formula for calculating the volume of a circular cone should bear some resemblance to that used for finding the volume of a pyramid. Imagine a circular cone and a circular cylinder that have the same height and congruent bases (Figure 11-39). If you filled the cone with colored water and poured the contents into the cylinder, you would find that the volume of

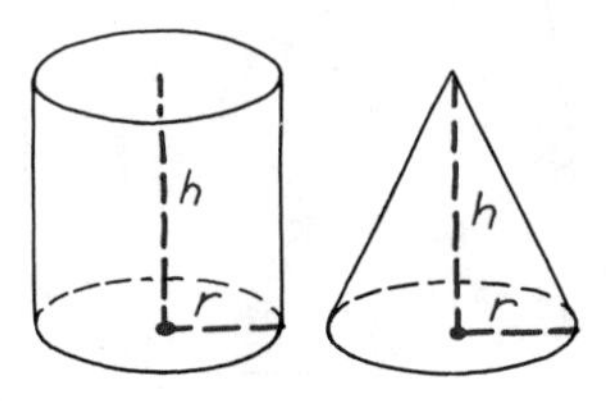

Figure 11-39

the cone is one-third the volume of the cylinder. We know how to find the volume of a circular cylinder:

$$V_{\text{circular cylinder}} = (\pi r^2) \cdot h$$

So the volume of the corresponding circular cone can be found as follows:

$$V_{\text{circular cone}} = \frac{1}{3} \cdot (\pi r^2) \cdot h$$

where r = measure of radius of base
h = measure of height

Volume of a Sphere

The final well-known shape we will consider is the sphere. How do we find its volume? It is obvious that the larger the radius of a sphere, the larger its volume. But can we find a formula that can be used to calculate the volume of a sphere if we know its radius?

We will have to do a little experiment to explore the problem more concretely. Take a solid rubber ball and a sharp knife, and cut out a piece of this spherical region as indicated in Figure 11-40. The piece cut out of the sphere looks very much like a pyramid with a curved base. Now cut out another "pyramid" with a smaller base in the same way. Note that the base of the "pyramid" with the smaller base is less curved (Figure 11-41).

Now suppose we could cut the sphere into many tiny "pyramids" whose height is approximately the measure of the radius of the sphere. The sum of the volumes of these tiny "pyramids" would be equal to the volume of the sphere. And we know how to find the

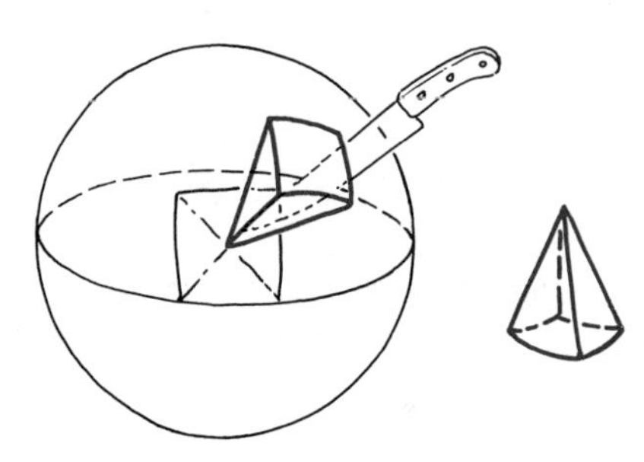

Figure 11-40

Figure 11-41

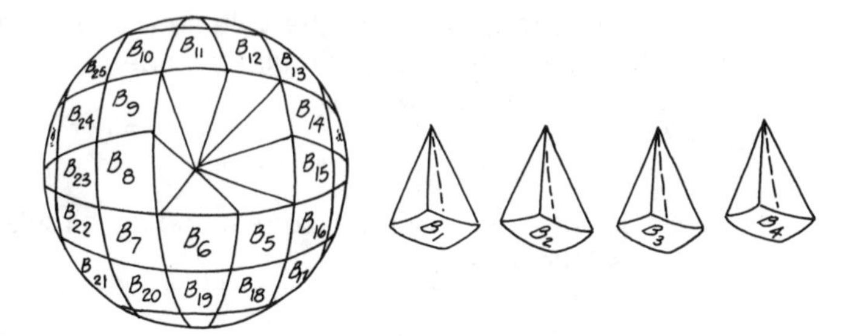

Figure 11-42

volume of a pyramid. We can use the formula

$$\underset{\text{pyramid}}{V} = \frac{1}{3} \cdot (B \cdot h) \qquad \text{or} \qquad \frac{1}{3}Bh$$

where B = measure of area of base
h = measure of height

If we express the measure of the areas of the bases of the "pyramids" as B_1, B_2, B_3, B_4, . . . , B_{10}, . . . , we can express the *sum* of the *volumes of all the "pyramids"* as

$$\underset{\text{"pyramids"}}{V} = \frac{1}{3}B_1h + \frac{1}{3}B_2h + \frac{1}{3}B_3h + \frac{1}{3}B_4h + \cdots + \frac{1}{3}B_{10}h + \cdots$$

The term *pyramid* appears with quotes because these shapes only *approximate* the shape of pyramids. They are *not* pyramids since the bases are curved. *But as we make the bases of the pyramids smaller and smaller,* these bases are *less and less curved* and the shapes look more and more like pyramids.

Imagine that we could make as many "pyramids" as we want (Figure 11-42). As we increase their number, the shapes more closely approximate actual pyramids and their height h gets closer to the measure of the radius r of the sphere.

Substituting r for h, the volume of the sphere can be expressed as

$$\underset{\text{sphere}}{V} = \frac{1}{3}B_1r + \frac{1}{3}B_2r + \frac{1}{3}B_3r + \cdots$$

Since ⅓ and r are factors of each term of this expression, we can write*

$$\underset{\text{sphere}}{V} = \frac{1}{3}r(B_1 + B_2 + B_3 + B_4 + \cdots)$$

But $(B_1 + B_2 + B_3 + B_4 + \cdots)$ represents the sum of the areas of the bases of the "pyramids," which is the same as the total

---

**We are using the distributive principle of multiplication over addition: for example, $(4 \times 5) + (4 \times 6) = 4 \times (5 + 6)$.*

surface area of the sphere. We know from our previous work that a sphere with radius r has a surface area of $4\pi r^2$. Therefore

$$B_1 + B_2 + B_3 + B_4 + \cdots = 4\pi r^2$$

and

$$V_{\text{sphere}} = \frac{1}{3}r(4\pi r^2)$$

$$V_{\text{sphere}} = \frac{4}{3}\pi r^3$$

where r = measure of radius of sphere

This tells us that if we have a spherical object like a ball with a radius of 6 cm, its volume is found (using 3.14 as an approximation for π) as follows:

$$V = \frac{4}{3}\pi r^3$$

$$\approx \frac{4}{3}(3.14)(6 \times 6 \times 6)$$

$$\approx \frac{4}{3}(3.14)(216)$$

$$\approx 904.32 \text{ cu cm}$$

Another Way to Measure Volume

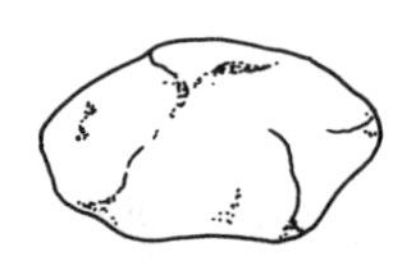

Figure 11-43

Sometimes we want to find the volume of an irregular shape like the stone pictured in Figure 11-43. How can it be done? We can find the volume indirectly by the *displacement* method. As you know, when a solid is immersed in water, the water level rises. Actually, the solid *displaces its volume in water.* By measuring the volume of the water displaced, we can find the volume of the solid.

Let us make a measuring jar and try to find the volume of the stone. Start with a clear jar like the one shown in Figure 11-44 and put some tape on the side as indicated. Find or make a small container whose volume you know. A container with a volume of 50 cu cm is used in this example. Fill the container with water and pour its contents into the jar. Mark the water level on the tape. Repeat the procedure to form a scale as shown in Figure 11-45.

Now we can measure the volume of the stone. The water level of the jar in Figure 11-46(a) indicates a volume of 300 cu cm. When the stone is placed into the jar, the water level rises to about 425

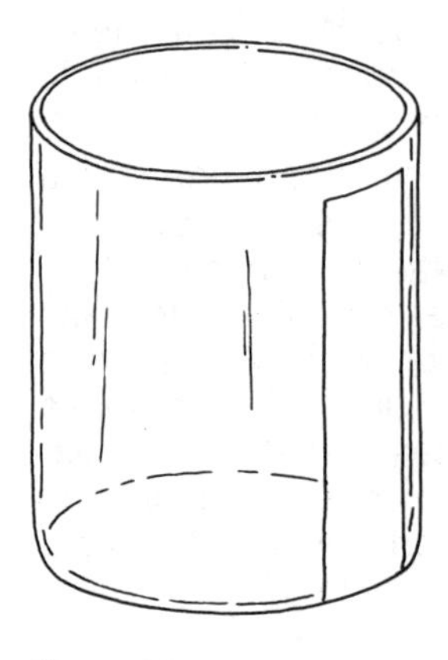

Figure 11-44

cu cm. Since the stone is displacing its volume in water, we find that the approximate volume of the stone is 125 cu cm (425 − 300).

This is a very useful method for estimating the volume of irregular objects. If an object does not sink, you can attach a weight so that it does. You will then need to deduct the measure of the volume of the weight from the total volume displaced to find the volume of the irregular object.

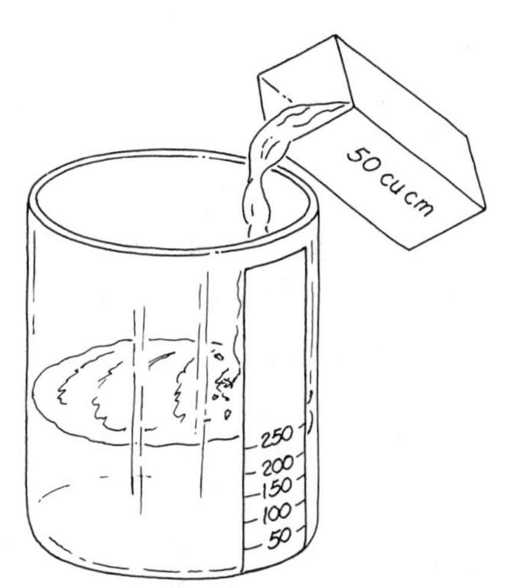

Figure 11-45

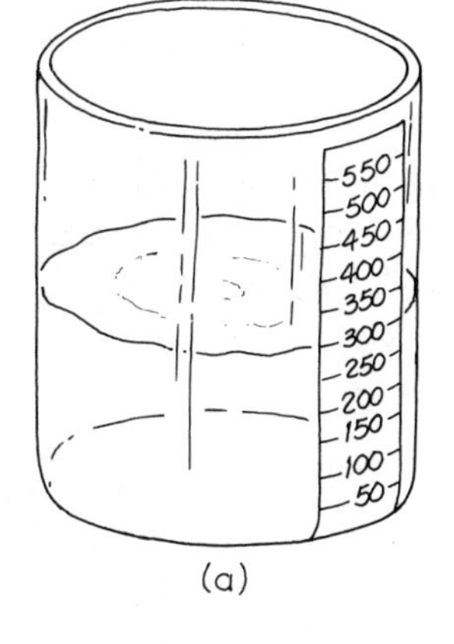

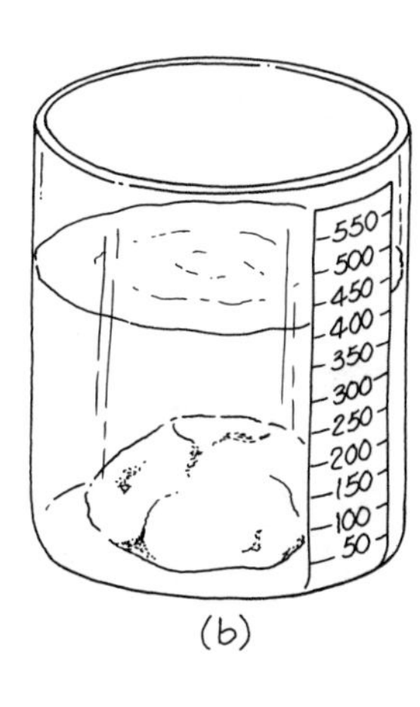

Figure 11-46

More Experiences

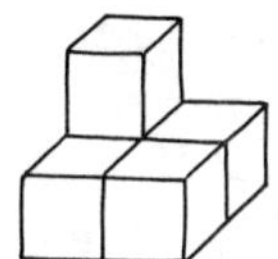

11-1. You will need some wooden cubes* for the following activities:

(*a*) Put some cubes together to make a larger cube. What is the least number of cubes you need to make the larger cube?

(*b*) If the length of an edge of a wooden cube is considered a unit of linear measurement, how many cubes does it take to make the model of a cube with an edge of 2 units? 3 units? Can you make larger cubes? What is the measure of the surface area for each cube formed? What is the measure of the volume?

Any commercially available wooden or plastic cubes will do.

Find the ratio of the surface area to the volume in each case. Keep a record of your results as follows:

Length of edge of cube (linear)	Total surface area of cube (area)	Volume of cube (volume)	Ratio of surface area to volume
1	6	1	$\frac{6}{1}$ or 6
2			
3			
.			
.			
.			
10			

When is the *number* of square units of surface area the same as the *number* of cubic units of volume (that is, ratio 1:1)? Explain why these numbers are the same. Will the ratio be 1:1 for other models of cubes? How is the ratio changing?

Can you make a graph showing the relationship between the length of an edge of a cube (x axis) and the ratio of surface area to volume for that cube (y axis)? What does the curve look like?

11-2. Take eight wooden cubes and make a structure, using all of them. What is the volume of your structure? What is its surface area?

How many different structures can you make using eight cubes? Some examples are shown at the left. What will the volume be for each of these structures? What will the surface area be in each case? Which has the greatest surface area? Which has the least surface area? Why?

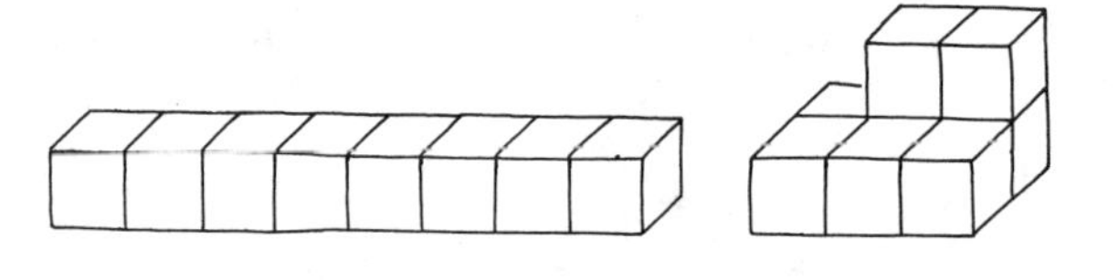

Try doing the same thing with 12 cubes.

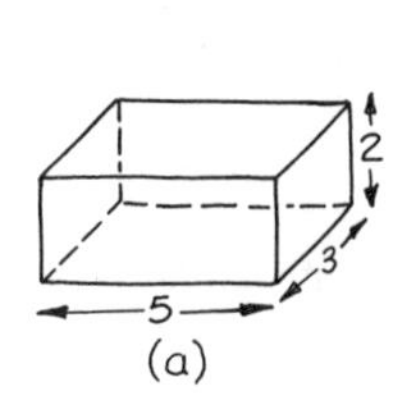

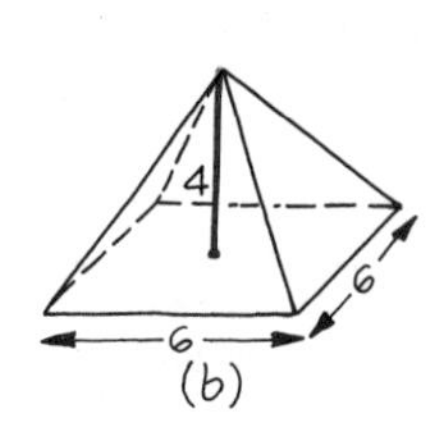

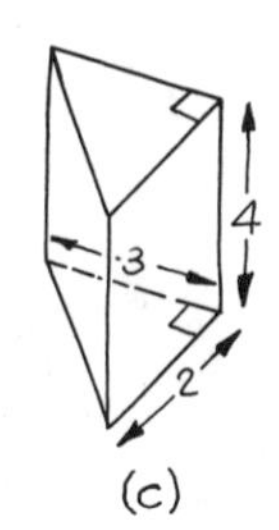

11-3. Find the surface area and the volume of each of the models shown at the left. (All measurements are in *yards.*)

Suppose each linear measurement of the figures is converted to feet. What is the measure of the surface area of each figure in square feet? What is the measure of the volume of each figure in cubic yards? Keep a record of all the information you collect as follows:

Figure	Surface area (in sq yd)	Surface area (in sq ft)	Volume (in cu yd)	Volume (in cu ft)
(a)				
(b)				
(c)				

If the surface area of a figure is expressed in square yards, what will be the surface area in square feet? If the volume of a figure is expressed in cubic yards, what will be the volume in cubic feet?

How would you find the surface area and the volume in square *inches* and cubic *inches,* respectively, for each figure? What relationships can you find between square inches, square feet, and square yards? Between cubic inches, cubic feet, and cubic yards?

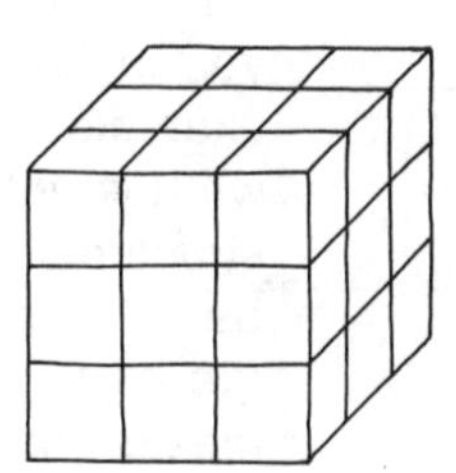

11-4. One wooden cube has a volume of 1 cubic unit. Can you use wooden cubes to make a model of a cube with a volume of 2 cubic units? Explain.

The model of the cube at the left has a volume of 27 cubic units. Can a model of a cube with twice that volume be made with the wooden cubes? Explain your results.

11-5. Suppose you wanted to make an open box out of a rectangular

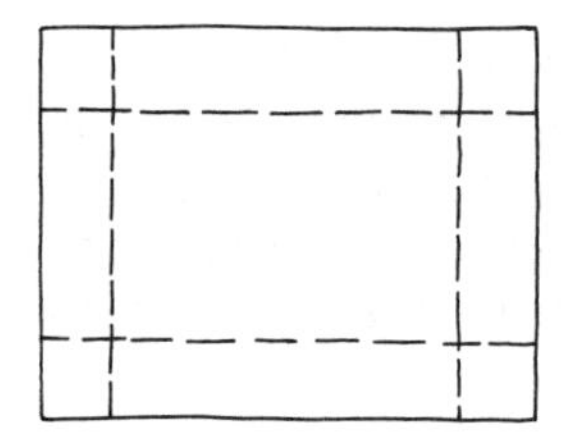

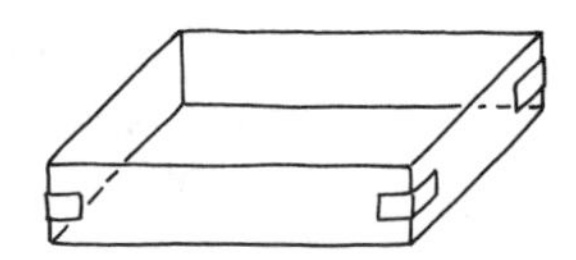

sheet of cardboard with dimensions 36×24 cm. You might cut squares at each corner, fold, and tape together as indicated in the diagram at the left.

The volume of the resulting open box will vary according to the size of the square cut out at each corner of the original rectangular sheet. How does the volume vary when the square is 1 cm on a side? 2 cm? 3 cm? 4 cm? etc.? When will the volume be the greatest? Why? Keep a record of your results. How is the surface area changing?

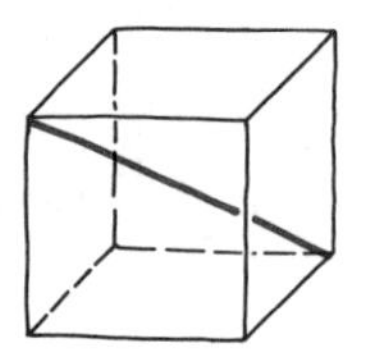

11-6. Suppose the edge of the cube at the left has a length of 5 in. Can you find the approximate length in inches of a diagonal of that cube? (*Hint:* How can you find the length of a diagonal of a square face?)

A rectangular solid has dimensions *a, b,* and *c,* where *a, b,* and *c* each represents a number of linear units. Can you develop a formula for finding the length of the diagonal of that solid?

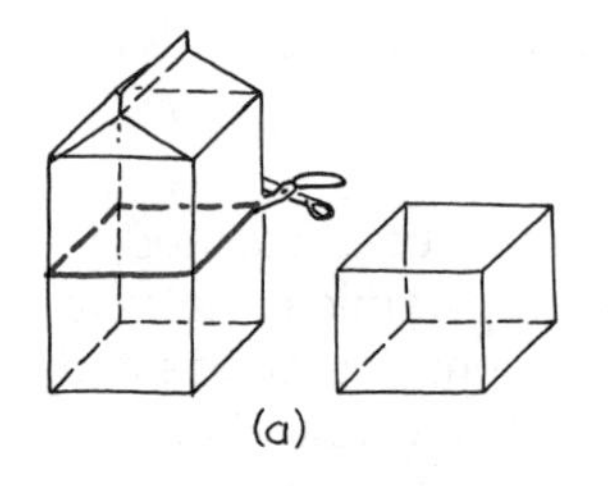

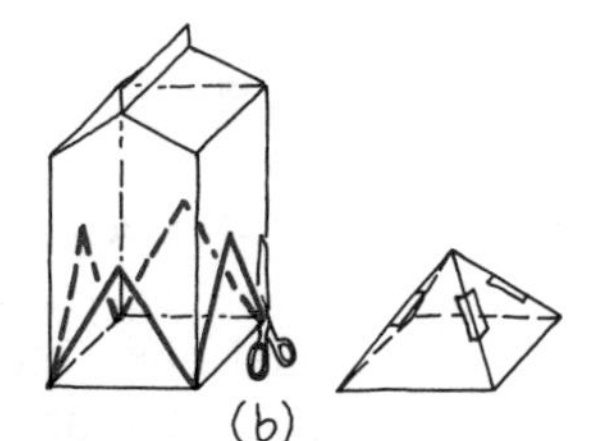

11-7. The diagram at the left shows how you can make models of an open cube (one face missing), which is labeled (a), and a square pyramid, labeled (b), by using empty milk cartons. Can you make these models so that both have the same height?

If the cube has an edge of 6 in., what should be the altitude of the triangles drawn on the faces of the carton in (b)?

These models have the same height and congruent bases. What is the relationship between their volumes? Make a hole in the base of the model of the pyramid and fill the model with water (completely). Then pour the water into the model of the cube. Repeat the process.How much larger is the volume of the cube than the volume of the pyramid?

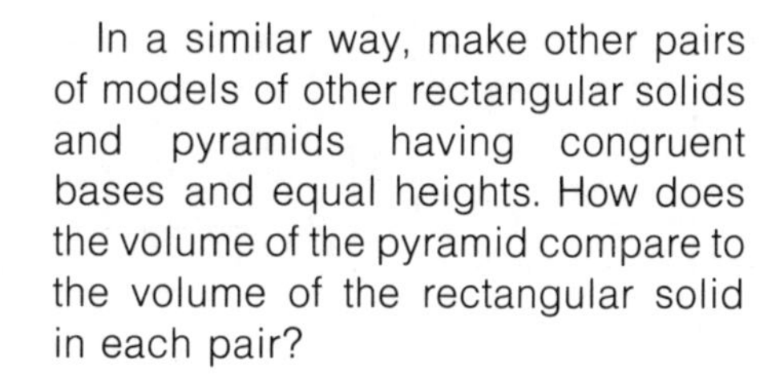

In a similar way, make other pairs of models of other rectangular solids and pyramids having congruent bases and equal heights. How does the volume of the pyramid compare to the volume of the rectangular solid in each pair?

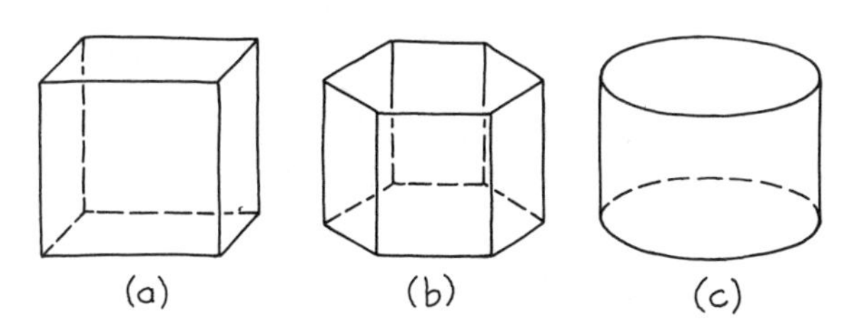

11-8. The sides, or lateral faces, of model (a) in the diagram at the left were made by folding a rectangular sheet of paper with dimensions 10×36 cm as shown below. How might a rectangular sheet of paper be used to make the sides of model (b)? Model (c)?

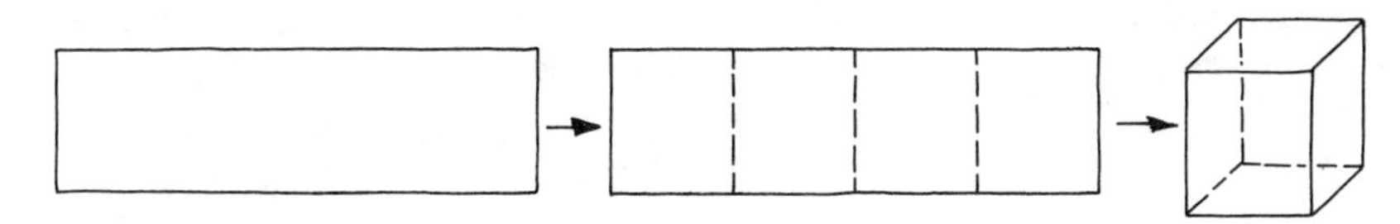

Each model has a base with a perimeter of 36 cm. (How do you know this must be true?) Can you find an approximate measure of the volume of each model? Is the measure of the volume the same for each model?

Why are cans of food cylindrical in shape?

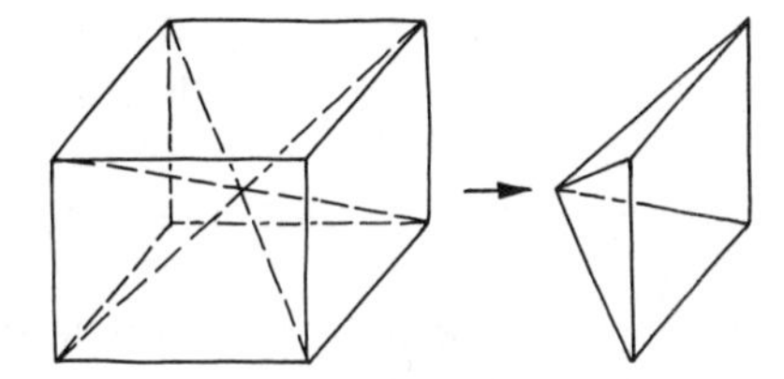

11-9. A cube can be dissected to form six square pyramids that are just alike (congruent), as suggested by the diagram at the left. Suppose an edge of the cube is 2 in. Find the volume of the cube. What would be the volume of one pyramid formed in this way?

Find the volume of the pyramid by using the formula developed in this chapter. (What is the height of the pyramid?) Do you get the same result?

11-10. A liter bottle has a volume the same as that of a cube with an edge of 1 dm (decimeter). How many cubic centimeters are in a liter? A bottle with a volume of 125 ml (milliliters) has the same volume as a cube whose edge is how many centimeters in length?

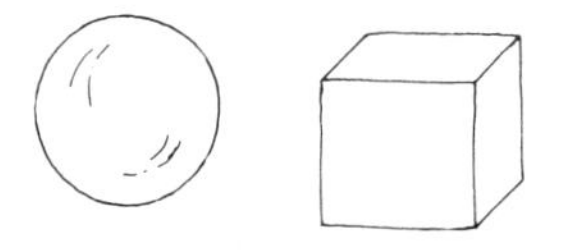

11-11. Take two pieces of plasticene that have the same weight. Make a model of a sphere with one piece and a model of a rectangular solid with the other. These models have the same volume. Why? Calculate the surface area of each model. (How will you find the diameter of the sphere?) Is the measure of the surface area the same for these shapes?

Why are bubbles spherical in shape?

11-12. Make a measuring jar (as described in this chapter) to measure the volume of objects by submerging them in water. Find objects with a shape that makes their volume easy to calculate by using basic formulas (shaped like cubes, rectangular solids, circular cones, spheres, cylinders). Calculate the volumes of these objects by using the appropriate formulas, and then use the measuring jar to find the volume of each. How accurate were your measurements?

Find some irregularly shaped objects or objects whose volume you cannot calculate with a formula. Find their volumes by using the measuring jar.

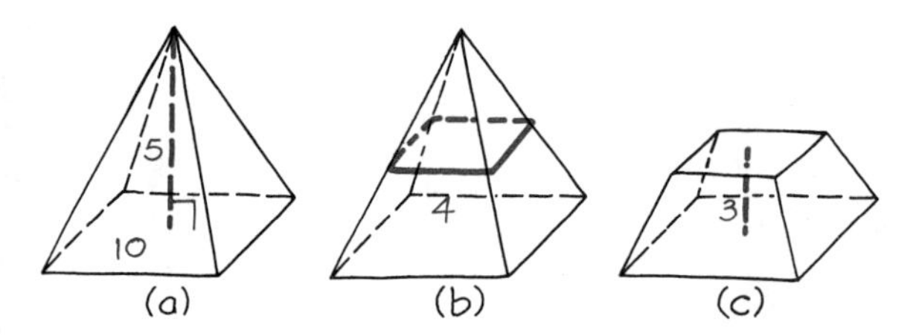

11-13. The solid shape in (c) was made by forming a square pyramid out of plasticene (a) and cutting off the top (parallel to the base) as in (b). Can you find the volume of the solid shape in (c)? This shape is called a **frustum** of a pyramid.

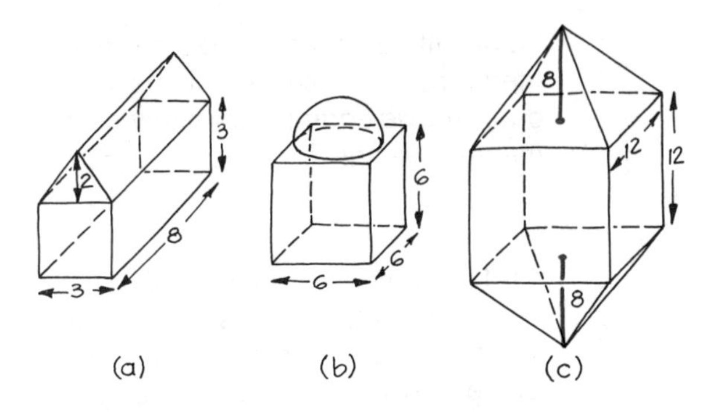

11-14. Find the surface area and volume of each solid figure shown at the left. (All measurements indicated are in centimeters.)

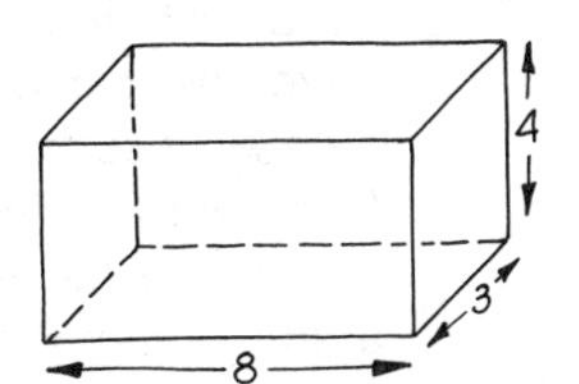

11-15. The box in the diagram at the left is shaped like a rectangular solid with dimensions 3, 4, and 8 in. What is the total surface area of the box?

How many different boxes can you imagine that have the same surface area as this box but different dimensions? List the dimensions of such boxes. What is the volume of each box? Keep a record of your findings. Are the surface areas and the volumes of all these boxes the same? Explain.

12 Five Special Solid Figures

The Platonic Solids

There are five solid figures that have special places in the world of geometry. These solids (Figure 12-1) were recognized for their unique characteristics centuries ago. In fact, they are known as the **Platonic solids,** after the famous Greek philosopher Plato.

Plato was so impressed with the extraordinary properties of these solid figures that he believed they played a fundamental role in the composition of the universe. Expanding upon the notion that all objects are composed of the four elements earth, air, fire, and water, Plato suggested that the smallest particles of earth have the form of a cube, those of air have an octahedron shape, those of fire have a tetrahedron form, those of water are shaped like the icosahedron, and finally the universe itself is shaped like a dodecahedron.

Why are these five solids so special? How are they unique? We will explore some answers to these questions in this chapter.

Polyhedra

Examine the diagrams of the Platonic solids in Figure 12-1 once more. What properties or characteristics of these solids do you notice?

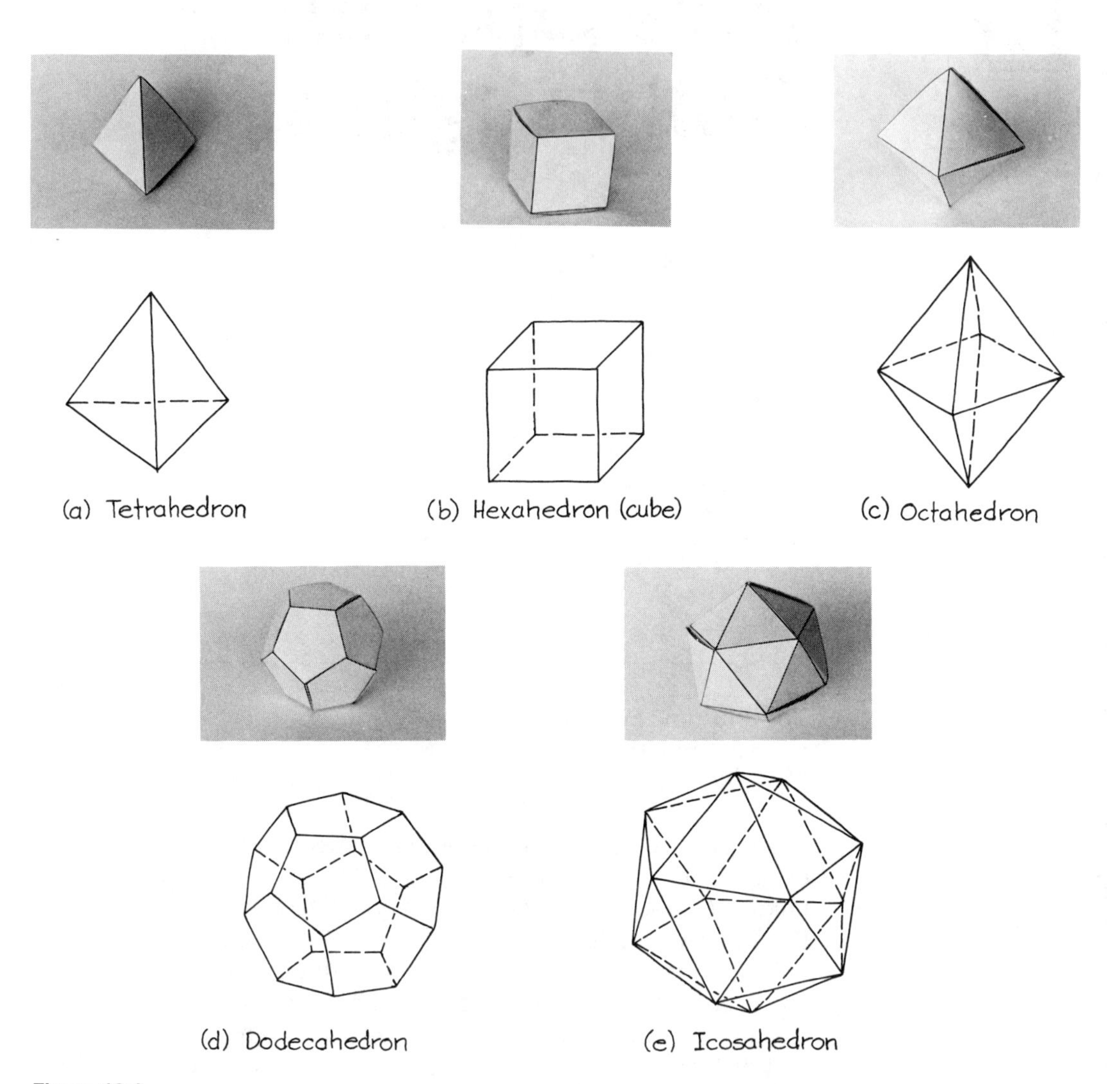

Figure 12-1
The Platonic solids

Each solid represents a simple closed surface that divides space into exactly three sets of points (points *inside, outside,* and *on* the figure). Also, all the faces of these figures are bounded by *polygons.* We call these solid figures **polyhedra** (singular: polyhedron). **Any simple closed surface whose faces are bounded by polygons is a polyhedron.**

The cylinder, cone, and sphere do not qualify as polyhedra. Why not? Do prisms and pyramids qualify? They are polyhedra because they are simple closed surfaces and all their faces are bounded by polygons.

When we discussed polygons, we found that a polygon can be considered convex if the extension of *any* side does *not* intersect with the interior (Figure 12-2). We can imagine the line determined by that side dividing the plane into two half planes. Then the con-

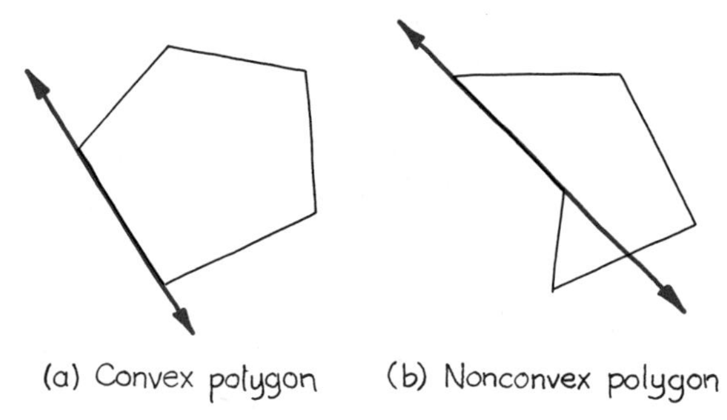

Figure 12-2

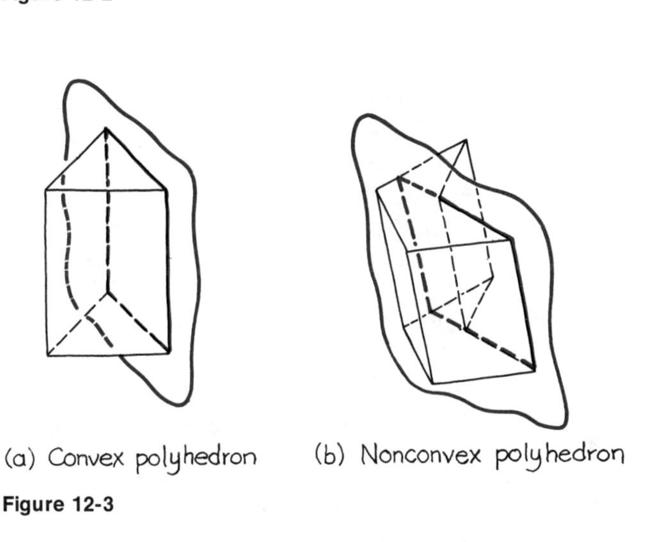

Figure 12-3

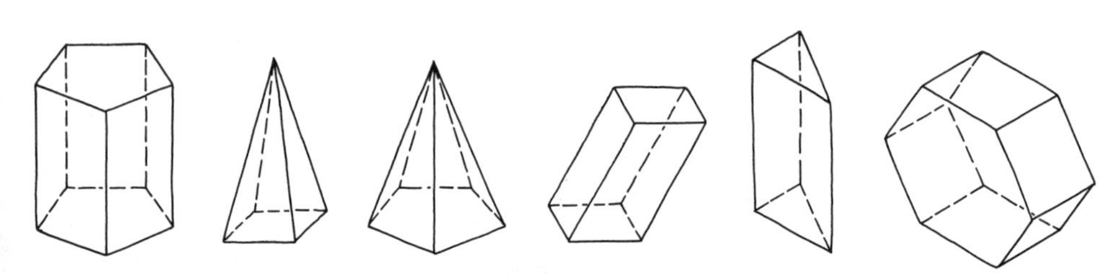

Figure 12-4

Other convex polyhedra

vex polygon is entirely in one half plane. As the diagram indicates, this is not true for a nonconvex polygon.

Analogously, a polyhedron can be considered convex if the extension of *any* face does *not* intersect with the interior of the polyhedron. We can think of the plane determined by that face dividing space into two half spaces. Then the convex polyhedron is entirely in one half space. As Figure 12-3 indicates, this is not true for a nonconvex polyhedron.

The five Platonic solids qualify as convex polyhedra. But many other solid figures qualify also, such as those shown in Figure 12-4.

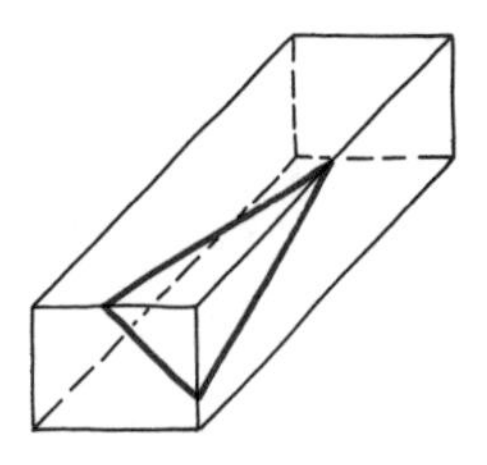

Figure 12-5

Vertex figure of a polyhedron

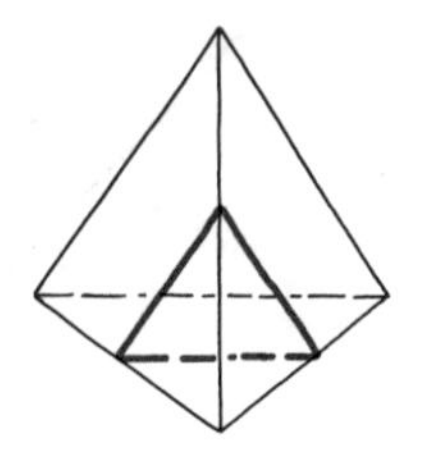

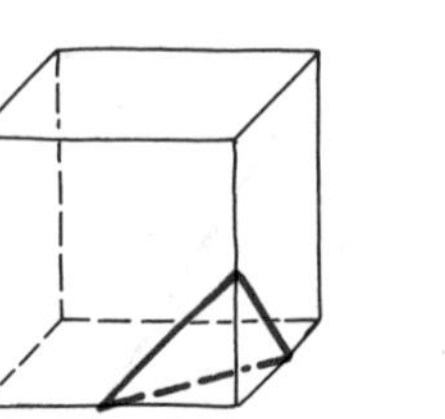

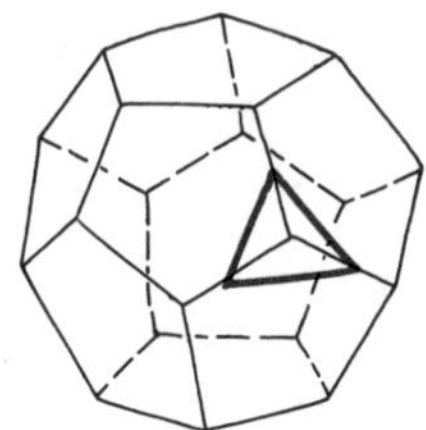

Figure 12-6

Vertex figures of tetrahedron, cube, and dodecahedron

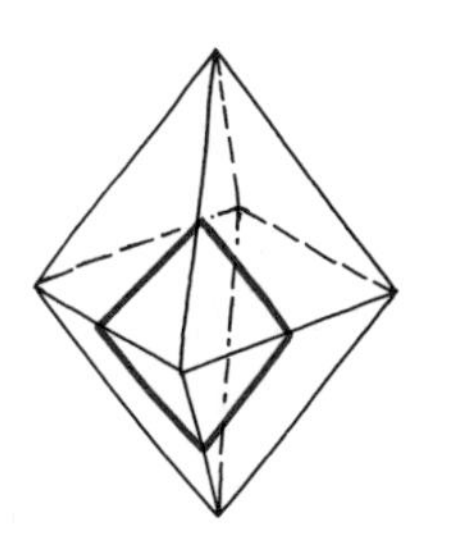

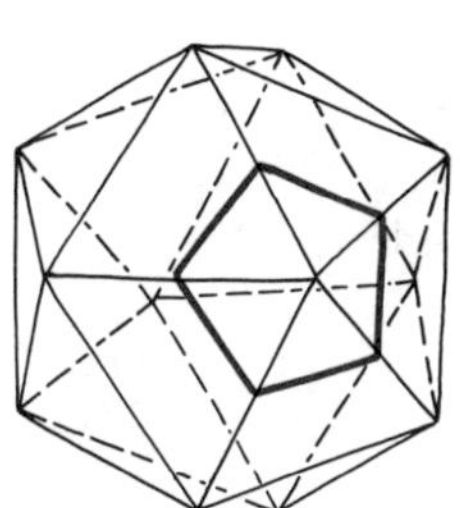

Figure 12-7

Vertex figures of octahedron and icosahedron

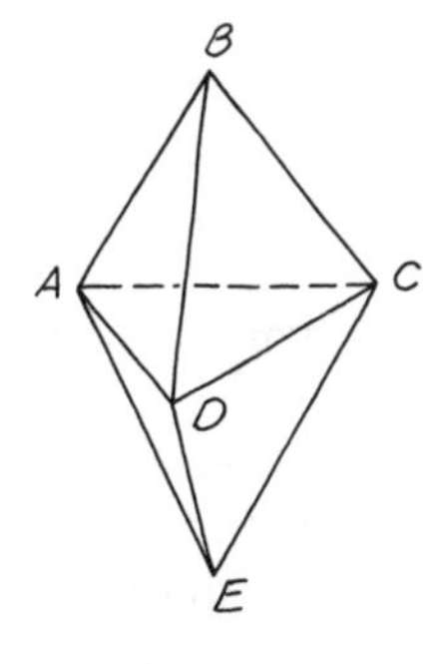

Figure 12-8

Each of these solid figures is a simple closed surface whose faces are convex and bounded by polygons (which makes them polyhedra). How are the five Platonic solids *special* convex polyhedra?

Each of these five solids has faces bounded by *congruent regular* polygons.* Three of the solids have all faces bounded by congruent equilateral triangles (tetrahedron, octahedron, icosahedron). One solid has all faces bounded by congruent squares (hexahedron) and one solid has faces bounded by congruent regular pentagons (dodecahedron).

The five Platonic solids have another interesting feature: **the same number of edges meet at each vertex.** Three edges meet at *every* vertex of the tetrahedron, hexahedron, and dodecahedron. Four edges meet at *every* vertex of the octahedron. How many edges meet at every vertex of the icosahedron?

Since all the faces are bounded by congruent regular polygons and the same number of edges meet at each vertex, these polyhedra have still another interesting feature. Imagine the polygon formed by joining the midpoints of consecutive edges about a vertex with line segments (Figure 12-5). This plane figure is called a **vertex figure** of the solid. Its sides lie on the faces of the solid. For the tetrahedron the vertex figures are always congruent equilateral triangles, as is true for the cube and the dodecahedron (Figure 12-6). What is the shape of the vertex figures for the octahedron and the icosahedron? As shown in Figure 12-7, the vertex figures for the octahedron are squares and the vertex figures for the icosahedron are regular pentagons. **For each of the five Platonic solids the vertex figures are congruent regular polygons.**

The solid illustrated in Figure 12-8 is a convex polyhedron. It looks like two tetrahedra with one set of faces joined together. All its faces are bounded by congruent equilateral triangles. Yet it is different from the Platonic solids. How is it different? Notice that the same number of edges do *not* meet at each vertex. Four edges meet at three of the vertices (*A, C, D*) and only three edges

---

**Recall from Chapter 1 that a regular polygon has congruent sides and congruent angles.*

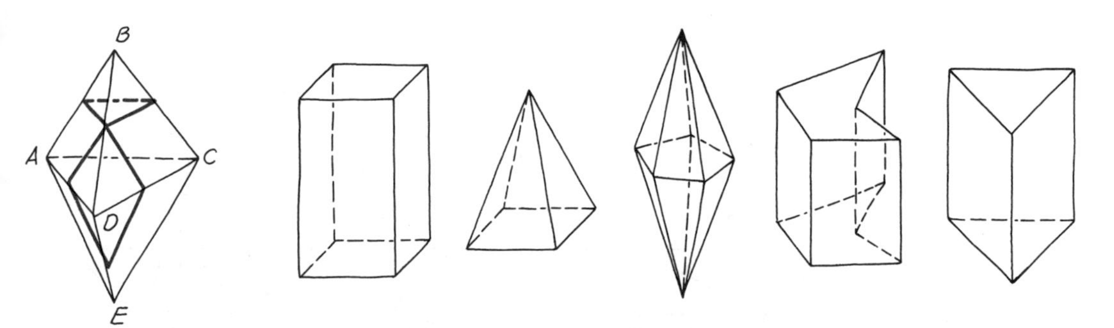

Figure 12-9

Figure 12-10

meet at the other two vertices (B, E). Consequently, the vertex figures are not the same (Figure 12-9).

Our five special solids can be called **regular polyhedra.** To qualify as a *regular* polyhedron, a solid figure must: (1) be a convex polyhedron, (2) have all its faces bounded by one kind of congruent regular polygon, and (3) have the same number of edges meeting at each of its vertices. The Platonic solids satisfy all three conditions. Why do the solid figures in Figure 12-10 *not* qualify as regular polyhedra?

Are There Other Regular Polyhedra?

As you know, the equilateral triangle, the square, and the equilateral pentagon are examples of *regular polygons.* It is possible to find as many different regular polygons as we wish (six-sided, seven-sided, eight-sided, etc.). All the polygons in Figure 12-11 are regular polygons—for each polygon all sides and all angles are congruent.

The five Platonic solids are regular polyhedra. Are there others? The astonishing fact is that **only five regular polyhedra are possible!** It would seem possible to describe any number of regular polyhedra (polyhedra with regular hexagon faces, regular septagon faces, etc.). Let us see why other regular polyhedra *cannot* exist.

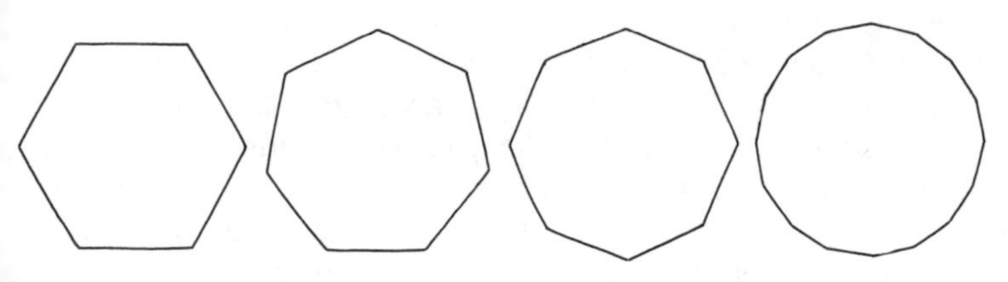

Figure 12-11

Some regular polygons

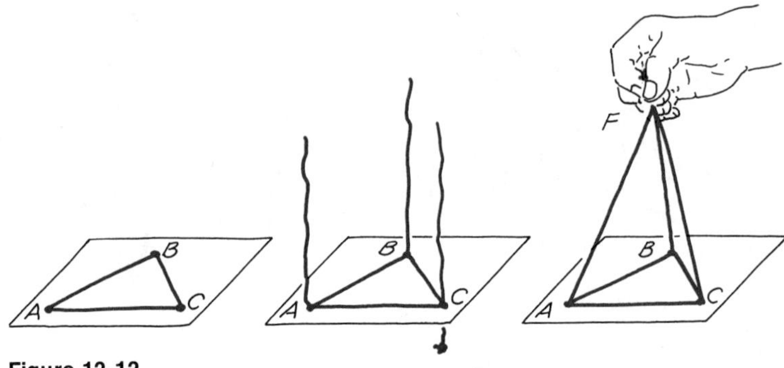

Figure 12-12

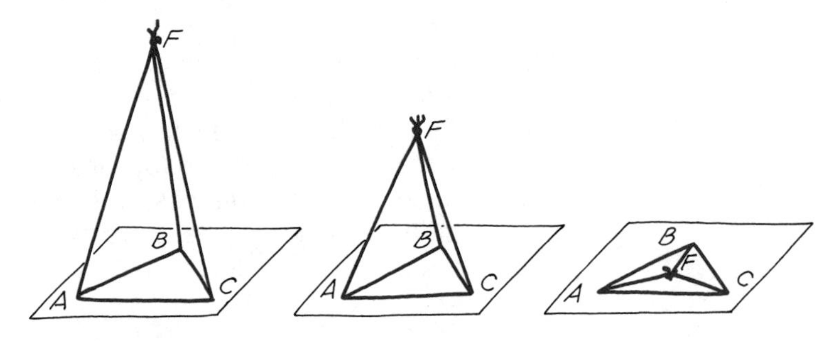

Figure 12-13

First of all, we need to examine some basic ideas about polyhedra. We know that all polyhedra are simple closed surfaces, all of whose faces are bounded by polygons. The intersection of any two faces is an *edge* of the polyhedron. The intersection of three or more edges is a *vertex* of the polyhedron.

We can discover an interesting fact about angles formed by the edges meeting at a vertex by making a model of a solid like the one pictured in Figure 12-12. Take a piece of wood or heavy cardboard and draw a triangle on it. Label the vertices *A, B,* and *C.* Make holes at the vertices of the triangle and insert a piece of elastic thread through each of the three holes (tie a knot at the end of each thread so it does not go through the hole). Bring the pieces of thread together to form a vertex (*F*) of a triangular pyramid.

Notice the three angles of the triangular faces of the pyramid with the common vertex *F.* Watch how they change as you transform this model. As you move vertex *F* away from the base, these angles get smaller and smaller. As you move vertex *F* closer to the base, they get larger and larger (Figure 12-13). In fact, as vertex *F* gets closer and closer to the base of the pyramid, the sum of these three angles gets larger and larger and approaches 360°. But the sum is never as large as 360° since when the sum of these three angles is 360°, vertex *F* is actually *on* the base of the model

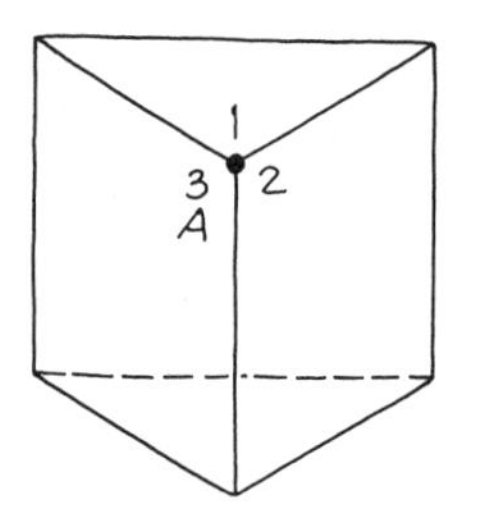

Figure 12-14

and no pyramid is formed. We can conclude the following: **the sum of the measures of the three angles of the faces of the pyramid (face angles) with vertex *F* must be less than 360°.**

We might make a similar model of a pyramid with a square, pentagonal, hexagonal, etc., base and make similar observations about the sum of the angles of the faces of these pyramids with a common vertex. But what does this have to do with our initial question of why there are only five regular polyhedra?

Consider the model of a polyhedron in Figure 12-14. Each vertex has corresponding face angles—that is, angles on the faces of the polyhedron with that common vertex and with edges of the polyhedron as sides. For example, vertex *A* has face angles 1, 2, 3. We know, intuitively, from the previous experience with the elastic thread model that **the sum of the measures of these angles must be less than 360°.** Vertex *A* is like the vertex of a pyramid.

We know that a regular polyhedron must have faces bounded by regular congruent polygons. Let us now list all the possible regular polyhedra.

Using Equilateral Triangles

If the faces of the polyhedron are bounded by equilateral triangles, it is possible to have *three* equilateral triangles meeting at a vertex (you need at least three faces to meet at a vertex), *four* equilateral triangles meeting at a vertex, or *five* equilateral triangles meeting at a vertex. The sum of the measures of the face angles at a vertex would be 180°, 240°, and 300°, respectively (Table 12-1). It is not possible to have *six or more* equilateral triangles meeting at a vertex since then the sum of the measures of the face angles at a vertex would be equal to or greater than 360°, and we know the sum must be less than 360°.

1. With three equilateral triangles at a vertex, a *regular tetrahedron* results.
2. With four equilateral triangles at a vertex, a *regular octahedron* results.
3. With five equilateral triangles at a vertex, a *regular icosahedron* results.

These are all the possible regular polyhedra with faces bounded by congruent equilateral triangles.

Using Squares

If the faces of the polyhedron are bounded by squares, only *one* polyhedron is possible. We need at least three faces to meet at a

Table 12-1

Face	Faces at vertex	Sum of measures of face angles at vertex	Regular polyhedron
Equilateral triangle	3	$3 \times 60° = 180°$	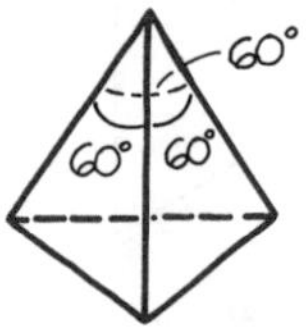 Tetrahedron
Equilateral triangle	4	$4 \times 60° = 240°$	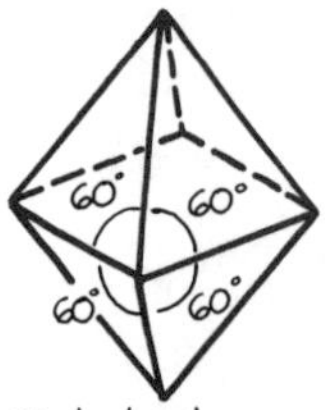 Octahedron
Equilateral triangle	5	$5 \times 60° = 300°$	 Icosahedron
Square	3	$3 \times 90° = 270°$	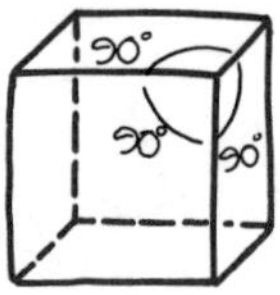 Hexahedron (cube)
Regular pentagon	3	$3 \times 108° = 324°$	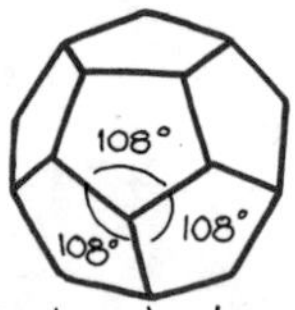 Dodecahedron

vertex. Each angle of a square has a measure of 90°. The sum of three face angles at a vertex would be 270° (Table 12-1).

> With three squares at a vertex, a *regular hexahedron* or *cube* results.

If four or more squares meet at a vertex, the sum of the measures of the corresponding face angles is equal to or greater than 360°. Therefore no other regular polyhedra with square faces are possible.

Using Regular Pentagons

Each angle of a regular pentagon contains 108°. If the faces of a regular polyhedron are regular pentagons, only one regular polyhedron is possible. When three regular pentagons meet at a vertex, the sum of the measures of the face angles is 324° (Table 12-1).

> With three regular pentagons at a vertex, a *regular dodecahedron* results.

If four or more regular pentagons meet at a vertex, the sum of the measures of the face angles is 432° or more. No such polyhedron is possible.

Using Other Regular Polygons

We might consider using other regular polygons as faces for a regular polyhedron as indicated in Table 12-2.

Table 12-2

Regular polygon	Measure of each angle
6 sides (hexagon)	120°
7 sides (septagon)	≈129°
8 sides (octagon)	135°
9 sides (nonagon)	140°
10 sides (decagon)	144°

It is not possible for a regular polyhedron to exist with faces bounded by any of these regular polygons: at least three faces must meet at a vertex. When you consider three or more hexagons meeting at a vertex, the sum of the measures of the face angles is 360° or more. Similarly, if you consider three or more of any other regular polygon in Table 12-2 meeting at a vertex, the sum of the measures of the face angles is greater than 360°. In fact, if you imagine regular polygons with more than 10 sides, each face angle will be greater than 144° and the sum of the measures of three face angles will always be greater than 360°. No other regular polygons can bound the faces of a regular polyhedron.

Realizing that the sum of the face angles at a vertex must always be less than 360°, we thus find that *only five regular polyhedra are possible.* Using equilateral triangles to bound the faces makes the *tetrahedron, octahedron,* and *icosahedron* possible. Using squares to bound the faces, the polyhedron becomes a *hexahedron* or cube. And with regular pentagons to bound the faces, it is possible to form a *dodecahedron.*

Euler's Formula and Duals

You can make sturdy models of the five regular polyhedra by using oaktag. Figure 12-15 shows patterns that can be used to make

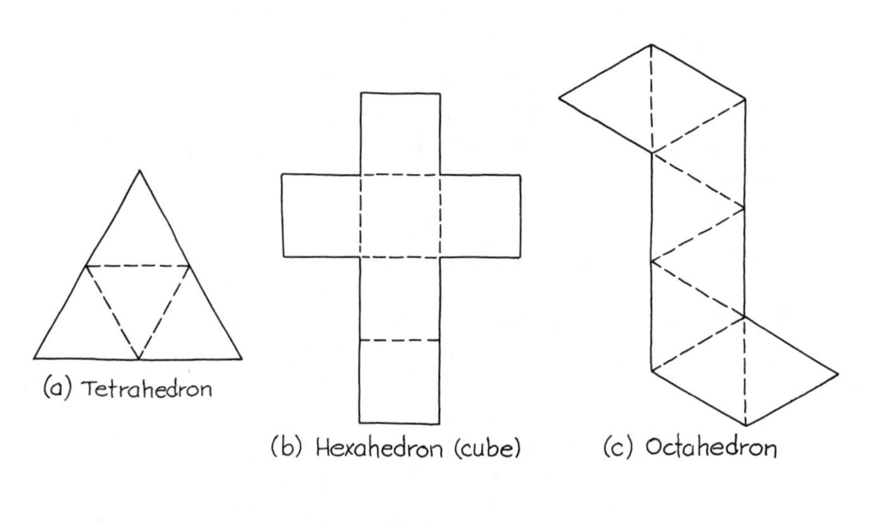

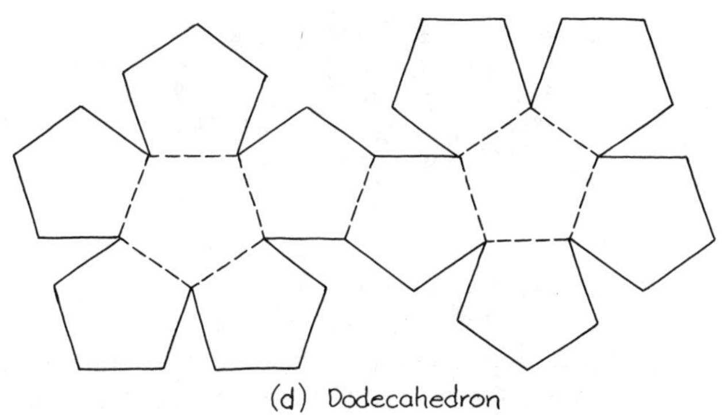

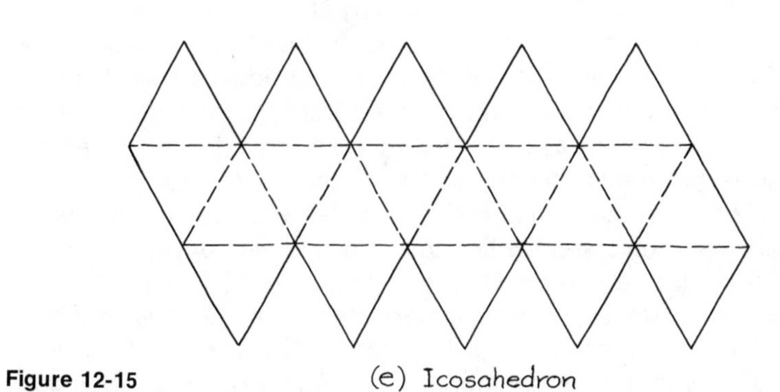

Figure 12-15

Patterns for the regular polyhedra

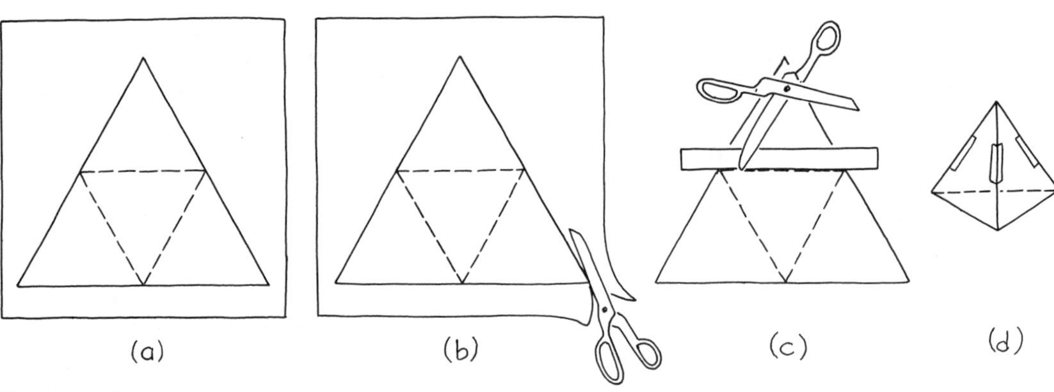

Figure 12-16

Making a polyhedron model from oaktag

these solids. In each case the pattern (or net) for the solid should be drawn on oaktag. Then the pattern is cut along the solid lines. Next, the pattern is scored along the dotted lines. Finally, the faces of the model are taped together along the edges. The procedure is illustrated in Figure 12-16.

If you count the number of faces, vertices, and edges of each of these solids, you can verify the information given in Table 12-3. It can be awkward trying to count the faces, vertices, and edges of the dodecahedron or icosahedron; numbering them right on the model as you count can make the job easier.

Table 12-3

Regular polyhedron	(1) Total number of faces F	(2) Total number of vertices V	(3) Total number of edges E
Tetrahedron	4	4	6
Hexahedron (cube)	6	8	12
Octahedron	8	6	12
Dodecahedron	12	20	30
Icosahedron	20	12	30

Examine Table 12-3 carefully. Do you see a relationship between the numbers in columns 1 and 2 and the corresponding number in column 3 for each figure? How are they related? The relationship is known as **Euler's formula,** after the eighteenth-century mathematician Leonard Euler. As an equation it can be stated as follows:

$$F + V = E + 2$$

That is, for each solid the sum of the number of faces and the number of vertices is equal to the number of edges increased by 2. You will see that this equation is valid for all five regular solids. For example, in the case of the tetrahedron where $F = 4$, $V = 4$, and $E = 6$,

$$4 + 4 = 6 + 2$$
$$8 = 8$$

It is interesting to note that this relationship is valid for *all* convex polyhedra, not just the five regular polyhedra. Try applying it to other convex polyhedra like a rectangular solid, a square pyramid, or a hexagonal prism to see that it is valid. You will find the relationship valid for *any* convex polyhedron you choose!

Take another look at the numbers in Table 12-3. Do you notice a kind of symmetry? The octahedron and the cube both have 12 edges; the octahedron has just as many faces (8) as the cube has vertices (8); and the cube has just as many faces (6) as the octahedron has vertices (6). Similarly, the dodecahedron and icosahedron have the same number of edges; the dodecahedron has just as many faces (12) as the icosahedron has vertices (12); and the icosahedron has just as many faces (20) as the dodecahedron has vertices (20). It would seem that these pairs of solid figures are related in some way.

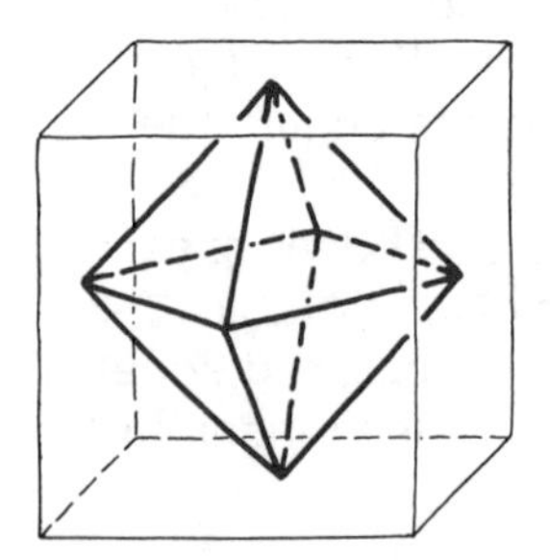

Figure 12-17

Octahedron inside a cube

In fact, they are related in a very definite way. The cube and the octahedron can be considered **duals** of each other. Figure 12-17 illustrates this. If you start with a cube, it is possible to construct an octahedron within the cube, with each of the six vertices of the octahedron at the center of one of the six faces of the cube. Similarly, it is possible to construct a cube within the octahedron, with each of the eight vertices of the cube at the center of one of the eight faces of the octahedron (Figure 12-18). You can imagine this process being carried on indefinitely, one figure within the other.

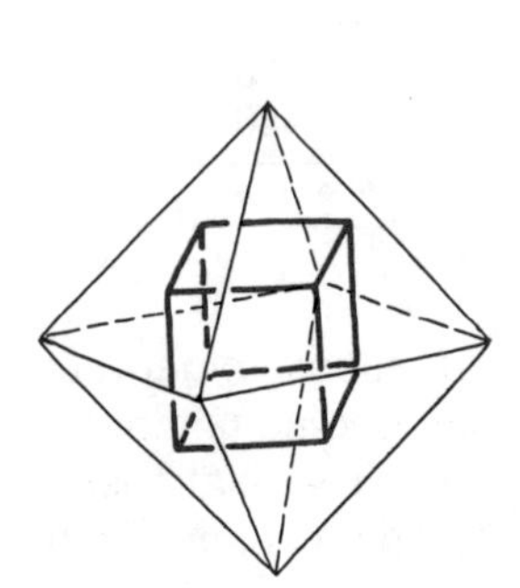

Figure 12-18

Cube inside an octahedron

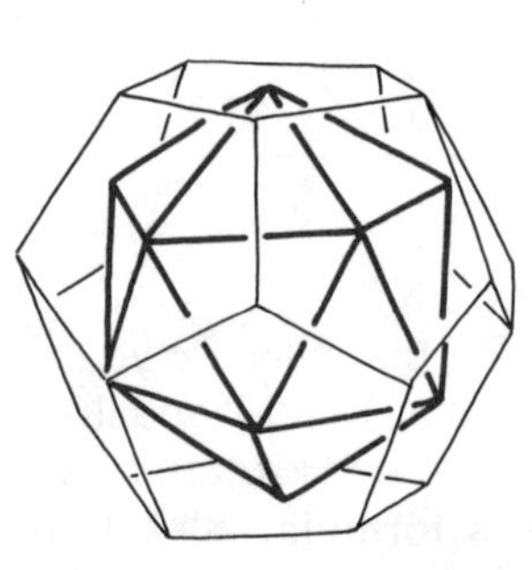

(a) Icosahedron inside a dodecahedron

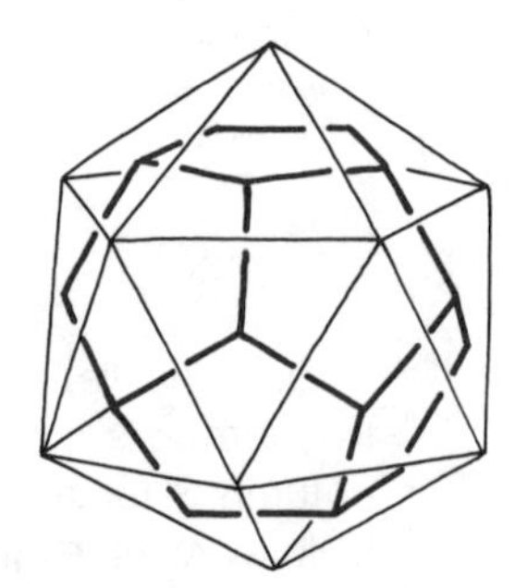

(b) Dodecahedron inside an icosahedron

Figure 12-19

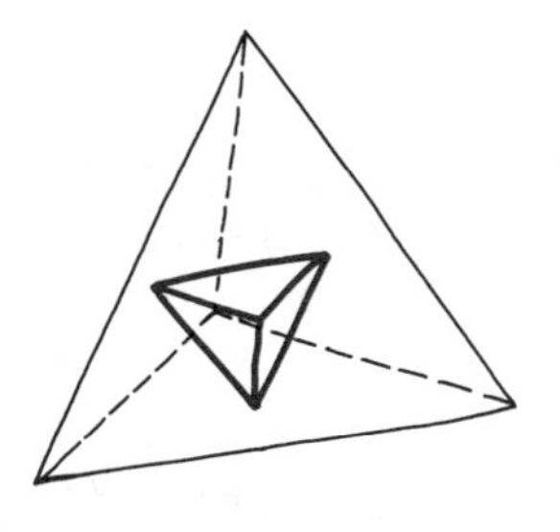

Figure 12-20

Tetrahedron inside a tetrahedron

The dodecahedron and the icosahedron are also duals of each other as shown in Figure 12-19. Again, you can imagine constructing another dodecahedron within the icosahedron, another icosahedron within that dodecahedron, and so on. Each time, the vertices of one figure correspond to the centers of the faces of the other.

Does the tetrahedron have a dual? It is the dual of itself. In the same manner used to make duals, we can construct a tetrahedron within a tetrahedron (Figure 12-20). This procedure can be carried on indefinitely.

The Kepler–Poinsot Polyhedra

There are four well-known polyhedra that are often associated with the regular polyhedra because they have most of the properties of regular polyhedra. These polyhedra are illustrated in Figure 12-21. They are *nonconvex* polyhedra, since the extension of a face *does* intersect with the interior of the polyhedron. Note, however, that all the faces of each polyhedron are congruent.

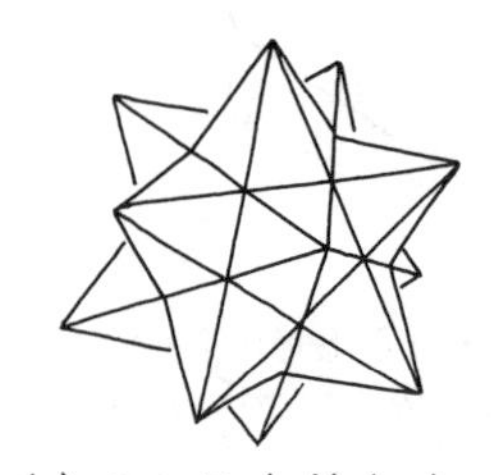

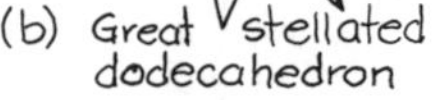

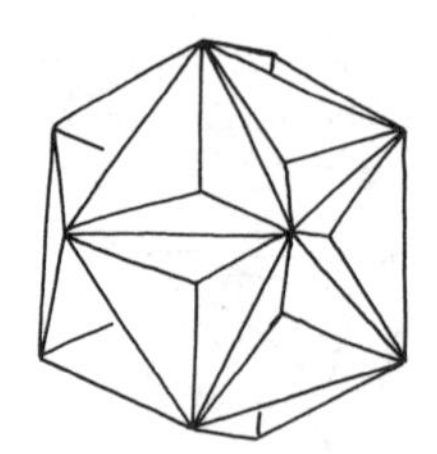

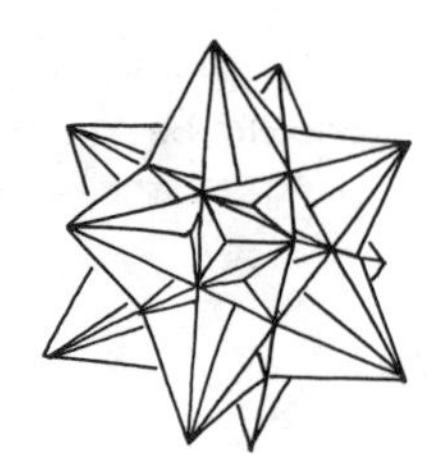

Figure 12-21

The Kepler–Poinsot polyhedra

These polyhedra are known as the **Kepler–Poinsot polyhedra,** named after the mathematicians who discovered them.* The small stellated dodecahedron, the great stellated dodecahedron, and the great dodecahedron are all considered dodecahedra (that is, solids with 12 faces) because their faces lie in 12 different planes.

**Johannes Kepler (1571–1630) discovered the small and great stellated dodecahedra. Louis Poinsot (1777–1859) later found the great dodecahedron and great icosahedron.*

Similarly, the great icosahedron is considered an icosahedron because its many faces lie in 20 different planes.

You may be familiar with the plane figure called a **pentagram.** A pentagram can be drawn by starting with a regular pentagon and extending its sides until they meet a nonadjacent side or by joining alternate vertices of the pentagon (Figure 12-22). Consider the outer boundary of the pentagram as a polygon. Let us call that polygon a *star polygon.* The star polygon is not a regular polygon, but it does have congruent sides and two sets of congruent angles.

If you take a closer look at the Kepler–Poinsot polyhedra, you will notice how often the star polygon is found as part of these solid shapes. For example, the star polygon is the shape of one of the 12 faces of the small and the great stellated dodecahedra. Do you see the star polygon shape in the great dodecahedron? In the great icosahedron?

Models of the Kepler–Poinsot polyhedra can be made out of oaktag. Patterns and directions for making them are presented in Figure 12-23. You can distinguish the different planes of the faces better if in the resulting models you paint all the faces in the same plane the same color.

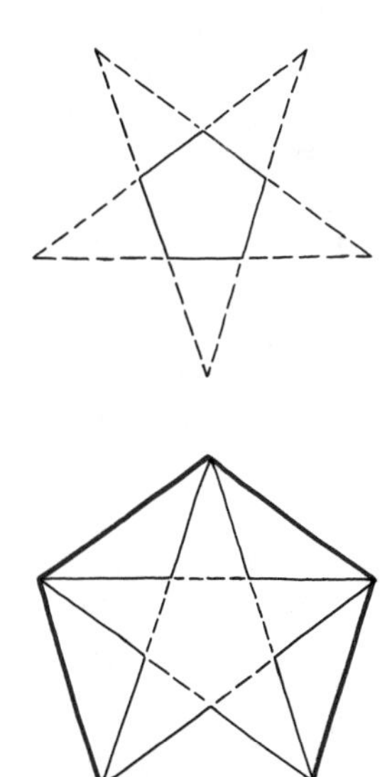

Figure 12-22

Making a pentagram from a regular pentagon

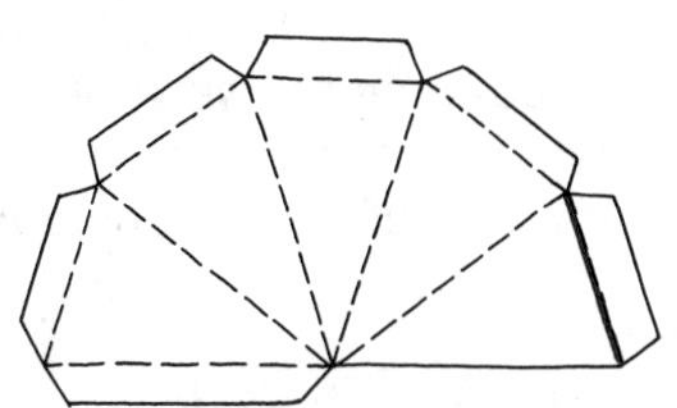

The length of the base of the triangle should be the same as the length of the edge of the dodecahedron.

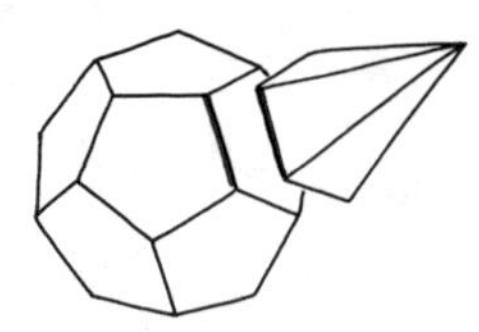

(a) Small stellated dodecahedron

1. First make a regular dodecahedron (Fig. 12-15).
2. Next make 12 pentagonal pyramid models, using the pattern given here.
3. Glue each pyramid onto a face of the dodecahedron, attaching the tabs to the face.
4. Paint all faces that are on the same plane the same color.

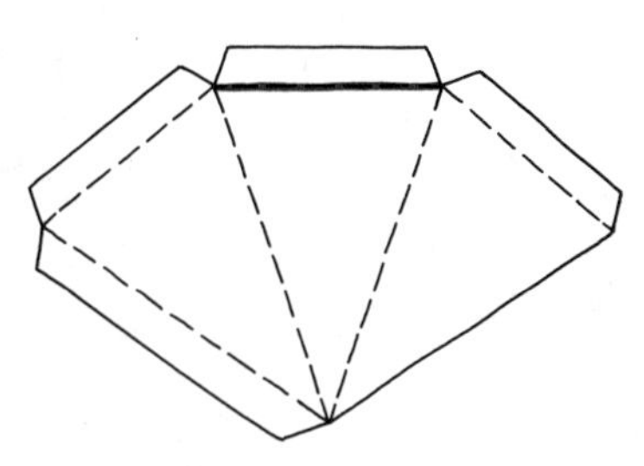

The length of the base of the triangle should be the same as the length of the edge of the icosahedron.

(b) Great stellated dodecahedron

1. First make a regular icosahedron (Fig. 12-15).
2. Next make 20 triangular pyramids, using the pattern given here.
3. Glue each pyramid onto a face of the icosahedron, attaching the tabs to the face.
4. Paint all faces that are on the same plane the same color.

Figure 12-23

Making the Kepler–Poinsot polyhedra

Semi-Regular Polyhedra

We found that all regular polyhedra are convex polyhedra which have (1) all faces bounded by the same kind of congruent regular polygon and (2) all vertex figures as congruent regular polygons. There are two types of convex polyhedra that are known as **semi-regular polyhedra** because they have *some* (but not all) of these properties.

One type of semi-regular solid is the **facially regular** polyhedron. This polyhedron has the following properties:

1. Each face is bounded by a regular polygon, but there can be two or more kinds or regular polygons.
2. The vertex figures are congruent but not necessarily regular polygons.

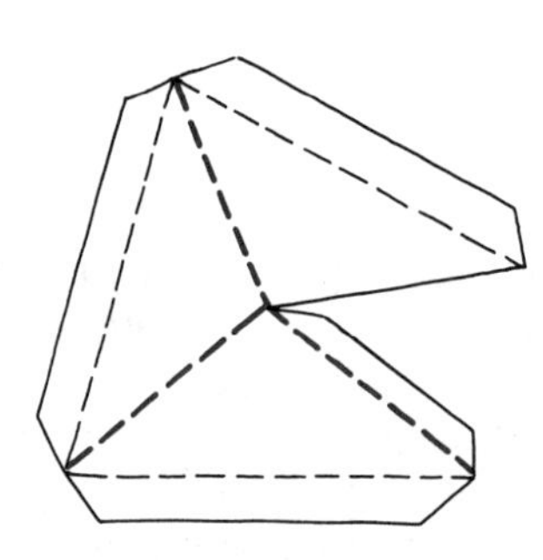

Score dashed dark lines on front and fold tabs backward; score dashed brown lines on back and fold forward.

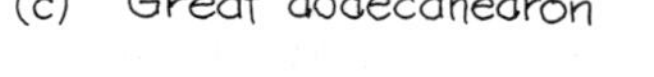

(c) Great dodecahedron

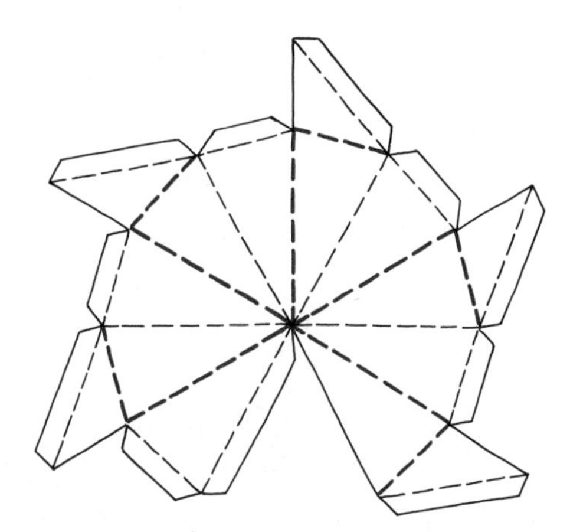

Score dashed dark lines on front and fold backward; score dashed brown lines on back and fold forward.

(d) Great icosahedron

1. Make 20 pyramids that are folded inward like a funnel using the pattern given here.
2. Glue the 20 parts together using the tabs. See Figure 12-21 for the completed polyhedron.
3. Paint all faces that are in the same plane the same color.

1. Make 12 star-shaped pyramids using the pattern given here.
2. Glue the 12 pyramids together using the tabs. See Figure 12-21 for the completed polyhedron.
3. Paint all faces that are in the same plane the same color.

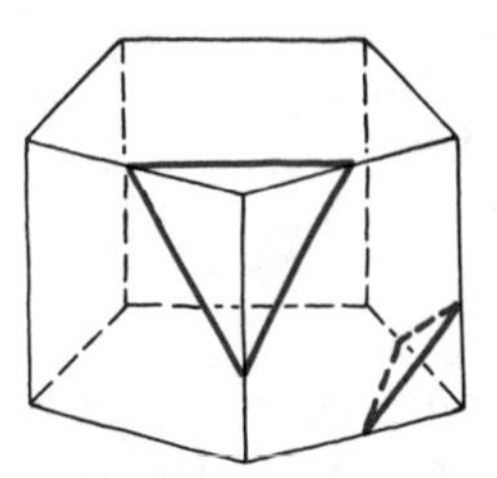
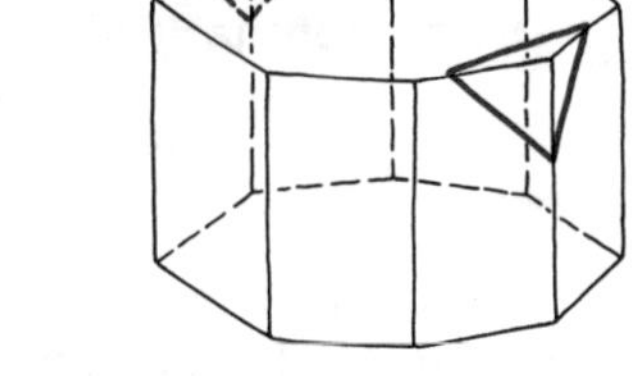

Figure 12-24
Facially regular prisms

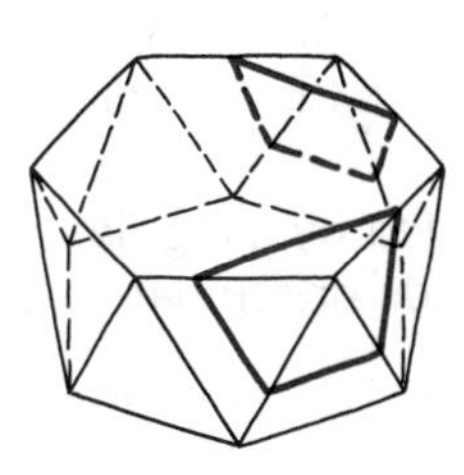
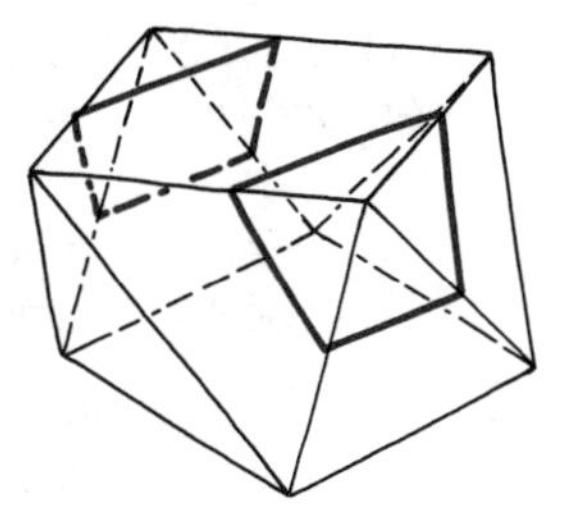

Figure 12-25
Facially regular antiprisms

All prisms with bases bounded by regular polygons and with square lateral faces like the ones in Figure 12-24 are examples of this type of facially regular solid. All the faces of the prisms are bounded by regular polygons of two or more kinds, and the vertex figures are congruent but not necessarily regular polygons. Similarly, *antiprisms* with bases bounded by regular polygons and with lateral sides bounded by equilateral triangles qualify as facially regular solids. Some examples are shown in Figure 12-25.

Aside from the infinite number of possible prisms and antiprisms that qualify as facially regular solids, only 13 other polyhedra qualify. They are known as the **Archimedean polyhedra.** Models of these polyhedra are pictured in Figure 12-26.*

The second type of semi-regular solid is the **vertically regular** polyhedron. This polyhedron has the following properties:

1. Each face is bounded by a congruent polygon, but that polygon is not necessarily regular.
2. The vertex figures are regular polygons (thus the name *vertically regular*) of two or more kinds.

**For a more detailed discussion of the Archimedean polyhedra and patterns for making them, see H. M. Cundy and A. P. Rollett, Mathematical Models, 2d ed. (New York: Oxford University Press, 1961).*

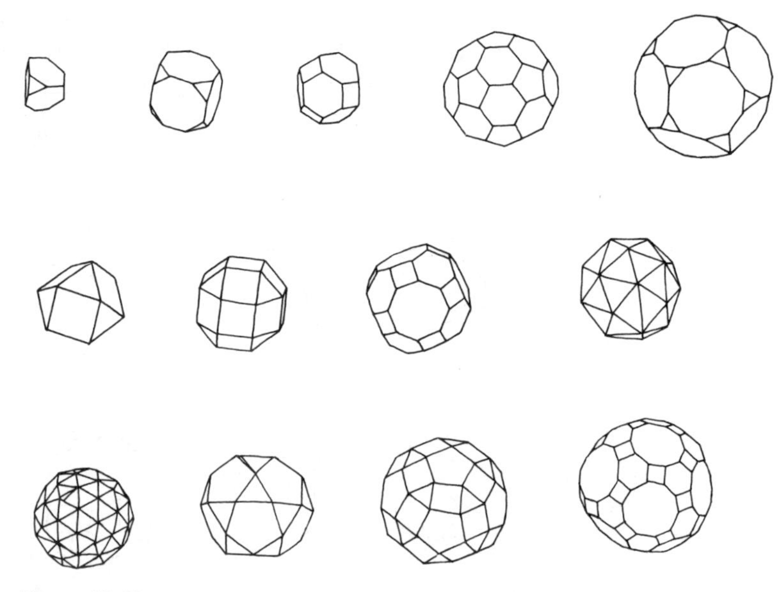

Figure 12-26

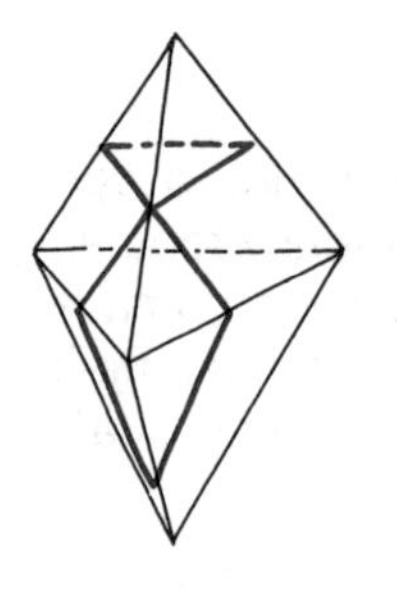

Figure 12-27

Triangular dipyramid (vertically regular)

The triangular dipyramid is an example of a vertically regular polyhedron. We have seen this polyhedron before. It looks like two regular tetrahedra with a common base (Figure 12-27). Note that each face is bounded by a congruent polygon (equilateral triangle) and the vertex figures are regular polygons (either an equilateral triangle or a square).

We can think of this triangular dipyramid as the dual of a triangular prism that qualifies as a facially regular polyhedron (Figure 12-28). In fact, we can imagine that for every prism which is facially regular, there exists a dual figure which is a vertically regular

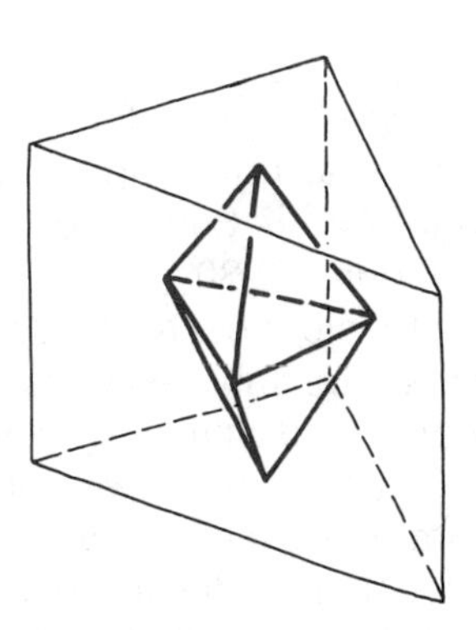

Figure 12-28

Triangular dipyramid (vertically regular) as the dual of a triangular prism (facially regular)

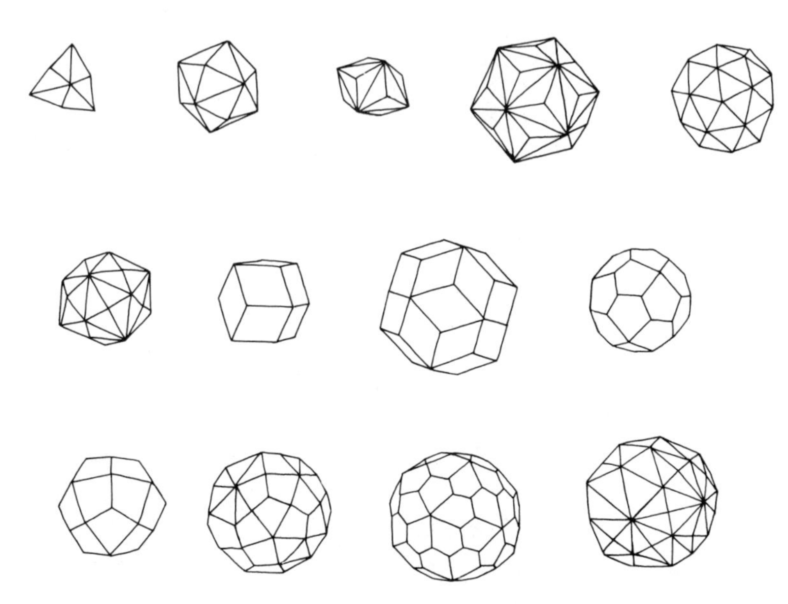

Figure 12-29

dipyramid. Similarly, every facially regular antiprism has a corresponding dual (called a trapezohedron) that is vertically regular.

As you might expect, all 13 of the Archimedean polyhedra have duals. While the Archimedean polyhedra are facially regular, their duals are vertically regular. Models of these Archimedean duals are pictured in Figure 12-29.

Regular Compounds

Imagine joining two regular polyhedra in such a way that their edges bisect each other at right angles. Such a polyhedron is called a **regular compound.** Some examples of regular compounds are shown in Figure 12-30.

Aside from their intrinsic beauty, these models show many geometric relationships. You can easily make the stella octangula by making a tetrahedron and attaching other tetrahedra to each face as indicated in Figure 12-31. If we consider each of the two intersecting tetrahedra represented by the model as a set of points, then the intersection of these sets is the polyhedron indicated by the brown lines in Figure 12-32. This intersection looks like the edges of what polyhedron? The regular octahedron! Now suppose we attach straws to the model so that the ends of the straws are attached to vertices of the model as indicated in Figure 12-33. What figure is formed? The cube! We might have

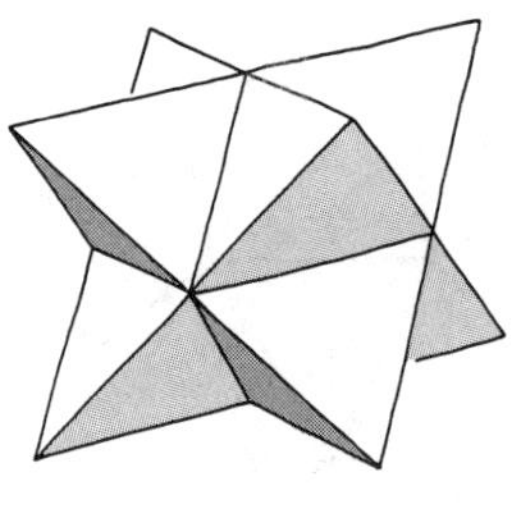

(a) Stella octangula shoded to show two tetrahedra

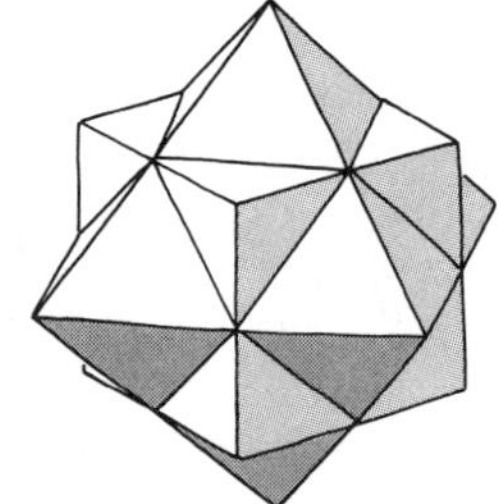

(b) Interpenetrating cube and octahedron

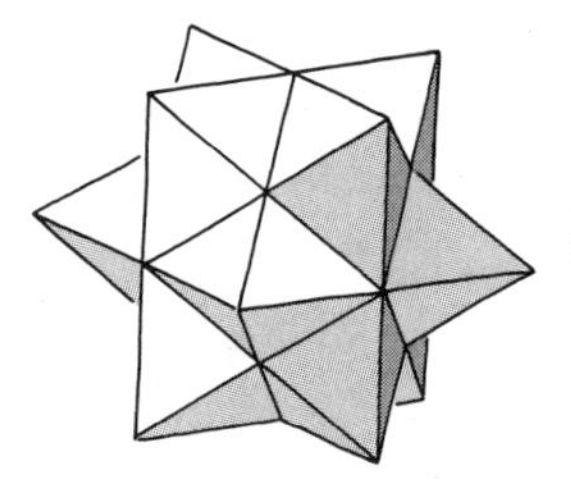

(c) Three interpenetrating octahedra

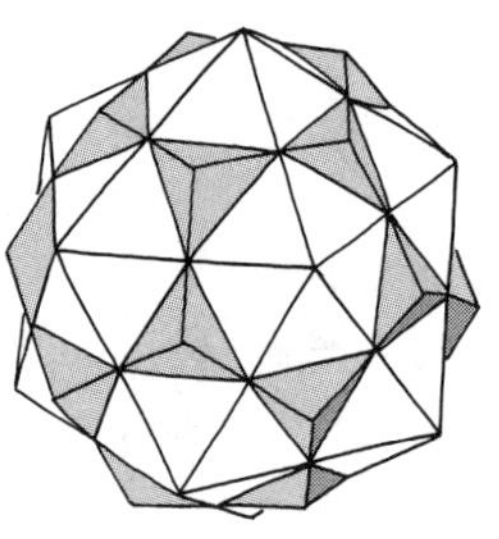

(d) Dodecahedron plus icosahedron

Figure 12-30

Some regular compounds

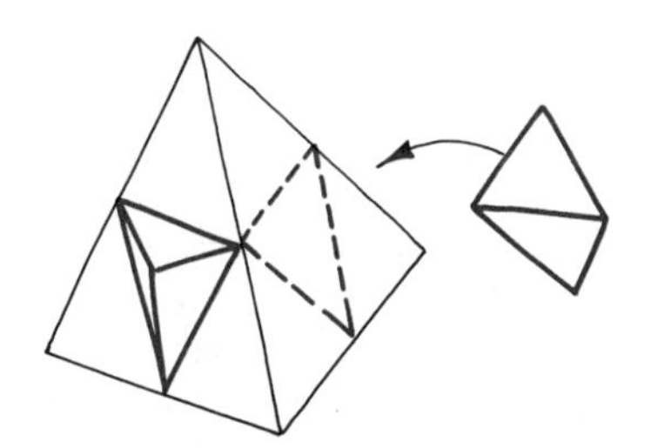

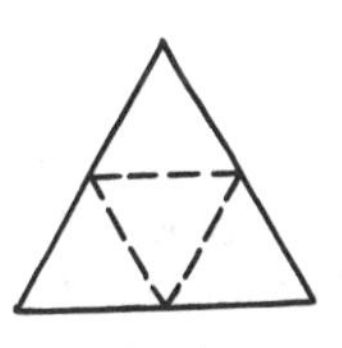

Figure 12-31

Making a model of two interlocking tetrahedra: the stella octangula

1. Length of edge of small tetrahedron is $\frac{1}{2}$ length of edge of large tetrahedron.
2. Attach one small tetrahedron to each face of large tetrahedron as indicated.

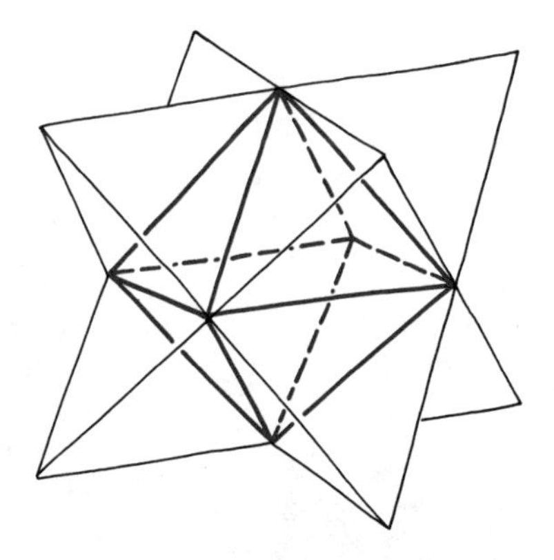

Figure 12-32

Octahedron as intersection of two interlocking tetrahedra

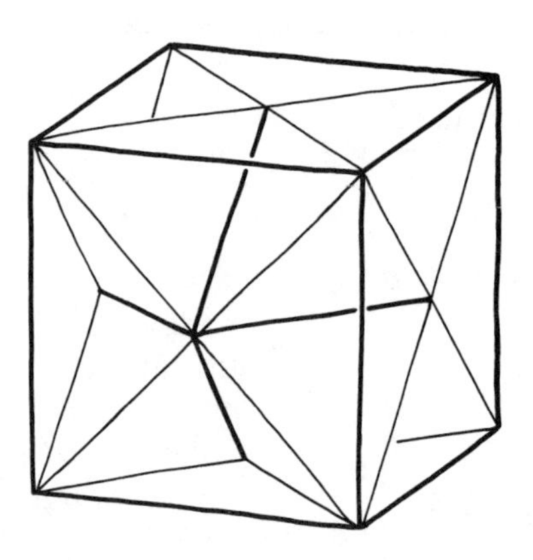

Figure 12-33

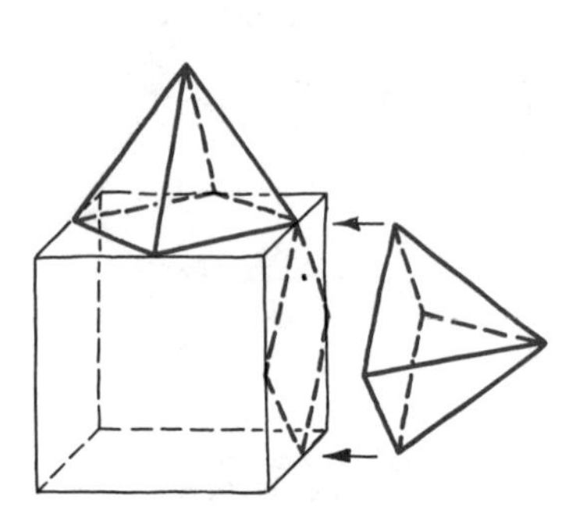

Figure 12-34

Making a model of an interlocking cube and octahedron

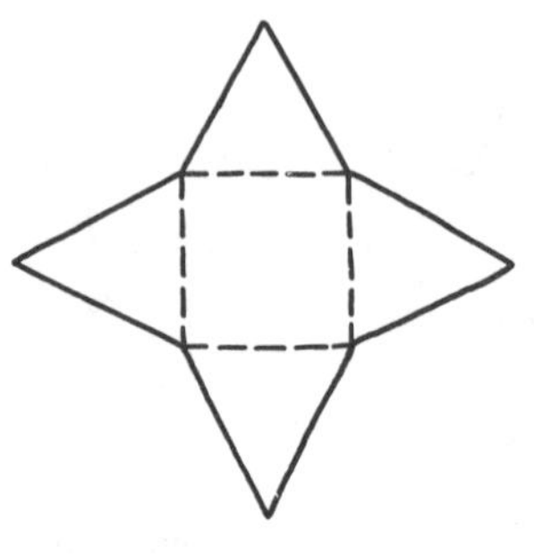

Length of edges of pyramids ≈ 7/10 length of edge of cube

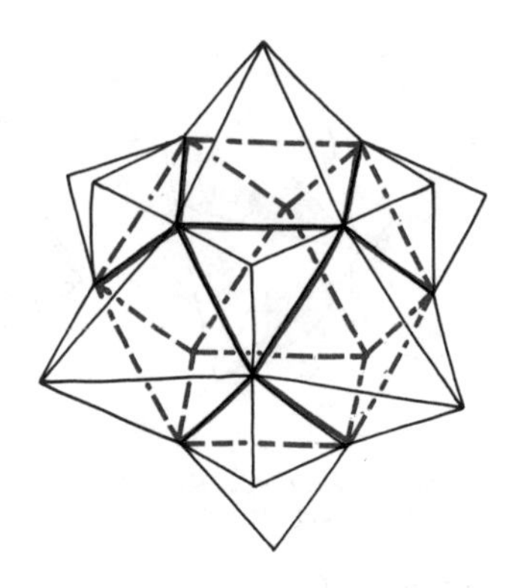

Figure 12-35

Cuboctahedron as intersection of a cube and an octahedron

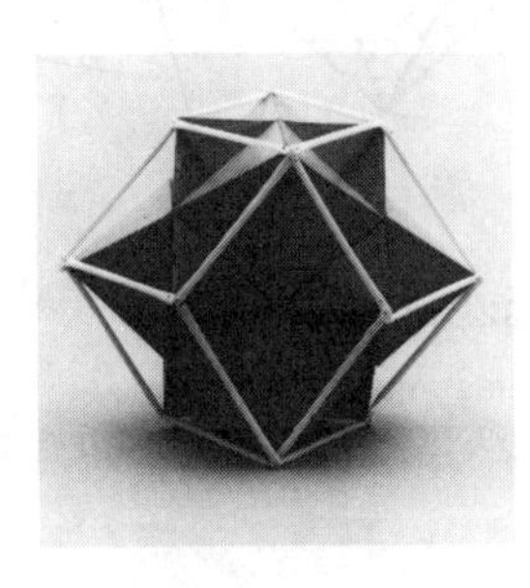

Figure 12-36

expected this, since we know that the cube is the dual of the octahedron. As the model of the stella octangula shows, the vertices of the octahedron meet the cube at the centers of its faces.

It is also easy to make a regular compound model of a cube and an octahedron by attaching one square pyramid to each of the six faces of a model of a cube as indicated in Figure 12-34. If you think of the cube as a set of points and the octahedron as a set of points, then the intersection of these sets will look like the polyhedron indicated by the brown lines in Figure 12-35. This polyhedron is called a **cuboctahedron;** it is facially regular and is one of the Archimedean polyhedra. As before, let us afix straws to the model so that their ends are attached to vertices of the original model as in Figure 12-36. The figure formed by the straws, a **rhombic dodecahedron,** is vertically regular and the Archimedean dual of the cuboctahedron.

More Experiences

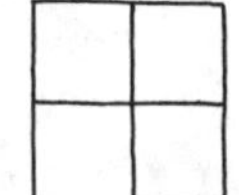
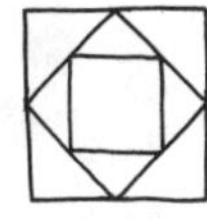
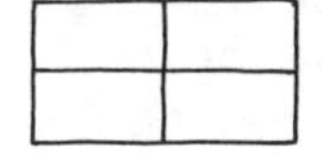

12-1. In Chapter 5, we found that certain plane shapes can be put together to cover a plane surface completely. Such an arrangement is called a **tessellation.** Some examples are shown at the left.

We can carry out a similar investigation in three-dimensional space by exploring the different solid

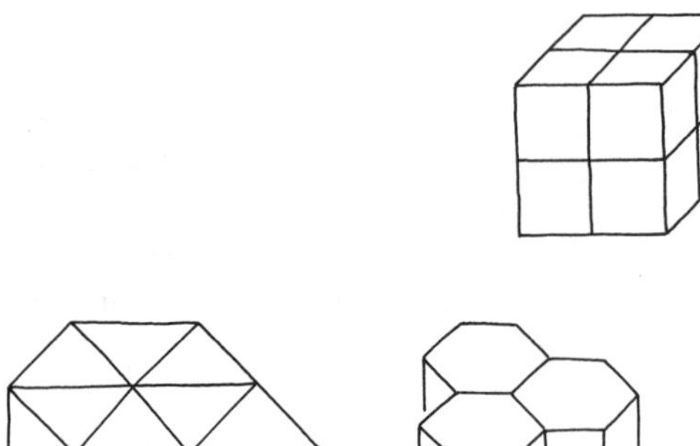

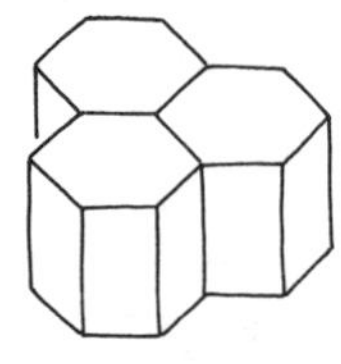

(a) Prisms with equilateral triangle base (b) Prisms with regular hexagonal base

shapes that can be put together or stacked to fill space completely. Cubes and certain prisms, for example, can be stacked to fill space completely (see diagrams at the left).

Can any prism with a regular polygonal base be used to fill space? How can you determine which prisms can fill space completely? Draw some that fill space.

Can any Platonic solids besides the cube fill space?

Can pyramids be used to fill space? What other solid shapes can fill space completely? You may have to make models to check your responses.

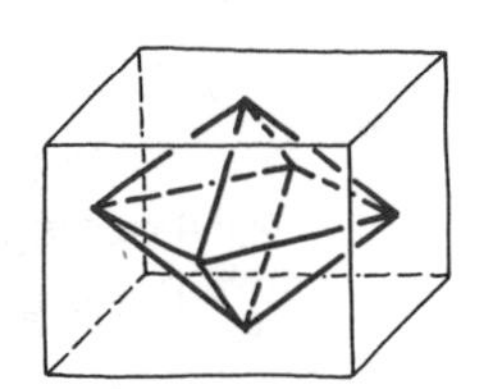

12-2. The cube and the regular octahedron are *duals* of each other. What relationship can you find between the axes of symmetry of the figures shown at the left?

12-3. Use the patterns in Figure 12-15 to make oaktag models of each of the Platonic solids.

12-4. In Chapter 10 we found that a cube has a total of nine planes of symmetry and a total of 13 axes of symmetry. Can you find the total number of planes of symmetry and the total number of axes of symmetry for each of the other Platonic solids? Keep a record of your results.

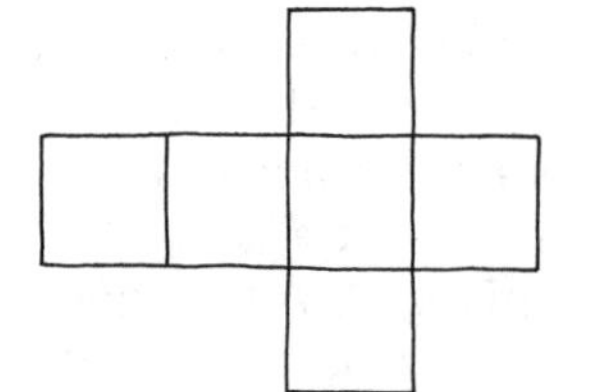

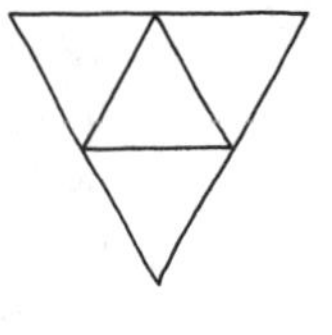

12-5. The arrangement of the six squares at the left can be used as a pattern (or net) for making the model of a cube. How many other arrangements of six squares can you make that will also form a cube?

The arrangement of four equilateral triangles at the left is the net of a regular tetrahedron. How many other arrangements of four equilateral triangles can you make that will also form a regular tetrahedron?

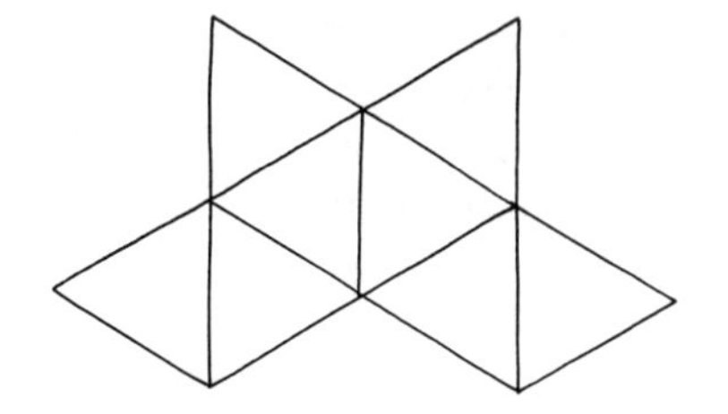

The arrangement of eight equilateral triangles at the left can be used as a net for a model of a regular octahedron. How many other arrangements of eight equilateral triangles can you make that can be used as a net for a model of a regular octahedron?

12-6. Find some examples of polyhedron shapes. Count the vertices, edges, and faces of each shape. Is Euler's formula valid for each of these polyhedra?

12-7. Starting with a triangular piece of paper or oaktag, it is easy to make a model of a polyhedron. Here is one way:

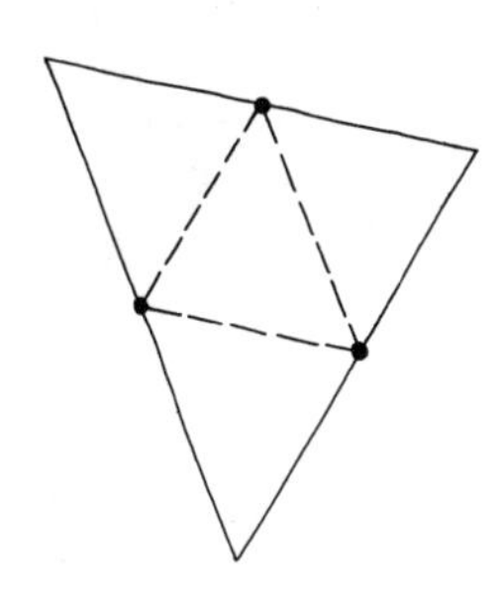

1. Cut out a triangular piece of oaktag or paper.
2. Locate the midpoint of each side.
3. Draw a line segment connecting the midpoints of adjacent sides.
4. Fold along the line segments drawn. Tape the model together.

What kind of polyhedron model results? Did you expect this? What do you know about the faces of this polyhedron? When will the resulting model be a regular polyhedron?

12-8. You know that a polygon has to have at *least* three *sides*. What is the *least* number of *faces* a polyhedron can have? Why?

12-9. Make a model of a cube out of oaktag. Examine the shadows as you hold it in different positions in sunlight. Can its shadow be a square? A rhombus? A triangle? A hexagon? Test its shadow by using an artificial light source. Try the same thing with models of other solids and keep a record of your results.

Appendix A

Patterns for Strips to Make Geometric Models

1. Reproduce these patterns.
2. Mount on oaktag.
3. Cut along outlines.
4. Punch holes where indicated, using a hole punch (preferably $1/8$-in. diameter)
5. Use brass fasteners ($1/2$-in. shank) to join the strips.

You will need many strips of each length. The number on each strip represents the distance (in centimeters) between the end holes.

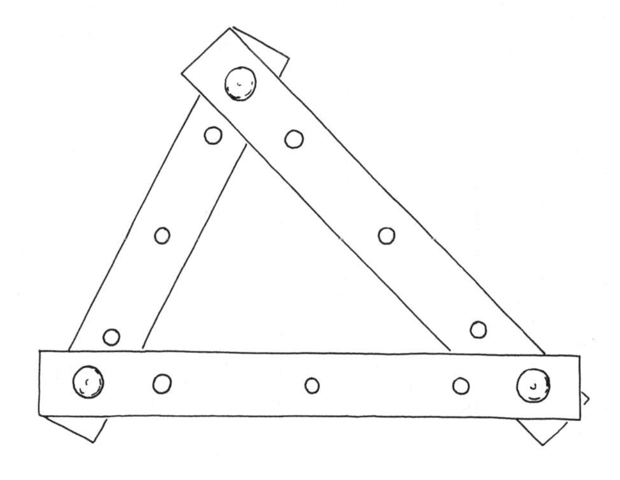

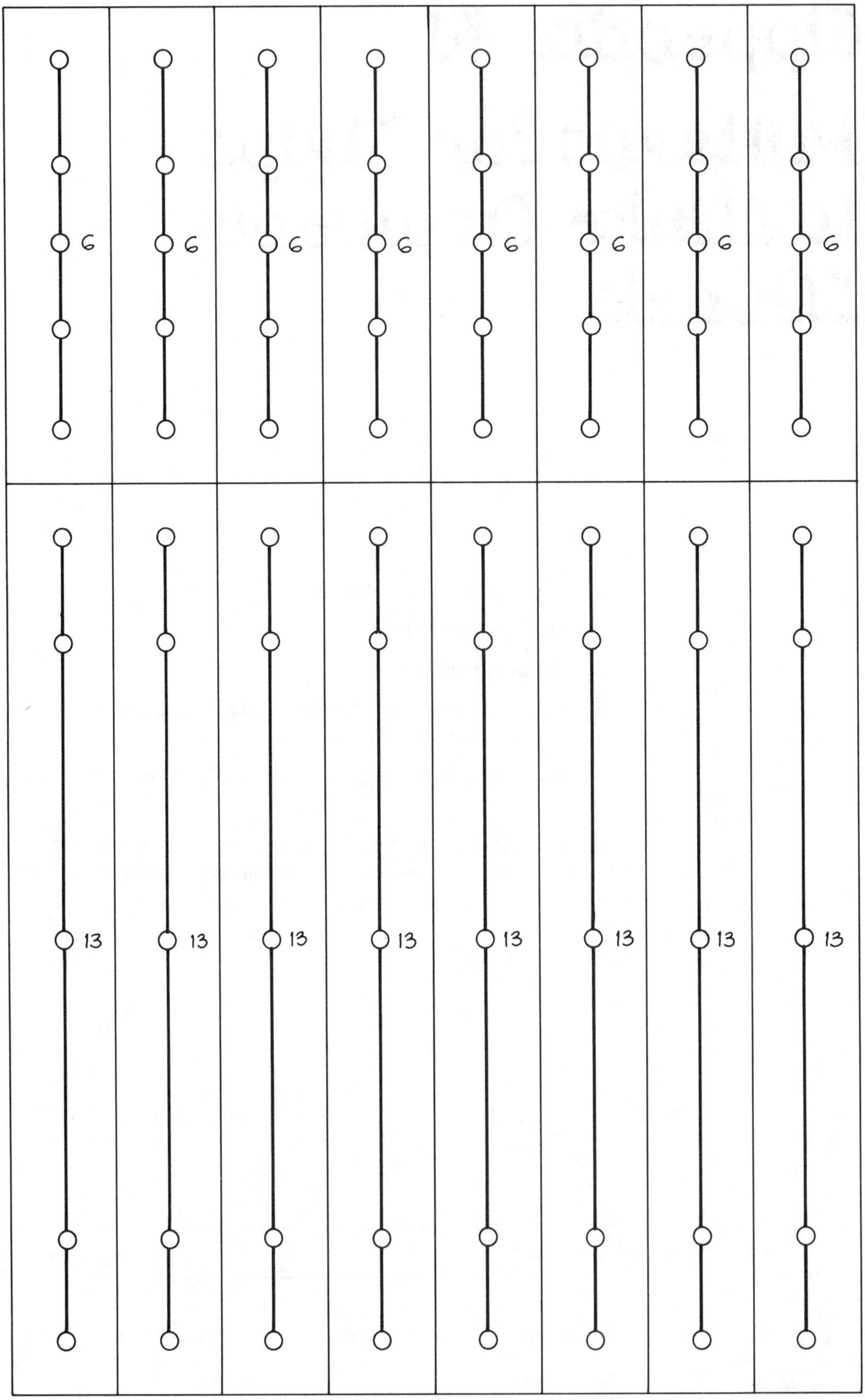

6 6 6 6 6 6 6 6

13 13 13 13 13 13 13 13

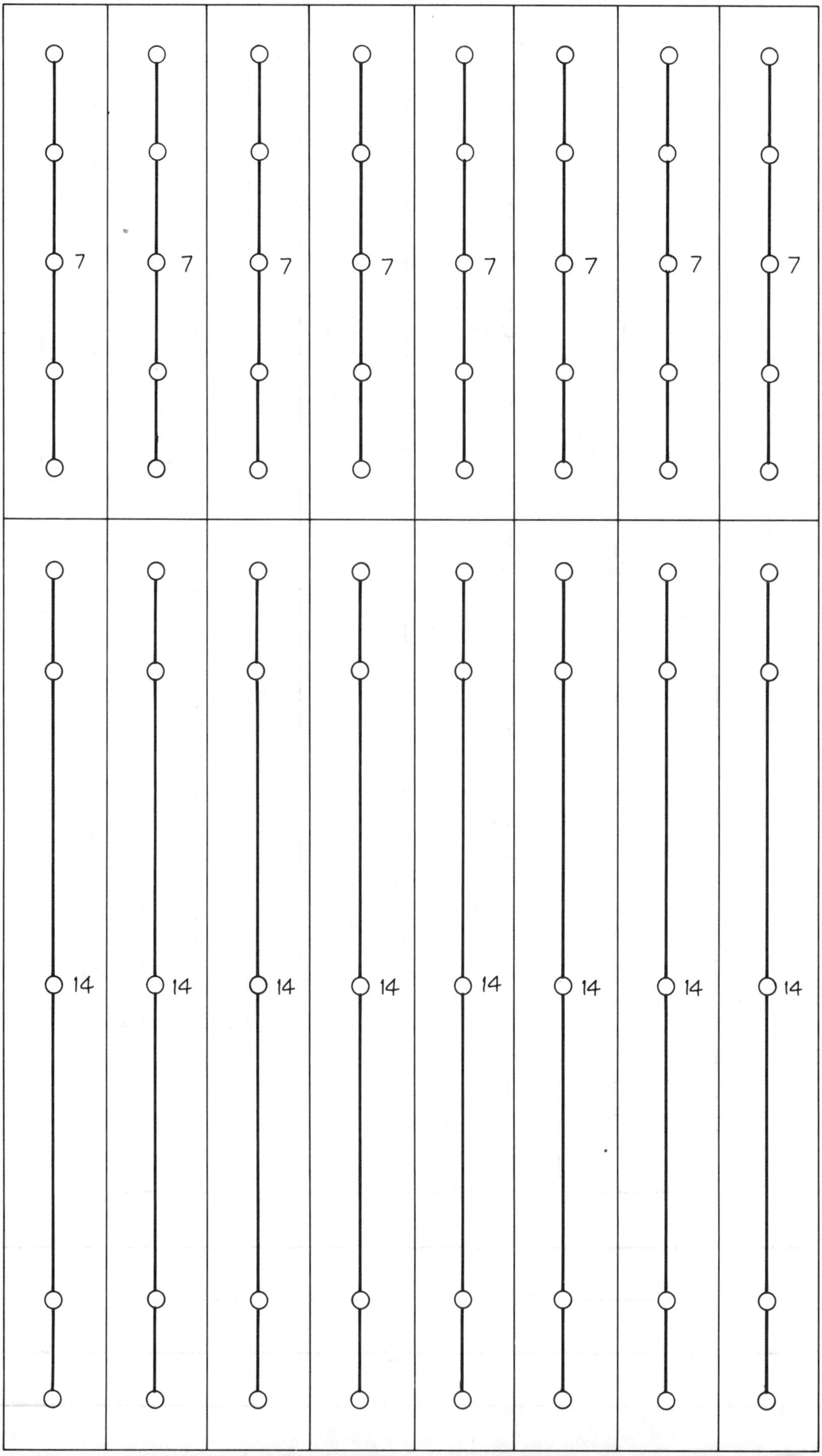

7 7 7 7 7 7 7 7

14 14 14 14 14 14 14 14

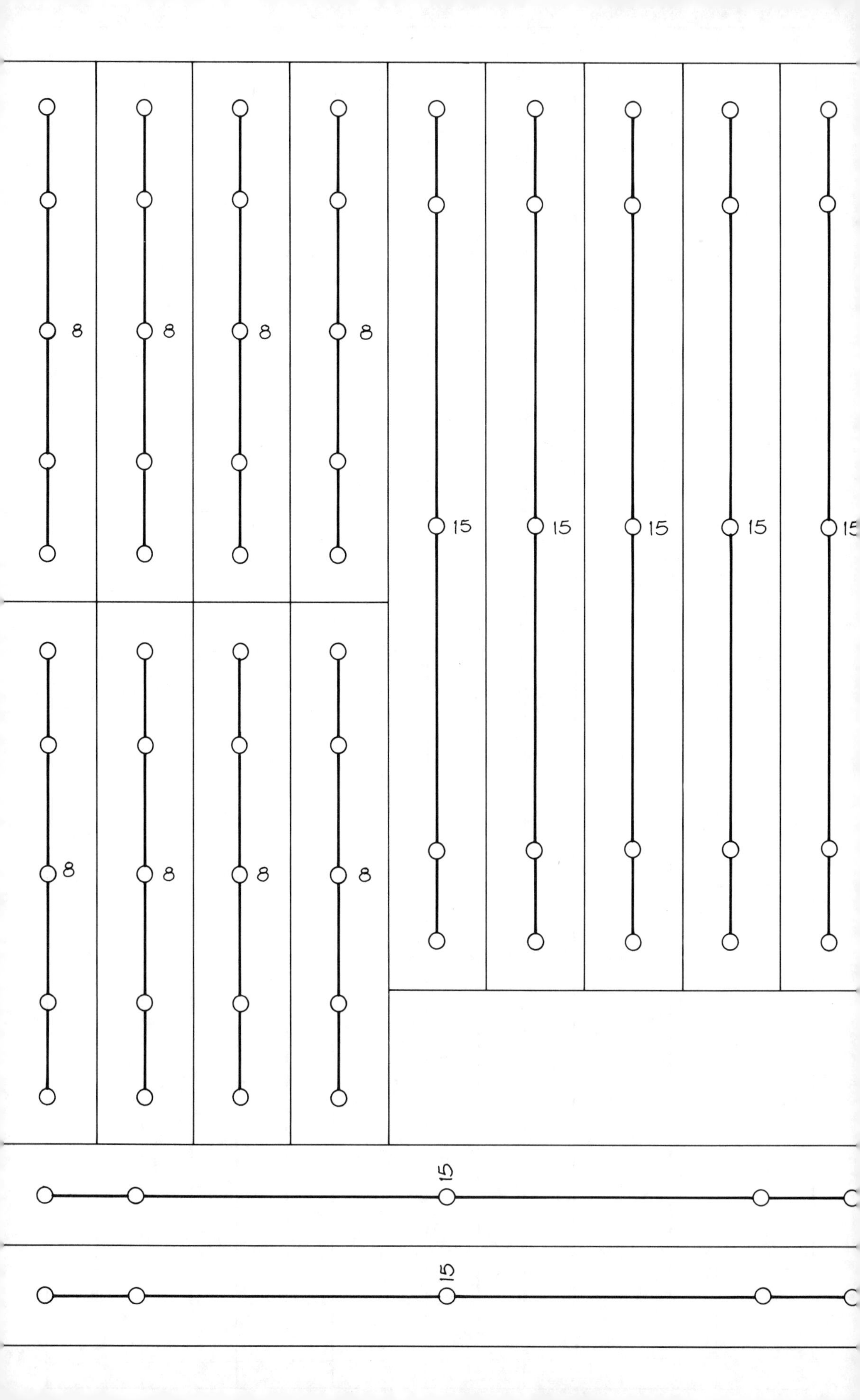

8
8
8
8
15
15
15
15
15
8
8
8
8
15
15

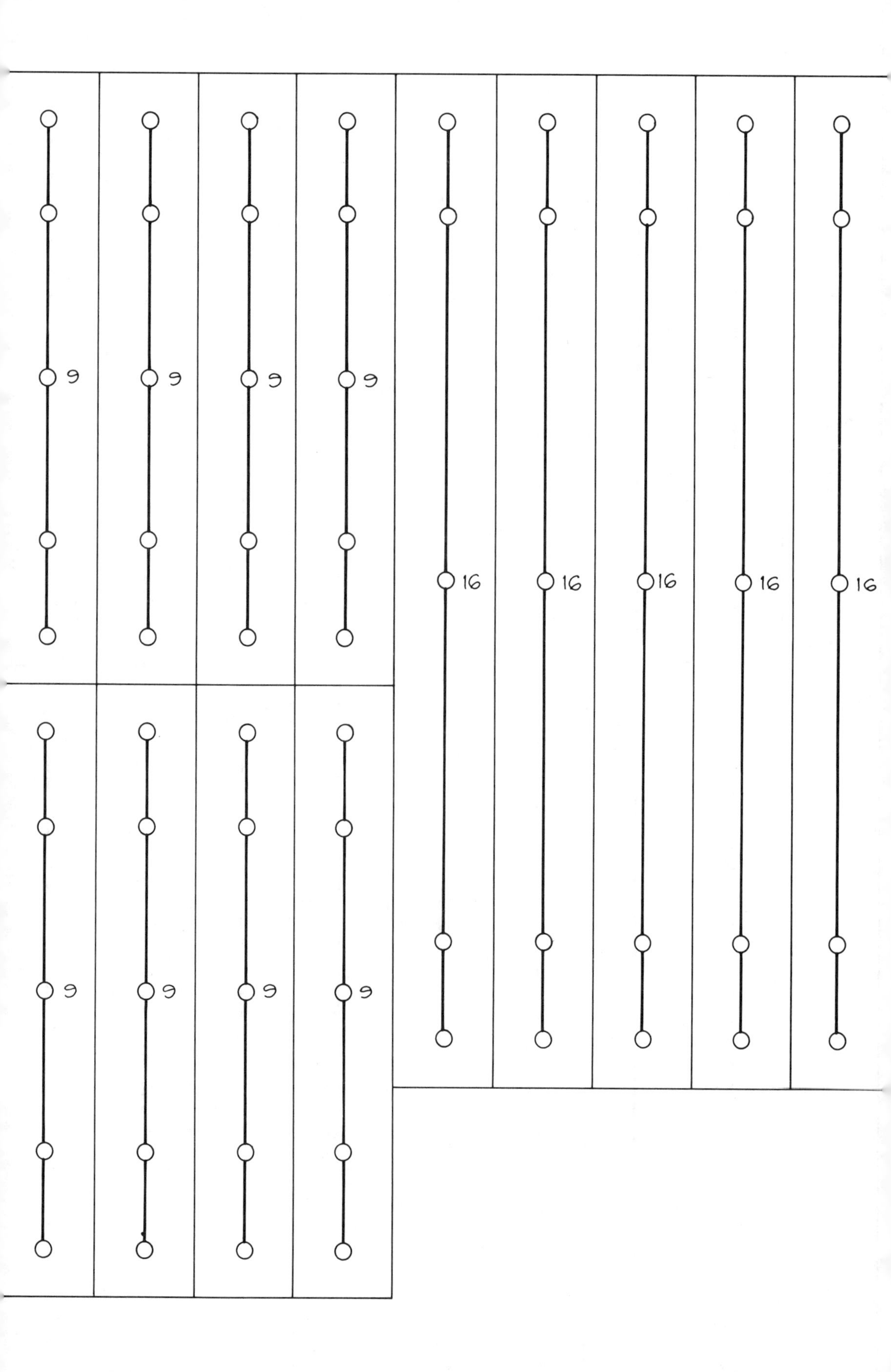
9
9
9
9
16
16
16
16
16
9
9
9
9

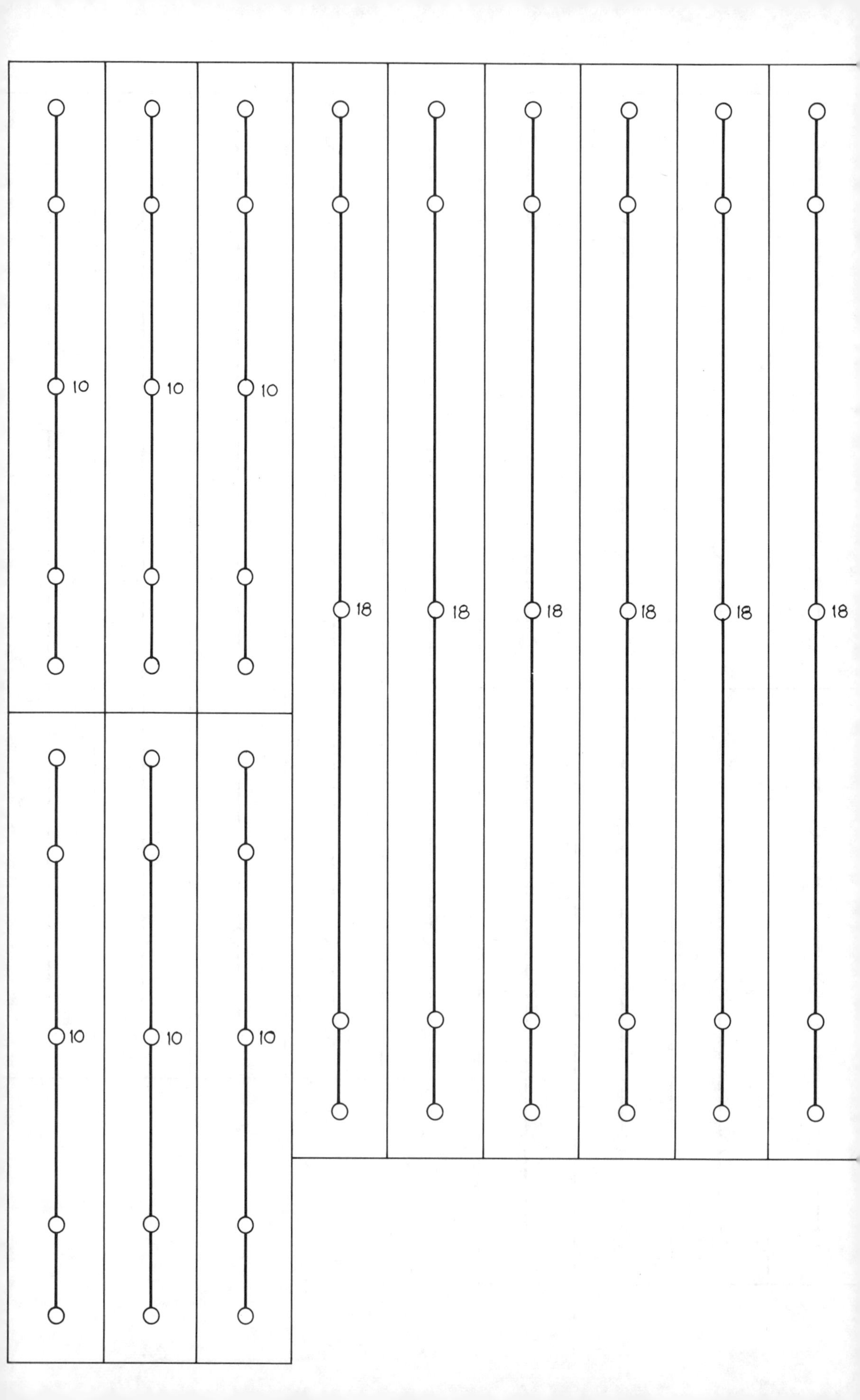
10
10
10
18
18
18
18
18
18
10
10
10

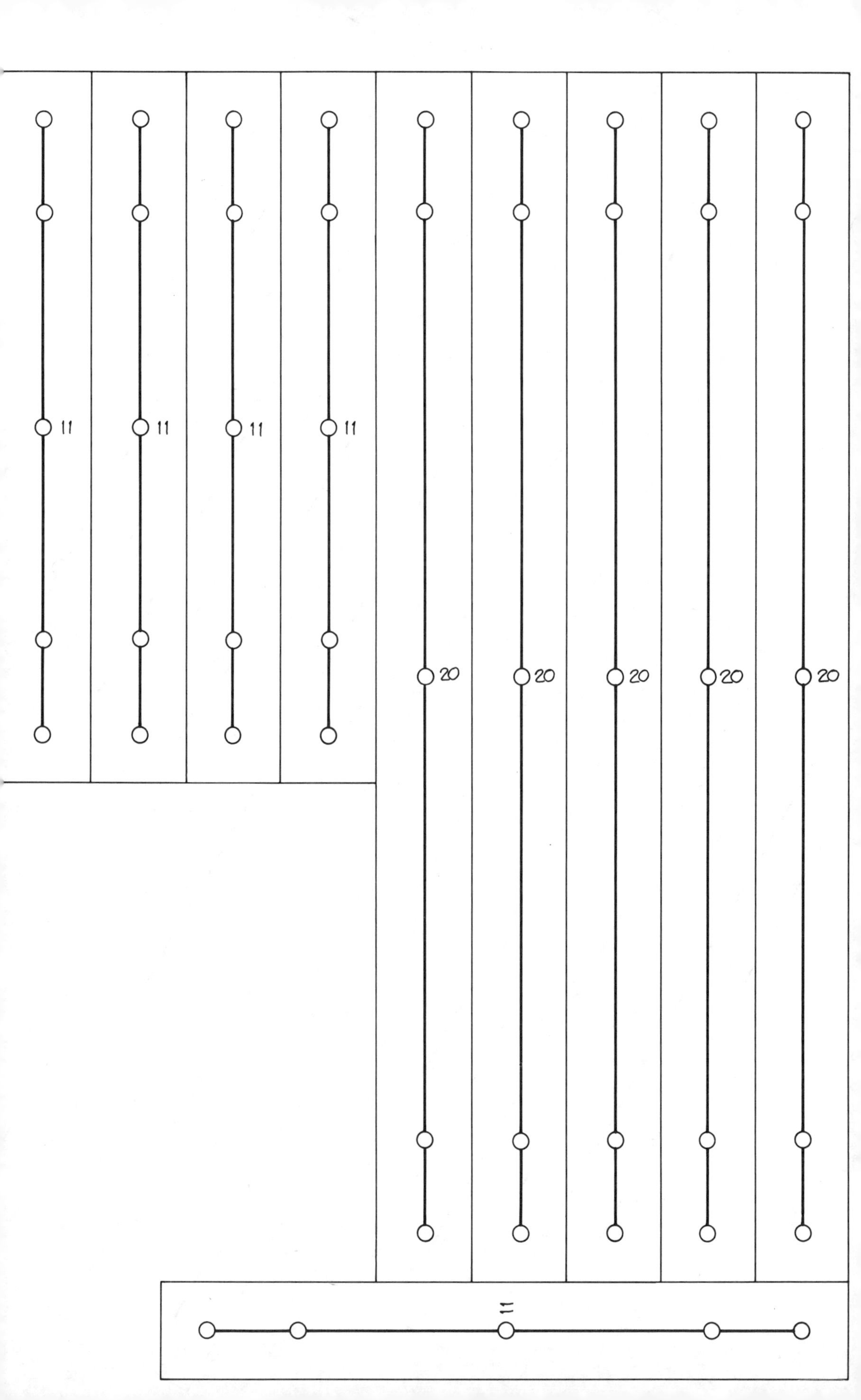
11
11
11
11
20
20
20
20
20
11

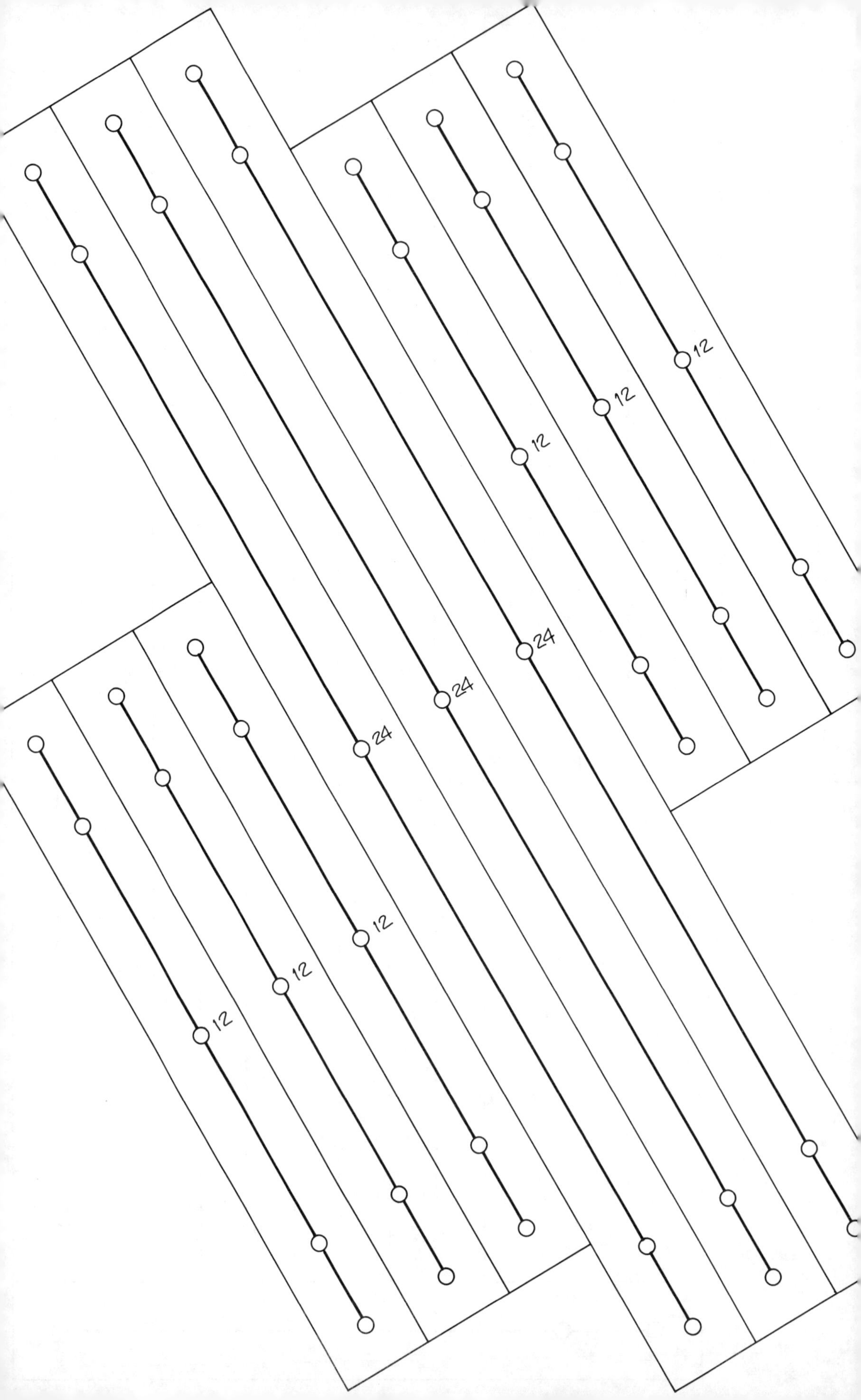
12
12
12
24
24
24
12
12
12

Appendix B
Patterns for Angle-Fixers

1. Reproduce these patterns.
2. Mount on oaktag.
3. Cut along outlines.
4. Punch holes where indicated—using a hole punch (preferably ⅛-in. diameter).
5. Use brass fasteners (½-in. shank) to join to strips.

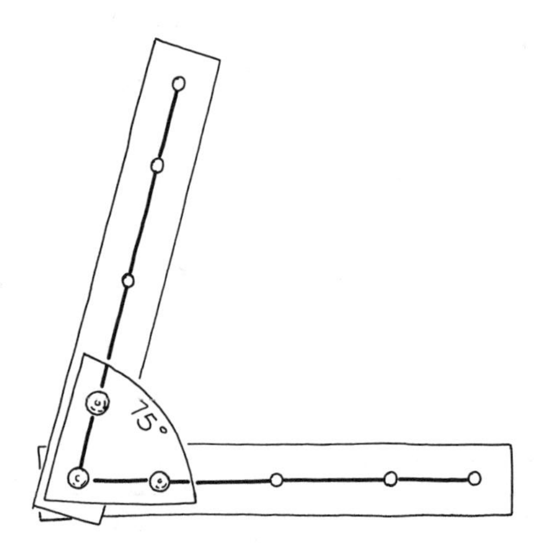

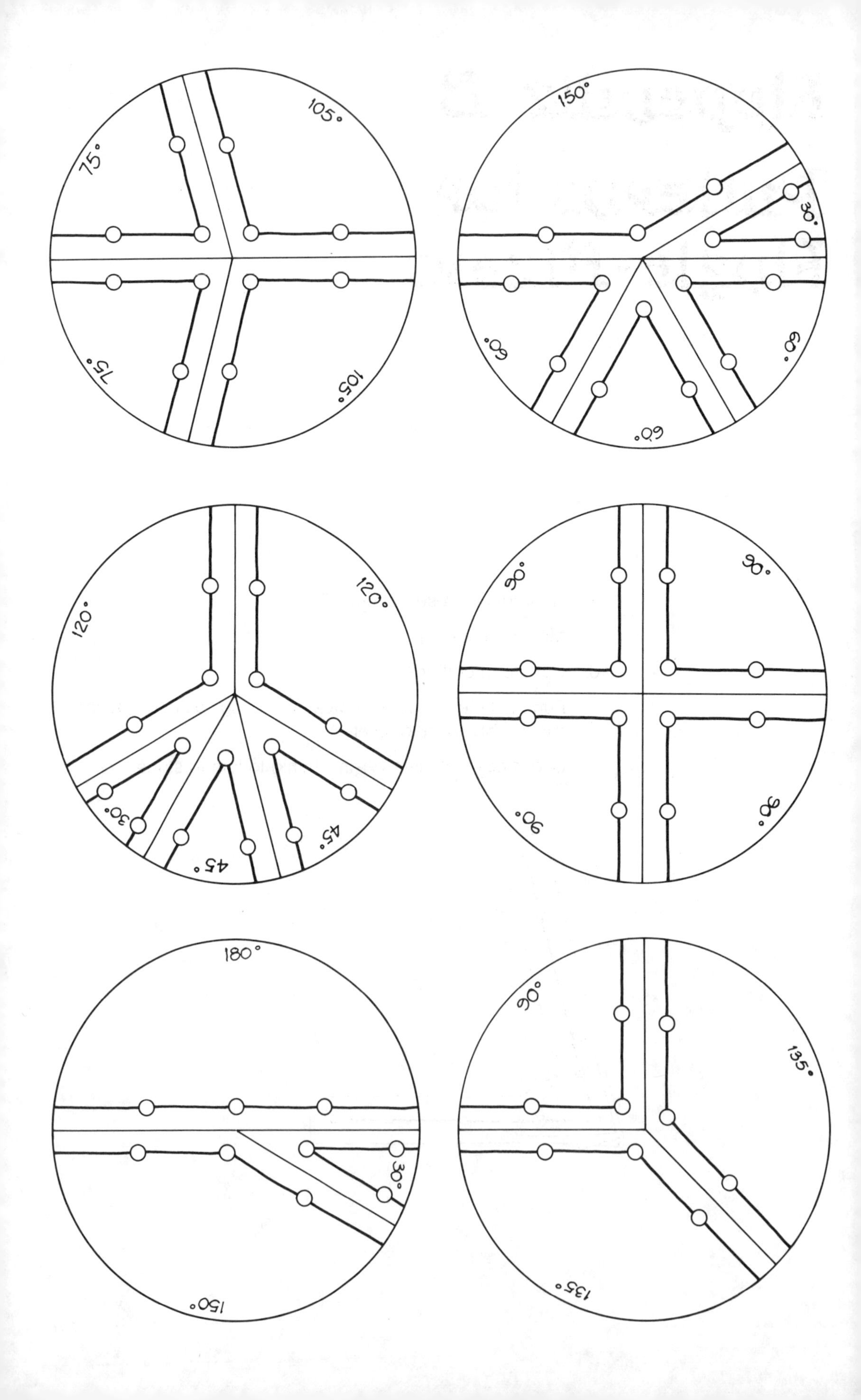

75°
105°
75°
105°
150°
30°
60°
60°
60°
120°
120°
30°
45°
45°
90°
90°
90°
90°
180°
30°
150°
90°
135°
135°

Bibliography

Cohen, D. *Inquiry in Mathematics Via the Geoboard.* New York: Walker Research Corp., 1967. 61 pp.

Cundy, H. M., and A. P. Rollett. *Mathematical Models.* 2d ed. New York: Oxford University Press, 1961. 286 pp.

DelGrande, J. J. *Geoboards and Motion Geometry.* Glenview, Ill.: Scott Foresman & Co., 1972. 122 pp.

Dienes, Z. P., and E. W. Golding. *Geometry of Congruence.* New York: Herder and Herder, 1967. 100 pp. Available through McGraw-Hill Book Company, New York.

Dienes, Z. P., and E. W. Golding. *Geometry of Distortion.* New York: Herder and Herder, 1967. 99 pp. Available through McGraw-Hill Book Company, New York.

Dienes, Z. P., and E. W. Golding. *Groups and Coordinates.* New York: Herder and Herder, 1967. 138 pp. Available through McGraw-Hill Book Company, New York.

Elliott, H. A., J. MacLean, and J. M. Jorden. *Geometry in the Classroom—New Concepts and Methods.* Toronto: Holt, Rinehart, and Winston of Canada, 1968. 266 pp.

Fielker, D. S. *Cubes.* Topics from Mathematics Series. New York: Cambridge University Press, 1969. 32 pp.

Fletcher, D., and J. Ibbotson. *Geometry One.* Edinburgh: Holmes-McDougall, 1965. 71 pp.

Fletcher, D., and J. Ibbotson. *Geometry Two.* Edinburgh: Holmes-McDougall, 1966. 98 pp.

Fletcher, D., and J. Ibbotson. *Geometry Three.* Edinburgh: Holmes-McDougall, 1969. 98 pp.

Fletcher, D., and J. Ibbotson. *Geometry with a Tangram.* Glasgow: W. & R. Holmes, 1965. 81 pp.

Fouke, G. R. *A First Book of Space Form Making.* San Francisco: GeoBooks, 1974. 64 pp.

Geometry. Experiences in Mathematical Discovery Series, Booklet Number 4. Washington, D.C.: National Council of Teachers of Mathematics, 1966. 98 pp.

Gillon, Jr., E. V. *Geometric Design and Ornament.* New York: Dover Publications, 1969. 66 pp.

Horemis, S. *Optical and Geometrical Patterns and Designs.* New York: Dover Publications, 1970. 92 pp.

Konkle, G. S. *Shapes and Perceptions—An Intuitive Approach to Geometry.* Boston: Prindle, Weber, & Schmidt, 1974. 248 pp.

Laycock, M. *Dual Discovery thru Straw Polyhedra.* Palo Alto, Ca.: Creative Publications, 1970. 41 pp.

Mold, J. *Circles.* Topics from Mathematics Series. New York: Cambridge University Press, 1967. 32 pp.

Mold, J. *Solid Models.* Topics from Mathematics Series. New York: Cambridge University Press, 1967. 32 pp.

Mold, J. *Tessellations.* Topics from Mathematics Series. New York: Cambridge University Press, 1969. 32 pp.

Mold, J. *Triangles.* Topics from Mathematics Series. New York: Cambridge University Press, 1971. 32 pp.

Nuffield Foundation. *Environmental Geometry.* Nuffield Mathematics Project Series. New York: John Wiley & Sons, 1969, 56 pp.

Ravielli, A. *An Adventure in Geometry.* New York: Viking Press, 1957. 117 pp.

Read, R. C. *Tangrams—330 Puzzles.* New York: Dover Publications, 1965. 334 pp.

Readings in Geometry from the Arithmetic Teacher. Washington, D.C.: National Council of Teachers of Mathematics, 1970. 121 pp.

Rowland, K. *The Development of Shape.* London: Ginn and Co., 1964. 128 pp.

Rowland, K. *The Shape of Towns.* London: Ginn and Co., 1966. 160 pp.

Sealey, L. G. W. *The Shapes of Things.* Oxford, England: Basil Blackwell & Mott, 1965. 45 pp.

Seymour, D. G., and R. A. Schadler. *Creative Constructions.* Palo Alto, Ca.: Creative Publications, 1974. 62 pp.

Seymour, D. G., L. Silvey, and J. Snider. *Line Designs.* Palo Alto, Ca.: Creative Publications, 1974. 80 pp.

Steinhaus, H. *Mathematical Snapshots.* 3d American ed. New York: Oxford University Press, 1969. 311 pp.

Taylor, A. *Math in Art.* Hayward, Ca.: Activity Resources Co., 1974. 60 pp.

Walter, M. I. *Boxes, Squares, and Other Things—A Teacher's Guide for a Unit in Informal Geometry.* Washington, D.C.: National Council of Teachers of Mathematics, 1970. 88 pp.

Wenninger, M. J. *Polyhedron Models for the Classroom.* Washington, D.C.: National Council of Teachers of Mathematics, 1966. 43 pp.

Index